Regional Geology Review

Series Editors

Roland Oberhänsli, Potsdam, Brandenburg, Germany

François M. Roure, Rueil-Malmaison, France

Dirk Frei, Department of Earth Sciences, University of the Western Cape, Bellville, South Africa

The Geology of series seeks to systematically present the geology of each country, region and continent on Earth. Each book aims to provide the reader with the state-of-the-art understanding of a regions geology with subsequent updated editions appearing every 5 to 10 years and accompanied by an online "must read" reference list, which will be updated each year. The books should form the basis of understanding that students, researchers and professional geologists require when beginning investigations in a particular area and are encouraged to include as much information as possible such as: Maps and Cross-sections, Past and current models, Geophysical investigations, Geochemical Datasets, Economic Geology, Geotourism (Geoparks etc), Geo-environmental/ecological concerns, etc.

More information about this series at http://www.springer.com/series/8643

Manuel Enrique Pardo Echarte
Editor

Geology of Cuba

Springer

Editor
Manuel Enrique Pardo Echarte
Exploration Scientific-Research Unit
Centro de Investigación del Petróleo
El Cerro, La Habana, Cuba

ISSN 2364-6438 ISSN 2364-6446 (electronic)
Regional Geology Reviews
ISBN 978-3-030-67800-5 ISBN 978-3-030-67798-5 (eBook)
https://doi.org/10.1007/978-3-030-67798-5

© Springer Nature Switzerland AG 2021
This work is subject to copyright. All rights are reserved by the Publisher, whether the whole or part of the material is concerned, specifically the rights of translation, reprinting, reuse of illustrations, recitation, broadcasting, reproduction on microfilms or in any other physical way, and transmission or information storage and retrieval, electronic adaptation, computer software, or by similar or dissimilar methodology now known or hereafter developed.
The use of general descriptive names, registered names, trademarks, service marks, etc. in this publication does not imply, even in the absence of a specific statement, that such names are exempt from the relevant protective laws and regulations and therefore free for general use.
The publisher, the authors and the editors are safe to assume that the advice and information in this book are believed to be true and accurate at the date of publication. Neither the publisher nor the authors or the editors give a warranty, expressed or implied, with respect to the material contained herein or for any errors or omissions that may have been made. The publisher remains neutral with regard to jurisdictional claims in published maps and institutional affiliations.

This Springer imprint is published by the registered company Springer Nature Switzerland AG
The registered company address is: Gewerbestrasse 11, 6330 Cham, Switzerland

Foreword

Books about the general theme of Geology of Cuba have been published in the course of the years, as a result of geological investigations and the collaboration of different specialists from diverse experiences, knowledge, and countries. It has been possible to expand the understanding of the regional geology, and the use of advanced technologies allows to provide new data every day.

In the last six decades, contributions that bring together different aspects of Cuban geology in a volume are known: *Geology of Cuba* (Furrazola et al. 1964), the first book to describe our geology, in very early periods of knowledge. Subsequently, and as a result of the geological mapping at scale 1:250000 of the entire national territory, carried out by the Academy of Sciences of Cuba in collaboration with the academies of sciences of Eastern European countries, a series of *Contributions to the Geology* of the different Cuban territories in the 80ths (Nagy et al. 1983, Albear et al. 1985, Pszczolkowski et al. 1987), which are still the foundation of stratigraphy and paleontology, as well as tectonics and metallogeny.

Geological investigations continued to develop, especially with the mapping works, at 1:50000 and more detailed scales, in almost half of Cuban territory, including the most potential areas for natural resource prospecting. Works such as the *Stratigraphic Lexicon of Cuba* (Franco et al. 1992), managed to combine in a synthetic monographic volume the lithostratigraphic units recognized and used in the resulting geological maps.

Ophiolites and volcanic arcs of Cuba (Iturralde Vinent Ed. 1996) is a compilation of different authors on the geochemistry, mineralogy, constitution, and genesis of oceanic crusts and volcanic arcs, with several contributions to Cuban tectonic, which meant a conceptual breakthrough of Cuban geology, integrating the data provided by several international projects.

Another monographic book, *Studies on Geology of Cuba* (Furrazola and Núñez, Ed. 1997), compiled a selection of articles on different aspects of Cuban geology, which generally covered all topics, including different degrees of studies, with an updated description of geology and stratigraphy of the national territory until that moment, differentiated by geological periods.

Several monographic volumes were published in different foreign journals, such as *Acta Geologica Hispanica* (Melgarejo and Proenza Ed. 1998), with a group of articles on the geology and metallogeny of Cuba, and *Geologica Acta* (Iturralde Vinent and Lidiak Ed. 2006), with different contributions on Caribbean plate tectonics, stratigraphy, magmatism, metamorphism, and tectonic events.

Compendium of the native land mammals of Cuba: living and extinct (Silva et al. 2007) is another monographic work that contains the distribution, taxonomy, and biogeography of the neo-quaternary mammals of Cuba.

Detailed descriptions of Cuban stratigraphy and its link with hydrocarbon prospecting are found in *Oilfields and manifestations of the Republic of Cuba* (Linares et al. 2011). *Petrophysical evaluation of Cuban gas-oil formations* (Castro 2017) characterizes the Cuban lithostratigraphic units from the petrophysical point of view and constitutes a methodological guide in these studies. *Classification system of Cuban rocks* (Collective of authors IGP 2018), is the most recent contribution to the standardization and nomenclature of Cuban rocks.

The Geology of Cuba and the Caribbean Compendium (Iturralde Vinent 2011–2019) is a compilation and author's update of all published information about Cuban geology, plate tectonics, paleontology, biogeography, lithostratigraphy, geological hazards, and among others, in digital version, and therefore, it keeps updating continuously.

It is also known that the advance in the geological knowledge of Cuba is deposited in the informative products, as a result of the events held in all these same years (Congresses and Conventions). The publication *Geosciences Memories, Works and Abstracts*, of the geoscience events in digital support, of the Cuban Geological Society, since 2005 has eight editions already, and stores an immense and invaluable information, all the topics of geological sciences. In these years also scientific journals were published such as the *Technological Journal* of the Industrial Minister (MINBAS), *Earth and Space Sciences* of the Cuban Academy of Sciences, Mining and Geology of the Moa University, and Yearbook of the Cuban Geological Society.

This edition is preceded by other significant ones in Springer, edited by Dr. Sc. Pardo Echarte, four volumes, which contain valuable and advanced information on several topics dedicated to Cuba: *Unconventional methods for oil & gas exploration in Cuba. The Redox Complex* (Pardo Echarte and Rodríguez Morán 2016); *Oil and gas exploration in Cuba. Geological-structural cartography using potential fields and airborne gamma spectrometry* (Pardo Echarte and Cobiella Reguera 2017; *Offshore exploration of oil and gas in Cuba using digital elevation models* (Pardo Echarte, Reyes Paredes, and Suárez Leyva 2018); and the most recent one *Non seismic and Non conventional methods for Oil & Gas Exploration in Cuba* (Pardo Echarte, Rodríguez Morán, and Delgado López 2019).

Now we are in front of another contribution to increase knowledge about the Geology of Cuba. It comes to complement or update essential aspects from the perspective of several authors with experience in their fields of work. Undoubtedly, in the scientific systematization of several themes advances are made, such as cartography, marine geology, fossil record, stratigraphy, tectonics, and other more specific, but no less important, which includes the ophiolites, the quaternary deposits, the metallogeny, and mineragenia of the Cuban territory.

The contribution about geological cartography in Cuba shows its history over 135 years and the most important, projection of the new geological mapping works at scale 1: 50 000 to be carried out throughout the national territory in the framework of the Geology Development Program until 2030.

Cuban marine geology is a significant subject, according to the greatest geological knowledge of its marine territory, which is covered entirely by research at different scales. There are in this chapter aspects that constitute new contributions such as morphogenetic complexes and relief forms, stratigraphic and sedimentological characteristics, neotectonic, geological evolution, and mineral resources of marine areas.

The Cuban Fossil Record constitutes the inheritance of the paleontological diversity exhibited by the current Cuban archipelago, the contribution offers abundant literature about Cuban fossils which allows us to know the varied degree of the study exhibited by the different fossil groups reported to this day. The richest fossiliferous lithostratigraphic units of the Triassic, Jurassic, Cretaceous, Paleogene, Neogene, and Quaternary periods participate in the geological structure of the Cuban archipelago.

Here, Precambrian rocks have been dated to singular outcrops in the southern area of Corralillo and La Teja in the provinces of Villa Clara and Matanzas, respectively. The stratigraphy of units from different periods is described by a large number of samples studied and the use and importance in the gaso-petroleum search works of different Cuban regions, especially in the Northwest Belt of Hydrocarbons of Cuba.

An overview of the Cuban tectonic is presented here, according to the author our territory is a geological mosaic positioned in the southeastern corner of the North American plate. Rocks from many different origins, with ages, from Proterozoic to Quaternary are present. The socle is divided into three major complexes: the Proterozoic-Paleozoic basement, the Mesozoic basement, and the Paleogene folded and thrust belt. The Eocene-Quaternary cover contains little disturbed beds, without magmatic or metamorphic rocks. It is evaluated that the nearness of SE Cuba to the Caribbean/North American plate boundary prints its cover with some special features.

Mesozoic ophiolites are an important feature of Cuban geology. They are distributed along the so-called "Northern" and "Eastern" Cuban ophiolite belts, and are strongly dismembered and intermingled mainly with Cretaceous volcanic arc rocks. The study of ophiolite associated basalts provides a comprehensive ophiolite classification according to the tectonic setting at which these ophiolites formed, then the studied lavas correspond mostly to subduction-related ophiolite.

In Cuba, twenty one lithostratigraphic units and five innominate carbonate and terrigenous deposits of Quaternary age are recognized. A stratigraphic subdivision scheme for these deposits applying relative dating methods, for classic subdivisions that are established for the Pleistocene, has been possible to compile. In its general conception, most formations correspond to marine transgressions that are closely related to the great sea-level glacio-eustatic transgressions that occurred during the Pleistocene in a number not less than seven, according to the majority of the world's researchers.

For Cuban metallogeny as a result of an analysis of several models of metallic mineral deposits, taking account descriptive—genetic type, together with the analysis of their spatial distribution and their relationship with the geology, allowed the identification and mapping of ten mineral systems, linked to the geodynamic environments present in the Cuban territory. In addition, some general inferences about Cuba's metallogeny are presented and several aspects that require new works are pointed out.

Our country has a wide variety of deposits, manifestations, and mineralized points of industrial minerals, and its origin has been summarized into seven genetic groups: magmatic, sedimentary, pyroclastic-sedimentary, metamorphic, skarn, hydrothermal, and supergene (residual). They appear in different geotechnical areas and geological contexts. This detailed approach shows the high geologic-economical Cuban potential.

The Cuban Geological Society has just completed its 40th anniversary; since its beginning it aims to include all geoscientists and in general, the specialists who deal with work related to geosciences in Cuba. Certainly, since its foundation, a cardinal goal is the scientific and technical improvement of its members, as well as dissemination of the results of their investigation, so we stimulate and highly value initiatives such as this, that complete our slogan *The Geosciences at the service of the Cuban Society and its development*.

My congratulations to the authors of this work, to Editor and Springer Publishing House, for the contribution to our geoscience.

Prof. Kenya Elvira Núñez Cambra M.Sc.
Presidente de la Sociedad Cubana de Geología
La Habana, Cuba

Preface

The monograph *Geology of Cuba* addresses different facets of Cuba's geological knowledge: cartography, marine geology, fossil record, stratigraphy, tectonics, classification of its ophiolites, quaternary deposits, metallogeny, and minerageny. The contributions are not linked by a common thread, being free versions according to the scientific positions of the authors; therefore, logically, there are contradictory opinions in many cases. However, it is a fabulous scientific work that puts in the hands of the most diverse readers all the wealth of geological knowledge of Cuba, essential for any type of application, exploration, teaching, research, or patrimonial. In itself, it constitutes a notable contribution to the advancement of knowledge of the Geology of Cuba.

The evolution of geological cartography in Cuba in its more than 135 years of history has been possible through the consultation of numerous archival reports, publications, maps, and personal interviews with different authors and geologists of vast experience. A brief critical analysis is made of the increase in the degree of geological knowledge of the country since the elaboration of the Geological Sketch of the Cuban Island at a scale of 1:2000000 (Fernández de Castro y Salterain P 1883), first of Cuba and of Ibero-America, until the most recent Digital Geological Map of Cuba at scale 1:100000 (Pérez Aragón et al. 2016). It's also briefly commented on the elaboration of some maps at detailed scales and the conditional surveys at scales 1:100000 and 1:50000 carried out in collaboration with the ex—Soviet Union and countries of the former Eastern European Socialist Field. Finally, the projection of the new geological mapping works at scale 1:50000 to be carried out throughout the national territory in the framework of the Geology Development Program until 2030 are briefly addressed.

Cuba is one of the countries with the greatest geological knowledge of its marine territory, which is covered in its entirety by research at different scales. It is considered as the geology of the marine territory of Cuba to that developed during the Pliocene-Quaternary in the areas occupied by the sea between the upper edge of the insular slope and the current coast. The investigations reveal that the deposits of this interval have different characteristics to those of the emerged territories of the islands of Cuba and of de la Juventud, determined by: (1) their biogenic-carbonate composition without or with very few terrigenous components; (2) distribution through the entire surface of the seabed and the keys, without the presence of pre-quaternary rocks; and (3) structure and morphology, typical of banks, bars, plains, and depressions. The following topics are addressed: (1) terms and definitions specific to the study territory or, conceptual-theoretical framework, to provide a better understanding of the composition, distribution, and genesis of seafloor and keys deposits, as well as its morphogenesis; (2) the main works, on which the chapter is based; and (3) aspects that constitute the contribution of this work (morphogenetic complexes and relief forms, stratigraphic and sedimentological characteristics, neotectonic, geological evolution, and mineral resources).

In general, the "Cuban Fossil Record", which covers approximately the last 200 million years of life on Earth, is rich in very varied fossils, witnessing a wide diversity of organisms, both animals and plants that inhabited the Antillean and Caribbean region and constituted the inheritance of the biological diversity exhibited by the current Cuban archipelago. The numerous literature about Cuban fossils allows us to know the varied degree of the study exhibited by the different fossil groups reported to this day. The irregular development of the investigations carried out so far reveals the possibilities of study that the Cuban fossil record still needs, and which points out the future path for new researchers interested in different topics on this paleontological richness, where there are many questions to be solved, even waiting to assess correctly, from modern work basis.

The lithostratigraphic units of the Triassic, Jurassic, Cretaceous, Paleogene, Neogene, and Quaternary periods participate in the geological structure of the Cuban archipelago. Precambrian rocks have been dated to singular outcrops in the southern area of Corralillo and La Teja in the provinces of Villa Clara and Matanzas, respectively. The formations of these periods are represented by sedimentary, volcanic, volcanic-sedimentary, and metamorphic rocks, which are registered in the subsoil or forming the depressions and mountainous areas of the national territory. The stratigraphy has been described through the Paleogeographic Domains: Synrift, North American Continental Margin, Cretaceous and Paleogenic Volcanic Arcs, and Superposed Basins with their corresponding Petrotectonic Sets. The metamorphic rocks, whose protolith were sedimentary rocks, and those of the Neogene and Quaternary systems have been briefly treated. Most of the mentioned lithostratigraphic units are registered in the Third Version of the Léxico Estratigráfico de Cuba (2013), but some do not appear. They are described by a large number of samples studied and the use and importance in the gaso-petroleum search works of different Cuban regions, especially in the Northwest Belt of Hydrocarbons of Cuba.

Cuba and its surroundings are a geological mosaic in the southeastern corner of the North American plate. Rocks from many different origins, with ages, from Proterozoic to Quaternary, belong to this puzzle, extended along the southern border of the plate, from northern Central America to the Virgin Islands. From the Eocene, this belt has been dissected by several great faults, related to the development of some great oceanic depressions (Cayman trough and Yucatan basin). Two main structural levels or stages can be distinguished in the geological architecture of Cuba. The lower stage is the socle, a great rock complex, mainly formed by Jurassic-Eocene rocks, unconformably resting below the cover. The socle is divided into three major complexes, according to their litho-structural features and rock age: (a) the Proterozoic-Paleozoic basement, (b) the Mesozoic basement, (c) the Paleogene folded and thrusted belt. The Eocene-Quaternary cover contains little disturbed beds, without magmatic or metamorphic rocks. The nearness of SE Cuba to the Caribbean/North American plate boundary prints its cover with some special features.

Mesozoic ophiolites are an important feature of Cuban geology. The ophiolites are distributed along the so-called "Northern" and "Eastern" Cuban ophiolite belts, and are strongly dismembered and intermingled mainly with Cretaceous volcanic arc rocks. Ophiolite associated basalts along the northern Cuban orogenic belt record magmatic history of the ophiolite formation from the Protocaribbean seafloor spreading to the subduction initiation stage. It has compiled geochemical data of oceanic basalt samples from previous works, together with data of analyzed samples during this study in order to discuss geochemical criteria based on immobile element proxies for fractionation indices, alkalinity, mantle flow and subduction addition, and field relationships, providing a comprehensive ophiolite classification according to the tectonic setting at which these ophiolites are formed. The results show then, that the studied lavas correspond mostly to subduction-related ophiolite and some of them have incorporated subduction component probably during the time when the Protocaribbean oceanic lithosphere downgo beneath the Caribbean plate. Some rare remnants of plume-MORB type ophiolite have been thought to also occur.

In Cuba, the existence of an important group of geological formations and innominate deposits of Quaternary age is recognized. They encompass both carbonate and terrigenous sequences and currently, twenty one of these lithostratigraphic units are recognized, as well as five innominate deposits. Applying relative dating methods, it has been possible to compile a stratigraphic subdivision scheme for these deposits that locates them specifically in each of the classic subdivisions that are established for the Pleistocene: Lower, Middle, and Upper. Six of the recognized units transited the Neogene/Quaternary limit, that is, they began to settle in the Upper Pliocene and concluded their deposition in the Lower Pleistocene. In the case of other units, it has been possible to better define their correspondence with the Middle Pleistocene, or with the Upper. Specifically with the Classical Upper Pleistocene corresponds one of the most extended formations of Cuba: the Jaimanitas Formation; and with the upper part of the Upper Pleistocene, in this case, the Oxygen Isotopic Stage (OIS) 3, there are currently several units of local character, but very important for their paleoclimatic peculiarities. In its general conception, most formations correspond to marine transgressions that are closely related to the great sea-level glacio-eustatic transgressions that occurred during the Pleistocene in a number not less than seven, according to the majority of the world's researchers.

As a result of the preparation of the Metallogenic Map at scale 1:250000 of the Republic of Cuba, forty-one models and eight sub models of metallic mineral deposits were identified. These models, of descriptive—genetic type, together with the analysis of their spatial distribution and their relationship with the geology, allowed the identification and mapping of ten mineral systems, linked to the geodynamic environments present in the Cuban territory. The work provides an up-to-date image of Cuba's metallogeny, with the incorporation of additional information obtained recently, presented through the description of each of the ten mineral systems identified. Special attention is given to the mafic—siliciclastic VMS—SEDEX—MVT mineral systems of the J-K1 distensive basin, which covers the Guaniguanico and the Cangre structural-tectonic units, Pinos and Guamuhaya terranes, and the supergene environment, this represented in a large part of the Cuban territory. The first hosts an important pyritic and polymetallic sulphide mineralization of Cu, Zn, Pb, Co, Au, and Ag, while the second contains the known Cuban lateritic deposits of Fe–Ni–Co, which also contain Au, PGE, Sc, and others metals. Finally, some general inferences about Cuba's metallogeny are presented and several aspects that require new works are pointed out.

The complex and prolific geological scenario of Cuba has led to geological-genetic conditions favorable for the formation of varied deposits, manifestations, and mineralized points of various industrial raw materials, decorative and for multiple practical uses, known worldwide as industrial rocks and minerals, or simply as industrial minerals. Thus, Cuba has large deposits of limestone, loam, dolomite, kaolin, gypsum and anhydrite, rock salt, marbles, sands and clays of different types, zeolites, peat, therapeutic peloids, and many more. There are manifestations of decorative and precious rocks such as jasper, jadeite, different varieties of quartz, and even xylopals. Its origin has been summarized in seven genetic groups: magmatic, sedimentary, pyroclastic-sedimentary, metamorphic, skarn, hydrothermal, and supergenic (residual). The different genetic groups are not governed, in general, by the same or similar geotectonic situations. They appear in different geotechnical areas and geological contexts, such as limestone rocks, granitoids, and zeolites. There are also tendencies, in some genetic types, to be located in specific environments such as peat, in the biogenic zones of the Miocene-Quaternary sedimentary cover and the kaolin, in the Mesozoic accreted terrane.

La Habana, Cuba   Manuel Enrique Pardo Echarte

Contents

1 **Geological Cartography of Cuba** 1
   Ramón Omar Pérez Aragón

2 **Geology of the Marine Territory of Cuba** 39
   Miguel Cabrera Castellanos and Betsy Vázquez Gainza

3 **Synthesis of Fossil Record of Cuba—A Bibliographic Compilation** 71
   Reinaldo Rojas Consuegra

4 **Stratigraphy of Cuba** 143
   Evelio Linares Cala and Dora Elisa García Delgado

5 **An Overview to the Tectonics of Cuba** 189
   Jorge Luis Cobiella Reguera

6 **Geochemical Fingerprinting of Ancient Oceanic Basalts: Classification of the Cuban Ophiolites** 219
   Angelica Isabel Llanes Castro and Harald Furnes

7 **Stratigraphy of the Quaternary Deposits in Cuba** 231
   Leandro L. Peñalver Hernández, Miguel Cabrera Castellanos, and Roberto Denis Valle

8 **Mineral Systems of Cuba: A Panoramic Vision of Cuban Metallogeny** 255
   Jorge Luis Torres Zafra and Xiomara Cazañas Díaz

9 **The Minerageny of Cuba** 289
   Waldo Lavaut Copa and Rolando Batista González

Abbreviations

AC URSS	Academia de Ciencias de la Unión de Repúblicas Socialistas Soviéticas
ACC	Academia de Ciencias de Cuba
APS	Alturas de Pizarras del Sur tectono-stratigraphic unit
Ar	Argon
BA	Back—Arc
BS	Boundary Sequences
CA	Calc—Alkaline
CAME	Consejo de Ayuda Mutua Económica
CARIS	Computed Aided Resources Information System
Cartageol 50K	Subprograma para la Cartografía Geologica de Cuba a escala 1: 50000
CC	Cayo Coco tectono-stratigraphic unit
CCT	Consejo Científico Técnico
CDM	Canyon Diablo meteorite troilite
CDT	Canyon Diablo troilite
CEEZ	Cuban Exclusive Economic Zone
CEINPET	Centro de Investigaciones del Petróleo
CFR	Cuban Fossil Record
CIDEM	Centro de Investigaciones de Estructuras y Materiales de UCLV (Cuba)
CIG	Centro de Investigaciones Geológicas
CIPIMM	Centro de Investigaciones para la Industria Minero-Metalúrgica (Cuba)
Cj	Camajuaní tectono-stratigraphic unit
C-MORB	Contaminated Middle Ocean Ridge Basalt
CNDIG	Centro Nacional de Información Geológica
CNFG	Centro Nacional del Fondo Geológico
COEB	Central Oriente Eocene Basin
CUJAE	Ciudad Universitaria José Antonio Echevarría (Cuba)
CUR	Comisión de Unificación y Redacción
CVA	Cretaceous Volcanic Arc
DGGG	Dirección General de Geología y Geofísica
DIGIAC	Dirección de Ingeniería Geotécnica Aplicada de Cuba
DSDP	Deep Sea Drilling Project
ECOB	Eastern Cuban Ophiolite Belt
E-MORB	Enriched Middle Ocean Ridge Basalt
FA	Fore—Arc
FAB	Fore Arc Basalt
FAR	Fuerzas Armadas Revolucionarias
Fm	Formation
FMI	Formation Microresistivity Image
FZ	Fault Zone
GDR	German Democratic Republic
GECEM	Grupo Empresarial del Cemento (Cuba)

GEODATO	Banco de Datos Geológicos de la República de Cuba
GEOMINERA	Empresas del Grupo de Geología-Minería-Sal (MINEM, Cuba)
GeoMinSal	Grupo Empresarial de Geología, Minería y Sal
GIS	Geographic Information System
GNL	Guacanayabo-Nipe Bay lineament
HPAL plant	High Pressure Acid Leaching plant
HT/MP	High Temperature/Mid Pressure metamorphism
IAT	Island Arc Tholeites
ICA	Instituto de Ciencias Agropecuarias
ICGC	Instituto Cubano de Geodesia y Cartografía
ICP-EOS	Inductively Coupled Plasma Optical Emission Spectrometer
ICP-MS	Inductively Coupled Plasma Emission Mass Spectrometer
ICRM	Instituto Cubano de Recursos Minerales
IGP	Instituto de Geología y Paleontología
IGT	Instituto de Geografía Tropical
IMR	Industrial Minerals and Rocks (nonmetallic)
INAA	Instrumental Neutron Activation Analysis
INFOYAC	Multimedia con banco de datos sobre yacimientos de Cuba
INQUA	Acronyms in English language of the International Union for Quaternary Research
IOCG	Iron Oxide Cupper Gold
ISMMM	Instituto Superior Minero Metalúrgico de Moa
ISPJAE	Instituto Superior Politécnico José Antonio Echeverría
IUGS	International Union of Geological Sciences
IVA	Insular Volcanic Arc
J3	Upper Jurassic
K	Potassium
K2	Upper Cretaceous
KGRT	Kiev Geo-Prospecting Technological Institute (in Russian)
KVA	Cretaceous Volcanic Arcs
KVAT	Cretaceous Volcanic Arcs Terrane
Ky	Kilo years
LCC	Low Carbon Cement
LOI	Lost On Ignition
MC	Mabujina Complex
MES	Ministerio de Educación Superior (Cuba)
MFS	Maximum Flood Surface
MGRI	Moscow Institute of Geological Prospecting (in Russian)
MICONS	Ministerio de la Construcción
MINBAS	Ministerio de la Industria Básica
MINEM	Ministerio de Energía y Minas
MMG	Ministerio de Minería y Geología
MNHNC	Museo Nacional de Historia Natural de Cuba
MOR	Mid Ocean Ridge
MORB	Mid Ocean Ridge Basalt
MTU	Main Tectonic Unit
MVT	Mississippi Valley Type
My	Million years
NAP	North America Paleomargin
NBHC	Northwest Belt of Hydrocarbon of Cuba
NC 622	Norma Cubana para la Cartografía Geológica
NCOB	Northern Cuban Ophiolite Belt

N-MORB	Normal Middle Ocean Ridge Basalt
NOB	Northern Ophiolitic Belt
N-Q	Neogene-Quaternary
OFZ	Oriente Fault Zone
OIB	Ocean Island Basalt
OIS	Acronyms in English language of Oxygen Isotopic Stage.
ONRM	Oficina Nacional de Recursos Minerales
P	Placetas tectono-stratigraphic unit
PAM	Passive Actives Mining
PBB	Piggyback Basins
PD(s)	Paleogeographic Domain(s)
PFB	Paleogene Foreland Basin
PFZ	Pinar Fault Zone
PG	Pan de Guajaibón
PgB	Paleogene Basin
PGE	Platinum Group Elements
PIA	Primitive Island Arc
PS(s)	Petrotectonic Set(s)
P-type	Plume type
PVA	Paleogene Volcanic Arc
R & D	Research and Development
R	Remedios tectono-stratigraphic unit
REE	Rare Earth Elements
RGB	Red-Green-Blue
SCG	Sociedad Cubana de Geología
SDB	Santiago Deformed Belt
SEDEX	Sedimentary Exhalative
SEM-EDS	Scanning Electron Microscopy/Energy Dispersive X-Ray Spectroscopy
SGC	Servicio Geológico de Cuba
SIGEOL	Sistema de Información Geológica
SMT	Southern Metamorphic Terranes
SO	Sierra de los Órganos tectono-stratigraphic unit
SR-APN-E	Sierra del Rosario-Alturas de Pizarras del Norte tectono-stratigraphic unit
SWB	Southwestern Cuban Basin
TAS	Total Alkali Silica
T-Cr	Turquino-Cayman Ridge volcanic arc
TFZ	La Trocha Fault Zone
TOC	Total Organic Carbon
TSU(s)	Tectono Stratigraphic Unit(s)
UCLV	Universidad Central Marta Abreu de Las Villas (Cuba)
UH	Universidad de La Habana
U-Pb SHRIMP	Uranium-Lead Sensitive High Resolution Ion Microprobe
U-Pb	Uranium-Plumb
UPR	Universidad de Pinar del Río Hermanos Saiz Montes de Oca (Cuba)
USA	United States of America
USSR	Union of Soviets Socialistic Republics
VA	Volcanic Arc
VMS	Volcanogenic Massive Sulphide
VSEGEI	Institute of Geological Researches of the Soviet Union (in Russian)
WSU	Washington State University
XRF	X-ray Fluorescence Spectrometry

List of Figures

Fig. 1.1	Geological Sketch of the Island of Cuba; made between 1869 and 1883 by the Spanish mine engineers Manuel Fernández de Castro Suero and Pedro Salterain Legarra. The first of its kind in Cuba and in Latin America [7].	5
Fig. 1.2	Geological Sketch of Cuba at 1:1000000 scale, made between 1941 and 1946 by a Forestry and Mines Technical Commission of the former Ministry of Agriculture of Cuba, led by Jorge Brödermann Vignier (Brödermann et al. [3].	6
Fig. 1.3	First Geological Map of Cuba at 1:1000000 scale [8].	7
Fig. 1.4	Reduced view (4 sheets) of the Geological Map of the Republic of Cuba at 1:500000 scale, from 1985. A fifth sheet located on the right (not shown) contains a legend of zonal type.	8
Fig. 1.5	Map of Metallic Mineral Deposits and Mineral Waters of the Republic of Cuba at scale 1:500000 (De los Santos Llerena et al. [84].	10
Fig. 1.6	Map of Deposits, Manifestations of Non-Metallic, and Fuel Minerals of the Republic of Cuba at 1:500000 scale (De los Santos Llerena et al. [85].	11
Fig. 1.7	Scheme of distribution of the areas to be mapped by the different brigades of the Academies of Sciences of the socialist countries for the preparation of the Geological Map of Cuba at 1:250000 scale. 1988. [2]	11
Fig. 1.8	Scheme of distribution of the areas covered by survey at 1:250000 scale of the former province of Pinar del Río by the (first) Cuban-Polish Brigade between 1972 and 1975.	12
Fig. 1.9	Distribution scheme of the areas covered by survey at 1:250000 scale of the old province of La Habana by the Cuban IGP brigade called the Habana-Matanzas Brigade, 1974–1977.	13
Fig. 1.10	Scheme of distribution of the areas covered by survey at 1:250000 scale of the old province of Matanzas by the (second) Cuban-Polish Brigade in 1977-1981.	13
Fig. 1.11	Scheme of distribution of the areas covered by survey at 1:250000 scale of the former province of Las Villas by the (first) Bulgarian-Cuban Brigade between 1969 and 1975.	14
Fig. 1.12	Scheme of distribution of the areas covered by survey at 1:250000 scales of the former province of Camagüey by the (second) Cuban-Bulgarian Brigade, between 1976 and 1981	15
Fig. 1.13	Scheme of distribution of the areas covered by survey at 1:250000 scale of the former province of Oriente by the Cuban-Hungarian Brigade between 1972 and 1976	16
Fig. 1.14	Distribution scheme of the areas covered by survey at 1:250000 scale of the former Isla de Pinos carried out by the Cuba-USSR Brigade between 1972 and 1976	16
Fig. 1.15	Geological Map of Cuba at 1:250000 scale. (Pushcharovsky et al. [12]	18
Fig. 1.16	Tectonic Map of Cuba at 1:500000 scale (Pushcharovsky et al. [13]	18

Fig. 1.17	New Geological Map of Cuba at 1:1000000 scale. Prepared by Francisco Formell Cortina and Alberto Morales Quintana, and published in the New National Atlas of Cuba of 1989	19
Fig. 1.18	Scheme of distribution of the polygons in which the national territory was divided for the execution of the projects of Updating and Generalization of the Geological Map of Cuba at 1:100000 scale per region	19
Fig. 1.19	Simplified facsimile of the Digital Geological Map of the Republic of Cuba at 1:100000 scale (unified format in the GIS CARIS version). Faults and annotations are not shown, as well as the geology of the marine-coastal zone	24
Fig. 1.20	Digital version of the Geological Map of the Vertientes sheet (4579-I). Result of the Complex Geological Survey of the Camagüey Polygon, sectors Ciego de Ávila-Vertientes at 1:50000 scale (CAME), carried out by a Cuban brigade between 1987 and 1991. (Piñero Pérez E et al. [52]. Digitized by Yadira Duran Cuervo, edited by Valentina Strazhevich, IGP, 2016	28
Fig. 1.21	Digital version of the Geological Map of the Central Zone, result of the Geological Survey at scale 1:50000 of the CAME polygon of the same name, located north of the Trinidad dome of the Escambray metamorphic massif, erected by a Cuban-Czech brigade between the years 1982 and 1986. (Dublan et al. [19]. Digitized by Melisa Texas Pita, edited by Valentina Strazhevich, IGP, 2016	29
Fig. 1.22	Scheme showing the polygons mapped at 1:50000 scale in the period 1961 and 1992. These surveys cover approximately 38% of the national territory (Pérez Aragón et al. [4]	29
Fig. 1.23	Diagram showing the polygons mapped at 1:100000 scale in the period 1961 and 1992. These surveys cover approximately 17% of the national territory (Pérez Aragón et al. [4]	30
Fig. 1.24	Scheme showing the polygons mapped at scales 1:50000 and 1:100000 in the period 1961 and 1992. Overall, these surveys cover approximately 55% of the national territory (Pérez Aragón et al. [4]	30
Fig. 1.25	Facsimile of the digital geological map at 1:25000 scale of the municipality of La Habana del Este, one of the 7 that make up the Digital Geological Map at said scale of the eastern and southern municipalities of Ciudad de La Habana (Pérez Aragón et al. [47]). Digitized Valentina Strazhevich, edited Ramón O. Pérez IGP, 2000	32
Fig. 2.1	Hypsometric regionalization of the marine-coastal territory and its bottom, according to its geological-geomorphological features	40
Fig. 2.2	Plan view of shelf (1-northwestern, 2-north central, 3-southeastern and 4- southwestern) and sectors without shelf (a- northwestern, b-northeastern, c-southeastern, d- south central, e- South of the Isla de la Juventud and f- southwestern	41
Fig. 2.3	Schematic representation of lithomorphogenetic zones from Cuba shelfs and their main forms of relief	41
Fig. 2.4	Geomorphological maps of the shelfs: A-northwestern, B-north central, C-southeastern and D-southwestern (based on [3, 13–15])	45
Fig. 2.5	Geological-geomorphological profile, generalized to the west of the northwestern shelf (based on [4, 17]). Note the existence of deep sublatitudinal fluvial channels in the calcareous substrate, in process of clogging and terraces in the vertical wall	45
Fig. 2.6	Morphology of the Gran Banco Buena Esperanza, seen in the upper vertical cut and geomorphological-geomorphological cut across the paleo valley of the Agabama or Manatí River (based on [4]; Avello and Pavlidis [19])	46

Fig. 2.7	Tidal channels: Left, in the sediments of Fragoso key of bank type (blue arrows), fragmenting homonymous deltas (black arrows); right, on the surface of reef limestones raised above the sea level, on Coco key (c) of the old reef type	47
Fig. 2.8	Fossil dunes, in Guillermo key, over 12 m altitude (left photo) and Coco key (right photo)	47
Fig. 2.9	Fossil bar crowned by storm ridges, Coco key (left photo) and tidal niches, Paredón Grande key (red lines, right photo)	48
Fig. 2.10	Typical structure of a bar key. Example: Guillermo key. The presence of a rock nucleus is not an obligatory condition for the formation of this type, but it is frequent.1. Reef limestone's, of the Jaimanitas Formation; 2. beach; 3. bar; 4. mangrove; 5. lagoons and 6. terrace	48
Fig. 2.11	Schematic representation of the typical structure of the recent reef keys on the southeastern shelf. They constitute a mangrove (swampy bottom), surrounded by a dead reef, which decant, with more than 30° at the bottom of the limestone of the Jaimanitas Formation	49
Fig. 2.12	Geological-geomorphological cuts through three karst depressions in the southwestern shelf, from top to bottom: La Broa, Siguanea and Batabanó (based on [2, 4, 19–22]	50
Fig. 2.13	Examples of intra-island bays: Santa Clara and Carahatas	51
Fig. 2.14	Linear Delta or fluvial-marine bar, known locally in Baracoa, province of Guantánamo, as the Tibaracón. It is typical of some coastal regions of the island of Cuba adjoining mountainous terrain, from which a steep river network is born capable of transporting large volumes of clastic material and depositing it on the coast. In times of drought when the flow decreases, the waves predominate, reworking the deposit and converting it into bar	52
Fig. 2.15	Geological map of the marine and coastal territory of Cuba at scale 1: 1,000,000 (based on [3, 7])	54
Fig. 2.16	Generalized stratigraphic columns of shelf. **a** northwestern Avello [3, 30–33]; **b** north central (based on Ionin and Pavlidis 1972; Ionin et al. [4, 34]; Cabrera et al. 1997; [35]); **c** southeastern (based on [4, 36, 19]; Cabrera and Batista [3]; and **d** southwestern (based on [37, 38, 21]b; [2, 3]	55
Fig. 2.17	Quaternary formations in Ocampo key, Cienfuegos bay. Jaimanitas Formation at the base of the cut covered by the La Cabaña Formation (from left to right, photos 1 and 2) and by the El Salado Formation (photo 3)	58
Fig. 2.18	Neotectonic map of Cuba [9]	59
Fig. 2.19	Paleogeographic outline of the territory of Cuba in the Upper Pliocene-Early Lower Pleistocene (Vedado transgression) 2.5–1.8 My)	60
Fig. 2.20	Paleogeographic scheme of Cuba during early Upper Pleistocene (130–75 Ky)	62
Fig. 2.21	Cuba's paleogeographic scheme during interstate Middle Wisconsin in the late Upper Pleistocene (24–36 Ky)	62
Fig. 2.22	Map of mineral deposits and manifestations of the current marine-coastal territory of Cuba	65
Fig. 3.1	View of a piece of furniture from the fossil invertebrate collection of the MNHNC; a true scientific—cultural treasure	75
Fig. 3.2	Fossil fern attributed to the Hymenophyllaceaea family of Lower–Middle Jurassic. San Cayetano Formation	77
Fig. 3.3	Ostreidos on ammonite shell, a fossil association of the Upper Jurassic. Jagua Formation	78
Fig. 3.4	Concretion of carbonate containing a gonoid fish of the Upper Jurassic, Oxfordian. Jagua and Francisco formations	80

Fig. 3.5	Perforations in wood (lower half of the image) and ammonite section (above) of the Upper Jurassic, Oxfordian. Jagua Formation	81
Fig. 3.6	Stem of wood in sandstones of the Lower Albian Cretaceous. Provincial Formation	82
Fig. 3.7	Shell of the rudists *Titanosarcolites giganteous* in marmorized limestones of the Upper Cretaceous, Maastrichtian. Cantabria Formation	86
Fig. 3.8	Ammonite of the Upper Cretaceous, Maastrichtian. Monos Formation. Courtesy of Alberto Arano	87
Fig. 3.9	The *aptychus* can be very abundant in some facies of the Lower Cretaceous, Barremian. Veloz Formation	88
Fig. 3.10	Internal mold and remains of the shell of an *Inoceramus* of the Upper Cretaceous, Turonian–Coniacian. Abreus Formation	89
Fig. 3.11	Elements of the crinoid *A. cubensis*, collected in Ciego de Ávila, west of the type locality. Presa Jimaguayú Formation	89
Fig. 3.12	Carapace of crutacean *Vegaranina rivae* of the Upper Cretaceous. Monos Formation. Photo courtesy of Alberto Arano	90
Fig. 3.13	Corals are abundant in the carbonate rocks of the Upper Cretaceous, Maastrichtian. Cantabria Formation. Photo courtesy of Carlos Borges and Alberto Arano	90
Fig. 3.14	Calcareous tubes of vermes on rudist shell of the Upper Cretaceous. Presa Jimaguayú Formation	91
Fig. 3.15	Macroforaminifera, abundant in the rocks of the Upper Cretaceous, Maastrichtian	91
Fig. 3.16	Coprolite found in clay-sandy facies of the Upper Cretaceous, Maastrichtian. Cantabria Formation	92
Fig. 3.17	Coprolite found in clay-sandy facies of the Upper Cretaceous, Maastrichtian. Cantabria Formation	93
Fig. 3.18	Endo-skeleton well preserved from an equinoid of the Paleogene	94
Fig. 3.19	Wood intensely bioturbed by a xylophagous, which produced a high concentration of perforations (inchnogenus Teredolites sp.). Monos Formation. Photo courtesy of Carlos Borges and Alberto Arano	97
Fig. 3.20	Abundant shells of gastropods in marmorized limestones of the Paleogene, Middle Eocene. Charco Redondo Formation	97
Fig. 3.21	Macroforaminifera in asphalt matrix of the Paleogene, Middle Eocene. Peñón Formation	98
Fig. 3.22	Traces fossils (*Scolicia* isp.) produced in sandy-clayey sediments of the Paleogene, Lower Eocene. Capdevila Formation	99
Fig. 3.23	Oxidized impressions of plants of the Upper Neogene, Pliocene. El Abra Formation	100
Fig. 3.24	Internal mold of a gastropod of the Neogene, Miocene. Güines Formation	101
Fig. 3.25	Decapod species *Euphylax* sp. from the Lower Miocene. Colón Formation	102
Fig. 3.26	Numerous specimens of cirripeds (*Balanus* sp.), adhered on an oyster shell. Lagunitas Formation	102
Fig. 3.27	Mold of the endocrane of a sirenium found in limestones of the Neogene. Colón Formation. Courtesy of Lázaro Viñola López	103
Fig. 3.28	Fossil teeth of Neogene sharks, Miocene. Güines Formation	104
Fig. 3.29	Fragment of bone skeleton of fish, from the Lower Miocene, collected in the region of Jaguey Grande, during the extraction of limestone blocks for construction. Colón Formation. Courtesy of Lázaro William Viñola López	104
Fig. 3.30	Very small perforations of a predator on bivalve shells *Lima* sp. Lagunitas Formation	105

List of Figures

Fig. 3.31	*Strombus giga*, a large gastropod common in the marine rocks of the Quaternary. It is easily seen on the colonial walls of Habana Vieja. Jaimanitas Formation. This species has current representatives (known as Cobo), which inhabit the shallow seas of Cuba	108
Fig. 3.32	The corals are diverse and abundant in the limestone rocks of the Quaternary. Jaimanitas Formation	108
Fig. 3.33	True coquina of diverse shells of terrestrial gastropods, formed in a Quaternary cave	109
Fig. 3.34	Large bony remains of sloths on a sandy asphalt matrix of the Quaternary Las Breas deposit in San Felipe, Matanzas	110
Fig. 3.35	Quaternary turtle bones preserved in karstic deposit. Courtesy of Lázaro Viñola López	111
Fig. 3.36	Various coprolites or pellets of rodents, mainly of "Jutias", preserved by the impregnation in asphaltite. Las Breas de San Felipe, Matanzas	112
Fig. 3.37	The vast majority of the fossiliferous vertebrate sites of the Quaternary in Cuba appear in the diverse manifestations of the extensive karst of their territory. Cave in Cayo Paredones, Sancti Spíritus	113
Fig. 3.38	Traces of plants in the Quaternary eolianits	114
Fig. 3.39	Wealth in the composition of the fossil record in Cuba for the five geological periods represented by its rocks	114
Fig. 3.40	A large scientific work has been carried out on the fossils of Cuba, although there are still several aspects in which to deepen the investigations	141
Fig. 4.1	**a** Scheme of paleogeographic domains and UTEs used for this chapter. **b** Legend of Fig. 4.1	145
Fig. 4.2	Folds lying in the San Cayetano formation. Km. 13 of the highway Consolación del Norte (La Palma) to Santa Lucía, province of Pinar del Rio	147
Fig. 4.3	Rocks from the upper part of the San Cayetano Formation, where studies of palynomorphs were carried out, south of Cinco Pesos, province of Artemisa	148
Fig. 4.4	Radiolarian silicites, clays, and interbedded tuffites. Santa Teresa formation in the locality of Sitio Morales, province of Pinar del Rio	153
Fig. 4.5	Basalts with pad structure, diabases with thin layers of carbonate rocks. El Sábalo Fm., mountain road, on El Sábalo hill	155
Fig. 4.6	Carbonate rocks well stratified in thin to medium layers, typical of the Artemisa Fm., Vegas Nuevas quarry, province of Pinar del Rio	156
Fig. 4.7	Micritical and marly limestones, violaceous, of the Ancón Fm., north of Valle del Ancón, province of Pinar del Rio. It has manganese minerals between the layers	160
Fig. 4.8	Policomponent Olistostrome of the Manacas Fm. in the vicinity of Rancho Mundito, province of Pinar del Rio. Observe large olistolites of limestones	160
Fig. 4.9	Original position of lithostratigraphic units of the North American continental margin and its substrate of synrift in placetas proto TSU (Precompresional)	162
Fig. 4.10	Radiolarian silicites, clays. and tuffites of the Santa Teresa Formation in the locality of La Sierra, south of Sierra Morena, province of Villa Clara	163
Fig. 4.11	Scheme of Placetas proto TSU and its orogenic coverage after the collision of the Caribbean plate with the North American one. Stratigraphic position of the Bacunayagua formation	164
Fig. 4.12	Carbonate rocks of the palenque formation in the Viet Nam Heroico quarry, sierra de Cubitas, Camagüey province	172

Fig. 4.13　Generalized stratigraphic columns of the PD North American continental margin and its coverage central Cuba-Havana. Current position of the TSUs 174

Fig. 4.14　Geological–structural model for block 17 (Puerto Padre-Gibara Region), provinces of Las Tunas and Holguín [40] 177

Fig. 4.15　Marmorized black limestones of the Mayarí Formation, homonymous locality, province of S. Spíritus 180

Fig. 4.16　Scheme with the location of the Central Basin of Cuba, provinces of S. Spíritus and C. de Ávila. Notice some of the oil and gas fields 182

Fig. 4.17　Basalts and andesite-basaltic rocks of El Cobre Group. Observe the columnar arrangement. Location Puerto de Moya, Carretera Central, near El Cobre, province of Santiago de Cuba 183

Fig. 4.18　Limestones of the Charco Redondo Formation, middle Eocene, on the Guantánamo-Santiago de Cuba highway, west of Guantánamo 184

Fig. 4.19　Organogenic limestones of the Cabo Cruz, Maya and Jaimanitas formations, in the terraces of Maisí to the south in the coastal zone, province of Guantánamo. Upper Pliocene—Early Pleistocene 186

Fig. 5.1　Major strike-slip faults of Cuba (red lines) and Precenozoic regional structures G: Guaniguanico Mountains, SR: Sierra del Rosario Mountains, Gm: Guamuhaya Massif, IY: Isla de la Juventud, V: Varadero lineament, P: Pinar fault zone, Y: Yabre lineament, T: La Trocha lineament, C: Camagüey lineament, O: Oriente fault zone, CEEZ: Cuban Exclusive Economic Zone in the Gulf of Mexico, MY: Mayabeque Province, LT: Las Tunas Province, GT: Guantánamo, F: Florida Peninsula 192

Fig. 5.2　The regional structures below the cover in western Cuba. Modified from Cobiella Reguera [19]. SO: Sierra de los Órganos TSU, APS: Alturas de Pizarras del Sur TSU; SR-APN-E: Sierra del Rosario-Alturas de Pizarras del Norte-Esperanza TSU; G: Pan de Guajaibón TSU; C: Cangre Belt TSU; NOB-KVAT: Northern Ophiolite Belt and Cretaceous volcanic arc terranes; PgB: Paleogene Basin (Cover); MM: Martin Mesa uplift (outlier of SR-APN-E); Cover: N-Q: Neogene–Quaternary deposits 193

Fig. 5.3　Lithostratigraphic columns of the Guaniguanico Cordillera TSU. An: Ancon Fm.; M: Moncada Fm.; Peña Fm.; Pons Fm.; A-T (El Americano, Tumbadero and Tumbitas members of Guasasa Fm.); SV: San Vicente Fm.; J: Jagua Fm.; SC: San Cayetano Fm.; Arroyo Cangre Fm.; Mr: Morena Fm.; Pa: Pinalilla Fm.; C: Carmita Fm.; ST: Santa Teresa Fm.; L: Lucas Fm.; Pl: Polier Fm.; Ar: Artemisa Fm.; ES: El Sábalo Fm. Modified from Cobiella Reguera [19]. See also Pszczolkowski [108] 194

Fig. 5.4　Outcrop of El Sábalo Fm., with a mix of mafic rocks (basalt and diabase-m) and xenoliths-x, of thin-bedded Upper Jurassic limestones and shales, containing complex synsedimentary folds 195

Fig. 5.5　Schematic map (**a**) and tectonic profile (**b**) along the central part of Guaniguanico Mountains, from Pinar fault zone to the northern coast, near Santa Lucia. Compare the flat lying nappes in APS and SO TSU with the north dipping linear structures in the SR-APN-E. **c** North dipping lowermost Cretaceous beds at Vegas Nuevas quarry, near La Palma, SR-APN-E 195

Fig. 5.6　Breccia in the San Adrian diapir, Mayabeque Province. According to Meyerhoff and Hatten [78], the main lithologies in the clasts are different types of limestones (those fine grained could contain radiolarian and Nannocunus), metamorphic rocks (marbles and quartz–mica schists), quartzose sandstones, and isolated metric lenticular blocks of serpentinite. Additionally, in the formation appear some grains of pyrite, tourmaline, and apatite 196

Fig. 5.7	Thin bedded chert and shales (dark rocks in the photo) in the vulcanogene-sedimentary member of the western Cuba ophiolites (Encrucijada Formation), south of Bahia Honda, Artemisa. Below, and probably in tectonic contact, rest deeply weathered basalts	198
Fig. 5.8	Vulcano-serpentinitic mélange at Loma Esmeralda, Matanzas Province. Serpentinite (S), strongly brecciated and with slickensides, contact with tuffs (T)	198
Fig. 5.9	Profile located near the eastern end of Cuba, from San Antonio del Sur, in the Caribbean coast, to Moa, in the Atlantic shore. NOB: G-gabbroids, S: serpentinites, and serpentinized ultramafic rocks, m-melánges. The KVAT is represented by Upper Cretaceous metavolcanics and some metasediments (the Sierra del Purial complex). It probably rests horizontally sandwiched between the ophiolites	199
Fig. 5.10	Upper Cretaceous conglomeratic and sandstone beds of La Trampa Formation, eastward Havana city. These sediments show that, contrary to the events in the Early Cretaceous, in the Late Cretaceous volcanic arc, some volcanic cones suffer the subaerial erosion	201
Fig. 5.11	Palinspastic reconstruction of the NW Caribbean–SW Gulf of Mexico region surrounding western and central Cuba at the Mesozoic/Cenozoic boundary (for details consult [20]). m: original area of Moncada-type deposits, c: original area of Cacarajícara-type deposits, a: original area of Amaro-type deposits, p: original area of Peñalver-type deposits, sc: original area of Santa Clara-type deposits	202
Fig. 5.12	Paleogeographic scheme at the Cretaceous/Paleogene boundary. Observe the complex regional relief in the area of the future western and central Cuba at the moment of the Chicxulub impact [20] modified from . m: original area of Moncada-type deposits, c: original area of Cacarajícara-type deposits, a: original area of Amaro-type deposits, p: original area of Peñalver-type deposits, sc: original area of Santa Clara-type deposits	203
Fig. 5.13	Early Paleogene tectonic profile across central Cuba, including the Turquino–Cayman Ridge arc, in its southern edge to the foreland basin, in the north	205
Fig. 5.14	Profile from Los Palacios Basin to the DSDP sites 97 and 540 in the SE Gulf of Mexico (modified from [19]). The profile integrates seismic and DSDP well data in the SE Gulf with data from inland wells and geological cartography in Pinar del Rio Province, Cuba	206
Fig. 5.15	Simplified tectonic map of the western end of the Sierra Maestra Mountains, supported on the interpretation of the Pushcharovsky [110] map and the unpublished Geological Map of Cuba from the Instituto de Geología y Paleontología [53]	207
Fig. 5.16	Basaltic columns at Puerto Moya, Central Road, Santiago de Cuba Province. The rocks belong to the middle–upper part of the axial strata of the Turquino–Cayman Ridge Paleogene arc	208
Fig. 5.17	Lower Eocene tuffs outcroping near the Caribbean coast at La Farola road, Cajobabo, Guantanamo.Th strata belong to a tectonic scale with T CR rocks (El Cobre Formation) and some slivers of serpentinite breccias (see Cobiella et al. [25], Iturralde- Vinent [54], and Cobiella-Reguera [23]). The scale (circa 5 sq. kilometers and perhaps 500 meters of maximum thickness) rests on the basal beds of the Central Oriente Intermontanious Basin	209
Fig. 5.18	Olistostrome of the basal beds of San Luis Formation at Cajobabo. The deposit is a pile of unsorted blocks, the biggest attaining several meters. The most abundant clast lithologiesare lapillitic tuffs and agglomerates with some porfiritic andesites.Obviously they were torn from a near T-CR	

	source rocks. Additionally, some greenschists clasts (from the Upper Cretaceous volcanic arc) and gabbro-pegmatites (from the NOB) are present.	209
Fig. 5.19	Stratigraphic distribution of the cover sub-stages in different localities. Whereas in western Cuba the oldest cover beds are almost of basal Lower Eocene age, in central Cuba the cover begins with Middle Eocene strata and in easternmost Cuba the lowermost cover rocks settled in the latest Late Eocene. In the figure, the different nature of the socle between eastern and central-western Cuba is emphasized.	210
Fig. 5.20	Unconformity between cover sub-stages C and D at Vía Blanca highway, La Habana. D: sub-stage D, represented by the cross-bedded Pleistocene Guanabo Formation, C: sub-stage C, represented by the shallow water, strongly weathered Lower–Middle Miocene Güines Formation.	210
Fig. 5.21	Distribution of uplifted areas and basins in the cover from Eocene to Quaternary. Included are the localities with salt diapirs in Cuba.	211
Fig. 5.22	Main structures in the Columbus strait (located between SE Cuba and Jamaica). The geologic setting is explained in the text.	212
Fig. 5.23	Southward prograding Neogene–Quaternary beds (toward the Caribbean Coast) at Macambo, Imías, Guantánamo Province. The study of the bedding reveals at least two sets: magenta and blue lines in the photo. A third (green) set is poorly defined.	212
Fig. 5.24	Simplified tectonic map of NE Matanzas and NW Villa Clara Province. The structural setting is explained in the text. Very important are the two small outcrops of Precambrian metamorphic rocks near Itabo (Matanzas) and Motembo (Villa Clara).	214
Fig. 6.1	Generalized geological map of Cuba (after Iturralde-Vinent 1998) with indication of the sampling sites. 1: ENC1, ENC2; 2: HAV5, SM1, SM2; 3: MGT1, MGT2, MT3, MT31; 4: SC1, SAG1; 5: CEN202; 6: MEL201; 7: M202.	221
Fig. 6.2	Diagram of volcanic rock classification from Pearce [27].	224
Fig. 6.3	Discrimination diagrams (A-C) of Pearce [29].	225
Fig. 6.4	Discrimination diagram of Saccani [35]. The Th and Nb values are 0.12 ppm and 2.33 ppm, respectively (taken from [30]).	225
Fig. 7.1	Scheme of stratigraphic correlation of the Quaternary of Western and Central Cuba. (alg – Alegrías Formation; vd – Vedado Formation; gne – Guane Formation; vs – Versalles Formation; gv – Guevara Formation; gnb – Guanabo Formation; vr – Villarroja Formation; js – Jaimanitas Formation; ro – Cayo Romano Formation; cmc – Camacho Formation; coc – Cocodrilo Formation; sf – Playa Santa Fe Formation; lcb – La Cabaña Formation; sdo – El Salado Formation; sgn – Siguanea Formation; lpi – Los Pinos Formation; mQ_2 – marine deposits; alQ_2 – alluvial deposits; pQ_2 – marshy deposits; bQ_2 – biogenic deposits)	235
Fig. 7.2	Stratigraphic correlation scheme of the Quaternary of Eastern Cuba. (rm – Río Maya Formation; by – Bayamo Formation; dt – Dátil Formation; gv – Guevara Formation; vr – Villarroja Formation; js – Jaimanitas Formation; ca – Cauto Formation; jam – Jamaica Formation; lcb – La Cabaña Formation; sdo – El Salado Formation; mQ_2 – marine deposits; alQ_2 – alluvial deposits; pQ_2 – marshy deposits).	235
Fig. 7.3	Satellite image of the Guanahacabibes Peninsula and adjacent areas. (pQ_2 – marshy deposits; bQ_2 – biogenic deposits; gne – Guane Formation; sgn – Siguanea Formation; vd – Vedado Formation; js – Jaimanitas Formation; gv – Guevara Formation).	237

Fig. 7.4	Satellite image of the Ciénaga de Zapata region and adjacent territories. (pQ$_2$ – marshy deposits; mQ$_2$ – marine deposits; bQ$_2$ – biogenic deposits; vd – Vedado Formation; vr – Villaroja Formation; js – Jaimanitas Formation; preQ – prequaternary rocks)	243
Fig. 7.5	Satellite image of the Cauto basin. (pQ$_2$ – marshy deposits; alQ$_2$ – alluvial deposits; ca – Cauto Formation; dt – Dátil Formation; by – Bayamo Formation; preQ – prequaternary rocks)	245
Fig. 7.6	Satellite image of the Isla de la Juventud. (pQ$_2$ – marshy deposits; mQ$_2$ – marine deposits; bQ$_2$ – biogenic deposits; gne – Guane Formation; gv – Guevara Formation; js – Jaimanitas Formation; ccl – Cocodrilo Formation; preQ – prequatenary rocks).	246
Fig. 8.1	**a–d** Spatial distribution of mineral Systems in Cuba. Maps and legend. The contours of the ten identified mineral systems, extracted from the Metallogenic Map of the Republic of Cuba at scale of 1:250 000 [29], are represented on a map of the administrative (municipal) political division of Cuba, shown in three separate maps to avoid overlapping and provide better visualization of them.	260
Fig. 8.2	Location of the outcrop areas of continental margin sequences in Cuba. In violet color the names of Cuban regions mentioned in the text. Numbers in red: 1. Yabre alignment, 2. La Trocha Fault and 3. Cauto—Nipe Fault	261
Fig. 8.3	Matahambre ore deposit. Fragmented pyrite (Py), allotriomorphic to hip idiomorphic aggregates (10x)	265
Fig. 8.4	Matahambre ore deposit. Pyrite (Py) with cataclastic texture with cracks filled with chalcopyrite and non-metallic minerals (20X)	265
Fig. 8.5	Longitudinal profile of the Guachinango prospect. As can be seen, the body is always concordant with the gently folded lay of the embedding rocks. Taken from Bolotin [16].	267
Fig. 8.6	Diffraction of RX of sample LISTE-3, taken in a greisen outcrop in Santa Elena, NW Isla de la Juventud. Symbols used: 1—U-uvite, 2—K-kaolinite, 3—G-goethite, 4—Q-quartz. The goethite here appears in small quantity. There is a relatively abundant phase of uvite (member of the solid solution dravite-uvite) of the tourmaline group, which is a relatively rare mineral and its presence is frequently related to evolved, volatile rich magmas and other characteristic elements of these	275
Fig. 8.7	**a–c** Polished sections of LID-4 sample. **a** Euhedral and idiomorphic crystals of arsenopyrite (asp) with presence of cataclastic in gangue minerals (gan) in section polished. Parallel Nicoles. **b** Thin section with parallel Nicoles. Quartz with abundant fluid inclusions and sulfide mineralization bordered by scaly mineral. **c** Thin section with crossed Nicoles. Quartz (Qz) with abundant fluid inclusions and sulfur bordered by scaly mineral, this also in veinlets	276
Fig. 8.8	Distribution Map of useful components in Lela. Legend. Circles: wells. In golden color Au; in light green Cu; in blue cyan Mo and in magenta W. Triangles: outcropping mineralization; coloring for Au, Cu, Mo, and W as in the case of wells. Red lines: faults. Dark green thick outline: Possible contour of intrusive body at 600–800 m depth. Dark blue thick contour: Possible contour of intrusive body at 400–600 m depth (in both cases deduced from the interpretation of the gravimetric data). The scale of the map does not allow to show the abundant faults present in the central area of the deposit.	277

Fig. 8.9	Results of the description of the polished section corresponding to a sample of ferricrete by schists of the. Cañada Fm. (J_{1-2}). **a** and **b** Hematite + goethite (hm + goe) with texture banded according to the direction of schistosity and Wolframite (wt) in gangue minerals (gan). Parallel Nicoles. **c** Grains of hematite and goethite intercepting wolframite. Parallel Nicoles	278
Fig. 9.1	Mineragenic and geo-structural map of Cuba	299
Fig. 9.2	Frontal view of Río del Callejón residual kaolin deposit	300
Fig. 9.3	View of the El Canal residual feldspar sand deposit	301

List of Tables

Table 1.1	Foreign countries and specialists participating in geological mapping of Cuba	33
Table 1.2	First results of the New Geological Map of Cuba at 1:50000 scale	33
Table 2.1	Types and subtypes of coast according to Decree—Law 212/2000	42
Table 2.2	Geochronological table of the Pliocene-Quaternary deposits of Cuba. Green: formations located in the coastal territory. Blue: those that appear in keys or in the submarine-coastal territory (Jaimanitas and Vedado). Black: they are exclusive of the mainland. The hiatus in the Lower-Middle Pleistocene appears with an interrogation, because it is only typical of the marine territory, without a clear explanation of its origin	53
Table 3.1	Some Cuban lithostratigraphic units with Jurassic—Early Cretaceous fossil record	76
Table 3.2	Some lithostratigraphic units with Lower Cretaceous to Early Upper Cretaceous fossil record	84
Table 3.3	Some lithostratigraphic units with Late Upper Cretaceous fossil record	85
Table 3.4	Some lithostratigraphic units with Paleogene to Neogene fossil record	96
Table 6.1	Whole-rock major- and trace-element data for basalts from NCOB and ECOB	222
Table 6.2	Summary of studied basaltic rocks from NCOB and ECOB with their inferred tectonic setting of formation and proposed ophiolite classification	226
Table 7.1	Scheme of the stratigraphic subdivision of the Quaternary of Cuba	234
Table 8.1	Distribution of metallic mineral deposits models in Cuba for geodynamic environments and tectonic units	258
Table 8.2	Content of organic matter in the distensive paleo-basin of $J-K_1$	262
Table 9.1	Mineragenic characteristics of Cuban industrial rocks and minerals	293

Geological Cartography of Cuba

Ramón Omar Pérez Aragón

Abstract

Through the consultation of numerous archival reports, publications, maps, and personal inquiries to different authors and geologists of vast experience, it has been possible to compile a large amount of data about the evolution of geological cartography in Cuba throughout its more than 135 years of history. A brief critical analysis of the degree of growing of the country's geological knowledge is made, from the elaboration of the Geological Sketch of the Island of Cuba at 1:2000000 scale [30], first of Cuba and of Latin America, until the latest Digital Geological Map of Cuba at scale 1:100000 [71]. It's also briefly commented on the elaboration of some maps at detailed scales and the conditional surveys at scales 1:100000 and 1:50000 carried out in collaboration with the ex—Soviet Union and countries of the former Eastern European Socialist Field. Finally, the projection of the new geological mapping works at scale 1:50000 to be carried out throughout the national territory in the framework of the Geology Development Program until 2030 are briefly addressed. Special tribute is given to all people, Cuban and foreigners, who have worked in the field of geological cartography in Cuba throughout its history.

Keywords

Geological cartography • Geological survey • Geological map • Geology • Scales • Degree of growing of geological knowledge

R. O. P. Aragón (✉)
Instituto de Geología y Paleontología, Servicio Geológico de Cuba
Vía Blanca No. 1002 Entre Río Luyanó y Prolongación de Calzada de Güines, Reparto Los Ángeles, San Miguel del Padrón, La Habana, Cp 11000, Cuba
e-mail: ramon@igp.minem.cu

© Springer Nature Switzerland AG 2021
M. E. Pardo Echarte (ed.), *Geology of Cuba*, Regional Geology Reviews,
https://doi.org/10.1007/978-3-030-67798-5_1

List of Abbreviations

ACC	Academia de Ciencias de Cuba
CAME	Consejo de Ayuda Mutua Económica
CARIS	Computed Aided Resources Information System
Cartageol 50 K	Subprograma para la Cartografía Geologica de Cuba a escala 1:50000
CCT	Consejo Científico Técnico
CEINPET	Centro de Investigaciones del Petróleo
CIG	Centro de Investigaciones Geológicas
CNDIG	Centro Nacional de Información Geológica
CNFG	Centro Nacional del Fondo Geológico
CUR	Comisión de Unificación y Redacción
CVA	Cretaceous Volcanic Arc
DGGG	Dirección General de Geología y Geofísica
DIGIAC	Dirección de Ingeniería Geotécnica Aplicada de Cuba
FAR	Fuerzas Armadas Revolucionarias
GDR	German Democratic Republic
GEODATO	Banco de Datos Geológicos de la República de Cuba
GeoMinSal	Grupo Empresarial de Geología, Minería y Sal
GIS	Geographic Information System
ICA	Instituto de Ciencias Agropecuarias
ICGC	Instituto Cubano de Geodesia y Cartografía
ICRM	Instituto Cubano de Recursos Minerales
IGP	Instituto de Geología y Paleontología
IGT	Instituto de Geografía Tropical
ISMM	Instituto Superior Minero Metalúrgico
ISPJAE	Instituto Superior Politécnico José Antonio Echeverría
KGRT	Kiev Geo-Prospecting Technological Institute (in Russian)
MGRI	Moscow Institute of Geological Prospecting (in Russian)
MICONS	Ministerio de la Construcción
MINBAS	Ministerio de la Industria Básica
MINEM	Ministerio de Energía y Minas
MMG	Ministerio de Minería y Geología
NC 622	Norma Cubana para la Cartografía Geológica
ONRM	Oficina Nacional de Recursos Minerales
PAM	Passive Actives Mining
PVA	Paleogene Volcanic Arc
R & D	Research and Development
SGC	Servicio Geológico de Cuba
SIGEOL	Sistema de Información Geológica
UH	Universidad de La Habana
USA	United States of America
USSR	Union of Soviets Socialistic Republics
VSEGEI	Institute of Geological Researches of the Soviet Union (in Russian)

1.1 Introduction

The geological map of a country or territory is a reflection of the degree of knowledge that we have about it, the types of rocks, and the main tectonic structures that built up the area. Geological knowledge, in turn, is the basis for optimal use and exploitation of a territory.

At present, the degree of geological knowledge of a country is considered one of the fundamental indicators of its progress and development, since it facilitates planning and attracts capital investments in productive activities vital to the economy.

The Republic of Cuba has a long history in the preparation of geological maps at different scales. From the publication in 1883 of the Geological Sketch of the Island of Cuba at 1:2000000 scale, the first one in Latin America, the Cuban territory has been successively covered with specialized geological and associated thematic maps (tectonic and ore deposits, among others) at scales ranging from 1:1000000 to 1:100000.

In addition, about 55% of the territory is covered by conditional geological surveys at scales 1:50000 and 1:100000, which have a large amount of semi-detailed information, and the rest are 1:250000 scale surveys.

Recently [2], the Instituto de Geología y Paleontología–Servicio Geológico de Cuba (IGP-SGC) has completed the updating and edition of the new Geological Map of Cuba at 1:100000 scale in digital format and in a Geographic Information System (GIS) environment, which covers the entire Cuban archipelago, including the marine-coastal zone. However, the geological cartography of the Republic of Cuba requires moving to a higher stage, which allows the collection and updating of all the scattered information at larger scales, to proceed with the systematization and completion of the geological cartography of the national territory at 1:50000 scale, including its marine part.

Taking into account the above, the IGP-SGC, with the support of the Geology Directorate of the Ministerio de Energía y Minas (MINEM), implemented, as part of the Development Program of Geology until 2030, the Subprogram of Geological Cartography of Cuba at 1:50000 scale (Cartageol 50 K). This establishes the policy of standardizing and completing the geological mapping of the country on this scale, thus completing one of the fundamental tasks of the IGP-SGC, which is the constant increase in the degree of study geological of the national territory.

The purpose of this chapter is to disseminate, in a summarized way, the history of the evolution of Geological Cartography in Cuba up to present. Also, to pay tribute to all those who directly or indirectly contributed and contribute even today, with their personal and collective effort, in the construction of the solid building that today constitutes the geological knowledge of Cuba, represented in its geological maps.

1.2 Materials and Methods

For the elaboration of this work, historical research techniques were used, carried out through the consultation of archival reports, personal inquiries with researchers and personnel from different specialties who worked in the activity of geological cartography during different stages of its development.

Among the materials consulted are dozens of reports and maps of the geological survey works at scales 1:50000, 1:100000, and 1:250000, stored in the archives and databases of the Oficina Nacional de Recursos Minerales (ONRM) and the Instituto de Geología y Paleontología–Servicio Geológico de Cuba (IGP-SGC). Also, a large number of monographs on different aspects of Cuba's geology, books, atlases, explanatory memoirs, and other textual and graphic materials resulting from the different geological works were carried out in different epochs and regions of Cuba, most of which are listed in the bibliography.

Of invaluable help were the personal communications obtained from numerous geologists and other specialists who worked in the activities of the geological surveys, or in the reception and critical review of the materials resulting from them.

1.3 Results

With a long tradition in the elaboration of geological maps, the cartography of this subject in Cuba has been characterized by the constant increase of the degree of study, which is evident, not only in the quantity and diversity of the studies carried out, but in the progressive increase of the scale and the degree of detail of the maps.

Cuba has, as we already know, the oldest map of Latin America: The Geological Sketch of Cuba at 1:2000000 scale, dating from 1883, made, as has also been said, by Spanish mine engineers Manuel Fernández de Castro Suero and Pedro Salterain Legarra, which constitutes a transcendental achievement of geological cartography in the nineteenth century.

In the first half of the twentieth century, what is known as the prerevolutionary stage, numerous works were carried out, mainly by foreign geologists (North Americans, Dutch, Germans, Swiss, Italians, and others), with their corresponding maps of different sectors and at different scales. The only general map of the Island, officially recognized, was the Geological Sketch of Cuba at 1:1000000 scale [3], edited and published by the Technical Commission of Forestry and Mines of the Ministry of Agriculture of that time.

Since the triumph of the Cuban Revolution in 1959, the country stood out for an unusual development in this aspect, becoming the head of the states of the region. It counting today with geological maps at scales 1:1000000, Instituto Cubano de Recursos Minerales (ICRM) 1962, IGP 1989 and 2014; 1:500000, Centro de Investigaciones Geológicas (CIG) 1985; 1:250000, IGP 1989; and 1:100000 in digital format, IGP 2016, of the entire national territory. The latter, is a result of the generalization and updating of a series of conditional geological surveys at scales 1:50000, 1:100000, and 1:250000, carried out over several years, from the early 1960s to the early 1990s of the past XX century.

It is known that from the decade of the sixties, the first conditional geological surveys at 1:50000 and 1:100000 scales began to be carried out in Cuba with companion searches at detailed scales (1:25000 and 1:10000) of the most perspective sectors. In this stage, between the years 1961 and 1982, close to 17 surveys were carried out [4].

From the end of the seventies until the mid-nineties of the last century, in the framework of collaboration agreements with the Consejo de Ayuda Mutua Económica (CAME) in the field of Geology and Mining, the last conditional geological surveys were carried out. The so-called "CAME polygons" had an approximate coverage of 38% of the national territory at 1:50000 scale and 17% at 1:100000 scale, which as a whole cover approximately 55% of the country (op. cit.).

These surveys were carried out in the regions that were considered as, and in fact constitute, the most prospective areas for the search of superficial and shallow mineral deposits. However, the disappearance of the USSR and the Socialist Field and the deep economic crisis that this historic event generated for Cuba, caused the interruption of all the works, leaving many planned areas without covering.

On the other hand, until 1959 the geological investigations of the submerged area of the national territory (marine-coastal areas) were minimal and dispersed. After the revolutionary triumph, exploration and geological cartography works were promoted for the development of various fields of the economy, obtaining a remarkable volume of information of these areas. In this, there were involved institutions of the Academia de Ciencias de Cuba (ACC), the Ministerio de la Construcción (MICONS), the Ministerio de la Industria Básica (MINBAS), and the Fuerzas Armadas Revolucionarias (FAR), with the cooperation of academic institutions of the former Soviet Union and Socialist Field [5].

Some of the works carried out in this part of the national territory are reported in different serial publications; the rest, along with many other investigations, are registered in numerous archival reports. Almost all, or the great majority of them, are reflected in the maps, including the Geological Map of the Republic of Cuba at 1:500000 scale [6] and the Digital Geological Map of Cuba at 1:100000 scale [2]. These are subject to analysis in the "Degree of Study of the Marine-coastal Territory", prepared for the program of Geological Map of Cuba at 1:50000 scale, called "Cartageol 50 K subprogram".

In the works of geological cartography in Cuba throughout its history, in addition to Cubans, hundreds of geoscientists have worked from more than 15 countries, among which the USSR, Bulgaria, Czechoslovakia, Hungary, the German Democratic Republic, Poland, and several North American, Dutch, and other geologists and geophysicists, representing, essentially, different oil companies, during the prerevolutionary stage.

Among the different specialists involved in geological mapping, geologists, geophysicists, geochemists, hydrogeologists, geomorphologists, paleontologists, micro paleontologists, petrographers, mineralogists, palynologists, and others stand out. Among the auxiliary or support personnel are geodetics, surveyors, drillers, drivers, translators, and even draughtsmen. All, with a greater or lesser degree of incidence, contributed to elevate the degree of geological knowledge, traveling a long and ascending way to the place of honor where the geological cartography in Cuba is at present.

1.4 Discussion

1.4.1 Geological Maps at General Exploratory Scales (1:2000000–1:1000000)

Geological sketch of the Island of Cuba at 1:2000000 scale

The Geological Sketch of the Island of Cuba (Fig. 1.1), made between 1869 and 1883 by the Spanish mine engineers Manuel Fernández de Castro Suero and Pedro Salterain Legarra, has the undeniable merit of being the first of its kind in Cuba and in Latin America [7]. It is unquestionable that it has an enormous historical value and constitutes, in addition, a monument to the stubbornness, courage, and perseverance of the geoscientists, personified values in the authors of this great scientific work.

It should not be ignored the historical conditions of the period (1869–1883) in which were made the necessary geological investigations, like sample collection, description of outcrops, and other fieldworks. It has to be remembered that this period coincided with the first Cuban wars to achieve its independence from the Spanish metropolis: The Ten Years War (1868–1878) and the so-called "Little War" (1879–1880). It is not difficult to imagine the adverse circumstances and the immense dangers, even for their lives, that these peninsular researchers had to face and avoid.

This map, although from a purely technical point of view, is a document that could be considered obsolete, due to the advance of geological knowledge. It does not stop being a work of great scientific merit and, at the time historical value for being, as it has already been declared, the first of its kind not only in Cuba, but also throughout Latin America.

From the geological point of view, however, the authors must be credited with having reflected relevant geological aspects, such as the correct layout of the structures in what is

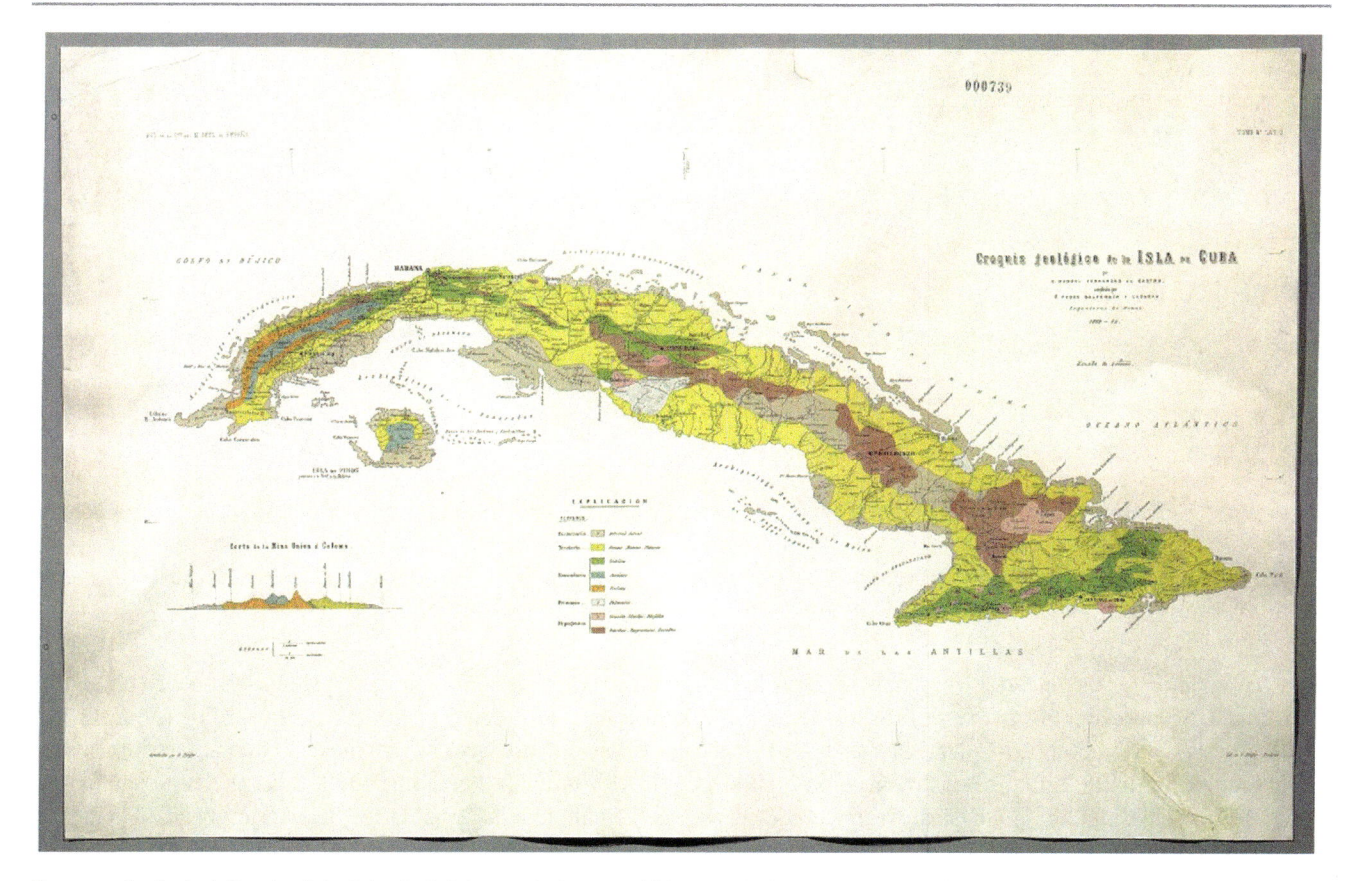

Fig. 1.1 Geological Sketch of the Island of Cuba; made between 1869 and 1883 by the Spanish mine engineers Manuel Fernández de Castro Suero and Pedro Salterain Legarra. The first of its kind in Cuba and in Latin America [7]

now known as the Cuban direction, that is, distributed throughout and parallel to the central axis of the island. With great success, they assigned Jurassic ages to the limestone rocks of the Cordillera de Guaniguanico and to the marbles and shales of the metamorphic massif of the Isla de Pinos (today Isla de la Juventud); they recognized the Cretaceous age of the rocks of the volcanic arc and colored the rocks of the Miocene sedimentary cover in yellow.

They also mapped the occurrence of intrusive rocks in the igneous terrains of Camagüey—Las Tunas, the "Manicaragua granitoids", as well as the bodies of oceanic crust rocks that make up the belt of the ophiolite association of Cuba, which they grouped under the generic name of "diorites, serpentines, and basalts".

The most significant mistake seems to be that of having assigned Triassic age to the slates of the north and the south of the mountain range of the west of Cuba, which is not very objectionable, because they were homologated with the oldest rocks, being in fact, and the oldest of Cuba. Still today, they are known to be from the Lower Jurassic to the Superior (Oxfordian), and are known as the San Cayetano, Castellano formations, and the Arroyo Cangre metamorphic fringe.

Another notable inaccuracy is to have considered from the Cretaceous the rocks of the Paleogene Volcanic Arc (PVA) of the Sierra Maestra, an error evidently induced when these rocks were considered as an extension of the Cretaceous Volcanic Arc (CVA), which appears in the rest of the Cuban territory.

Geological sketch of Cuba at 1:1000000 scale

During the first half of the twentieth century, most of the geological work was carried out by foreign technicians, mainly North Americans and, to a lesser extent, Dutch, Germans, Italians, etc., who made explorations for oil and gas, with their corresponding maps of different sectors and at different scales. The only general map of the island, officially recognized, was the Geological Sketch of Cuba at 1:1000000 scale [3], made in 1941–1946. It was published that year by the Forestry and Mines Technical Commission of the Ministry of Agriculture of the time, with a second edition by the Technical Commission of Mining and Geology of the same ministry in 1955. (Figure 1.2).

When comparing the 1946 Cuban Geological Sketch, with its predecessor, it is possible to notice many differences between them. First, the scale is doubled, which constitutes a

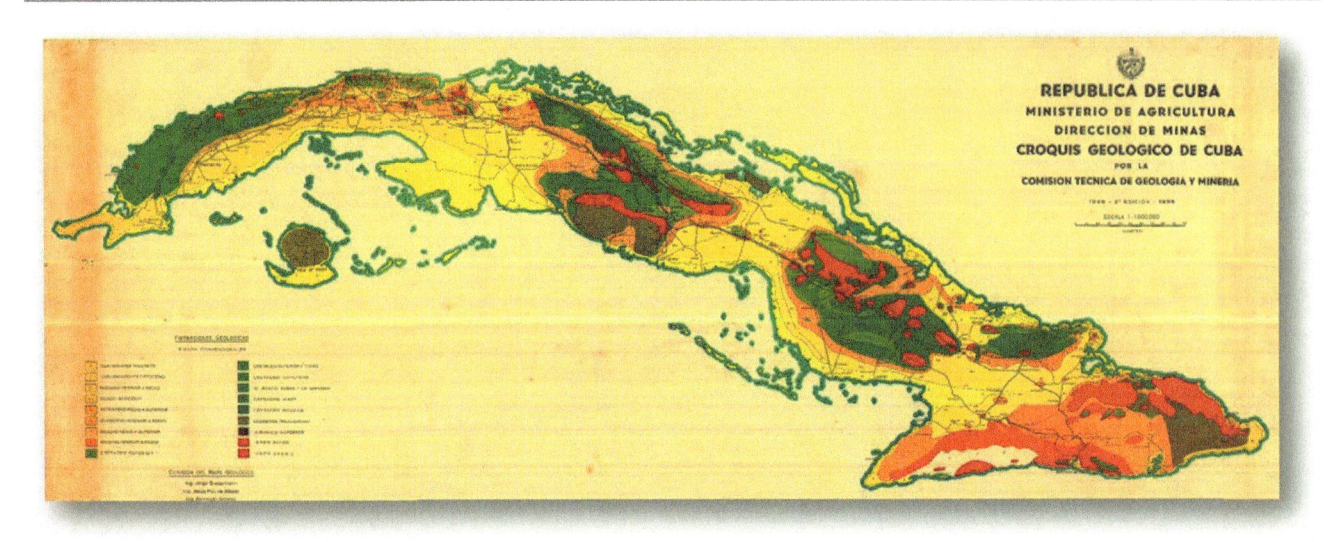

Fig. 1.2 Geological Sketch of Cuba at 1:1000000 scale, made between 1941 and 1946 by a Forestry and Mines Technical Commission of the former Ministry of Agriculture of Cuba, led by Jorge Brödermann Vignier (Brödermann et al. [3])

notable advance in the increase of detail and geological knowledge. Second, and this is one of the most remarkable, this map was made entirely by Cuban specialists, led by the Civil Engineer, Architect, and paleontologist Jorge Brödermann Vignier; the doctor in Physical-Chemical Sciences, Civil Engineer, paleontologist, and stratigraphist Jesús Francisco de Albear Fránquiz; and by the engineer-topographer and amateur geologist-surveyor Armando Andreu, without having data on other participants.

From the geological point of view, the greatest success seems to be the implicit recognition of the second arc of volcanic islands (PVA), when the volcanic and volcano-sedimentary rocks of the Sierra Maestra are located in the Paleogene. Although they are still treated as intrusive rocks, there is a better outline of the ophiolites of Eastern and Central Cuba. In this last region, an excellent cartography of the "Manicaragua granitoids" is made, although these are not yet separated from the "Mabujina amphibolites". The "metamorphic massif of the Escambray" is delimited with great accuracy, as well as the others from the eastern end of Cuba and the Isla de Pinos (Isla de la Juventud). In this last territory, the rocks of the Cretaceous Volcanic Arc are still not mapped. The salt domes of the north of central Cuba are mapped for the first time, to which they are assigned not Jurassic but the Upper Cretaceous age.

In the Cordillera de Guaniguanico, however, the authors corrected the error of mapping as the Triassic the so-called "slates" of the north and south (formations that we know today as Arroyo Cangre, Castellano, and San Cayetano). However, they commit another, perhaps worse, by rejecting the Jurassic age correctly assigned prior to the carbonated formations of this region (Artemisa, Jagua, Viñales formations, among others) and assigning the Cretaceous age to the entire north-western mountain system. Another step back, so to speak, is to have assigned a yellow color to recent formations (quaternary), which makes it difficult to differentiate them from those belonging to the Miocene carbonate cover.

Highlights the qualitative leap that means having an updated topographical base for the time, emphasizing the roads, rural roads, and railroads networks.

As a summary, it is worth saying that, with successes and mistakes, this work meant in its time a great leap forward in the geological knowledge of the Island of Cuba and its surrounding keys.

Geological Map of Cuba at 1:1000000 scale

After the triumph of the Cuban Revolution, on January 1, 1959, the country was characterized by an unusual development of geological activity in general and geological cartography in particular, becoming at the head of the states of the region. Appointed Commandant Ernesto Guevara de la Serna (Che), at the head of the Ministry of Industries, making use of the futuristic vision and the advanced economic thought that characterized him, Che conceded great importance to the development of the geological-mining activity. He led the process of nationalization of all companies, including the foreign, that operated in Cuba and of all the geological information that laid in their archives.

Collaboration in this area was agreed with the countries of the so-called Socialist Field, especially with the Union of Soviet Socialist Republics (USSR), Czechoslovakia, Hungary, Poland, Bulgaria, Romania, and the German Democratic Republic (GDR). There they began to be sent the first students to get prepared in these specialties and from where they began to arrive specialists and researchers of high scientific level, in quality of advisers and consultants for the ordering and processing of all the available geological information and to develop up this activity in the country. The valuable collaboration of a group of Hungarian specialists, led by the Vegh spouses, in the organization of the Centro Nacional del Fondo Geológico (CNFG), now the Oficina Nacional de Recursos Minerales (ONRM), should be highlighted here.

The results of this collaboration were obtained immediately. As early as 1964, the first book of Geology of Cuba, prologued by Che himself, became known. However, two years earlier, in 1962, the first Geological Map of Cuba had been published at 1:1000000 scale (Fig. 1.3). It was prepared in the Instituto Cubano de Recursos Minerales (ICRM) of the then Ministry of Industries, by a team of Cuban and Soviet scientists led by the geographer and Captain of the Rebel Army, Antonio Núñez Jiménez, the engineer-topographer Armando Andreu, and the Soviet geologists' academics Boris Bogatiriov, Ivan Novajatsky, and Konstantin Judoley.

It is worth noting that since that time the tradition was established of putting in the credits of the maps, as main editors and/or responsible editors, the administrative and scientific leaders, under whose responsibility the work of cartography was carried out. The degree of direct participation of the same in the tasks of compilation, generalization, and interpretation of the information, as well as in the writing and edition of the maps, was subordinate. This practice, taken from the rules governing cartographic works in the USSR and other socialist countries, sometimes makes it difficult to know the real authorship or the degree of participation of the specialists in the different works.

Of which we are concerned, it can be said that it constitutes the first Geological Map of Cuba—the previous ones were practically diagrams (sketches)—elaborated with a scientific methodology, in accordance with the Soviet norms of geological cartography. If compared to its predecessors, the qualitative leap is evident and substantial.

Among the aesthetic improvements that come to the fore is the correct application of conventional international signs, referred to the age (colors) of rock formations and complexes. In this way, the domains of the ages of the recognized formations in Cuba are recognized immediately: Jurassic (blue), Cretaceous (green), Paleogenic (orange and brown), Miocene (yellow), Quaternary (gray), and so on.

In addition, the domains of the intrusive rocks (red) are separated for the first time, from those corresponding to the oceanic crust (magenta), which constitutes not only an aesthetic improvement, but also a great step forward in the geological knowledge of the national territory, due mainly to the metallogenic implications of this differentiation. In addition to the Oriente and Centro regions, the ophiolitic rocks are mapped in the Habana-Matanzas region, and the Cajálbana massif at the western end, thus completing the so-called ophiolitic belt of Cuba.

Continuing in the analysis of the geological improvement, the "Mabujina amphibolites" are separated from the "Manicaragua granitoids". The mapping of the rocks of the CVA is perfected, separating this from the rocks of the PVA in the Sierra Maestra, accepting the presence in this zone of two superimposed arches of islands, although the extension of the first domain is still exaggerated.

Among the aspects to be overcome in this map are, besides the aforementioned exaggeration of the CVA

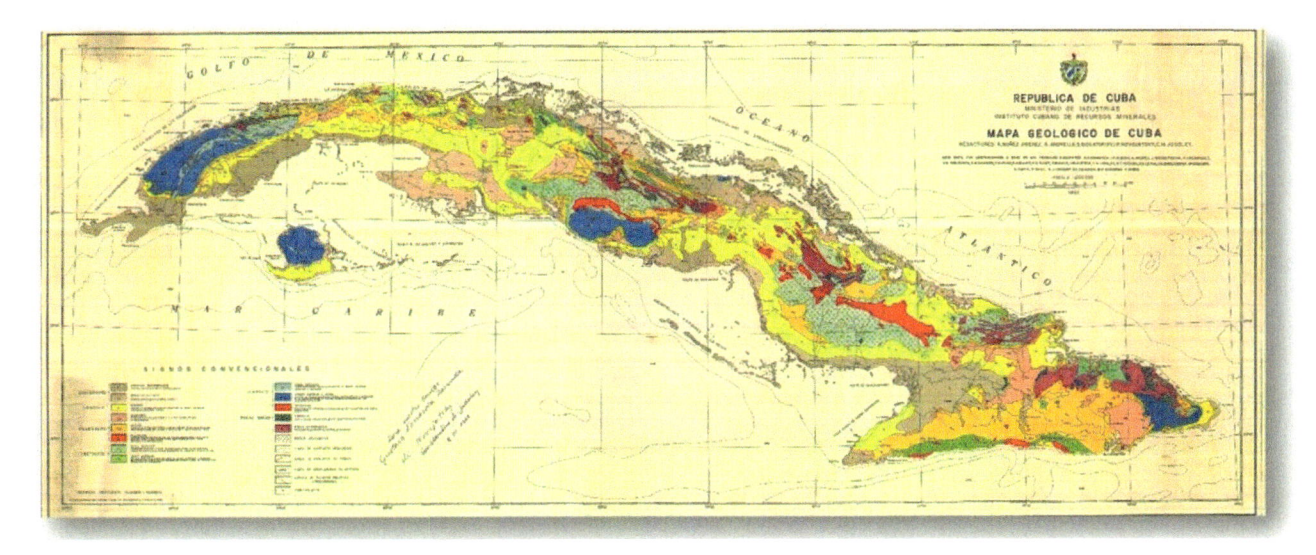

Fig. 1.3 First Geological Map of Cuba at 1:1000000 scale [8]

domains in the Sierra Maestra, there is the omission of the presence of the same in the north of the Isla de Pinos and Pinar del Río. Also, the absence of the Paleogenic olistostromic deposits (Vieja and Manacas formations) of the Guaniguanico Mountain Range.

Despite these inaccuracies and some others that escape these brief comments, the geological map of Cuba at 1:1000000 scale is one of the first great scientific achievements of the nascent Cuban Revolution. In addition, it was the first in the field of geosciences, which marked in Cuba a before and an after in the history of geological sciences in general, and of geological cartography in particular.

1.4.2 Geological Maps at Regional Recognizance Scales (1:500000–1:250000)

Geological Map of the Republic of Cuba at 1:500000 scale

Since the ends of the 1970s, a group of Cuban and Soviet geologists working at the CIG of the MINBAS, entities that had replaced the previous institutions Dirección General de Geología y Geofísica (DGGG) and Ministerio de Minas y Geología (MMG), respectively, had begun to make a new geological map of Cuba at 1:500000 scale (Fig. 1.4). It would double the scale of the previous one, also made by the Soviets and Cubans [8].

To achieve this, it had begun to collect and generalize the preexisting geological information, including the one recently generated by the 1:250000 regional conditional surveys carried out by the IGP of the ACC in collaboration with the Academies of Sciences of the USSR, Poland, Hungary, Bulgaria, and GDR. In addition, it considers the first surveys with companion searches at scales 1:50000 and 1:100000, carried out by brigades of Cuban and Soviet geologists in the territories of the Mining Companies of the MMG.

In the credits of the map, Jesús Pérez Othón, at the time Deputy Minister of Geology of the MINBAS and V. Yarmoliuk, chief and representative of the mission of Soviet collaborating specialists before said ministry, appear as Chief Editors. As Responsible Editors appear the geologist engineer Elio de los Santos Llerena, then Director of the CIG, the licensed geologist Jesús Hernández Fernández, Deputy Director of Geology, and the leading Soviet specialists I. N. Tijomirov and A. M. Zagoskin.

The drafting board of the map consisted of the Soviet specialists A.V. Dovbnia, P.G. Osadchiy, K.M. Judoley, Y. B. Evdokimov, B.A. Markovskiy, V.A. Trofimov, and A.L. Vtulochkin, and of the Cuban geologists Evelio Linares Cala, Dora García Delgado, Alberto Zuazo Alonso, Gustavo Furrazola Bermúdez, and Amelia Brito Rojas.

Among the authors of the map is an extensive collective of Cubans and Soviets. On the Cuban side, the geological engineer Evelio Linares Cala headed the work group, from the beginning. It was integrated by Santa Gil, Luis García, Rafael González, Carbeny Capote, Jesús Hernández, and Francisco Vergara, marine geologist Miguel Cabrera (Brigada Geomar), geologists (doctors) Jorge Luis Cobiella (of the Instituto Superior Minero Metalúrgico (ISMM)), Lilavatti Díaz de Villalvilla, Eugenia Fonseca, Alfredo Norman, and Mireya Pérez, the geophysical engineer Verania Bello, and the geographer licensed Leandro Peñalver (from the IGP). Also, it was integrated by Elio de los Santos, Marlene

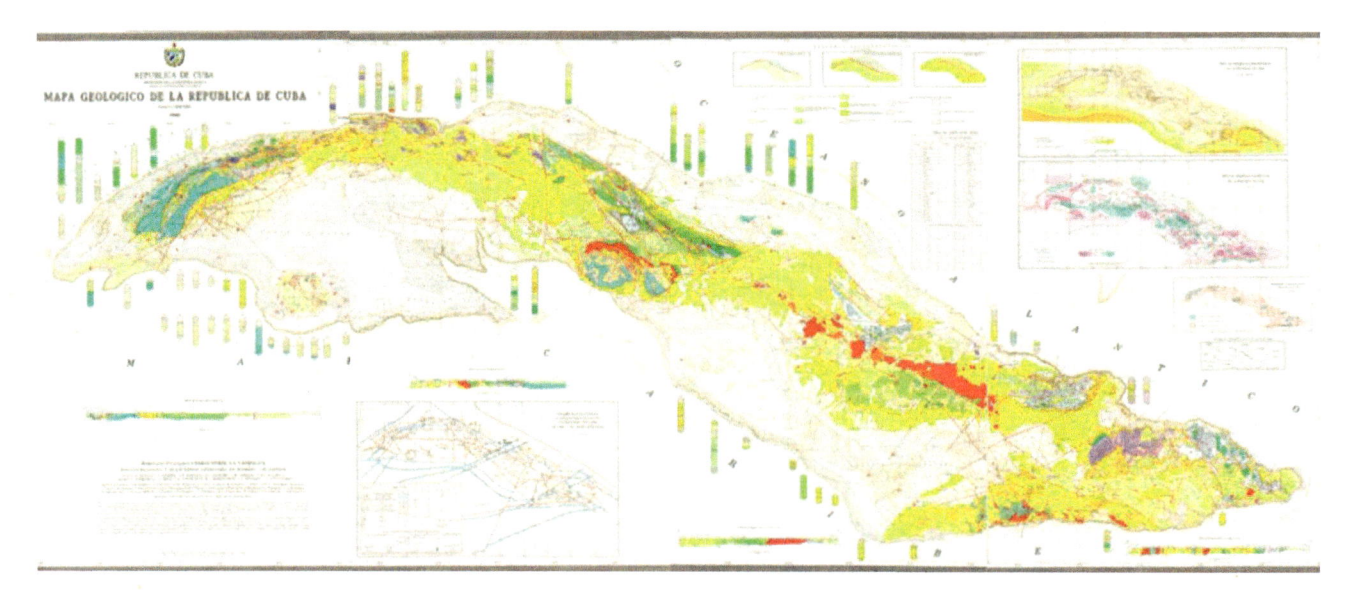

Fig. 1.4 Reduced view (4 sheets) of the Geological Map of the Republic of Cuba at 1:500000 scale, from 1985. A fifth sheet located on the right (not shown) contains a legend of zonal type

Dilla, Raisa Delgado, Gustavo Echevarría, Ileana García, Margarita Heredia, Roberto Morales, Ernesto Milián, Ana María Recio, Jorge Sánchez (CIG), as well as Manuel Iturralde, Guillermo Millán and Domingo González (IGP) and also José Oro, Omara Avello, Carlos Suyí (Brigada Geomar), and many others.

Leading the team of Soviet consultants and collaborators was P.G. Osadchiy, while the group of specialists who worked directly on the map was led by the geologist A.V. Dovbnia, and integrated by W.A. Busch, O.I. Eguipko, Y.B. Evdokimov, K.M. Judoley, L.A. Kondakov, B.A. Markovskiy, I.N. Tijomirov, V.A. Trofimov, A.L. Vtulochkin, A.M. Zagoskin, and V.N. Zelepuguin, among others.

All the works of data compilation and drafting of the Geological Map of the Republic of Cuba at 1:500000 scale were carried out in Cuba by the appointed staff and innumerable technicians, specialists, and drafters of the CIG.

Among the main materials used are the maps of the geological surveys at 1:250000 scale, prepared by the Academies of Sciences of Cuba, Bulgaria, Hungary, and Poland. It considers the Cuban authors J. F. de Albear, M. A. Iturralde, D. P. Coutin, and L. L. Peñalver; the Bulgarians I. L. Kantchev, I. Boyanov, D. Tchounev, E. Belmustakov; the Hungarian E. Nagy; the Polish K. Piotrowska, A. Pszczolkowski, and others. The unified geological map was also used at 1:250000 scale, prepared by the academies of sciences mentioned above, in conjunction with the Academy of Sciences of the USSR.

The maps of different surveys at 1:50000 and 1:100000 scales and the materials and data contributed by numerous authors were also used. Among these are the Soviets A.F. Adamovich, V.D. Chejovich, K.P. Astajov, A.M. Alioshin, V.A. Burov, O.I. Eguipko, I.N. Garapko, L.M. Golovkin, R. I. Ismagulov, K.M. Judoley, G.N. Kuzovkov, M.L. Somin, A.V. Maksimov, A.A. Nikolaev, I. Kartashov, I.A. Shevchenko, A.V. Dovbnia, V.S. Shein, G.B. Simakov, A.A. Teperin, A.N. Varvarov, and many others.

The final works of editing and printing were carried out in the Institute of Geological Research of the USSR (nowadays of Russia) "AP Karpinski" (VSEGEI, in the Russian language), in the city of Leningrad (now St. Petersburg), in 1985.

Among the main advantages of this map with respect to the previous ones, in addition to doubling the scale, as already noted, is that, despite being a surface geological map, the deep structure of the country is well sketched in four regional transversal cuts, that cross the Island from south to north, perpendicular to the Cuban structure. In addition, around 65 columns of wells reflect the composition of the subsoil. The map is accompanied by several boxes with diagrams at 1:2500000 scale, where valuable additional information is exposed, such as maps of magnetic and gravimetric anomalies, as well as a tectonic interpretation scheme.

One of the most outstanding innovations of this map is that, for the first time, it includes the mapping of the lithological composition of the sediments of the seabed between the coastline and the upper edge of the insular slope. To this purpose, all the data provided by previous works of different Cuban entities of the ACC, MICONS, MINBAS, and the FAR were used, with the cooperation of academic institutions of the former Socialist Field.

A decisive role in the contribution of data, the compilation and writing of the geology of this part of the map played the members of the extinct "Brigada Geomar", attached to the then Union de Geología of the MINBAS (today Grupo Geominsal of the MINEM), and directed by geologists José Oro Alfonso and J.R. Rodríguez García.

Among the members of this brigade were the geological engineers specialized in marine geology, Miguel Cabrera, Raciel González, and Victor Estrada. Also, the geologists Carlos Hernández, Rolando Batista Leyva, Jorge L. Álvarez, Jorge Ríos, Violeta Ramos, Magaly Sánchez, Rafael Rodríguez, Jorge Nápoles, María Eugenia, and Aidee Díaz. Then, the geophysical engineers Carlos Suyí, Luis Moreno, and Rubén Corrada; the geographers Rolando González, Carlos Ayla, Luis Arquímedes, and Bernardo Columbié; the specialized divers Pedro Tito, Roberto Castellanos, Castor A. Rodríguez, and Porfirio Pérez; the drillers Ángel Fonseca, Nelson Gutiérrez, Heriberto Venegas, and Raymundo Garcés; the biologists Jorge Puentes and Omara Avello, as well as the surveyor José M. García, among others.

This map, in addition to the advantages already mentioned of duplicating the scale, served as a geological basis for the preparation of a series of thematic maps at 1:500000 scale, namely: The Map of Metallic Mineral Deposits and Mineral Waters and the Map of Deposits and Manifestations of Non-Metallic and Fuel Minerals of the Republic of Cuba.

Map of Metallic Mineral Deposits and Mineral Waters of the Republic of Cuba at 1:500000 scale

Prepared and edited in 1986–1988, on the map are reflected, with an appropriate legend, all the deposits and manifestations of metallic minerals known to date, as well as the most important mineral water sources. It also shows some 20 cross-sections of the main deposits based on geophysical and drilling data (Fig. 1.5).

Among the main editors of the Map of Metallic Mineral Deposits and Mineral Waters, appear the director of the CIG Elio de los Santos, the Cuban specialists Leandro Peñalver and Rustin Cabrera of the IGP, Rafael Lavandero, and Jesús Hernández from CIG, the Bulgarian A.I. Krivtsov, and the Soviets Igor N. Tijomirov and Victor A. Trofimov.

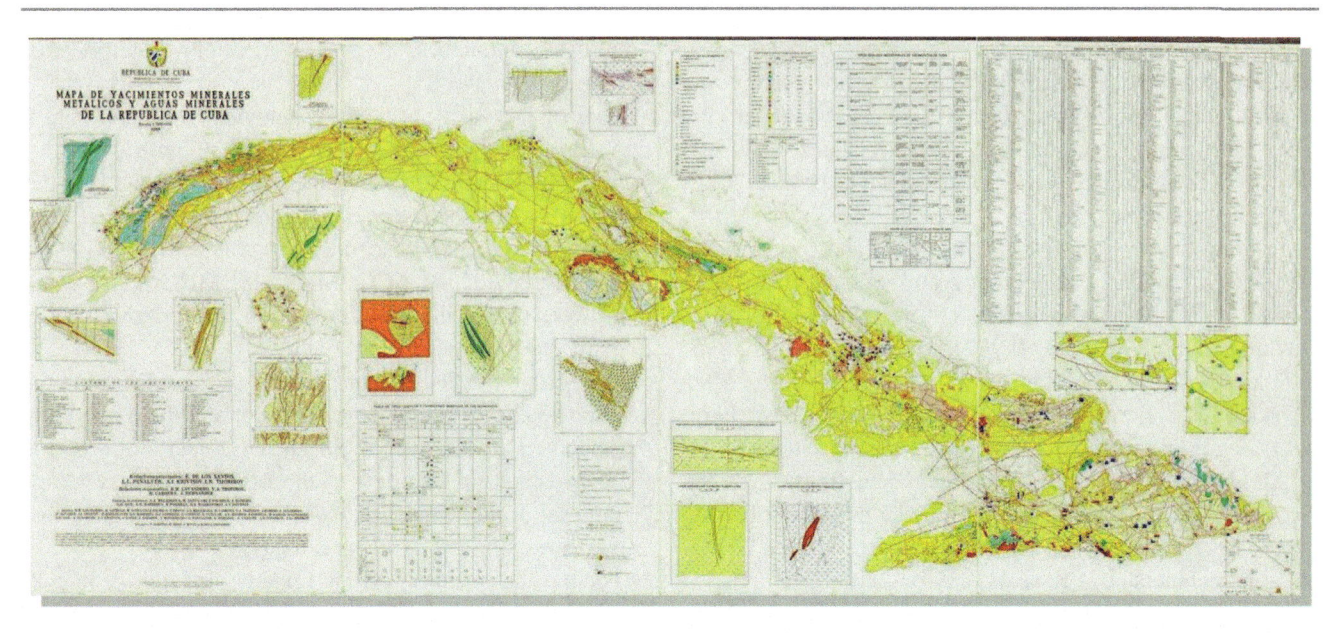

Fig. 1.5 Map of Metallic Mineral Deposits and Mineral Waters of the Republic of Cuba at scale 1:500000 (De los Santos Llerena et al. [84]

The group of authors consists of a long list of Cuban, Soviet, and Bulgarian specialists, namely: Cuban engineers Mario Estrugo, María Santa Cruz-Pacheco, Félix Bravo, Rustin Cabrera, Julio Romero, Idenia Altarriba, Pauxides Álvarez, Dalia J. Carrillo, Xiomara Cazañas, Julio Montenegro, Guillermo Pantaleón, Francisco Formell, and Jesús López. The technicians Miguel García, Octavio Vázquez, Domingo González, and Néstor Cuéllar. While the Soviet Group was made up of the geologists A.A. Melnikova, V.A. Trofimov, I.L. Aniatov, A.V. Dovbnia, A.N. Barishev, G.G. Gue, L.J. Krapiva, A.M. Zagoskin, and A.Y. Zhidkov, and the Bulgarian team were I. Lozanov, B. Badamgavin, A.I. Krivtsov, A. Janchivin, and N. Stefanov.

Map of Deposits, Manifestations of Non-Metallic, and Fuel Minerals of the Republic of Cuba at 1:500000 scale

This map was developed in parallel and proceeded analogously to the previous one, only that, the deposits and manifestations known to date of non-metallic raw materials and the main deposits of hydrocarbons reported until then in Cuba are reflected with an appropriate legend. The marginal information includes several boxes, where enlargements of the areas with important reserves or perspectives of non-metallic mineral raw materials are shown. It also includes a list of the names and main characteristics of the mapped deposits (Fig. 1.6).

As main editors of the map, appear Elio de los Santos, director of the CIG; the Cuban specialists Leandro Peñalver (IGP), Jesús Martínez, and Zulema González (CIG); the Bulgarian A.I. Krivtsov; the German geologist Lottar Lippstreu; and the Soviets I.N. Tijomirov, A.M. Zagoskin, B.A. Markovskiy, A.V. Dovbnia, and K.A. Klischov.

The group of authors consists of numerous Cuban, German, Czechoslovakian, Bulgarian, and Soviet specialists, namely: Cuban engineers Donis P. Coutin, Jesús Martínez, Zulema González, Blanca Delgado, Juan G. López, Gustavo Echevarría, Rodobaldo Rodríguez, the geographer licensed Mario Barea, and the technicians Eugenia Rodríguez and Miguel García. The group of foreign geologists consisted of the German geologist Lottar Lippstreu, the Czech M. Marek, the Bulgarians A. Kamensky, J. Tabak, R. Cerny, and the Soviet A.V. Dovbnia.

Geological Map of Cuba at 1:250000 scale

One of the first tasks addressed by the IGP of the ACC, after its foundation in 1967, was the development of the Geological Map of Cuba at 1:250000 scale. International collaboration with the Academies of Sciences of the member countries of the former Eastern European Socialist Field was used for this purpose.

In the introduction of the Explanatory Report of the geological map of the province of La Habana at 1:250000 scale [9], a constituent part of this great project, the story of the beginning of the collaboration that led to this great cartographic work is expressed better than anywhere else. In said document, it is expressed as verbatim:

> The Geological Meeting of La Habana, held in February 1968, constitutes the real beginning of the activities of the geological mapping of Cuba at 1:250000 scale… In this meeting, the representations of the Academies of Sciences of Bulgaria,

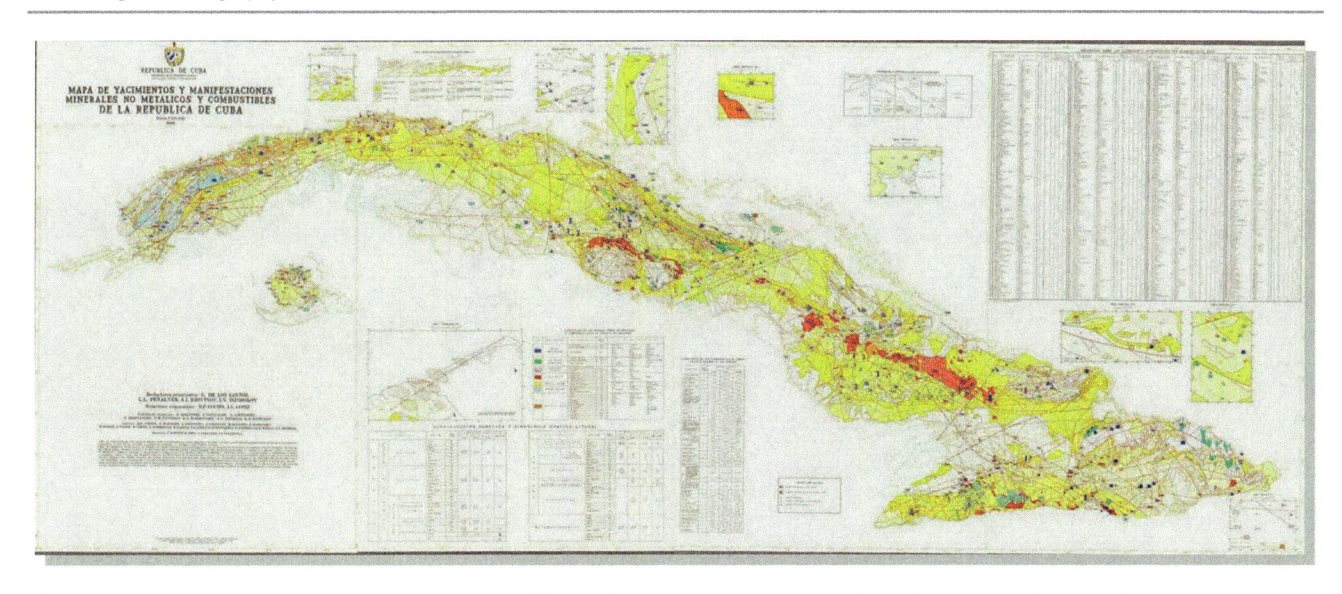

Fig. 1.6 Map of Deposits, Manifestations of Non-Metallic, and Fuel Minerals of the Republic of Cuba at 1:500000 scale (De los Santos Llerena et al. [85]

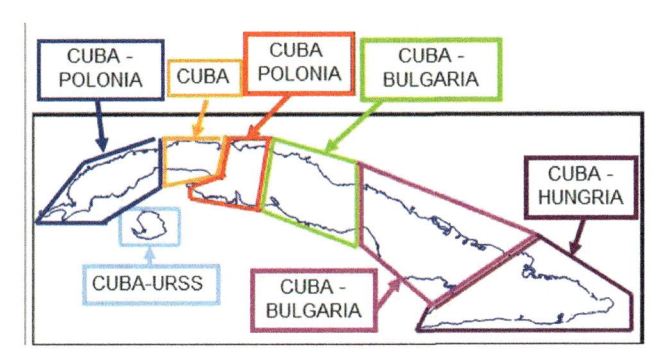

Fig. 1.7 Scheme of distribution of the areas to be mapped by the different brigades of the Academies of Sciences of the socialist countries for the preparation of the Geological Map of Cuba at 1:250000 scale. 1988. [2]

Czechoslovakia, Hungary, Poland, Romania and the Union of Soviet Socialist Republics, with a high spirit of internationalist confraternity, decided to establish a multilateral collaboration. So, the current Instituto de Geología y Paleontología of the ACC could face the great task that was he entrusted him with carrying out the fundamental research and the corresponding scientific and thematic studies that would allow obtaining a representative and updated cartographic instrument of the geological problems of our entire national territory.

As a result of the agreements taken at the aforementioned meeting, a survey program was prepared at the indicated scale, for which the national territory was divided into several polygons, each corresponding to one of the six provinces existing in the country before the political-administrative division of 1976 (Fig. 1.7).

In such a manner, the territories of Pinar del Río and Matanzas would be mapped by two brigades in which specialists from the Academies of Sciences of Cuba and Poland would collaborate. Other joint brigades of Bulgarian and Cuban specialists would be in charge of the mapping of the Las Villas and Camagüey territories. A Cuban-Hungarian brigade would map the territory of the former province of Oriente. Cubans and Soviets were in charge of cartography of the metamorphic terrains of the then Isla de Pinos and the Escambray Massif, as well as the insular territories of the archipelagos of the Canarreos, Jardines del Rey, and Jardines de la Reina. Cuban technicians, led by the experienced Jesús Francisco de Albear, mapped only the polygon corresponding to the current provinces La Habana, Mayabeque, and part of Artemisa.

The objective of this collaboration between the IGP of the ACC and the rest of the aforementioned academies of the former socialist republics of Eastern Europe was the geological cartography of the respective territories and, the transmission of knowledge and theoretical-practical skills to the assigned Cuban personnel, generally technicians of middle level and engineers of little experience.

Geological Survey of the Province of Pinar del Río at 1:250000 scale

A so-called First Cuban-Polish Brigade made the geological survey works at 1:250000 scale from the former province of Pinar del Río, which included the western part of the today province of Artemisa (Fig. 1.8). These works were carried out in 1972–1975 by a mixed brigade of Polish and Cubans specialists, belonging to the Institute of Geological Sciences of the Academy of Sciences of Poland and the IGP of the ACC.

The results of this collaboration were the geological map of the territory and a report under the name of "Explanatory text to the Geological Map at 1:250000 scale from the

Fig. 1.8 Scheme of distribution of the areas covered by survey at 1:250000 scale of the former province of Pinar del Río by the (first) Cuban-Polish Brigade between 1972 and 1975

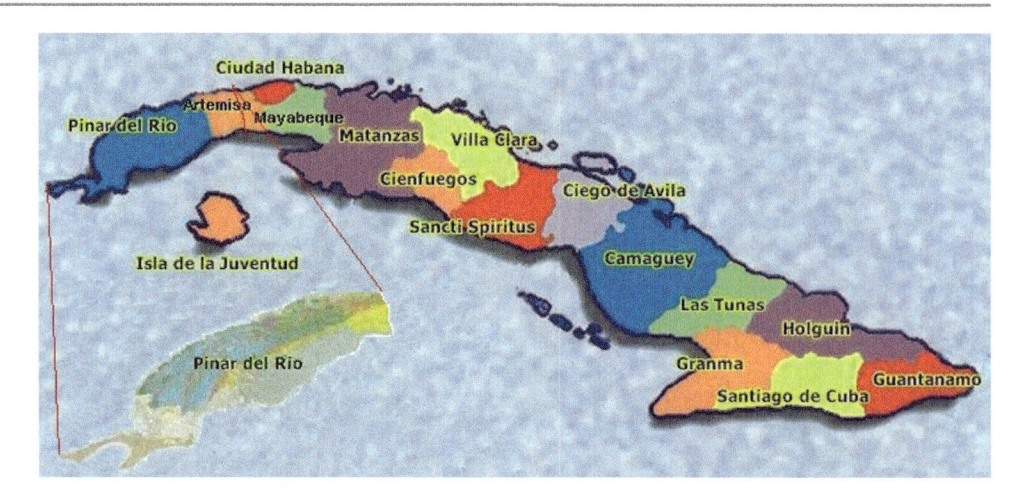

province of Pinar del Río". This was prepared in its entirety by the Polish specialists Andrzej Pszczolkowski, Krystyna Piotrowska, Ryszard Myczynski, Jerzy Piotrowski, Andrzej Skupinski, Jerzy Grodziccki, Daniel Danilewski, and Grsegorz Haczewski. The team for the fieldwork was made up of the cartographer geologists Daniel Danilewski, Jerzy Grodziccki, Grsegorz Haczewski, and Andrzej Pszczolkowski (group leader); by the geologist cartographer and tectonist Krystyna Piotrowska, the petrographer of sedimentary rocks Bozena Lacka, the biostratigraph Ryszard Myczynski, and the geologist, cartographer, and petrographer Andrzej Skupinski. In addition, the doctors in biostratigraphy of the Institute of Basic Geology of the University of Warsaw, Andrzej Wiezzbowski and Jan Kutek, collaborated.

On the Cuban side, the petrographer Gustavo Carrassou, the geologist and metamorphist Guillermo Millán, the micro paleontologist Alfredo de la Torre, the petrographer Marla Muñoz, and the biostratigraph María Luisa de la Nuez participated as specialists in petroleum geology and deposits. The auxiliary geologists Bernardo Machin, Esteban Díaz, and Eustaquio Gallardo also collaborated.

Geological Survey of the Province of La Habana at 1:250000 scale

At the beginning of 1974, all the territories of the country had begun their respective surveys at 1:250000 scale, except the provinces of La Habana and Matanzas. Faced with this worrying situation, the directors of the ACC and the IGP, proposed the task of undertaking work in this territory with a group (Habana-Matanzas Brigade) of Cuban specialists, with their own efforts and resources.

The works projected for the simultaneous surveying of both provinces began in May of 1974 and lasted throughout that year with great difficulties due to the scarcity of human and material resources, especially transportation. In 1975, due to the persistence of the listed problems, it was decided to restrict the works that were being executed, and to continue them only in the province of La Habana (Fig. 1.9).

The so-called Brigada Habana-Matanzas was made up of a small group of Cuban professionals and technicians, headed by Civil Engineer and Architect Jesús Francisco de Albear, with vast experience—perhaps the only one—in the regional geological survey work, since his participation in the making of the Geological Sketch at 1:1000000 scale [3]. The rest of the participants was a group of young geoscientists from the first group of geologists graduated from the Universidad de Oriente. Among them, graduates in geology specialized in petrography Gustavo Carrassou and Marlene Dilla, geologists Manuel A. Iturralde and Juan G. López, and the geographers Nestor Mayo and Leandro L. Peñalver. The paleontology team was led by the already veteran Dr. Alfredo de la Torre. The young biologist Consuelo Díaz, as well as the techniques Caridad Fuentes and Rafaela and Ernestina Pérez composed it. The technician Teresita Soto carried out the petrographic study of some samples.

The rest of the personnel dedicated to the field survey work consisted of a small troop of middle-level technicians in geology, composed by Arsenio Barrientos, José Ferrer, Julieta Maya, Hilda de la Paz, Isabel Salazar, Octavio Vázquez, as well as the geology assistants José A. Barros, Juan A. Estrada, Pascual Peregrín, and Angel Pérez.

The Brigade also had the support of some specialists from the DGGG of the then MMG, who collaborated in the paleontological determinations. This group was headed by engineers Manuel Marrero and José L. Iparraguirre, and integrated by Dr. on Natural Sciences and paleontologist Gustavo F. Furrazola; the licensees Gena Fernández and Silvia Blanco, specialists in foraminifera; Dr. Jorge Sánchez, who was in charge of the determination of ostracods and Emilio Flores, who dedicated himself to the study of radiolarians.

Fig. 1.9 Distribution scheme of the areas covered by survey at 1:250000 scale of the old province of La Habana by the Cuban IGP brigade called the Habana-Matanzas Brigade, 1974–1977

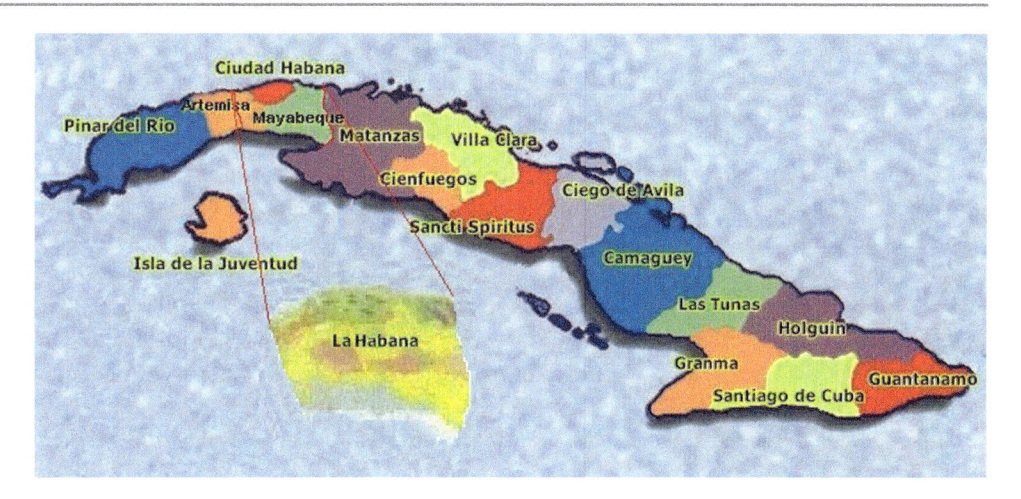

They also collaborated in the supply of hydrogeological and lithostratigraphic data from wells, several colleagues from the Instituto de Hidroeconomía, headed by the engineer Alfredo Álvarez. The engineers Roberto Martínez and Enrique Guillen, from DIGIAC, provided important data on geological and geotechnical engineering.

Because of these works, the geological map of the former province of La Habana was obtained at 1:250000 scale and a report called "Explanatory Memory of the Geological Map at 1:250000 scale of the Province of La Habana", whose authors of the fundamental texts are the engineers Jesús F. de Albear and Manuel A. Iturralde. Gustavo Carrassou, Nestor A. Mayo, and Leandro L. Peñalver also appear as authors of the auxiliary texts.

Geological Survey of the Province of Matanzas at 1:250000 scale

As has been described before, the first attempts to make the geological survey of the province of Matanzas, at 1:250000 scale, were carried out by a brigade of Cuban IGP specialists since 1974, which was initially called the Habana-Matanzas Brigade. Work in this last region was abandoned since the beginning of 1975 due to difficulties presented with material and human resources.

In April 1976, when the survey of the province of Pinar del Río by the first Cuban-Polish Brigade was completed, a protocol of collaboration between the Academies of Sciences of Cuba and Poland was signed in La Habana. In it was established that the geological survey of the province of Matanzas (Fig. 1.10) would be carried out by the staff of the Geological Sciences Institutes of Poland and the IGP between 1976 and 1980, for which another joint brigade of Cuban and Polish specialists would be formed, the second Cuban-Polish Brigade. The investigations would not begin, however, until September 1977, extending fieldwork until June 1979. Cabinetwork lasted until the end of 1980, so the stay in Cuba of Polish collaborators was delayed by this date, because all these works were made in the offices of the IGP in La Habana.

The results of this collaboration, consisting of the Geological Map of the province at 1:250000 scale and its textual and graphic annexes, are shown in the "Explanatory Text for the Geological Map at 1:250000 scale of the province of Matanzas ", presented in May 1981. The Polish and Cuban authors prepared it: Krystyna Piotrowska, Andrzej

Fig. 1.10 Scheme of distribution of the areas covered by survey at 1:250000 scale of the old province of Matanzas by the (second) Cuban-Polish Brigade in 1977-1981

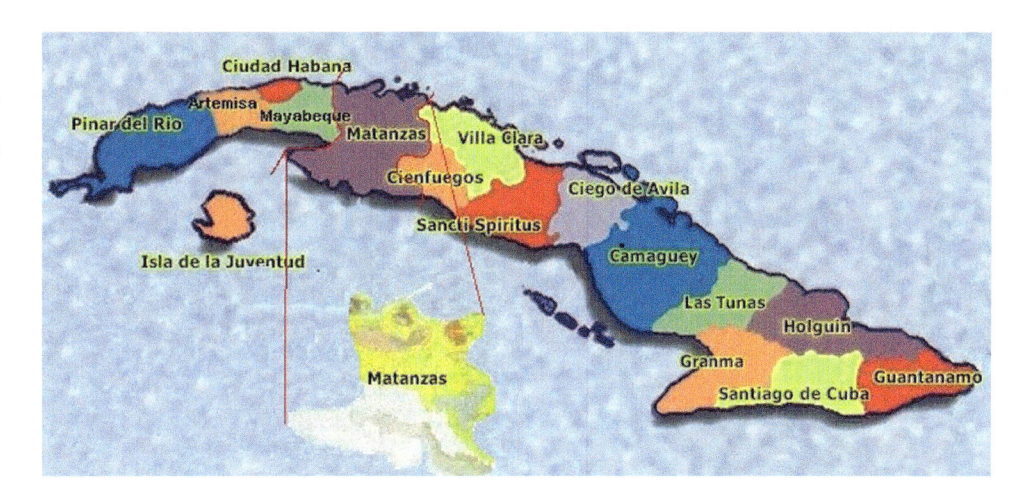

Pszczolkowski, Jerzy Piotrowski, Ryszard Myczynski, Jan Rudnicki, Michal Kuzniarski, Leandro Peñalver, Guillermo Franco, Nelson Pérez, and Jesús F. de Albear. Alfredo de la Torre, Stanislaw Dzulynski, Maciej Hakenberg, Consuelo Díaz, Danuta Peryt, Caridad Fuentes, Ernestina Pérez, and Gustavo Carrassou also collaborated in the writing of the different chapters.

The Polish geologist-doctors Andrzej Pszczolkowski, Jerzy Piotrowski, Krystyna Piotrowska, the Cuban engineer Nelson Pérez, and the Polish technician A. Szafranski carried out the fieldwork for the cartography of pre-quaternary sediments. Dr. Jan Rudnicki and the Cuban specialist Leandro Peñalver carried out the cartography of the quaternary sediments. The geologist stratigraphist, specialist in ammonites Ryszard Myczynski, carried out stratigraphic studies while Dr. Alfredo de la-Torre and Danuta Peryt, as well as the geologist technicians Ernestina Pérez and Caridad Fuentes carried out micro paleontological studies, with the assistance and supervision of the paleontologist Consuelo Díaz.

The Cuban specialist Dr. Guillermo Franco carried out Neogene stratigraphy studies, while as consultants participated Dr. Jesús F. de Albear, from Cuba, and from Poland Prof. Dr. S. Dzulynski and Dr. Maciej Hakenberg. The Polish Licentiate E. Janiak also carried out hydrogeological works, while the Polish Licentiate R. Kuzniarski and the also licentiate, but Cuban, Gustavo Carrassou, carried out the petrographic studies.

Geological Survey of the Province of Las Villas at 1:250000 scale

The geological survey of the territory of the former province of Las Villas was carried out between 1969 and 1975 by the so-called Bulgarian-Cuban Brigade, which as its name indicates was composed of specialists from the Institute of Geology of the Bulgarian Academy of Sciences and young specialists of the IGP of the ACC.

The results of this work are the "Geological Map of the Province of Las Villas at 1:250000 scale", which includes the territory of the current provinces of Cienfuegos, Villa Clara, and Sancti Spiritus (Fig. 1.11) and, the corresponding report, entitled "Geology of the Province of Las Villas. Result of the Investigations and Geological Survey at 1:250000 scale, carried out during the period 1969–1975", authored by I. Kantchev, I. Boyanov, N. Popov, R. Cabrera, A. Goranov, N. Volkichev, M. Kanazirski, and M. Stancheva [10]

The Bulgarian team consisted of around 25 specialists, led by geologists Emil Belmustakov and Ilia Kantchev. It was integrated by Liliana Dodekova, Nikolai Popov, Vladimir Schopov, Vassil Zlatarski, Alexandar Gomov, Nikolai Volkichev, Milko Kanazirski, María Stancheva, Maleschko Yordanov, Rumiana Racheva-Arnaudova, Ivan Dgyanov, the paleontologists Emilia Kojundgieva, Parankova Tzanova, Yanka Vaptzarova, and the hydrogeologists Delcho Molov, Todor Kojayov, Stefan Zvatkov, Boyan Alexiev, Zlatka Daikova, and Nikolina Ruskova.

For its part, the Cuban team was made up of about 20 people, including technicians of higher and middle level, as well as assistants and administrative personnel. Among them were the geologists Rustin Cabrera, Francisco Formell, Guillermo Millán, and Manuel Iturralde; the paleontologists Lenia Montero, María Luisa de la Nuez, Sara Arruti, Primitivo Borro, and Ángel García; the geologist technicians Jesús Triff, Manuel Domínguez, and Ernestina Pérez; the assistants Orestes Alfonso, Celso Cárdenas, Hugo Cordero, Luis Martín, Antonio Ortiz, and Dagoberto Paz. He also collaborated with the geophysical engineer Juan Antonio Valdés Bombino, from the DGGG of the then MMG.

It should be noted that the metamorphic terrains of the Escambray massif, located in this territory, was not studied

Fig. 1.11 Scheme of distribution of the areas covered by survey at 1:250000 scale of the former province of Las Villas by the (first) Bulgarian-Cuban Brigade between 1969 and 1975

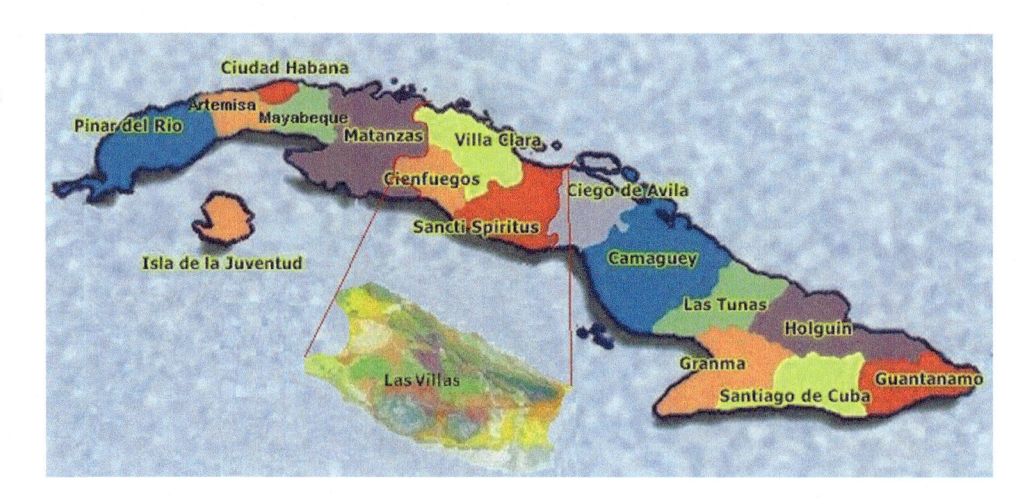

by the Bulgarian-Cuban Brigade, but was mapped by a mixed team of metamorphists composed of the Cuban engineer Guillermo Millán and the Soviet doctors Mark L. Somin, Gennady E. Nekrasov, and Sergei D. Sokolov. They carried out, in addition, the recognition and mapping of the metamorphites of this massif, as well as of the Mabujina Amphibolites that surround it from the north, and wrote an explanatory report for the Map of this complicated structure in the southern part of the territory of Central Cuba.

Geological Survey of the Province of Camagüey at 1:250000 scale

The geological survey at 1:250000 scale of the old province of Camagüey, territory that today encompasses those of Ciego de Avila, Camagüey, and Las Tunas (Fig. 1.12), would be carried out after the previous one (Las Villas), between the years 1976 and 1981. It was carried out by another team of Bulgarian and Cuban specialists of the Institute of Geology of the Academy of Sciences of Bulgaria and the IGP of the ACC. This time, the mixed brigade would receive the name of Cuban-Bulgarian Brigade.

The results of this survey are included in the report entitled "Geology of the Ciego-Camagüey-Las Tunas territory. Result of the investigations and geological survey at 1:250000 scale", whose scientific writing was in charge of the geologists Manuel Iturralde, Dimiter Tchounev, and Rustin Cabrera. In the preparation of the chapters of this report, participated specialists from the Bulgarian side: Emil Belmustakov, Elena Dimitrova, Milcho Ganev, Ivan Haydoutov, Vasil Kostadinov, Slavtcho Ianev, Jovka Ianeva, Emilia Kojumdjieva, Evgenia Koshujarova, Nikolai Popov, Vladimir Shopov, Plamen Tcholakov, and Tzanko Tzankov, while Rustin Cabrera, Consuelo Díaz, Manuel Iturralde, and Fidel Roque, did it for Cuba.

Slavcho Ianev, Ivan Haydoutov, Vasil Kostadinov, Vladimir Shopov, Plamen Tcholakov, Ivan Velinov, Ivan Lotov, Stoian Dakov, Yanka Vaptzarova, Iovka Ianeva, and Gueorgui Radev participated also in the survey work on the Bulgarian side. On the Cuban side, Guillermo Millán, Nelson Pérez, Arsenio Barrientos. Numerous IGP geologist technicians also worked, including Néstor Enríquez, Fernando Fleites, Gabino Blanco, Juan A. Estrada, Raimundo Vargas, Hiray Arzuaga, Domingo González, Jorge Díaz, and Julieta Maya.

Geological Survey of the Province of Oriente at 1:250000 scale

The geological survey works at 1:250000 scale in the territory of the former province of Oriente, encompassing what we know today as Guantanamo, Holguín, Santiago de Cuba, Granma, and part of Las Tunas (Fig. 1.13). It was carried out between 1972 and 1976 by a mixed group of Cuban specialists from IGP of the ACC and foreign, mainly from the State Institute of Geology of Hungary, the so-called Cuban-Hungarian Brigade, with the additional collaboration of some Soviet and Romanian specialists.

The group of foreign specialists of this survey was formed by the Hungarian geologists Elemer Nagy, Josef Ando, Karoly Bresznyansky, Pal Gyarmaty, Peter Jacus, Laslo Korpas, Gyula Radocz, M. Baldine-Beke, M. Rajos, G. Noskene-Fazekas, L. Ravaszne-Beranyai, and Imra Vato; the paleontologists Josef Bona and Pal Gozcan; and the geophysicist Ivan Polcz. The Soviet specialists Aleksandr K. Cherniajovski and V. Erotchev-Chak, as well as the Romanians M. Bratu and G. Popescu, collaborated at different times in the development of the survey and map preparation work.

Among the Cuban Geologists, they were Donis P. Coutin, Francisco Formell, Guillermo L. Franco, Nelson Pérez, the geologist technicians José Oro, Guillermo Pantaleón, Pablo Varela, the petrographer Amelia D. Brito, the paleontologist Alfredo de la-Torre, the geophysicists Carlos Sacasa, Pablo Lledías, and others.

As a result of this survey, the geological map of the province of Oriente and its corresponding report was

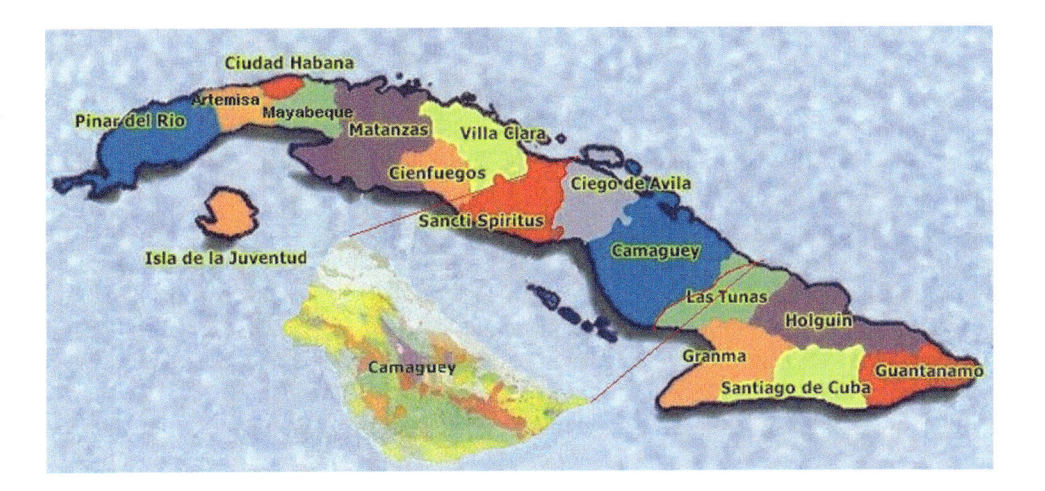

Fig. 1.12 Scheme of distribution of the areas covered by survey at 1:250000 scales of the former province of Camagüey by the (second) Cuban-Bulgarian Brigade, between 1976 and 1981

Fig. 1.13 Scheme of distribution of the areas covered by survey at 1:250000 scale of the former province of Oriente by the Cuban-Hungarian Brigade between 1972 and 1976

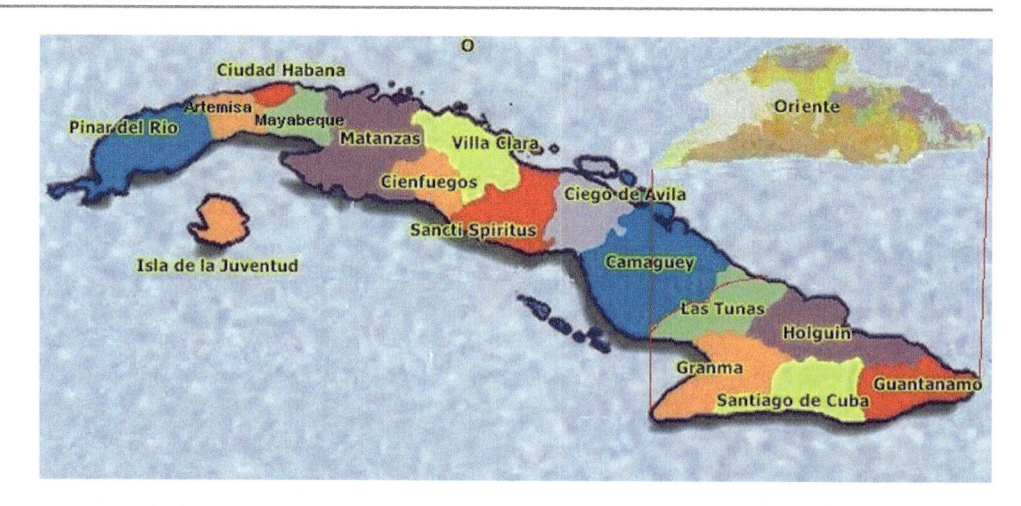

Fig. 1.14 Distribution scheme of the areas covered by survey at 1:250000 scale of the former Isla de Pinos carried out by the Cuba-USSR Brigade between 1972 and 1976

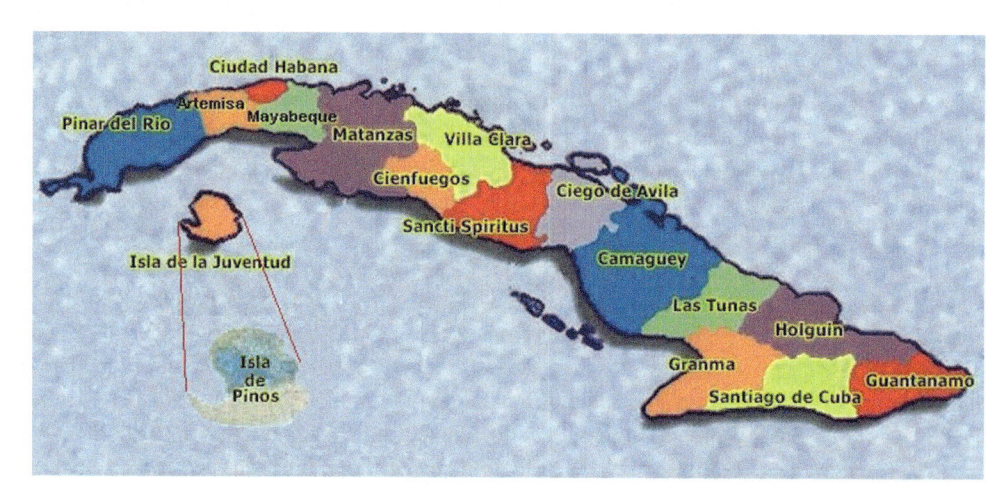

obtained with the title: "Explanatory Text of the Geological Map of the East at 1:250000 scale. Mapped and made for the Cuban-Hungarian Brigade between 1972 and 1976". As authors of this report appear E. Nagy, K. Bresznyansky, A. Brito, F. Formell, G. Franco, P. Gyarmaty, P. Jacus, G. Radocz, F. Abello, D.P. Coutin, J.P. Lledías, J. Oro, G. Pantaleón, P. Jakus, N. Pérez, I. Polcz, C. Sacasa, R. Suárez, and R. Varela.

Geological survey of the Isla de Pinos at 1:250000 scale

The geological survey at 1:250000 scale in the territory of the then Isla de Pinos (since 1976 Isla de la Juventud, Fig. 1.14) was made possible thanks to the collaboration of the USSR Academy of Sciences and the IGP of the ACC between 1972 and 1976. For this, a mixed brigade of specialists from both institutions was created, which was called the Cuba-USSR Brigade.

On the Cuban side, the geologist and metamorphic rock petrographer Guillermo Millán participated, while the doctors Mark L. Somin and Aleksandr Cherniajovski took part for the Soviet Union. These specialists would be in charge of the study and cartography of the rocks of the folded basement and the Cretaceous volcanic arc that emerge in the central and northern part of the island, while Doctor Guillermo L. Franco Alvarez studied the rocks of the Miocene and Quaternary coverage that emerge mainly in the south.

The Soviet A. Adamovich and the Cuban Francisco Formell carried out the study of the Cayo Largo del Sur key and the other small islands that together with the one of the Isla de Pinos make up the Canarreos Archipelago, as part of the completion of the 1:250000 scale survey from this part of southern Cuba [11].

Geological Map of Cuba (unified) at 1:250000 scale

This map [12] arose because of the unification of the regional maps obtained from the survey works executed through the collaboration agreements between the Academy of Sciences of Cuba—represented by the Institute of Geology and Paleontology— with several institutions of the Academies of Sciences of many other country members of the extinct Eastern European Socialist Field. As we already know, the collaboration in the field of geological cartography became effective within the framework of agreements signed with the Academies of Sciences of the USSR, Poland, Bulgaria, Hungary, and Romania. In practice, the Romanian Academy of Sciences participated with only a few specialists.

In the credits of this map, the editors' category was given to the directors of the participating institutions, which appear in alphabetical order, without giving less or greater weight to its participation on the work, nor of their respective institutions. The list is headed by Dr. M. Borkowska, of the Institute of Geological Sciences of the Polish Academy of Sciences, followed by G. Hamor, of the State Institute of Geology of Hungary. Then, the academic Dr. Yuri M. Pushcharovsky, from the Institute of Geology of the Academy of Sciences of the USSR; the engineer Julio Suárez Morales, Director of the IGP of the ACC; and finally, Dr. I. Velinov, from the Institute of Geology of the Bulgarian Academy of Sciences.

The map was edited and printed in the "Kartfabrica" of the Institute of Geological Research of the USSR "A. P. Karpinski (VSEGEI, in Russian language)" of Leningrad, under the direction of the Doctors in Sciences, S. Sokolov, A. Mossakovskiy, and G. Nekrasov, who served as principal editors.

In the unification, redaction, and editing of the map, a large number of specialists from the different institutions already mentioned participated, for which it was necessary to create a Commission of Unification and Redacting (CUR), integrated, according to the credits of the finally published map, by the following specialists:

For the IGP of the ACC: Jesús Francisco de Albear, Rustin Cabrera, Bienvenido Echevarría, Francisco Formell, Guillermo Franco, Manuel Iturralde, Guillermo Millán, José Oro, and Leandro Peñalver. For the Academies of Sciences of the socialist countries participated: A. Adamovich, V. Chejovich, I. Kartachov, and M. Somin from the USSR. R. Myczynski, K. Piotrowska, J. Piotrowski, A. Pszczolkowski, and J. Rudnicky, from Poland. I. Boyanov, I. Haydotov, I. Kantchev, V. Kostadinov, and D. Tchounev, of the Institute of Geology of Bulgaria, and K. Bresznyansky, E. Nagy, and J. Radocz, of the State Institute of Geology of Hungary.

The unified Geological Map of Cuba at 1:250000 scale, which is shown in Fig. 1.15, consists of 40 sheets according to the format and nomenclature of the topographic charts of that scale, prepared by the former Instituto Cubano de Geodesia y Cartografía (ICGC), nowadays known as Geo-Cuba. This includes in its marginal information a legend and a map showing the location of 34 generalized columns of wells, distributed throughout the country, which show the deep geological constitution of the Island. In his time (1988), the work was a milestone in the knowledge of the geology in Cuba and throughout the region, since, at that years, few countries in the area had a map of that degree of detail for all its territory.

Regarding its usefulness, it should be noted that this map had an immediate insertion in social practice and led to the appearance of new thematic maps such as the first Tectonic Map of Cuba at 1:500000 scale and a new geological map at 1:1000000 scale.

Tectonic Map of Cuba at 1:500000 scale

One of the immediate practical applications of the Geological Map of Cuba at 1:250000 scale was its use, by a group of Cuban specialists and the member countries of the CAME, as a geological basis for the creation of the Tectonic Map of Cuba at 1:500000 scale (Fig. 1.16). It was also published in the USSR in 1989, that is, barely a year after the geological one [13].

The authors of this tectonic map also include a large group of Cuban specialists in collaboration with Polish, Hungarian, Bulgarian, and Soviet geologists, led by the main editors, the USSR Academicians Y.M. Pushcharovsky, A.A. Mossakovskiy, and the mining engineer Julio Suárez Morales, at that time Director of the IGP of the ACC. Among the Cuban technicians are the doctors Francisco Formell, Rustin Cabrera, and Manuel Iturralde; the geological engineers Raul Flores, José Oro, Alberto Morales, and Guillermo Pantaleón; and the then geologist technician Leonel Pérez. As the Academies of Sciences of the socialist countries participated also the Soviets G.E. Nekrasov, S.D. Sokolov; the Bulgarians: I. Boyanov, I. Haydotov, V. Kostadinov, I. Kantchev, and D. Tchounev; the Polish A. Pszczolkowski; and the Hungarian K. Bresznyansky.

This tectonic map has the merit of being the first of its kind on this scale. It contains abundant and novel information about the main tectonic structures, folds and disjunctive deformations, the direction of the main thrusts, axes of the folds, and others.

It also has a zonal legend where the age and succession of the different tectonic events are represented; in a box the name of the most relevant (80) structures present in the map is shown. Three profiles or cross-sections are shown, where the interpretation of the deep tectonic structure of the Cuban territory in the western, central, and eastern regions is revealed.

On the other hand, as part of the marginal information, this work has two interpretative schemes at 1:2500000 scale. The first one, types of land crust of Cuba and its adjacent regions, shows the distribution and tectonic relationship of the different types of rocks and structures that make up the substrate of Cuba and its Caribbean environment. This map is of the total authorship of the Soviet specialists A.L. Vtulochkin, A.A. Mossakovskiy, G.E. Nekrasov, and S.D. Sokolov. The second scheme is presented under the title of

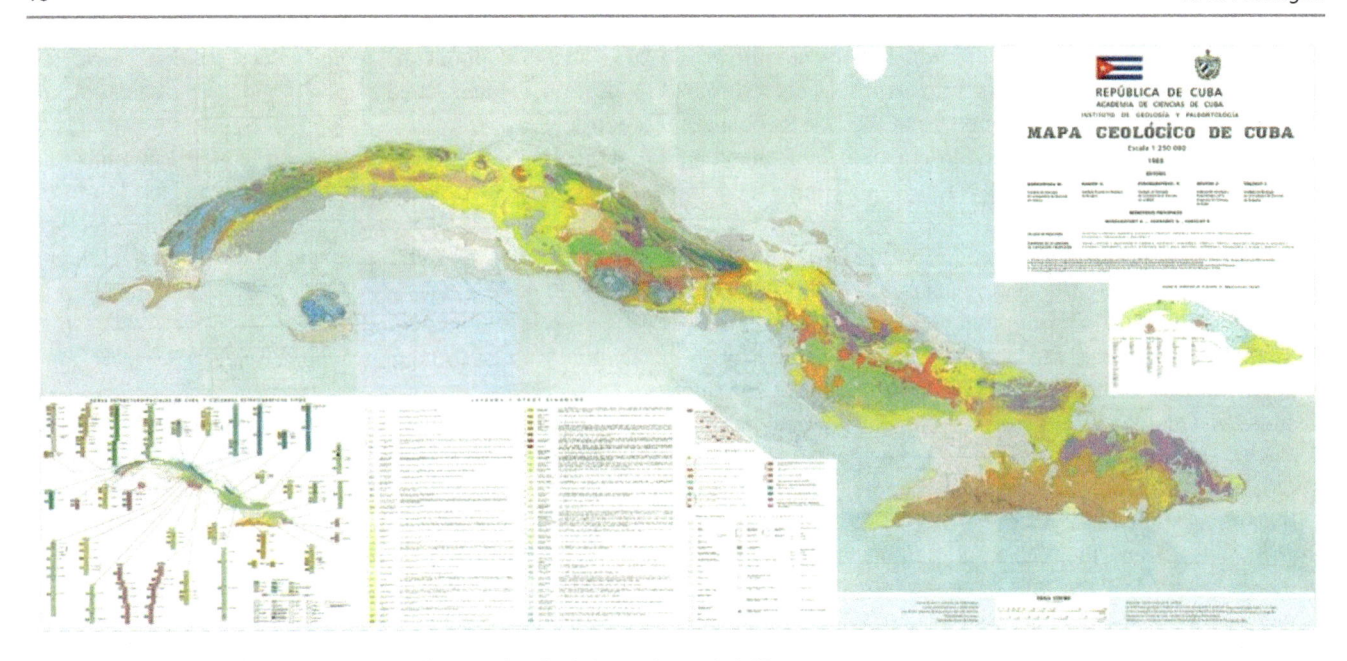

Fig. 1.15 Geological Map of Cuba at 1:250000 scale. (Pushcharovsky et al. [12]

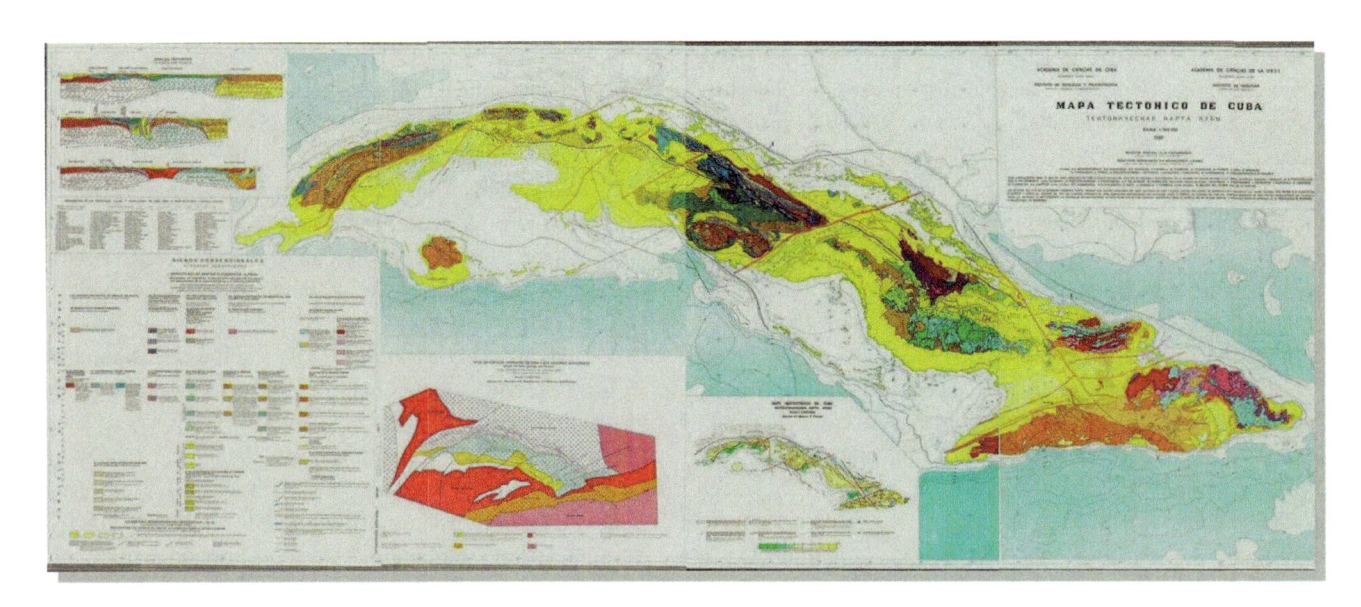

Fig. 1.16 Tectonic Map of Cuba at 1:500000 scale (Pushcharovsky et al. [13]

Neotectonic Map of Cuba and is a joint interpretation of the Soviet Researcher V.I. Makarov and the Cuban geologist engineer Dr. Francisco Formell Cortina.

New Geological Map of Cuba at 1:1000000 scale

Another direct product of the Geological Map of Cuba at 1:250000 scale was its namesake, but at 1:1000000 scale (Fig. 1.17), made in the IGP by Dr. Francisco Formell Cortina for its inclusion in the New National Atlas of Cuba [14].

If it compares this map—known colloquially among geoscientists as the "Atlas Geological Map" or simply as the "Formell's Map"—with its predecessor [8], it highlights the evident advantage of its greater degree of updating and detail, precisely because it is a suitable version and reduced to a smaller scale, from the most detailed map to date. In its marginal information, this geological chart has a general legend for the geological formations and four regional cross-sections that show the deep structure in an equal quantity of profiles drawn in different areas of the national

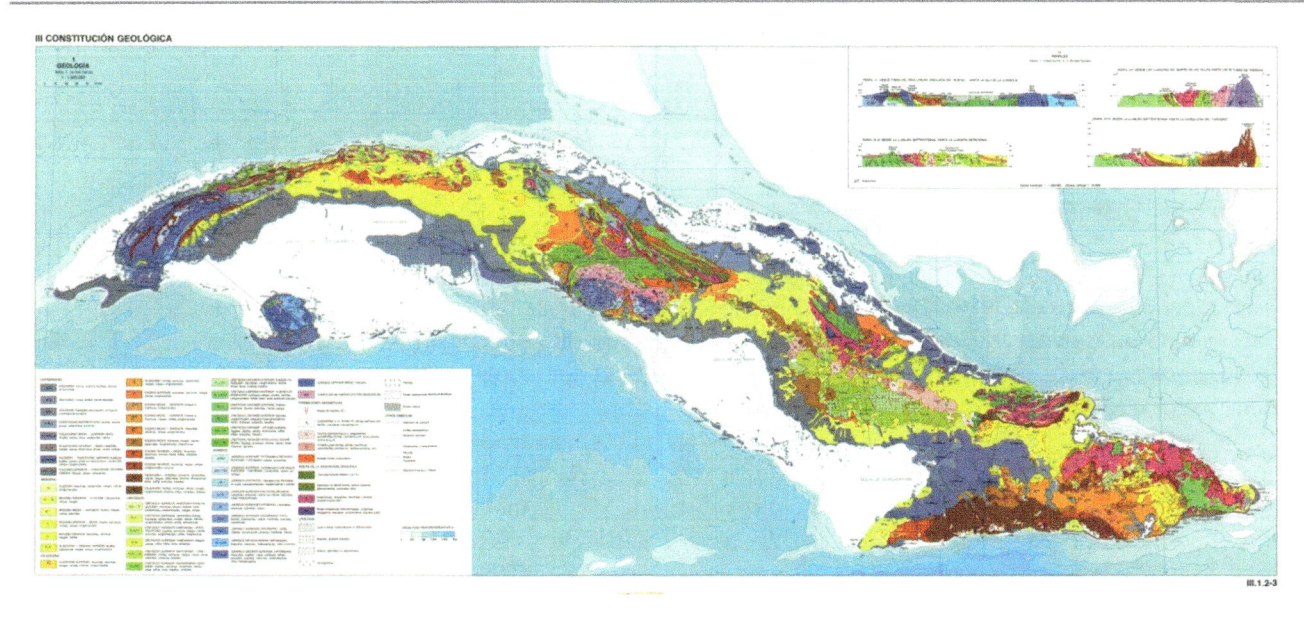

Fig. 1.17 New Geological Map of Cuba at 1:1000000 scale. Prepared by Francisco Formell Cortina and Alberto Morales Quintana, and published in the New National Atlas of Cuba of 1989

territory. The geologist Alberto Raul Morales Quintana made these cuts.

1.4.2.1 Geological Maps at Semi-detailed Scales (1:100000–1:50000)

Digital Geological Map of Cuba at 1:100000 scale

As of 1989, the national economy would suffer one of the hardest blows in its history. The disintegration of the USSR and the collapse of the so-called Socialist Field, on which 85% of the commercial exchange with the exterior depended, plunged the Island into what would be known as the Special Period.

The geological activity also suffered, no less than the rest, the effects of the recession. Financial assistance was interrupted with the different companies and institutions, including the Academies of Sciences of the socialist countries, which collaborated with Cuba through bilateral and multilateral agreements, within the framework of the CAME.

There was a drastic reduction in research projects, the logistical infrastructure created (brigades, camps, laboratories, etc.) was lost, so exploration-prospecting work was reduced to almost zero and, of course, geological mapping was not an exception. Several geological survey projects at semi-detailed scales that were carried out in different polygons were canceled. Such was the case of Complex "Geological Survey at 1:50000 scale of the Camagüey`s Polygon III, Sector Loma Jacinto" [15] or the "Geologic Survey 1:50000 of the northern part of Villa Clara, polygon Esperanza-Santo Domingo" [16], to mention just a couple of examples.

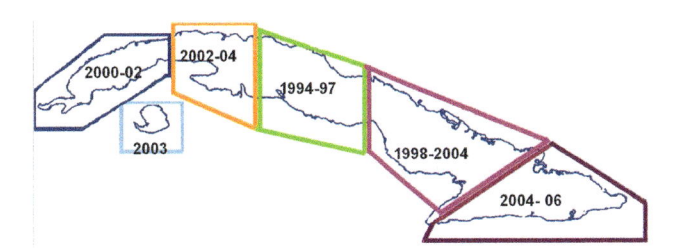

Fig. 1.18 Scheme of distribution of the polygons in which the national territory was divided for the execution of the projects of Updating and Generalization of the Geological Map of Cuba at 1:100000 scale per region

Because of this situation, many geoscientists were re-profiling or devoting themselves to other specialties inside or outside the geological activity. In the middle of the hard economic crisis, a part of the researchers of the IGP were dedicated to the collection, unification, and updating of all the preexisting information in the archives, texts, and maps of the reports of surveys at different scales in format analog, going to new technologies for digitization and conversion on a single scale. The chosen scale was 1:100000.

This is how the series of Research and Development (R & D) projects was born, under the general title of "Project for the Generalization and Updating of the geological cartography of (such or such a region) at 1:100000 scale". Similar was done for the map 1:250000. Then, the national territory was divided into several polygons (Fig. 1.18) and digital maps were obtained at 1:100000 scale of the different regions, only this time; all the authors would be the Cuban researchers of the IGP, already by then attached to the MINBAS.

Generalized Geological Map of Central Cuba at 1:100000 scale (provinces of Cienfuegos, Villa Clara, and Sancti Spiritus)

The first "Project of Generalization and Updating of Geological Cartography at 1:100000 Scale", was the one corresponding to the Central Cuba region, understood as such, the provinces of Cienfuegos, Villa Clara, and Sancti Spiritus. This project was conceived and directed by geologist Dora E. García Delgado and was carried out between 1994 and 1997 [17].

The team was made up mainly of the members of the Regional Geology Group of the IGP. It considers the geological engineers Raisa Delgado and Yamirka Rojas, who were responsible for the stratigraphy; the Dr. Guillermo Millán, a specialist in metamorphic rocks; the geologist Luis R. Bernal, who attended the tectonics; and the geographer Leandro L. Peñalver, who took care of the cartography of the Quaternary deposits.

The doctors in the geological-mineralogical sciences Lilavatti Díaz de Villalvilla and Kustrini Sukar Sastroputro (an Indonesian specialist placed in Cuba), as well as the geologist Ileana García, would be in charge of the study of the magmatic rocks, while the engineer Angélica I. Llanes took care of the study and cartography of the rocks of the Ophiolitic Association. Dr. Gustavo F. Furrazola, the biologist Consuelo Díaz, and the geologist Iliana Delgado formed the team responsible for the paleontological description of thin sections and washes.

Dr. Carlos Manuel Pérez was in charge of the remote sensing and image processing work, while Dr. Manuel E. Pardo and the geophysical engineer Valia Suárez carried out the geophysical interpretation work. The geologist technician Estela Duani made dissimilar auxiliary works.

Because of this project, the geological maps were obtained at 1:100000 scale from each of the 21 topographic sheets of the GeoCuba nomenclature that covers this territory, with the corresponding legend of all the units mapped in them and the conventional symbols. The sheets were edited in AutoCAD system, applying the symbology approved by the special project "Design of the Geological Information System of Cuba at 1:100000 scale" (SIGEOL).

A report was also written with the name of "Explanatory Text to the Geological Map of Central Cuba at 1:100000 scale (Cienfuegos provinces, Villa Clara and Sancti Spiritus)". In the writing of the chapters participated Dora E. García Delgado, Carlos M. Pérez, Raíza Delgado, Consuelo Díaz, Guillermo Millán, Gustavo Furrazola, Lilavatti Díaz de Villalvilla, Ileana García, Kustrini Sukar, Iliana Delgado, Luis Bernal, Manuel Pardo, Yamirka Rojas, Valia Suárez, and Estela Duani.

For the preparation of this map were used, as geological basis, the graphic materials and reports of preexisting geological surveys at scales 1:100000 and 1:50000. Among these, the following works stand out by region: [16, 18, 19, 10, 20, 21, 22, 23, 24, 25, 26, 27].

Generalized Geological Map of Pinar del Río at 1:100000 scale

The "Project of Generalization and Updating of the Geological Map of Pinar del Río at 1:100000 scale", which includes the territories that today occupy the provinces of Pinar del Río and part of Artemisa, was carried out between 2000 and 2002 and it was directed by the geologist Dora Elisa García Delgado [28].

The stratigraphy team consisted of the geological engineers Santa Gil, Raisa Delgado, and Yamirka Rojas, while Dr. Guillermo Millán, as metamorphist, and the licensed geographer Leandro Peñalver, specialist in Quaternary deposits, integrated the regional geology team. The marine geologist Miguel Cabrera was in charge of the studies of the areas of the marine-coastal territory. The collaboration of Roberto Denis, from the Geominera Pinar del Río, was valuable.

In charge of the photointerpretation and the digital processing of aerial and satellite images were the engineers Ramón O. Pérez and Dr. Carlos M. Pérez. The engineers Jorge L. Chang, Magaly Fuentes, and Valia Suárez integrated the team of geophysics, while the paleontologists Dr. Gustavo Furrazola, Consuelo Díaz, and Juana R. Pérez, all from the IGP Paleontology Group, carried out the paleontological descriptions of thin sections and washes.

A team of Petrography, integrated by Dr. Lilavatti Díaz de Villalvilla and Ileana García, carried out the study and description of the igneous rocks, while Mercedes Torres and Angélica I. Llanes attended the description of the volcanic and ophiolitic rocks, respectively.

All the digitization works, both of the materials used and of the resulting map, were in charge of the geologist technician Valentina Strazhevich, a Soviet citizen of Belarusian nationality, based in Cuba, with the supervision of the engineer Ramón O. Pérez. Different works to support the cartography were carried out by geologist Ramón Rivada and geologist technician Roberto Morales.

At the end of the project, the generalized and updated geological maps at 1:100000 scale were obtained from each of the 17 topographic sheets of the GeoCuba nomenclature for this territory, with the corresponding legend and the conventional symbols of all the units mapped in them. The maps and the cross-sections by regional profiles were edited in the AutoCAD system, applying the symbols approved by SIGEOL.

An explanatory text was also elaborated under the title of "Generalization and Geological Update of the province of Pinar del Río at 1:100000 scale". In the writing of the chapters worked Dora E. García Delgado, Santa Gil, Raíza

Delgado, Guillermo Millán, Leandro L. Peñalver, Miguel Cabrera, Roberto Denis, Jorge L. Chang, Magaly Fuentes, Consuelo Díaz, Valia Suárez, Angélica I. Llanes, Ramón O. Pérez, Mercedes Torres, Carlos Pérez, and Lilavatti Díaz de Villalvilla.

For the preparation of this map, the graphic materials and reports of the preexisting geological surveys at scales 1:100000 and 1:50000 were used as a geological base. Among these, the following works stand out by region: [29, 30, 31, 32, 33, 34, 35, 36, 37, 38], and Vologdin et al. (1963).

Generalized Geological Map of the Isla de la Juventud at 1:100000 scale

The "Project for the Generalization and Updating of the Geological Map of the Isla de la Juventud (Isla de Pinos before 1976) at 1:100000 scale" began and was completed in the course of 2003. At the head of it was the Dr. in Geological and Mineralogical Sciences and specialist in metamorphic rocks, Guillermo Millán Trujillo. He directed and assumed the cartography of the outcrop of the folded substrate, the Cretaceous volcanic arc, and the intrusive rocks that make up the complex geology of this small portion of the Cuban territory.

For his part, the geographer Leandro Peñalver Hernández assumed the cartography of the rocks of the Neogene—Quaternary coverage of the south and the coastal areas of the island. The submerged areas of the surrounding marine-coastal territory were in charge of the geological engineers Miguel Cabrera and Roberto Denis.

The geological interpretation of aerial and spatial images was in charge of Dr. Carlos M. Pérez. The geologists Cecilia Ugalde and Adriana Rosa, supervised by the geologist Guillermo Pantaleón, carried out the digitization work of the previous information used, as well as the newly generated map.

As a result of this project, the Geological Map of the Isla de la Juventud was obtained at 1:100000 scale, whose revision and the final digital edition were in charge of the geologist Ramón O. Pérez and the geologist technician Valentina Strazhevich.

For the preparation of this map were used as geological basis the graphic materials and reports of preexisting geological surveys at scales 1:250000, 1:100000, and 1:50000, among which the following works stand out by region: [39, 40], Millán et al. (1981), and [41].

Among the others, the main critic this map received is that it is more similar to the version of the geological map at 1:250000 scale of Pushcharovsky et al. [12], than to the conditional surveys at scales 1:100000 of Garapko et al. [40] and 1:50000 of Babushkin et al. [39], carried out in this territory.

Generalized Geological Map of Habana—Matanzas at 1:100000 scale (provinces of Artemisa, La Habana, Mayabeque, and Matanzas)

As in the westernmost region of the Island, the geologist Dora E. García Delgado, directed the "Project for Generalization and Updating of the Geological Map of the Habana—Matanzas region at 1:100000 scale" during the period from 2002 to 2004 [42]. In this opportunity, the geological engineers Raisa Delgado and Yamirka Rojas also integrated the stratigraphy team.

For the group of Regional Geology, Dr. Guillermo Millán, who was responsible for the geology of metamorphic rocks, the graduate geographer Leandro L. Peñalver, specialist in geology of the Quaternary, and the engineer Miguel Cabrera, specialist in marine geology, were in charge. The engineer Ingrid Padilla carried out the geophysical studies.

The paleontological descriptions were made by the graduate in Biology Consuelo Díaz, the geologist engineer Ana Ibis Torres, and the graduate geographer Juana R. Pérez, directed by Dr. Gustavo F. Furrazola. The geological engineer Angélica I. Llanes was in charge of the improvement of the oceanic crust rocks cartography, while the engineering geologists Mercedes Torres and Ileana García led the studies of the igneous and effusive rocks intrusive, respectively.

The geological engineer Ramón O. Pérez and Dr. Carlos M. Pérez were in charge of the remote sensing and image processing works, while the engineer Rolando Batista González was in charge of the accuracy of aspects related to the location and distribution of mineral deposits in the mapped territory.

The Cartography Group, headed by Ramón O. Pérez, carried out the digitization work of the graphic material used and the resulting maps. This group was integrated by the geological engineers Jesús Triff and Cecilia Ugalde, as well as by the geologist technician Valentina Strazhevich. Geologist technicians Roberto Morales, Luisa Rodríguez, Diana Sosa, and Estela Duani carried out different works in support of cartography.

An explanatory text was prepared under the title of "Generalization and Geological Update of the Habana—Matanzas region at 1:100000 scale". In the writing of the chapters worked Dora E. García Delgado, Raisa Delgado, Yamirka Rojas, Guillermo Millán, Leandro Peñalver, Miguel Cabrera, Ingrid Padilla, Consuelo Díaz, Ana I. Torres, Gustavo Furrazola, Angélica Llanes, Mercedes Torres, Ramón O. Pérez, Luis Bernal, and Roberto Morales.

The primary materials used in this project were, first, the data of the 1:50000 scale survey [36] and the 1:250000 scale surveys of [9 and 23]. Also used as a geological basis for the preparation of this map are the graphic materials and reports of the preexisting geological surveys at 1:100000 and 1:50000 scales. Among these, the following works stand out by region: [43, 44, 45, [36, 46].

Of great use for this work were the geological maps at 1:25000 scale of the municipalities of east and south of La Habana, made between 1994 and 1998 by Pérez Aragón et al. [47], within the framework of the first projects of Environmental Geology made in the country.

Generalized Geological Map of Ciego-Camagüey-Las Tunas at 1:100000 scale (provinces of Ciego de Ávila, Camagüey, and Las Tunas)

The "Project of Generalization and Updating of the Geological Map of the Ciego-Camagüey-Las Tunas provinces, at 1:100000 scale", was carried out between 1998 and 2004, led by Dr. in Geological Sciences Carbeny Ramiro Capote Marrero and integrated mainly by geologist María Santa Cruz-Pacheco and geologist Domingo González.

Specialists in the mineralogy of intrusive and effusive igneous rocks were Dr. Lilavatti Díaz de Villalvilla and the geological engineers Daysi de la Nuez, Ariadna Suárez, and Miriela Ulloa. The geographer Leandro L. Peñalver was in charge of the cartography of the Quaternary deposits. The geologist technician Valentina Strazhevich, the licensed geographer María R. Santos, and the geologist Guillermo J. Pantaleón edited the final map of this project.

It should be noted that, because both the management and the composition of the group of technicians who carried out this work was different from those that executed the "update-generalization" projects of the rest of the national territory, for its insertion in the final general map, it was necessary to carry out a reedition work. It was carried out by geologist Ramón O. Pérez, based strictly on the critical remarks and recommendations of Dr. Manuel A. Iturralde Vinent of the ACC and, in minor degree, of the engineer-geologist Enrique C. Piñero Pérez of the Geominera Camagüey.

For the preparation of this map, the graphic materials and reports of the preexisting geological surveys at scales 1:100000 and 1:50000 were used as a geological base. Among these, the following works stand out by region:[18, 48, 49], Kantchev et al. (1976), [50, 51, 52, 15, 13, 53, 25, 26].

Generalized Geological Map of the Oriente province at 1:100000 scale (provinces of Las Tunas, Holguín, Granma, Santiago de Cuba, and Guantánamo)

The "Project for the Generalization and Updating of the Geological Map of the Eastern Provinces at 1:100000 scale", which included the territories of the current provinces of Holguín, Granma, Santiago de Cuba, Guantanamo, and part of Las Tunas, was the last goal. It had several changes in its leadership due to various causes. The same began to run at the end of 2004 and was completed in 2007. In the beginning, it was developed and directed by the engineer Dora E. García Delgado. Later it was managed by the engineer Ramón O. Pérez and, finally, directed by the geological engineer Kenya E. Núñez and the geological engineer Dr. Bienvenido T. Echevarría.

The composition of the team also suffered several difficulties during the execution. Initially, the members of the Regional Geology Group of the IGP, the stratigraphists Santa Gil, Yamirka Rojas, and Raisa Delgado, integrated it. In the end, worked Bienvenido Echevarría, Kenya Núñez and Arelis Núñez of the Geoprocessing Group, as well as, Guillermo Pantaleón, Denyse Martín, María R. Santos, and Valentina Strazhevich of the Cartography Group. Dr. Guillermo Millán was in charge of the metamorphism and the licentiate Leandro L. Peñalver, of the cartography of the Quaternary deposits.

The specialists of the Geominera Santiago de Cuba, the engineers Iris Méndez and Ramona Rodríguez, specialists in intrusive rocks, integrated the magmatic rock petrography team this time. The geological engineers Mercedes Torres and Angélica I. Llanes from IGP dealt with the volcanic and ophiolitic rocks, respectively.

Consuelo Díaz and Juana R. Pérez commissioned the paleontological studies. The geophysical engineer Jorge L. Chang carried out the work related to this specialty.

The digitization of data and the editing of maps was carried out by the geologist technician Valentina Strazhevich. The geological-geophysical engineer Dalia J. Carrillo and the geologist technicians Luisa Rodríguez and Roberto Morales did several cartography support projects.

For the preparation of this map, the graphic materials and reports of the preexisting geological surveys at 1:100000 and 1:50000 scales were used as a geological base. Among these, the following works stand out by region: [54, 55, 56, 57, 58, 59, 60, 48, 49, 61, 62, 63, [50], [64, 51, 65, [66].

Generalized Geological Map of the Marine-Coastal Zone at 1:100000 scale

The work of the geological cartography of marine-coastal zone included the region between the coastline and the upper edge of the insular slope, considering all the islands and keys of the Cuban Archipelago. The geological engineer and specialist in marine geology Miguel Cabrera Castellanos led it.

For the updating and generalization works of these territories, the data of the Geological Map of Cuba at 1:500000 scale [6] and the geological cartography at 1:250000 scale of the Canarreos Archipelago [11] were used. The Geological Maps of the Southern and Northern Platform of Camagüey at 1:250000 scale [67] and others also were used.

Geological engineers Guillermo J. Pantaleón and Cecilia Ugalde carried out the digitization and final editing of this information, as well as its connection with the rest of the map. The final revision was in charge of the engineer Ramón O. Pérez, as well as the geologist technician Valentina Strazhevich.

For the preparation of this map, the graphic materials and reports of the preexisting geological surveys at scales 1:100000 and 1:50000 were used as a geological base. Among these, the following works stand out by region: [68, 69–71, 72, 73, 74, 75, 67, 76, 77, 78].

Geological Information System (SIGEOL) and NC622

Parallel to the work of generalization and updating of the geological cartography at 1:100000 scale and in support of this, the project "Design of the Geological Information System of Cuba at 1:100000 scale" (SIGEOL) was carried out. It was led by the geologist Dr. Enrique A. Castellanos Abella with the participation of numerous geological specialists and engineers: Kenya E. Núñez, Arelis Núñez, Cecilia Ugalde, andDr. Bienvenido T. Echevarría; the geological-geophysical engineer Dalia J. Carrillo; the geological engineers Guillermo J. Pantaleón and Jesús Triff; the licensed geographer Denyse Martín; and the geophysical technician William Alfonso.

As a result of this project, a series of procedures was obtained that regulate the content and design of the graphic symbols for the execution of the geological map at 1:100000 scale, which was the one that was being executed at that time. These procedures, after their critical review by different geoscience specialists and their approval by the Consejo Científico Técnico (CCT) of the IGP, gave way to the preparation and publication of the Cuban Standard for Geological Cartography NC 622 [79].

Digitization and Edition

All the digitization and graphic edition works of the 120 pages of the Geological Map of Cuba at 1:100000 scale, in this first stage, were carried out in AutoCAD and were directed by the engineer geologist Ramón O. Pérez Aragón. He worked in front of the IGP Cartography Group. In these works, the members of the group which participated were: the engineers Guillermo Pantaleón, Jesús Triff, Cecilia Ugalde, and Adriana Rosa; the licentiate geographers María R. Santos and Denyse Martín; and the technicians Valentina Strazhevich, Geraldine Quintana, Lourdes Fernández, and Orlando Rodríguez.

Great participation and influence in this work had specialists from other areas of the IGP such as the engineer Raisa Delgado and the geologist technician Luisa Rodríguez, from the Regional Geology Group, as well as, the engineer Arelis Núñez and the technicians William Alfonso, Melisa Tejas, and Katherine Hernández, from the Geoprocessing Group. Technician Domingo González, of the Mineral Deposits Group, also participated in an outstanding way.

Digital Geological Map (unified) of Cuba at 1:100000 scale (first version)

Once all the update-generalization projects for the territories described above were completed, they were unified in a single map, for which a new R & D project was created, called "Generalization and Updating of the Digital Geological Map of Cuba at 1:100000 scale".

This project consisted of the reissue of the 120 sheets, with their respective legends, of the maps already generalized by regions and their coherent connection in their limits, to obtain a unified map. Then, its conversion from the AutoCAD system to a Geographic Information System (GIS) environment. For this purpose, through the payment of the corresponding licenses, the Canadian CARIS system was acquired, in order to evade the US economic blockade that prevented the acquisition of other North American systems more known and used internationally for these purposes, such as MapInfo and ArcGIS.

In its beginnings (2007–2008), the project was led by the Regional Geology Group and commanded by engineer Dora E. García Delgado, who headed a large group of specialists among which were the engineers Santa Gil, Yamirka Rojas, and Raisa Delgado. For different reasons, these specialists leave the IGP, including the project manager, so, from 2009 to 2011, the execution of the same was transferred to the area of Cartography and led by engineer Ramón O. Pérez Aragón.

As a result of this project, the first variant of a Digital Geological Map was obtained for the entire territory (emerged and marine-coastal) of the Republic of Cuba, at 1:100000 scale, in a GIS environment. It was updated with the information closing of 2007 and in concordance with the Stratigraphic Lexicon published in 2002. This fact constitutes a great advance step in the geological cartography, without precedents, since it is the first geological map for all Cuba in digital format.

The finished version was submitted to the critical review of different geologists and specialists of various geological entities from different territories. They made a group of critical statements and recommendations, which were added to a number of shortcomings known by the authors. However, that could not be resolved in the temporal and financial framework of the project, at the end of its execution, remaining pending to be corrected in subsequent works.

Digital Geological Map of Cuba at 1:100000 scale (latest version)

As a result of the critical evaluation to which it was submitted and the accumulation of signs and insufficiencies detected during the process between 2012 and 2016 where

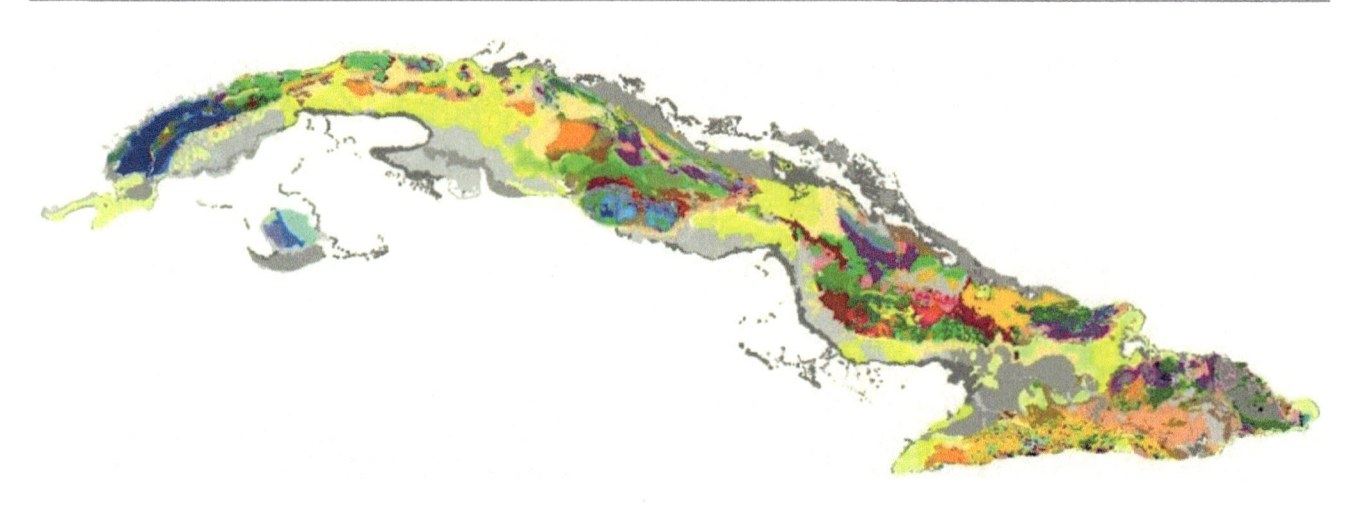

Fig. 1.19 Simplified facsimile of the Digital Geological Map of the Republic of Cuba at 1:100000 scale (unified format in the GIS CARIS version). Faults and annotations are not shown, as well as the geology of the marine-coastal zone

the referenced R & D project was executed, then emerged the final (last) variant of this geological map in digital format (Fig. 1.19).

The complete map, understood as the 120 sheets that make up the nomenclature of the topographic maps elaborated by GeoCuba on this scale, was subjected to review, taking into account the critical statements made by numerous opponents and users. It was detected that many of the indicated errors were introduced automatically during the conversion of the AutoCAD system to CARIS. Some were human errors committed during the digitization process and, others, the least, were deficiencies of the original maps that were mechanically reproduced.

In the elaboration of this map, from the initial projects of generalization and updating to the revision and final edition, a relatively small number of Cuban specialists participated, most of the IGP. A minimum, but an effective collaboration of specialists was from other institutions in character of critics and advisers. These were the geologists Dr. Manuel A. Iturralde, from the ACC and the engineers Enrique Piñero, from the Geominera Camagüey; Aldo Fernández, from the Geominera Oriente; Marcelino Arce and Julio Blanes, from the Geominera Pinar del Río; and Emilio Milián, from the Geominera Center, among others.

It is necessary to highlight that this map, despite being a collective work, which reflects the work of hundreds of geologists of different generations and nationalities, responds to the efforts of a group of researchers, specialists, and technicians of the IGP. They worked in a selfless way during the hard years of the "special period" with all kinds of deprivations. It is difficult therefore to define an authorship for it, so in the "Main Sheet" that serves as the cover, we do not speak of authors, but of editors. However, it should be noted the outstanding participation of geologist Eng. Dora Elisa García Delgado, who, although she did not work until the end, was in charge of the initiative and direction of the first generalization and updating projects that led to obtaining this result.

The category of Principal Editors of the latest version of the Geological Map of the Republic of Cuba at 1:100000 scale was granted to: geological engineers Ramón O. Pérez; Kenya E. Núñez; Miguel Cabrera; Angélica I. Llanes; Carbeny R. Capote; Arelis Núñez; Cecilia Ugalde; Guillermo J. Pantaleón; Jesús Triff. Also, the geographer licentiates Leandro L. Peñalver and Denyse Martín and the geophysical technician William Alfonso.

The map had an Editorial Board composed of research geologists R.O. Pérez (Auxiliary Researcher), K.E. Núñez (Attache Researcher), L.R. Bernal (Attache Researcher), M. Cabrera (Auxiliary Researcher), Dr. C.R. Capote (Senior Researcher), R.A. Denis (Attache Researcher), A.I. Llanes (Attache Researcher), A. Núñez (Attache Researcher), L.L. Peñalver (Auxiliary Researcher), and Dr. C.M. Pérez (Senior Researcher).

All the digitization works of the past information as well as of the new maps and edition of the final maps were directed by engineer Ramón O. Pérez, in his role as project manager and head of the Cartography Group of the IGP. He also participated directly in the digitization and editing of maps with: the geological engineers Cecilia Ugalde, Guillermo J. Pantaleón, Jesús Triff, Adriana Rosa, Arelis Núñez, and Raisa Delgado; the licentiates geographers María R. Santos, and Denyse Martín; the geophysical technician William Alfonso; the geologist technicians Valentina Strazhevich, Domingo González, and Luisa Rodríguez; the computer technicians Katherine Hernández, Melisa Tejas, Geraldine Quintana, Orlando J. Rodríguez, as well as the drafter Lourdes Fernández.

The Geological Map of Cuba at scale 1:100000 in its latest version [2] has several advantages over its predecessors. One of them, perhaps the most significant, is to have achieved the first geological map on a semi-detailed scale of the entire territory of the Cuban archipelago, including the marine-coastal zone, in digital format. In addition, it can be consulted on separate sheets (120 in total) or unified by regions and even one whole at a time.

The projects of generalization and updating by regions that gave rise to the final variant were the first step to achieve unification, stratigraphic correlation, and coherent splicing of the different lithostratigraphic units within the boundaries of the various polygons. It allows a close interrelation with the Léxico Estratigráfico de Cuba (LEC), defining the main problems to solve in the necessary task of constantly updating this governing document.

One of the biggest shortcomings of this work is the lack of a unified explanatory text for the entire map. As an explanatory memory of the same are the texts prepared for the different projects updating–generalization, which exist for the corresponding regional polygons to which reference has been made.

Although a guiding document was drawn up (Annex of Representation) that contains all the lithostratigraphic units, rock complexes, types of contacts, and faults, a unified general legend for all of Cuba was never finished. Nevertheless, each one of the 120 sheets has its own legend.

Another weakness of the map is tectonics. The lack of specialists in this branch led to an insufficient study of the types of faults, their directions, ages, and magnitude of the displacements, a task that must be faced during other future updating work and during geological mapping at 1:50000 scale, a goal for the years to come.

In addition to the inherent deficiencies that remain to be determined and solved, this work constitutes a detailed, updated, generalized, and complete document. It compiles and summarizes all the previous geological information, accumulated over dozens of years of dedicated physical and intellectual work of hundreds of Cuban and foreign geoscientists. It is, therefore, the most reliable reflection of the superficial geological constitution of the national territory. It is, also, an excellent new point of departure to undertake future works of detail in order to continue increasing and improving the degree of study and geological knowledge of the Cuban territory.

This map has had a great reception and immediate application in several branches and spheres of national science and economy, being fully or partially demanded by other institutions such as the Centro de Investigaciones del Petróleo (CEINPET), the Instituto de Geografía Tropical (IGT), Instituto de Ciencias Agrícolas (ICA), and the different territorial Empresas Geomineras of the Grupo GeoMinSal. It has also been acquired by different foreign companies (China, Russia, Canada, Australia, etc.), interested in mining investment, including the hydrocarbon exploration.

The map has also been used as a geological basis for the preparation of new thematic maps such as the Metallogenic Map of Cuba at 1:250000 scale [80], the Tectonic-Structural Sketch of Cuba at 1:250000 scale [81], and the Geological Map of Cuba at 1:1 000000 scale [82], prepared especially for the Geological Map of the World Program at that scale.

One of the most important applications of the Digital Geological Map of Cuba at 1:100000 scale, perhaps the main one, is that it constitutes the substrate or geological foundation for the Geographic Information System (GIS) that supports the Geological Data Bank of the Republic of Cuba (GEODATO).

Conditional geological surveys at scales 1:100000 and 1:50000

We cannot talk about geological cartography in Cuba without referring to geological surveys at semi-detailed scales (1:100000 and 1:50000) that were carried out over three decades (1961–1992), a period extending from the revolutionary triumph until the very collapse of the Soviet Union and the Socialist Field.

At the behest of Commandant Ernesto Che Guevara, since he took office as Minister of Industries, he began a collaboration with the USSR and with almost all the socialist countries of Eastern Europe.

One of the collaboration modalities was the sending to Cuba of specialists of experience in geological surveys at scales 1:50000 and 1:100000, mainly geologists and geophysicists, with the purpose of detecting new mineral prospective sectors. In these, prospections would be made to that were called "companion searches", at more detailed scales (1:25000 and 1:10000).

From each of these works, the main product that remained was the geological map at the scale of the survey in question, together with a series of specialized maps such as the geomorphological, hydrogeological, mineral deposits, and others. Their characteristics and explanations were described in voluminous texts, sometimes of several volumes, with their graphic and textual annexes, called "reports".

This type of collaboration can be divided into two stages. The first, which we have called Pre-CAME, is between the years 1960 and 1982 and was specifically with Soviet specialists from the 15 socialist republics that integrated the USSR. At this stage, according to the records of the Oficina Nacional de Recursos Minerales (ONRM)—which at the time was called the Centro Nacional del Fondo Geológico (CNFG)—17 "conditional" surveys were carried out, two of them at 1:100000 scale and the rest [33] at 1:50000 scale.

All these works were executed according to Soviet standards, which translated into English mean, "Directory of aggregate estimation standards for geological exploration works" (Fig. 21). These standards required a high degree of detail and the realization of large volumes of sampling, drilling, and field descriptions, among others, to achieve compliance with the "conditions" required for each scale, hence the expression "conditional surveys".

The second stage is that of the CAME surveys, named for being the result of bilateral collaboration agreements that were established with the member countries of the Consejo de Ayuda Mutua Económica. This entity grouped and favored commercial and economic exchange among the countries of the socialist field, and of which Cuba became an effective member since 1971, after the failure of the so-called "Ten Millions Mill", benefiting from a very advantageous program of exchanges. Within this framework, a series of bilateral agreements in the sphere of mining and geology were signed in La Habana, selecting the areas where the surveys would be executed, which were distributed among the different signatory countries of the agreement.

The geological mapping works with accompanying searches in the so-called CAME polygons began to be executed in 1982 and extended until 1992. In these 10 years, according to the ONRM archive data, 26 surveys were carried out, of which 16 are at scales 1:50000 and 10 to 1:100000. In them participated geoscientists of Bulgaria, Czechoslovakia, Hungary, GDR and the USSR, and of Cuba, the host and beneficiary country.

To carry out these works, a guiding document was drafted (original in Russian), which, as a rule, would set the guidelines for the execution of survey work by the different "binational brigades" in their respective sectors. This document, translated into Spanish, without distinction with the titles: "Instruction for the execution of the geological survey at 1:50000 scale" and "Standard for the execution of the geological survey at 1:50000 scale" consists of two parts or volumes. The first is a textual relation of the steps to execute during the works, while the second part collects the symbols to be used in the preparation of the maps and their graphic annexes.

These documents do not count (at least in the copies that could have been consulted) with a list of authors, nor the year of elaboration, so is it impossible to cite it formally. It only refers that it is a document prepared by the Ministerio de la Industria Básica (MINBAS) of the Republic of Cuba.

General characteristics of the conditional surveys at scales 1:50000 and 1:100000

As it is known, since the first years of collaboration in the fields of mining and geology with the countries of the former Socialist Field, several mining companies (now Geominera) were created in the country, which were distributed as follows:

- Geominera Pinar del Río, located in Santa Lucia, Pinar del Río. It attended the Pinar-Habana-Matanzas region.
- Geominera Isla de Pinos, located in La Fe, Isla de la Juventud.
- Geominera Las Villas (Centro), located in Santa Clara, Villa Clara. It served the entire region of Cienfuegos, Villa Clara, and Sancti Spiritus.
- Geominera Camagüey, in the city and province of that name. Attending the Ciego-Camagüey-Las Tunas region.
- Geominera Oriente, located in Santiago de Cuba. It attended all the vast regions of the provinces of Guantanamo, Santiago de Cuba, Holguín, Granma, and almost all of Las Tunas.

Generally, the conditional surveys at the scales indicated above, carried out in the Republic of Cuba between 1960 and 1992, corresponded to a stage of the regional geological prospecting of the national territory. With the exception of the first ones, carried out in collaboration with Soviet geologists, most of them were carried out following the geological surveys at 1:250000 scale.

For this, temporary camps or brigades were created, located in the regions of influence of the active mining companies (Geominera). They remain in the surroundings or within the prospective areas for the revelation of useful mineral deposits, as well as for the satisfaction of other needs of the national economy, such as construction, agriculture, and water resources.

These regions were classified by the conditions of execution of the works, for which several factors were taken into account. Among these, as the most important, were considered the degree of outcrop and complexity of the geological structure, as well as the geological interpretation in aerial photos and satellite images, the physical-geographic conditions of the territory, and accessibility.

For the selection of the area to be mapped in each project, i.e. the size of the polygons, the factors listed above were taken into account. In addition, the presence of useful minerals, the degree of complexity of the geophysical fields, as well as the material possibilities of simultaneously investigate a whole territory. However, as a rule, the area of the selected polygons oscillated between 2 and 6 topographic blankets at 1:50000 scale and very rarely exceeded these dimensions.

The 1:100000 scale surveys were carried out in those areas, where the characteristics of the relief (especially mountainous regions) made it difficult to comply with the conditionality required for 1:50000 scale. Also, for the application of the whole complex of methods, mainly the drilling footage, as is the case of the Sierra Maestra and the Guamuhaya massif (Escambray). Other times, they were reserved for areas that are not very complex from the geological point of view, but which did not require a more detailed scale. This is the case of the Pinar Sur survey at 1:100000 scale [32].

The work required the so-called methods of advancement that included the geological photointerpretation of aerial and/or satellite images for the preparation of a photo geological scheme or preliminary map (pre-map). This reflected the main structures of the chosen area and traced the reconnaissance itineraries to be carried out by transversal regional profiles to the main geological structures.

The survey required the geological mapping of the entire territory at the chosen scale and the preliminary prospecting at scales 1:25000 and 1:10000 in certain sectors of mineral interest revealed during the execution of the works.

To comply with the conditionality of the surveys, the density of the observation network considered as "optimal" was 2 linear km of geological itineraries for each km^2 for the scale 1:50000 and 1 km for the works at 1:100000 scale. In the first case, the description of an outcrop every 500 meters was also made, apart from the continuous description of the observed between points throughout the itinerary.

The drilling in regional profiles, in the surveying sectors, and in the rest of the polygon included up to four types of wells:

- Parametric (up to 600 m);
- Structural (up to 400 m);
- Hydrogeological (until cutting the water table mirror);
- Mapping and search wells (up to 60 m).

The use of morphometric methods of study of the relief and the making of a geomorphological map of the study area were mandatory. The hydrogeological survey was also carried out, with observations of the hydrogeological regime during the drilling, test pumping in the wells drilled and "criollo wells", the taking of water samples for chemical and spectral analysis of the dry residue.

The works of pedestrian geophysics comprised a wide complex of methods that included:

- Gravimetry (by areas and in regional profiles);
- Magnetometry (regional profiles, in support of mapping and in search sectors);
- Radiometry in conjunction with geological itineraries, in regional profiles, in support of mapping and search sectors, terrestrial verification of gamma-spectrometric anomalies, discrete gamma logging in the mapping wells;
- Electrical methods (induced polarity, medium gradient profiling, vertical electric sounding in regional profiles);
- Physical properties of the rocks (density and magnetic susceptibility).

On the other hand, within the complex of geochemical works, the following were carried out:

- Sampling of fragments of outcrops;
- Metalometric sampling;
- Lithogeochemical sampling of drill core;
- Lithogeochemical sampling in mining works.

As for the types of analysis, the following were carried out:

- Paleontological (washing and thin sections);
- Petrographic;
- Mineralogical (sampling of jagua);
- Mineralogical;
- Chemicals;
- Semi-quantitative spectral (for 18 and 32 elements);
- Granulometric;
- Technological tests;
- Plasticity of the sands.

A notable difference between the first surveys and those that have already taken place in more recent years is that while maps of the first were made by separate sheets with legends in the form of a geochronostratigraphic column and their own geological sections (Fig. 1.20), the second were made to a whole polygon. That included several complete sheets of the nomenclature of the scale or fragments of them, with the legend—and often the geologic cuts—in one sheet (Fig. 1.21).

As has already been said previously, with the end of the USSR and the so-called Eastern European Socialist Field, the CAME ended and with it, all the economic collaboration with these countries. Consequently, the aid in the geological-mining-oil sphere ceased and several polygons that had been planned were left undone, some even had to be canceled. However, the balance of cooperation in the field of geological mapping was largely positive: around 38% of the territory of the Cuban Archipelago was mapped at 1:50000 scale (Fig. 1.22), while at 1:100000 scale, it was mapped about 17% (Fig. 1.23). From the above, it can be inferred that, as a result of the conditional surveys carried out over three decades in Cuba, 55% of the national territory is mapped at semi-detailed scales (Fig. 1.24).

Reflecting on the quantity, density, and quality of the data provided by these geological works, it can be said that they are well above the standards currently used in the world for the aforementioned scales. If we also take into account that the surveys were carried out in the most favorable areas for the occurrence of mineral resources, it is not exaggerated when it is asserted that close to 100% of the most prospective areas are raised at semi-detailed scales (1:50000

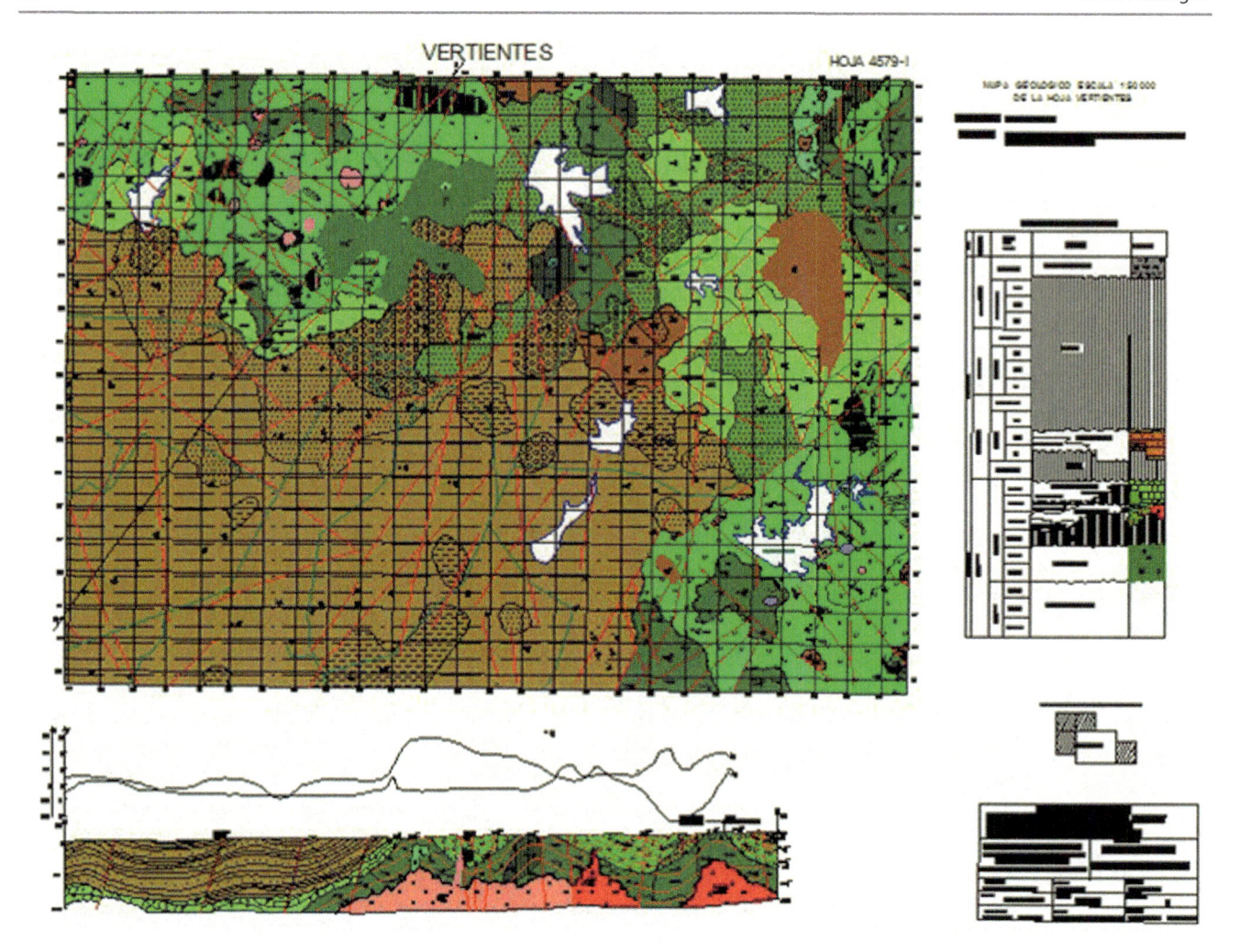

Fig. 1.20 Digital version of the Geological Map of the Vertientes sheet (4579-I). Result of the Complex Geological Survey of the Camagüey Polygon, sectors Ciego de Ávila-Vertientes at 1:50000 scale (CAME), carried out by a Cuban brigade between 1987 and 1991. (Piñero Pérez E et al. [52]. Digitized by Yadira Duran Cuervo, edited by Valentina Strazhevich, IGP, 2016

and 1:100000), with a large number of detailed sectors at scales 1:25000 and 1:10000).

From this period, date the discoveries of new deposits and manifestations of solid metallic and non-metallic minerals. From the first, the findings of iron, copper, gold, tungsten, bauxite, polymetallic, and others stand out, many of which have been exploited and others are still in research phases. In the sphere of non-metallic, numerous deposits of stones for construction were discovered: limestone for cement, clays, kaolin, sands (aggregates in general), decorative and ornamental rocks, zeolites, different ores for agriculture uses, etc.

1.4.2.2 Geological Maps at Detailed Scales (1:25000–1:10000)

Throughout the history of geological cartography in Cuba, which, incidentally, already has more than 135 years, some attempts have been made of detailed geological studies of certain areas. In fact, it would be very difficult to mention all of them since, as we know, many of these studies were done within the conditionally mapped areas at semi-detailed scales. We refer to the sectors where the so-called "companion searches" were carried out, where the degree of detail reached 1:25000, 1:10000, and up to 1:2000.

However, some examples can be related, due in some cases to their historical and scientific interest; in others, because they are independent studies, specially designed at small scales.

Geological map of La Habana and its surroundings at 1:25000 scale

In the first case, we can mention the maps resulting from a geological study of the territory of the Cuban capital and its surroundings carried out between 1957 and 1959, and published in the form of a monograph in 1963 [43]. In this invaluable work, which became a treaty of inescapable consultation, with several paleontological discoveries and

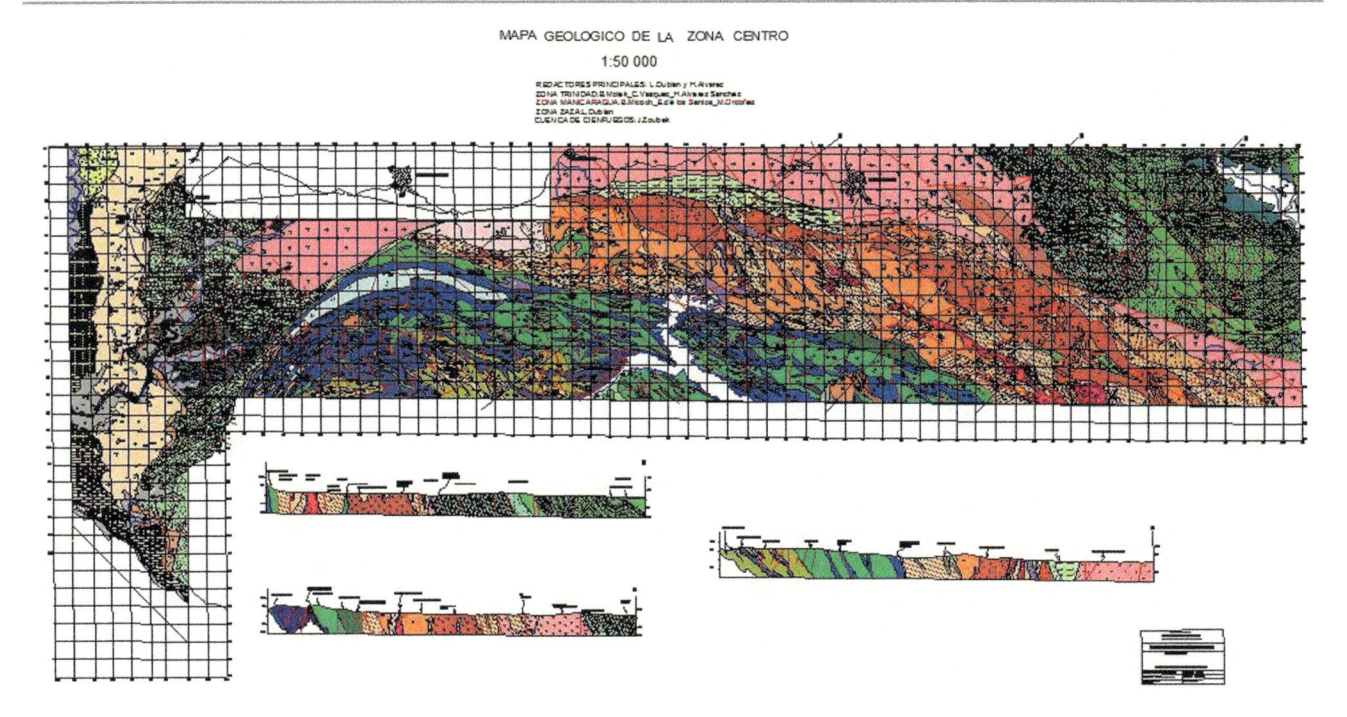

Fig. 1.21 Digital version of the Geological Map of the Central Zone, result of the Geological Survey at scale 1:50000 of the CAME polygon of the same name, located north of the Trinidad dome of the Escambray metamorphic massif, erected by a Cuban-Czech brigade between the years 1982 and 1986. (Dublan et al. [19]. Digitized by Melisa Texas Pita, edited by Valentina Strazhevich, IGP, 2016

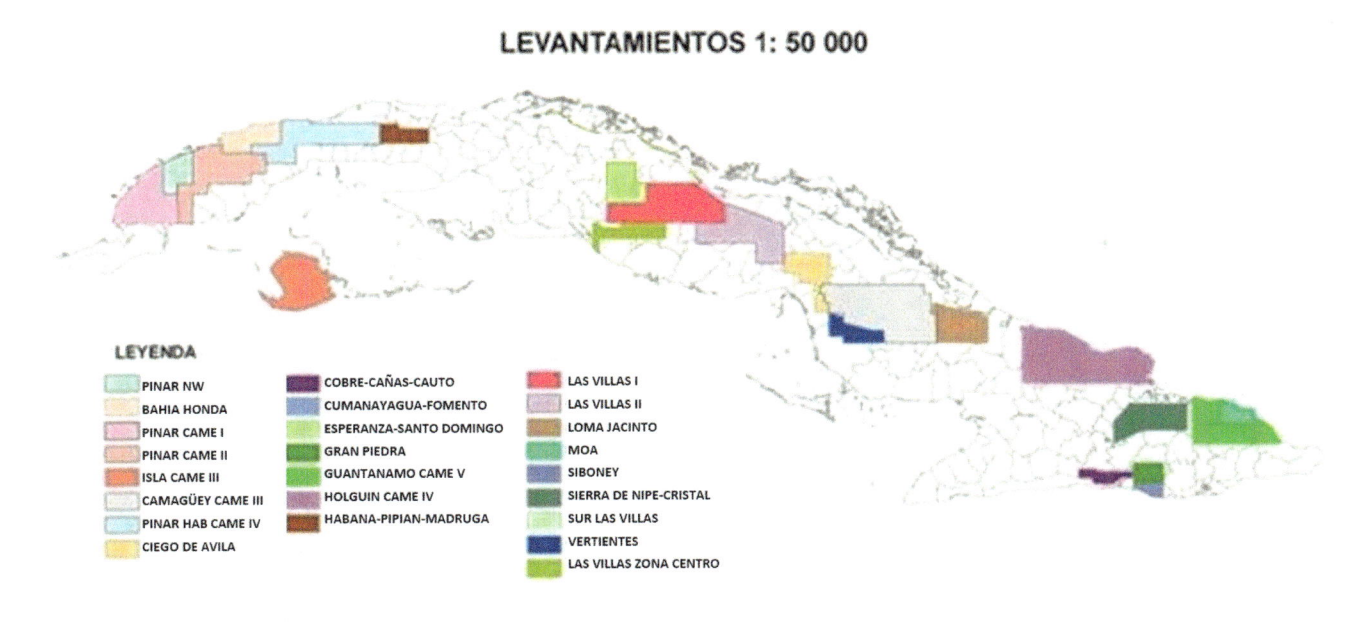

Fig. 1.22 Scheme showing the polygons mapped at 1:50000 scale in the period 1961 and 1992. These surveys cover approximately 38% of the national territory (Pérez Aragón et al. [4]

contributions to the Léxico Estratigráfico de Cuba, the Swiss researcher Paul Brönnimann and the Italian Danilo Rigassi published two geological maps (Plate II and Plate III) of obligatory consultation for any student of the geology of the capital region. The first was published on a scale of 1:100000 and the second one on a scale of 1:25000.

For the realization of these studies, the authors themselves claim to have had the good fortune to carry out their fieldwork in a time of great constructive activity. Roads (the tunnel of the Bay of La Habana, the Monumental avenue, the Via Blanca) and several residential deals to the south of the Capital (Capdevila, Alcazar, etc.) meant access to

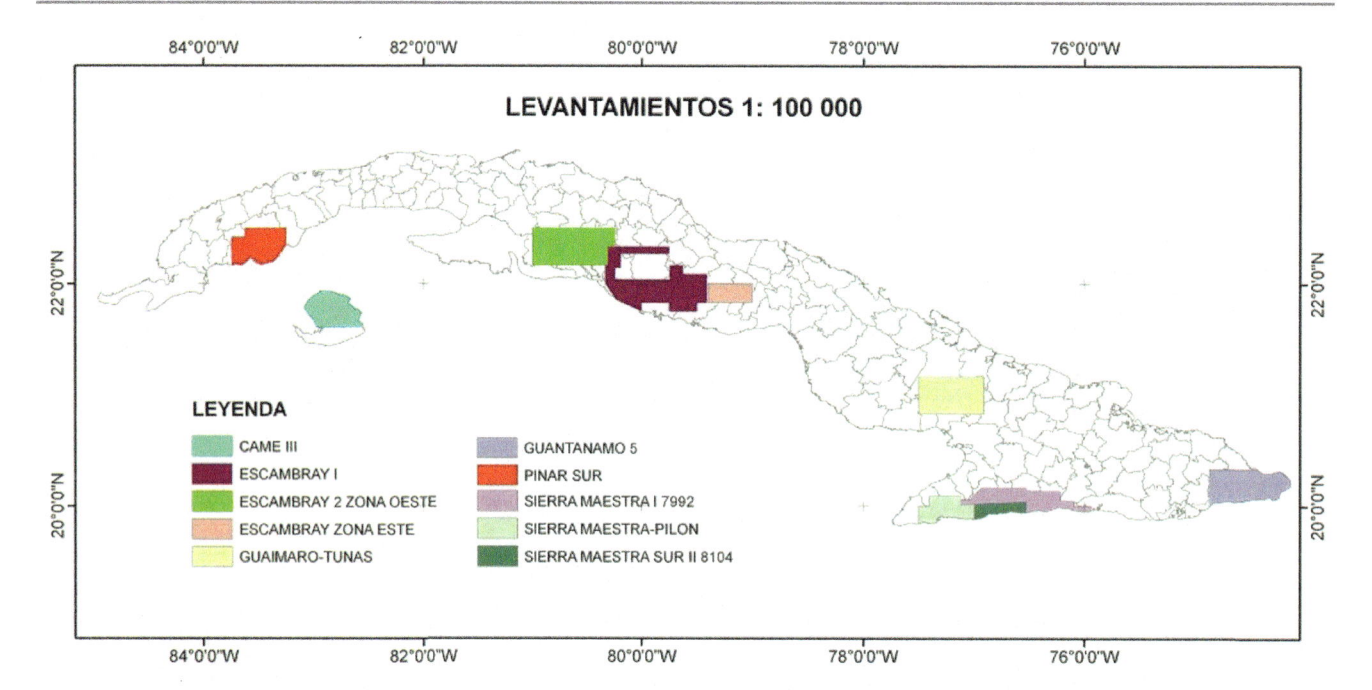

Fig. 1.23 Diagram showing the polygons mapped at 1:100000 scale in the period 1961 and 1992. These surveys cover approximately 17% of the national territory (Pérez Aragón et al. [4]

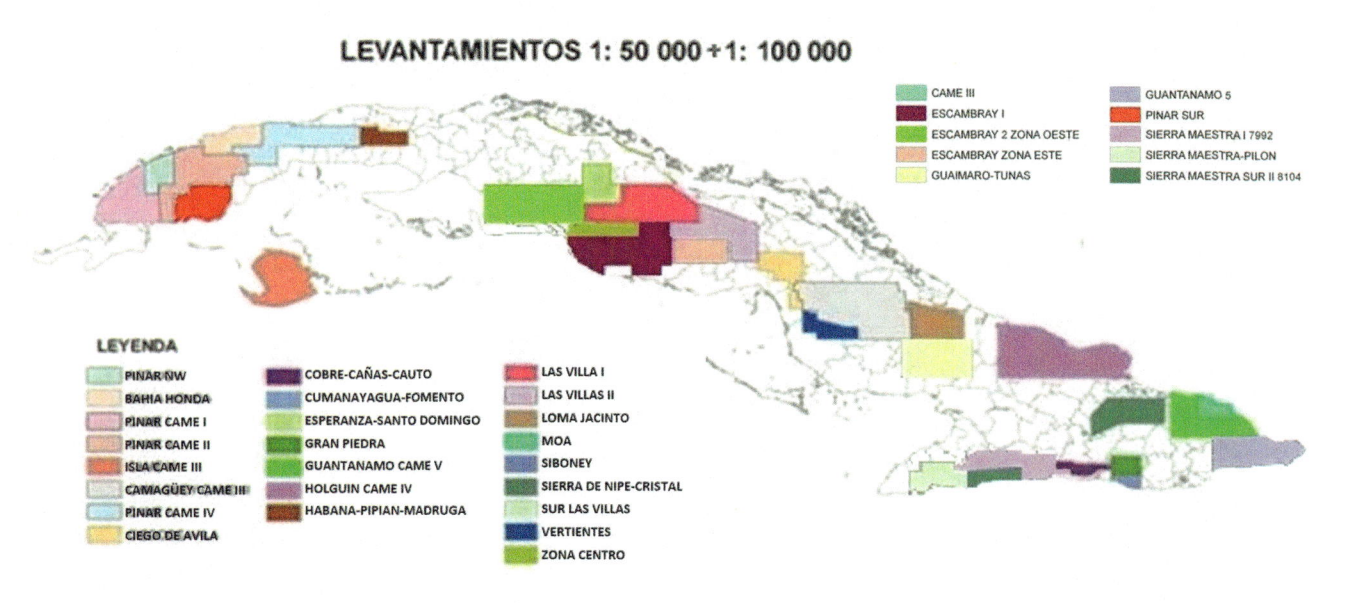

Fig. 1.24 Scheme showing the polygons mapped at scales 1:50000 and 1:100000 in the period 1961 and 1992. Overall, these surveys cover approximately 55% of the national territory (Pérez Aragón et al. [4]

unbeatable outcrops, and some of which even gave name to new lithostratigraphic units described by them. Such are the cases of: Via Blanca, Capdevila, Alcazar, Via Tunel conglomerates, Rio Piedras conglomerates, Bacuranao limestone, Cacahual limestone, Habana group, Maríanao group, and others.

At the same time, they refer to the difficulty of not having access to many outcrops "due to the political situation in the country" in those years of full revolutionary effervescence against the bloody dictatorship of Fulgencio Batista. This reminds us of what was lived a century before by Fernández de Castro and Salterain Legarra during the elaboration of their Geological Sketch of the Island of Cuba in the middle of the war of independence [1].

Geological map of the El Cobre deposit and its surroundings at 1:10000 scale

Also for the year 1963, a couple (in science and in life) integrated by Soviet geologists Y. Bogdanov and V. Bogdanova, together with Cuban geologist technicians Manuel Miralles, Reynol Sosa, and Mario Estrugo [57], made a detailed geological mapping in the area of the El Cobre cooper deposit, in the former province of Oriente (today, Santiago de Cuba).

The study was carried out with the objective of specifying the stratigraphic, tectonic, and volcanogenic data of the El Cobre mineral zone, which, according to the opinion of the authors, was very poorly studied from the geological point of view. There was only a schematic geological map at scale 1:50000 and an outline of the geological structure of the Great Mine.

As a result, the geological map was obtained at 1:10000 scale in an area of 11 km^2, containing four geological sections that show the deep structure of the deposit and the mineralized areas surrounding it.

In addition, an explanatory report was presented under the title "Report on the results of the survey and search works in the area of the El Cobre deposit (Oriente Province)". Both the report and the geological map presented and constituted a great step forward in the knowledge of the internal structure, genesis, and mineral associations of one of the oldest deposits in Cuba. This allowed to project, with a scientifically argued vision, new areas of exploration, exploitation, and extension of the useful life of the mine.

Geological map of the Mariel-Cojímar area at 1:20000 scale

According to work in the archives of the ONRM, with the code or inventory number 896, toward the year 1969, the "Report of the geological survey, Mariel—Cojímar area at 1:20000 scale" was approved and deposited in that institution. Its authors are the Soviet specialist Ivan Garbuz and the Cuban geologists Evelio Linares Cala and Carlos Álvarez.

Among the graphic annexes of this report appears the Geological Map of the Mariel—Cojímar area, at the aforementioned scale, which reflects a wide coastal strip that extends between the towns cited, belonging to the current provinces of Artemisa and La Habana, respectively.

Although among the objectives of this survey was fundamentally the study of the manifestations of hydrocarbons, especially asphaltites, present in the rocks of this region, at the end, a geological map is presented where the deposits and superficial lithostratigraphic units are mapped with the detail of the scale.

Geological map of the municipalities of the south and east of La Habana at 1:25000 scale

In 1994, in the midst of the economic crisis that followed the collapse of the Socialist Field and the dismemberment of the USSR, partly because it was in tune with the currents of the moment and also as a way of shoveling the lack of research projects and source of employment of the researchers, the group of Environmental Geology was created in the IGP.

Among the first tasks of this group, was the execution of a series of projects called "Comprehensive environmental geology of the municipality (such or more) at 1:25000 scale". These projects, as it is deduced from the name, were made for the municipalities of the then province of Ciudad de La Habana, today simply, La Habana, and were partially financed by the beneficiaries, that is, the municipal governments.

A multidisciplinary group that included several specialties carried out territorial geological-environmental studies and, as a result, a series of thematic maps at 1:25000 scale were obtained. Such is the case of pollution by industrial waste, use and occupation of the territory, geomorphological, slopes, soils, vegetation, among others. Also, they consider a geological map with the distribution of the occurrences of resources or raw materials of mineral origin of possible use in the development of the municipalities studied.

The first municipality evaluated would be San Miguel del Padrón, because this is the headquarters of the IGP, and because of lack of fuel for transportation, almost all itineraries could be done only on foot or by bicycle. Then, the studies of La Habana del Este (Fig. 1.25), Guanabacoa, Regla, Cotorro, Arroyo Naranjo, and, finally, Boyeros will be made, using resources, especially the fuel provided by the municipal governments. Similar studies that were carried out in the basins of the Luyanó and Martín Pérez rivers would come to provide additional data on the territory of the Diez de Octubre and San Miguel del Padrón municipalities.

As basic materials for the geological mapping of these territories, data from two fundamental studies would be useful. The first, at scales 1:100000 and 1:25000 [43] and the second, at scale 1:250000 [9], relying on photointerpretation of aerial and satellite images and abundant field verification itineraries.

An IGP team integrated by Ramón O. Pérez, Miguel A. García Saborit, and Valentina Strazhevich made all maps. Once finished the seven maps on mentioned municipalities, they were generalized and unified in only one. Finally, the map was cut according to the frames of the 13 topographic sheets of the nomenclatures corresponding to the scale 1:25000, being currently available in all three formats.

This map has had an immediate application in several branches of the economy and society. Such was the case of

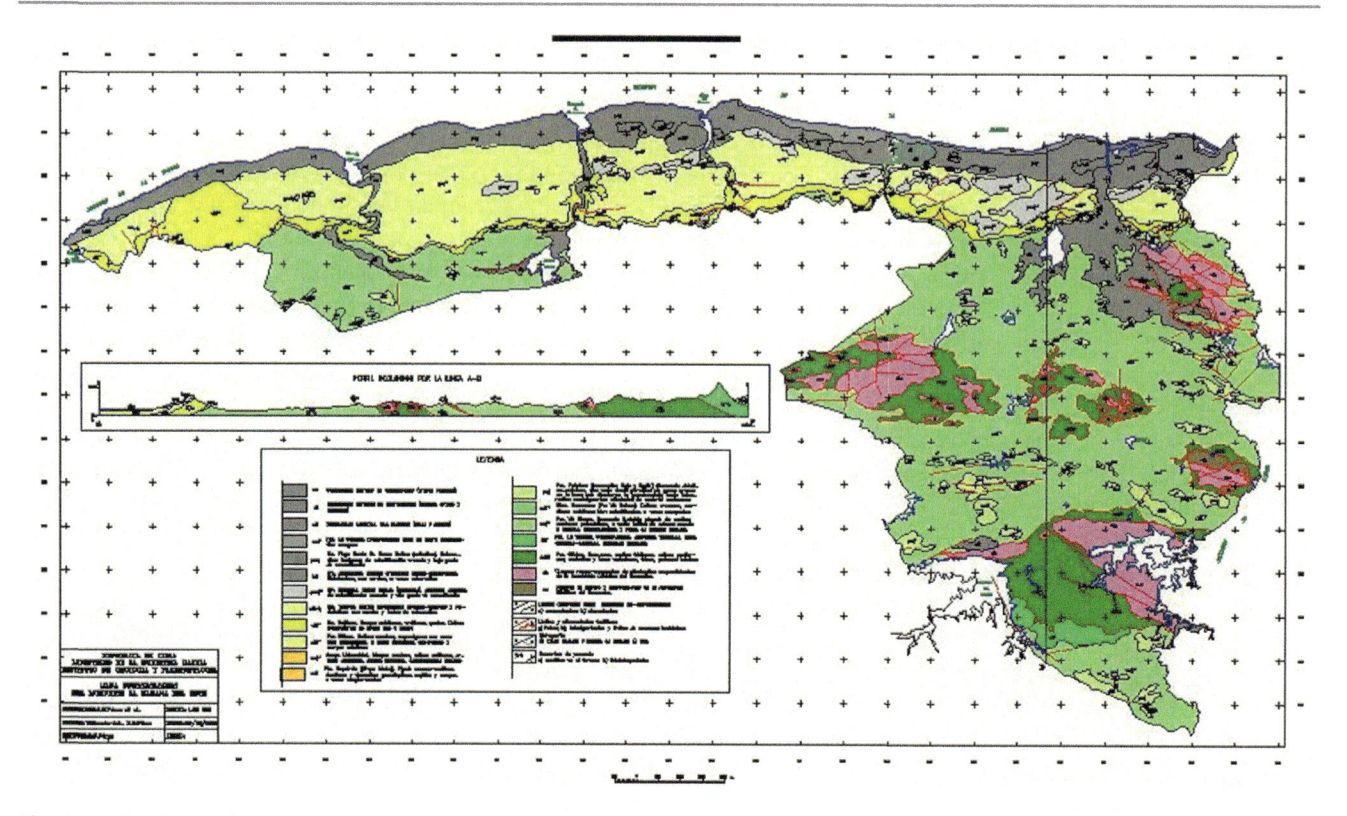

Fig. 1.25 Facsimile of the digital geological map at 1:25000 scale of the municipality of La Habana del Este, one of the 7 that make up the Digital Geological Map at said scale of the eastern and southern municipalities of Ciudad de La Habana (Pérez Aragón et al. [47]). Digitized Valentina Strazhevich, edited Ramón O. Pérez IGP, 2000

the location of deposits or accumulations of raw materials for construction. They, without reaching large deposits, can serve to alleviate the needs of the territories for the repair or construction of low-cost housing, sand, gravel, clays, and others. Other was the environmental study of mining environmental liabilities of Passive Mining Actives (PMA). It consisted in making an inventory of mines, quarries, and abandoned loans and make proposals to use other treatment to mitigate the impacts caused to the environment by mining activity. Finally, were the geo-environmental studies of the basin and sub-basins of the Almendares River and its tributaries, with the objective of detecting and eliminating the sources of contamination of the water resources and the responsible and ecological management of them.

1.4.3 The Human Resources and Some Interesting Statistics

In the elaboration of the different generations of geological maps of all the scales that cover, totally or partially, the territory of the Cuban Archipelago, hundreds of scientists and specialists from different branches of geosciences have participated directly or indirectly. It includes geologists, paleontologists, mineralogists, petrologists, palynologists, hydrogeologists, geophysicists, geochemists, geographers, geomorphologists, surveyors, drillers, miners, and others.

Many architects, civil engineers, biologists, and even doctors in pharmacy changed their specialties and became notable geologists, tectonics, paleontologists, stratigraphers, etc., and made their more or less remarkable contributions to paleontology, stratigraphy, and finally, to the geological cartography of Cuba.

Likewise, it would not have been possible to carry out geological surveys, nor to prepare maps, without the direct or indirect participation of hundreds of workers. It includes: mining digging workers, drilling engineers and assistants, machete-cutters of topography trails, samplers, backpackers, operators of bulldozers, cranes and other heavy equipment, drivers, boat captains, divers, drafters, translators, brigade administrators, typists, camp-chefs, and all kinds of support personnel.

It is interesting to note that the geological cartography of Cuba is not the unique heritage of the country's nationals (Table 1.1); on the contrary, as a result of the consultation of reports, books, documents, and maps, we can see that there are hundreds of foreign specialists. These are citizens of more than one dozens of countries, from several continents, that have worked in Cuba. Among these, at list, appear citizens of Bulgaria, Chile, Czechoslovakia, Germany

1 Geological Cartography of Cuba

Table 1.1 Foreign countries and specialists participating in geological mapping of Cuba

1		Czechoslovak Socialist Republic	71
2		Federal Republic of Germany	3
3		French Republic	1
4		German Democratic Republic	43
5		Italian Republic	2
6		Kingdom of Holland	6
7		Kingdom of Spain	3
8		People's Republic of Bulgaria	81
9		People's Republic of Hungary	79
10		People's Republic of Poland	19
11		Republic of Chile	4
12		Republic of Guatemala	1
13		Republic of Indonesia	1
14		Republic of Peru	1
15		Socialist Republic of Romania	2
16		Swiss Confederation	1
17		Union of Soviet Socialist Republics	361
18		United States of America	51
Total			731

(Federal and Democratic), Guatemala, France, Italy, Holland, Hungary, Indonesia, Peru, Poland, Romania, Spain, Swiss, USA, and especially from the former Soviet Union.

The Cuban participation in the geological cartography, during the prerevolutionary period, that is, before 1959, was rather scarce and limited to some isolated personalities: famous naturalists, prominent specialists, and some enterprising executives in the sphere of mining, which would total a couple of dozens (Table 1.2).

In the second stage, that is, after the triumph of January 1959, thanks to the clear educational policy undertaken by the revolutionary government, the graduation of specialists in geosciences would reach an unusual peak. It adds several hundreds of mid-level and superior technicians, graduates of

Table 1.2 First results of the New Geological Map of Cuba at 1:50000 scale

No.	Executing entity	Name of the sheet	Number of the sheet
1	Geominera Pinar del Rio	Consolación del Sur	3483-I
2	Geominera Isle of Juventud	Nueva Gerona	3681-IV
3	Geominera Centro	Camajuaní	4283-II
4	Geominera Camagüey	Cascorro	4779-IV
5	Geominera Santiago de Cuba	Gran Piedra	5076-II
6	GeoEM	Cayo Ines de Soto	3484-III

different schools, and faculties that were created in the country, as well as at institutions of different countries of the former European socialist field.

Hundreds of young Cubans were trained as mid-level technicians in geology, geophysics, and topography in technological schools created in different periods and places. Such are: the School of Geological Assistants of El Cobre, in Santiago de Cuba; the "Félix Corzo" School of Technicians in Geology and Geophysics, addressed in Linea and 8, Vedado, La Habana, which worked since 1962; the Military Technological School "Comandante Vitalio Acuña", located in Cotorro, La Habana; and the Technological Geological School of La Carlota, in Cumanayagua, Cienfuegos. About four dozens of geologists, hydrogeologists, and geophysicists graduated from the Kiev Geo-Prospecting Technological Institute (KGRT, in Russian language) in the beautiful Ukrainian capital between 1970 and 1980.

The first 26 licensed geologists, graduated by the Revolution, egressed in 1966 from the first School of Geology. This was attached to the Faculty of Sciences of the University of La Habana (UH), as part of a course, inaugurated in 1962, under the direction of the first Cuban geologist engineer, graduated in USA, Gustavo Echevarría Rodríguez and the rectory of Dr. Juan Marinello Vidaurreta. The cloister of this school [83] was a true example of internationalism. It was composed of Czechoslovak specialists: Frantisek Cech (Geological survey), Milan Mishic (Sedimentology), Javelka Bojuslav (Inorganic geochemistry), Vladimir Tyls (Geology for engineers), Yuri Kralik (Mineralogy), V. Jladic (Geophysics), and Y. Sorkovski (Structural geology). From the USSR participated Irina Shirokova (Petrography) and Vladimir Sacedatle (Geophysics for minerals, Gravimetry, Magnetometry, and Electrical methods). From Peru, Guillermo Cox (Petrology). From Guatemala Jorge García Calderon (Petroleum geology). Also, the Italian Amadeo Sikoski (General geology) and the Cuban-Spanish Rafael Segura Soto, who taught Petrography.

New licentiates and engineers would graduate in the School of Geology of the University of Oriente, in Santiago de Cuba, in the University of Matahambre, Pinar del Río, in the Higher Metallurgical Mining Institute (ISMM) of Moa, Holguín, and in the University Hermanos Saíz from Pinar del Río.

Great influence has also had the School of Geophysics of the Ciudad Universitaria José Antonio Echevarría (CUJAE) of La Habana, providing hundreds of specialists in this branch. This without demerit of other UH faculties, such as those of Geography and Biology, from where geographers and biologists graduated. They have traditionally later converted into geomorphologists and paleontologists who have given their contribution to geological cartography.

Engineers' geologists, geophysicists, miners, and others, graduated from several universities of the USSR. Such are: the Red University of Kiev, Ukraine; the State University and the Gorniy Institute, Leningrad; the University of the Peoples Patricio Lumumba, the State University M. V. Lomonosov, the Institute of Geological Prospecting (MGRI, in Russian language), all of them in Moscow; and other Universities in Uzbekistan and Kazakhstan. In other countries of the former Socialist Field such as Bulgaria, Czechoslovakia, and Romania, several dozens of young higher level specialists in the various branches of geosciences between 1960 and 1980 would also graduate.

In the years of maximum activity of the geological researches in Cuba (decades from 60 to early 90), the Cuban specialists, among mid-level, and high-level technicians, came to be counted by hundreds. They were distributed by the different brigades dedicated to the prospecting, exploration, and surveys in the six territorial companies (Pinar del Río, Habana-Matanzas, Isla de la Juventud, Villa Clara or Centro, Camagüey, and Oriente or Santiago de Cuba) and in research centers and middle-level and superior teaching units, already mentioned. In these years, the activity of geological cartography, according to the data collected for this chapter, brought together human resources in numbers that greatly exceed 1080 workers from various specialties related to geology.

1.4.4 Cartageol 50 K, the New Geological Map of Cuba at 1:50000 Scale

As it's known, recently, the Instituto de Geología y Paleontología–Servicio Geológico de Cuba (IGP-SGC) has finished the updating and editing of the last Geological Map of Cuba, at 1:100000 scale, in digital format and in Geographic Information System (GIS) environment [2], which covers the entire Cuban archipelago. This is the most updated and detailed map that covers the entire national territory.

However, the geological cartography of the Republic of Cuba requires moving to a higher stage, which allows the collection and updating of all the scattered information at larger scales. This is in order to proceed with the systematization and completion of the geological cartography of the national territory at 1:50000 scale, including, of course, its marine-coastal zone.

Taking into account the above, the IGP-SGC, with the support of the Directorate of Geology at the MINEM, implemented, as part of the Development Program of Geology until 2030, the Subprogram of Geological Cartography of Cuba at 1:50000 scale (Cartageol 50 K). It establishes the policy of standardizing and completing the geological mapping of the country at this scale, thus completing one of the fundamental tasks of the IGP-SGC, which is the constant increase in the geological degree of study of the national territory.

In line with the above and, with the evolution experienced by Geology worldwide in recent decades, there is a need to introduce substantial changes in the methodologies used so far in the realization, edition, and publication of geological maps. This is in order to undertake the mapping of the 420 sheets at 1:50000 scale, which will make up the new Geological Map of the Republic of Cuba on that scale. Also, to establish the rules that guarantee a homogeneous product quality and a rapidly available database.

To carry out this task, was created a **Methodological Instruction** [4]. It determines the main criteria of the organization of the work of geological cartography, the bases, and principles of the methodology of its realization. Also it considers the main requirements of the content of the maps and of the materials for the elaboration of their **Explanatory Memory**. This Instruction must constitute the governing document for all the entities that will carry out the works for geological cartography at 1:50000 scale in the territory of the Republic of Cuba.

Since January 2018, six so-called "executing entities" have been implementing the same number of projects for the execution of the cartography of geological sheets in the new format designed for this purpose. The execution of these "pilot projects" aims to validate, in practice, the aforementioned Methodological Instruction, currently in β version, so that it can be perfected and converted into the rule that governs the development of the remaining sheets in the coming years. In such a manner, it is expected that by 2019, the first six geological maps will be ready at 1:50000 scale in the new proposed format. It includes the map itself, an explanatory report in the form of a book, and an Interactive CD that includes all the information used, as well as that resulting from the process of preparing the cartography.

The Cartageol 50 K Subprogram represents a challenge that will put to the test all the experience of the few cartographer geologists who remain active. Also, it will require the preparation and training of new young specialists in the realization of geological maps with the help of new technologies. The work is just in the beginning and is hard, but achievable. The results will undoubtedly be the subject of a new article.

1.5 Conclusions

As a result of the researches and inquiries carried out for the execution of the R & D Project "Methodological Design of the Geological Map of Cuba at 1:50000 scale", in the period elapsed from 2016 to the present, a large amount of information about the history and development of geological cartography in Cuba has been collected and compiled. These data have been of inestimable value for the preparation of the new Methodological Instruction, but it has also revealed some paradigms about the work carried out to this day in this branch of geological work. The same, in a manner of conclusions, could be summarized in a few fundamental aspects:

- Cuba's experience in geological mapping is supported by a tradition that goes back more than 135 years.
- The development of geological cartography in Cuba has been characterized by the constant and sustained increase in the degree of study and detail of geological information.
- Because of the foregoing, the entire territory of Cuba is covered by geological maps at scales of 1:1000000; 1:500000; 1:250000, and 1:100000.
- The carrying out in Cuba of Conditional Surveys with accompanying searches at 1:50000 scale in 38% and at 1:100000 scale in 17% of the national territory, whose results and primary data are kept in the archives of the ONRM, make the country to have a large and valuable volume of data of about 55% of its territory. This must be used and reinterpreted, in light of the new technologies, for the completion of geological mapping at 1:50000 scale.
- The remaining 45% of the territory of Cuba has 1:250000 high-quality surveys, so in these areas, geological mapping at 1:50000 scale can be carried out and completed with the execution of small volumes of additional works. These, mainly include the interpretation of remote sensing data and the reinterpretation of aero geophysical data.
- Despite serious financial difficulties and lack of resources of all kinds, including human resources, the IGP-SGC is willing and determined to accept the challenge posed by the new geological cartography of Cuba at 1:50000 scale. It will be in a modern format, adapted to current requirements and according to the new Methodological Instruction designed for that purpose.

Acknowledgments To my friend Dr. Manuel Enrique Pardo Echarte, for so kindly invited me to participate in this important book.To my dear colleague and friend Nyls Gustavo Ponce Seoane, for his patient and accurate review that helped improve the texts and for his enthusiastic praise that conveyed confidence and desire to follow ahead.To my dear friend Dr. Evelio Linares Cala, for opening, disinterestedly, the archives of his prodigious memory.To Roberto Denis Valle and Luis Bernal Rodríguez, for providing valuable unpublished information and graphic memory.To Kenya Núñez Cambra, Iris Méndez Calderón, Juan Guerra Tasé, Rosendo Oña Álvarez, Tomás Martínez Drake, Rey Carral Chao, Aurora Borja Suárez, Ana Rita Díaz Vera, and many others, who helped to dust off old names and surnames lost in time.To the more than 1840 geoscientists of several generations and more than 20 nations that helped directly or indirectly to write the true story of the geological cartography in Cuba.

References

1. Fernández de Castro M, Salterain P (1883) Croquis Geológico de la Isla de Cuba a escala 1:2000000. Bol. Map Geol, España, Madrid, p 8
2. Pérez Aragón RO, García Delgado DE et al (2016) Mapa Geológico Digital de Cuba a escala 1:100000. Archivo IGP-SGC, Inédito
3. Brödermann J, De Albear JF y Andreu A (1946) Croquis Geológico de Cuba a escala 1:1 000000. Comisión Técnica de Montes y Minas del Ministerio de Agricultura. La Habana Cuba. Primera edición
4. Pérez Aragón RO y otros (2017) Instrucción Metodológica para el Mapa Geológico de Cuba a escala 1:50000. IGP-SGC. Inédito
5. Cabrera Castellanos M (2016) Grado de estudio del Territorio Marino-Costero de Cuba. Instituto de Geología y Paleontología, Inédito
6. Linares Cala E, Osadchiy PG y otros (1985) Mapa Geológico de la República de Cuba a escala 1:500000. Centro de Investigaciones Geológicas (CIG). Printed on the "Kart Fabrica" of the Institute of Geological Research of the USSR "A. P. Karpinski" (VSEGI in Russian) Leningrad. USSR
7. Wikipedia (2017). Digital Enciclopedia of free content
8. Núñez Jiménez A, Andreu A, Bogatiriov BS, Novajatsky IP, Judoley KM et al (1962) Mapa Geológico de Cuba a escala 1:1000000. Instituto Cubano de Recursos Minerales, Ministerio de Industrias, La Habana
9. De Albear JF, Iturralde Vinent MA, Carrassou G, Mayo NA, Peñalver LL (1977) Memoria Explicativa del Mapa Geológico escala (sic) 1:250000 de las Provincias de La Habana, Informe. Inventario 2819, ONRM. Inédito
10. Kantchev I, Boyanov I y otros (1981) Geología de la Provincia de Las Villas. Resultado de las Investigaciones y Levantamiento Geológico a escala 1:250000, Realizados Durante el Período 1969–1975. Informe. Archivo ONRM. Inventario 2434. Inédito
11. Adamovich AF, Formell F et al (1980) La cartografía geológica a escala 1:250000 del Archipiélago de los Canarreos. Archivo IGP-SGC, Inédito
12. Pushcharovsky YM, Mossakovskiy AA y otros (1988) Mapa Geológico de Cuba a escala 1:250000. Instituto de Geología y Paleontología. Printed on the Cart Factory of the Institute of Geological Research of the USSR "A. P. Karpinski" (VSEGI), Leningrad. USSR
13. Pushcharovsky YM, Mossakovskiy AA y otros (1989) Mapa Tectónico de Cuba a escala 1:500000. Instituto de Geología y Paleontología. Printed on the Cart Factory of the Institute of Geological Research of the USSR "A. P. Karpinski" (VSEGI), Leningrad. USSR
14. Colectivo de Autores (1989) Nuevo Atlas Nacional de Cuba. Academia de Ciencias de Cuba. Editorial Científico Técnica, La Habana, Cuba
15. Piñero Pérez EC y otros (1992) Informe sobre los resultados del Levantamiento Geológico Complejo. Polígono Camagüey III, Sector "Loma Jacinto". Archivo ONRM. Inventario 04191. La Habana. Inédito
16. Arcial Carratalá F, Milián E, Rodríguez S y Bueno I (1994) Informe levantamiento geológico 1:50000 parte norte de Villa Clara "Esperanza-Santo Domingo". Archivo ONRM. Inventario 04311. La Habana. Inédito
17. García Delgado DE et al (1998) Texto Explicativo al Mapa Geológico de Cuba Central (provincias Cienfuegos, Villa Clara y Sancti Spíritus) a escala 1:100000. Archivo IGP-SGC, Inédito
18. Belmustakov E y otros (1981) Geología del Territorio Ciego – Camagüey—Las Tunas. Resultados de las investigaciones y levantamiento geológico a escala 1:250000. Archivo ONRM. Inventario 02892. La Habana. Inédito
19. Dublan L, Álvarez H y otros (1987) Informe del Levantamiento Geológico 1:50000 Zona Centro. Archivo ONRM. Inventario 03562. La Habana. Inédito
20. Lobik I, Dostal D, Zimmerhall P, Rodríguez R, Darias JL y Fernández J (1986) Informe Final del Levantamiento Geológico 1:100000 Escambray II Zona Este 1985–1986. Archivo ONRM. Inventario 03515. La Habana. Inédito
21. Pavlov I y otros (1970) Informe sobre los trabajos búsqueda-levantamiento a escala 1:50000, realizados en 1969–70 en el área comprendida entre las ciudades de Cumanayagua y Fomento (Provincia de Las Villas). Archivo ONRM. Inventario 01299. La Habana. Inédito
22. Maksimov A, Grachev G y Sosa R (1968) Geología y Minerales Útiles de las pendientes noroccidentales del sistema montañoso Escambray. Informe sobre los trabajos búsqueda-levantamiento a escala 1:50000, realizados en la parte Sur de la provincia de Las Villas, en 1966–1967. Archivo ONRM. Inventario 01289. La Habana. Inédito
23. Piotrowska K, Pszczolkowski A y otros (1981) Texto Explicativo para el Mapa Geológico en la escala (sic) 1:250000 de la provincia de Matanzas, Informe. Archivo ONRM. Inventario 3423. Inédito
24. Stanik E y otros (1981) Informe de los levantamientos geológico, geoquímico y trabajos geofísicos, realizados en la parte sur de Cuba Central en las provincias de Cienfuegos, Sancti Spíritus y Villa Clara. Archivo ONRM. Inventario 02882. La Habana. Inédito
25. Vasiliev E y otros (1989) Informe Levantamiento Geológico 1:50000 y Búsqueda Norte Las Villas II Jíbaro-Báez. Archivo ONRM. Inventario 03879. La Habana. Inédito
26. Vázquez C y otros (1993) Informe Levantamiento Geológico 1:50000 y Búsqueda Norte Las Villas III. Archivo ONRM. Inventario 04239. La Habana. Inédito
27. Zelenka P y otros (1991) Informe Levantamiento Geológico Escambray II 1:100000 Zona Oeste. Archivo ONRM. Inventario 04509. La Habana. Inédito
28. García Delgado DE, Gil S y otros (2003) Informe del Proyecto 228. Generalización y Actualización Geológica de la Provincia de Pinar del Río a escala 1:100000. Archivo IGP-SGC. La Habana. Inédito
29. Abakumov B, Stepanov V y Hernández A (1967) Estructura geológica y minerales útiles de la región Viñales en la provincia de Pinar del Río. Informe sobre el Levantamiento Geológico en escala 1:50000 y la Búsqueda Detallada en escala 1:10000 en la parte central de la provincia de Pinar del Río efectuados en 1965–67. Archivo ONRM. Inventario 00138. La Habana. Inédito
30. Astajov K et al (1981) Trabajos de levantamiento geológico a escala 1:50000 en la parte NO de la provincia de Pinar del Río (Hojas -3484-III, 3483-IV y 3483-III-A) Archivo ONRM. Inventario 02971. La Habana. Inédito
31. Biriukov B, Messina V, Ponce N y Navarro N (1968) Informe sobre los trabajos Búsqueda y Levantamiento a escala 1:50000, realizados en los años 1967–1968 en la parte oriental de la provincia de Pinar del Río (región de La Palma). Archivo ONRM. Inventario 00143. La Habana. Inédito
32. Barrios E y otros (1988) Informe de Levantamiento Geológico a escala 1:100000 y Búsqueda Acompañante "Pinar-Sur". Archivo ONRM. Inventario 03659. La Habana. Inédito
33. Burov V y otros (1988) Informe sobre los trabajos de Levantamiento Geológico a escala 1:50000 realizados en la parte Occidental de la provincial Pinar del Río (hojas 3382-III, IV; 3383-I, II, III; 3482-IV-a, c; 3483-III-c) en los años 1981–1985. Archivo ONRM. Inventario 03563. La Habana. Inédito

34. Cherepanov VM, Cuéllar A, Glebov ON y otros (1971) Informe de los trabajos de Búsqueda y Levantamiento a escala 1:50000 realizados en la parte noroeste de la provincia de Pinar del Río. Archivo ONRM. Inventario 00154. La Habana. Inédito
35. Martínez D, Fernández R y otros (1988) Informe sobre los resultados del Levantamiento Geológico y Búsqueda a escala 1:50000 en la parte Central de la provincia de Pinar del Río. Archivo ONRM. Inventario 03642. La Habana, Inédito
36. Martínez D y otros (1991) Informe sobre los resultados del Levantamiento Geológico y Prospección preliminar a escala 1:50000 Pinar-Habana. Archivo ONRM. Inventario 04002. La Habana. Inédito
37. Maksimov A, Mediakov I y otros (1981) Informe sobre los resultados de los trabajos de levantamiento geológico a escala 1:50000 en la zona de Bahía Honda (planchetas 3584-I, 3584-III parte norte) y (3584-IV). Archivo ONRM. Inventario 02867. La Habana, 1968. Inédito
38. Pszczolkowski A, Piotrowska K y otros (1975) Texto explicativo al Mapa Geológico a escala 1:250000 de la provincia de Pinar del Río. Informe. Archivo ONRM. Inventario 02430. Inédito
39. Babushkin V y otros (1990) Informe de los Trabajos de Levantamiento Geológico-Geofísico a escala 1:50000 y Búsqueda Acompañante en el municipio especial Isla de la Juventud en colaboración con la URSS. Archivo ONRM. Inventario 03880. La Habana. Inédito
40. Garapko I y otros (1974) La Composición Geológica y los minerales útiles de Isla de Pinos. Informe sobre el levantamiento geológico y las búsquedas a escala 1:100000 en los años 1971–1974. Provincia de La Habana. Archivo ONRM. Inventario 02719. La Habana. Inédito
41. Millán G (1997) Estudios sobre la Geología de Cuba. Instituto de Geología y Paleontología. CNDIG. ISBN 959-243-002-0. pp 243–259
42. García Delgado DE, Delgado R y otros (2005) Informe del Proyecto 216. Generalización y Actualización Geológica de la Región Habana-Matanzas a escala 1:100000. Archivo IGP-SGC. La Habana. Inédito
43. Brönnimann P, Rigassi D (1963) Contribution to the geology and paleontology of the area of the city of Havana and its surroundings. Ecologae Geologicae Helveticae. 56(1)
44. Kovaliov BM, Zaitsev VI, Mederos P, Panasenko A, Fernández R y Rodríguez A (1982) Informe del Levantamiento a escala 1:50000 en la región de Güines-Madruga-Pipián (hojas 3784-I y 3884-IV). Archivo ONRM. Inventario 03047. La Habana. Inédito
45. Garbuz I, Linares Cala E y Álvarez C (1969) Informe del levantamiento geológico, área Mariel—Cojímar a escala 1:20000. Archivo ONRM. Inventario 0896. La Habana. Inédito
46. Linares Cala E, García R, Garriga D (1981) Levantamiento Geológico. Editorial Pueblo y Educación. La Habana, Cuba
47. Pérez Aragón RO, García Saborit MA y Strazhevich VP (1998) Mapa geológico a escala 1:25000 de los municipios del este y sur de Ciudad de la Habana. IGP
48. Iturralde Vinent MA, Tchounev D y otros (1981) Geología del territorio Ciego-Camagüey-Las Tunas. Resultado de las investigaciones y levantamiento geológico a escala 1:250000. Informe. Archivo ONRM. La Habana Inventario 2892. Inédito
49. Iturralde Vinent MA, Thieke HU y otros (1987) Informe Final sobre los resultados del Levantamiento Geológico y Búsquedas Acompañantes a escala 1:50000 en el Polígono CAME III, Camagüey, 1981–1987. Archivo ONRM. Inventario 03539. La Habana. Inédito
50. Nagy E, Bresznyansky K y otros (1981) Texto Explicativo del Mapa Geológico de Oriente a Escala a escala 1:250000 Levantado y Confeccionado por la Brigada Cubano-Húngara entre 1972 y 1976. Archivo ONRM Inventario 02808. Inédito
51. Pentelenyi L, Garcés E y otros (1991) Informe final sobre los resultados del Levantamiento Geológico y Búsquedas Acompañantes a escala 1:50000 en el Polígono IV CAME-Holguín, 1983–88. Archivo ONRM. Inventario 04524. La Habana. Inédito
52. Piñero Pérez EC y otros (1991) Informe sobre los resultados del Levantamiento Geológico Complejo del Polígono Camagüey sectores Ciego de Ávila-Vertientes. Archivo ONRM. Inventario 03949. La Habana. Inédito
53. Shevchenko I y otros (1979) Informe final del levantamiento búsqueda a escala 1:100000 en las zonas de la parte sur del anticlinorio Camagüey (Región Guáimaro – Victoria de las Tunas) Archivo ONRM. Inventario 02985. La Habana. Inédito
54. Adamovich AF, Chekhovich VD y otros (1963a) Estructura Geológica y Minerales Útiles de la zona de Moa, provincial de Oriente. Informe sobre el levantamiento geológico en escala 1:50000 realizado en 1962. Archivos ONRM. Inventario 00340. La Habana. Inédito
55. Adamovich AF, Chekhovich VD y otros (1963b) Estructura Geológica y Minerales Útiles de los macizos montañosos de las Sierras de Nipe y del Cristal. Informe sobre el levantamiento geológico a escala 1:50000 realizado en 1961–1962. Archivo ONRM. Inventario 01640. La Habana. Inédito
56. Alioshin V y otros (1982) Informe final sobre los trabajos de Levantamiento Geológico y búsqueda a escala 1:100000 ejecutados en las montañas de la Sierra Maestra en la provincia de Santiago de Cuba. Sierra Maestra Nororiental I. Archivo ONRM. Inventario 02980. La Habana. Inédito
57. Bogdanov YV, Bogdanova V, Miralles M, Sosa R y otros (1963) Informe sobre los resultados de los trabajos Levantamiento y Búsqueda en la Zona del yacimiento "El Cobre" (Provincia de Oriente). Archivo ONRM. Inventario 01647. La Habana. Inédito
58. Golovkin L, Sviridov A, López L, Sojo E y otros (1983) Informe del Levantamiento 1:100000 Parte Oeste de la Sierra Maestra durante los años 1976–1977. Archivo ONRM. Inventario 03045. La Habana. Inédito
59. Grechanik T, Norman A y otros (1970) Informe sobre los trabajos Geológicos de Levantamiento a escala 1:50000 que se realizaron en la Cuenca de los ríos Cobre, Cañas y Cauto en 1967–1969 Provincia Oriente. Archivo ONRM. Inventario 01736. La Habana. Inédito
60. Gyarmaty P, Leye J y otros (1989) Informe de los trabajos de Levantamiento Geológico a escala 1:50000 y de las prospecciones acompañantes en el Polígono V Guantánamo. Archivo ONRM. Inventario 04001. La Habana. Inédito
61. Kinev T y otros (1968) Informe sobre los trabajos búsqueda para cobre a escala 1:100000, realizados en la parte meridional de la provincia Oriente en el área interfluvial Buey-Guamá (Sierra Maestra) en 1967 – 1968. Archivo ONRM. Inventario 01299. La Habana. Inédito
62. Kuzovkov G et al (1982) Informe final sobre los trabajos de levantamiento 1:100000 realizado por la brigada Sierra Maestra en la pendiente sur del Pico Turquino, parte oeste Provincia de Santiago de Cuba, años 75–76. (Sierra Maestra Sur) Archivo ONRM. Código 03044. La Habana. Inédito
63. Kuzovkov G, Zinchenko V y otros (1988) Informe sobre el levantamiento Geológico a escala 1:50000 y Búsquedas Acompañantes en el área de la "Gran Piedra" al este de Santiago de Cuba en los años 1983–1987. Archivo ONRM. Inventario 03613. La Habana. Inédito
64. Nikolaev A y otros (1981) Informe Geológico sobre los trabajos de levantamiento búsqueda a escala 1:100000 y los resultados de los trabajos de búsqueda a escala 1:50000 y 1:25000 ejecutados en la parte este de la provincia de Guantánamo. Planchetas 5276, 5376. Año 1977–81. Archivo ONRM. Inventario 02895. La Habana, Inédito

65. Millán G, Somin M, et al. (1980) Texto Explicativo del Mapa Geológico de la Isla de Pinos a Escala a escala 1:250000 Levantado y Confeccionado por la Brigada Cuba-URSS entre 1972 y 1976. Informe. ONRM. Unpublished
66. Velázquez M, Zalay P, Quiñones L, García M y Correa B (1991) Informe sobre los resultados de la Búsqueda Acompañante a escala 1:25000 en el área del complemento al proyecto Levantamiento Geológico a escala 1:50000 del Polígono CAME-HOLGUÍN. Archivo ONRM. Inventario 03924. La Habana, Inédito
67. Iturralde Vinent MA (1980) Mapa geológico de la plataforma norte y sur de Camagüey a escala 1:250000. Archivo IGP-SGC, Inédito
68. Cabrera Castellanos M y Peñalver Hernández LL (2001) Contribución a la estratigrafía de los depósitos cuaternarios de Cuba. Rev. C. & G., 15 (3–4): 37-49, © SEG. AEQUA. GEOFORMA Ediciones
69. Estrada Sanabria V et al (1987) Informe de los trabajos de reconocimiento geológico-evaluativo preliminar en el shelf oriental de Cuba y la franja costera adyacente, Archivo IGP. La Habana, Inédito
70. Estrada Sanabria V y otros (1989) Informe sobre la prospección de arenas marinas para la construcción en el tramo costero Santa Fe-bahía de Santa Lucía (plataforma noroccidental). Escala 1:50000, Archivo IGP, La Habana. Inédito
71. Estrada Sanabria V et al (1992) Informe del levantamiento geólogo-geofísico de la ensenada la Broa, 1:100000. Archivo IGP. La Habana, Inédito
72. González D et al (1981) Investigaciones complejas geólogo-geofísicas marinas en el extremo occidental del archipiélago Sabana-Camagüey y la península de Hicacos para establecer el balance de sedimentos, 1:25000. Archivo ONRM. La Habana, Inédito
73. Hernández CE et al (1985) Informe sobre los trabajos regionales de apoyo a la geología de la plataforma marina suroccidental de la República de Cuba para minerales sólidos. Archivo ONRM. La Habana, Inédito
74. Hernández CE, Ramos V, Sánchez M, Rodríguez R y Corrada R (1988) Informe sobre los trabajos de Levantamiento Geológico y Búsqueda de minerales sólidos en el Shelf de la Isla de la Juventud. Archivo ONRM. Inventario 03638. La Habana. Inédito
75. Ionin AS y otros (1977) Geología del shelf de Cuba, AC. URSS, Ed. Naúka, Moscú, p 277
76. Iturralde Vinent MA y Cabrera Castellanos M (1998) Estratigrafía de los cayos del archipiélago Sabana-Camagüey entre Ciego de Ávila y Las Tunas, t. 1, Memorias III Congreso Cubano de Geología y Minería, La Habana, pp 319–322
77. Ortega R, Rodríguez C y otros (1991) Informe sobre el Levantamiento Geológico para la ubicación de reservas pronóstico de variedades de mármol Isla de la Juventud. Archivo ONRM. Inventario 04444. La Habana. Inédito
78. Suyí Ruiz C et al (1981) Informe sobre los resultados de los trabajos de reconocimiento pronóstico-evaluativo para minerales sólidos en el shelf noroccidental, a escala 1:100000. Archivo ONRM. La Habana, Inédito
79. Oficina Cubana de Normalización (2012) Norma Cubana de Cartografía Geológica. NC 622-12: 2012. La Habana
80. Cazañas Díaz X, Torres JL et al (2017) Mapa Metalogénico de la República de Cuba, a escala 1:250000. Instituto de Geología y Paleontología, Inédito
81. Cobiella Reguera JL, Capote C, Martín D, Rivada R, Núñez A (2017) Esquema Tectónico –Estructural de Cuba a escala 1:250000. Instituto de Geología y Paleontología, Inédito
82. Núñez Cambra KE, Iturralde Vinent MA (2017) Mapa Geológico de Cuba a escala 1:1000000. Instituto de Geología y Paleontología, La Habana Inédito
83. Linares Cala E, García Delgado D, Delgado O, López J y Strazhevich V (2011) Yacimientos y manifestaciones de hidrocarburos de la República de Cuba. Centro de Investigaciones del Petróleo. Palcograf editions. ISBN 978-959-7117-33-9. La Habana
84. De los Santos Llerena E, Peñalver LL y otros (1988a) Mapa de Yacimientos Minerales Metálicos y Aguas Minerales de la República de Cuba a escala 1:500000. CIG. Printed on the Cart Factory of the Institute of Geological Research of the USSR "A. P. Karpinski" (VSEGI in Russian) Leningrad
85. De los Santos Llerena E, Peñalver LL y otros (1988b) Mapa de Yacimientos y Manifestaciones de Minerales No Metálicos y Combustibles de la República de Cuba a escala 1:500000. CIG. Printed on the Cart Factory of the Institute of Geological Research of the USSR "A. P. Karpinski" (VSEGI in Russian) Leningrad
86. Cabrera Castellanos M (1998) Estudio geólogo-ambiental del ecosistema Sabana-Camagüey. Archivo IGP. La Habana, Inédito

Geology of the Marine Territory of Cuba

Miguel Cabrera Castellanos and Betsy Vázquez Gainza

Abstract

It is considered as geology of the marine territory of Cuba to that developed during the Pliocene-Quaternary in the areas occupied by the sea between the upper edge of the insular slope and the current coast. The investigations reveal that the deposits of this interval have different characteristics to those of the emerged territories of the islands of Cuba and of de la Juventud, determined by: (1) their biogenic-carbonate composition without or with very few terrigenous components; (2) distribution through the entire surface of the seabed and the keys, without the presence of pre-quaternary rocks and (3) structure and morphology, typical of banks, bars, plains and depressions. Cuba is one of the countries with the greatest geological knowledge of its marine territory, which is covered in its entirety by research at different scales. Its results are recorded in numerous reports and publications, among which the geological maps of the Republic of Cuba at scales 1: 500,000 and 1: 100,000, the neotectonic map at scale 1: 500,000 and 2 monographs stand out. The purpose of this chapter is to contribute to the dissemination of the results on the geology of the marine territory of Cuba. The following topics are addressed: (1) terms and definitions specific to the study territory or, conceptual-theoretical framework, to provide the better understanding of the composition, distribution and genesis of seafloor and keys deposits, as well as its morphogenesis. For example: shelf, sectors without shelf, coastal zone, coasts and insular slope, as well as the predominant lithomorphogenetic processes and factors (biogenic, hydrogenic, tectonic, paleoclimatic, eolic, chemogenic, lithological and anthropogenic); (2) the main works, on which the chapter is based and (3) aspects that constitute the contribution of this work: morphogenetic complexes and relief forms, stratigraphic and sedimentological characteristics, neotectonic, geological evolution and mineral resources. The text is widely illustrated with geological, geomorphological, neotectonic, paleogeographic and mineral resource maps. In addition, abundant geological-geomorphological sections, chronostratigraphic columns and photos.

Keywords

Marine geology • Insular slope • Shelf • Seabed • Coastal zone • Keys • Lithomorphogenetic • Morphogenesis

Abbreviations

N-Q	Neogene-Quaternary
Ky	Kilo years
My	Million years

2.1 Introduction

It is considered as geology of the marine territory of Cuba to that developed during the Pliocene-Quaternary in the areas occupied by the sea between the upper edge of the insular slope and the current coast. The investigations reveal that the deposits of this interval have different characteristics to those of the emerged territories of the islands of Cuba and of de la Juventud, determined by their composition, distribution, structure and morphology.

This geological interval is distinguished by the presence of a series of lithomorphogenetic factors and processes, which constitute an active natural laboratory in a territory of 70,000 km^2 (about 70% of the territory of the Cuban archipelago), 5746 km of coasts and near 5000 keys, based

M. C. Castellanos (✉) · B. V. Gainza
Instituto de Geología y Paleontología-Servicio Geológico de Cuba, Vía Blanca no. I002 y Carretera Central. San Miguel del Padrón CP, La Habana, 11 000, Cuba
e-mail: miguel@igp.minem.cu

B. V. Gainza
e-mail: betsy@igp.minem.cu

on which can be reconstructed the history of a series of transgressive and regressive events occurred and the potential of mineral resources.

Currently the country has a high degree of geological study of its marine territory, as a result of the policy established by the Cuban State to elevate and implement the knowledge of this specialty. Testimony of this are the results of numerous basic and applied geological-geophysical investigations carried out under the scientific-technical direction of prestigious Cuban and foreign institutions.

The purpose of this chapter is to briefly update the main geological-geomorphological features of the marine territory of Cuba, as well as the prospects of carrying out new investigations.

2.2 Theoretical Framework

In the field of marine geology, in general, controversial concepts and definitions still exist. This is due to: (a) the youth of their geological investigations and (b) the complexity of the processes and factors (planetary, regional and local). These are responsible for the genesis, evolution and development of the deposits and their relief in the seas and coasts of the different latitudes of the planet. For the purposes of this chapter it is important to refer to the terms and definitions which determine the geomorphological regionalization for the study territory. For example: shelf, sectors without a shelf, types of coastal zone, coasts and insular slope (Figs. 2.1 and 2.2), since they have been treated indistinctly in different investigations [1–4]. It is also necessary to mention and evaluate lithomorphogenetic processes and factors.

Shelf. Submerged part of the coastal plains, shallow waters surrounding the islands of Cuba and de la Juventud. Among the coast and the upper edge of the insular slope (isobaths of 10–50 m). It is composed of two lithomorphogenetic zones: external and internal (Fig. 2.3).

Sectors without shelf. Abrasive step, slightly inclined towards the depths of the sea. Its frontal cliff is the insular slope and its rear cliff the adjacent emergent land. Its width varies between hundreds of meters and the first kilometres.

Coastal zone and types of coast. The general line of coast and the extension form the coastal zone, both in the shelfs and in the sectors without a shelf, inland as far as the tides influence the low coasts or where it reaches the highest surf on the raised coasts. There is a wide range of classifications of the coasts both for the world and for Cuba [3], which do not always result from practical application. To standardize the information on the subject, the classification established in Decree-Law 212/2000 (Table 2.1) has been adopted nationally.

Insular slope. It is the unevenness between the surface of the bottom of the neritic seas (shelfs and sectors without a shelf) and the abyssal depths. Constitute extensive structural-tectonic steps, formed by faults and bends, with a slope of up to more than 45°. Its upper edge is located at varying depths between 10 and 50 m.

Processes and lithomorphogenetic factors. The different deposits of the marine territory of Cuba and their forms of relief are polygenic, due to the intervention of a large number of lithomorphogenetic factors and processes of a physical, organic and chemical nature, as a consequence of changes in the atmosphere, hydrosphere and lithosphere, in addition to those of anthropogenic nature. Among the main ones are the following: biogenic (fauna and flora), wind, hydrogenic, paleoclimatic, chemogenic, tectonic and anthropogenic.

Biogenic. Within the **fauna** that intervenes in the lithomorpho-lithogenesis, there are the tropical hermatypic stone corals or builders of thick reefs, that in biocenosis with a set of species (foraminifera, sponges, urchins, calcareous algae, molluscs, echinoderms, gastropods and others). They form coral reefs, which are an important source of bio clasts for the creation of cumulative forms of coastal relief (keys, beaches, bars and storm ridges) and marine relief (banks, terrace Zero and others). When fossilized, they constitute lithostratigraphic units (Río Maya, Vedado and Jaimanitas), which constitute an important element in the paleogeographic analysis for the reconstruction of geological development.

Within the **flora** are marine grasses, with a great development in the soft bottom of the shelfs (sands, silts and clays). These biotopes are distributed discontinuously and

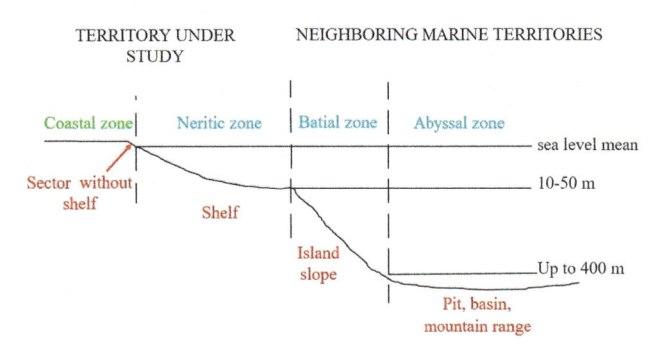

Fig. 2.1 Hypsometric regionalization of the marine-coastal territory and its bottom, according to its geological-geomorphological features

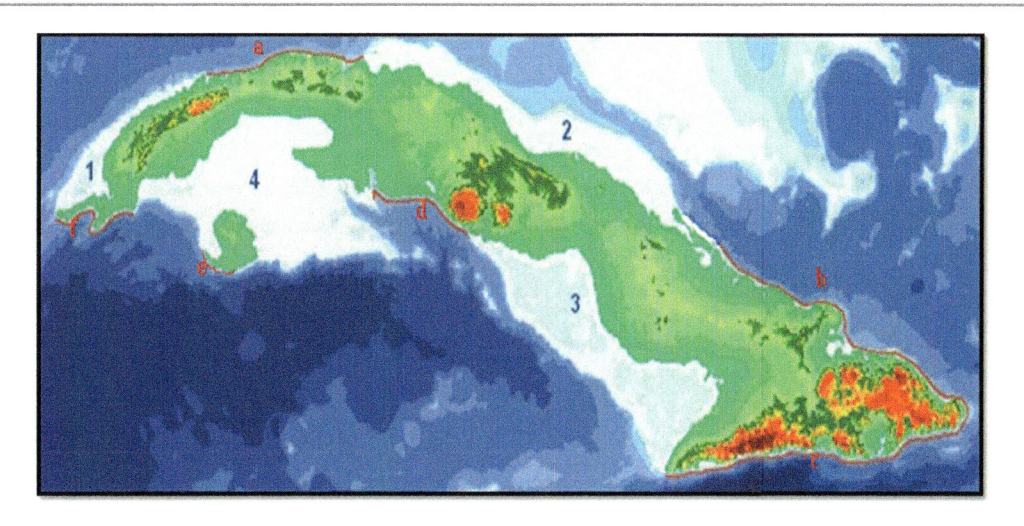

Fig. 2.2 Plan view of shelf (1-northwestern, 2-north central, 3-southeastern and 4- southwestern) and sectors without shelf (a- northwestern, b-northeastern, c-southeastern, d- south central, e- South of the Isla de la Juventud and f- southwestern

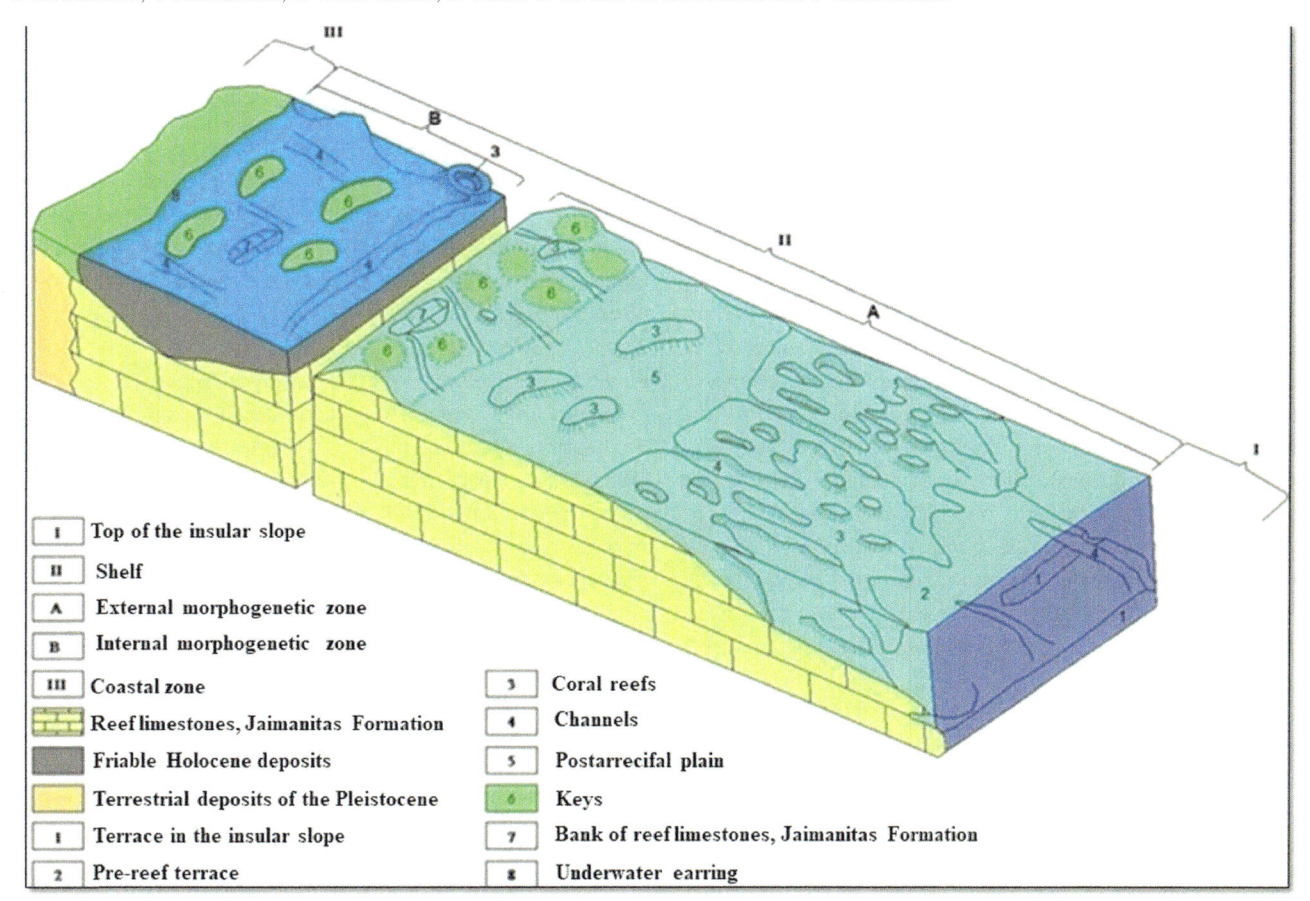

I	Top of the insular slope
II	Shelf
A	External morphogenetic zone
B	Internal morphogenetic zone
III	Coastal zone
	Reef limestones, Jaimanitas Formation
	Friable Holocene deposits
	Terrestrial deposits of the Pleistocene
1	Terrace in the insular slope
2	Pre-reef terrace
3	Coral reefs
4	Channels
5	Postarrecifal plain
6	Keys
7	Bank of reef limestones, Jaimanitas Formation
8	Underwater earring

Fig. 2.3 Schematic representation of lithomorphogenetic zones from Cuba shelfs and their main forms of relief

with variable density in the different territories, occupying more than 50% of their area, predominating in the southwestern and northwestern shelfs. They are sometimes called "ecosystem engineers", because they partly create their own habitat: the leaves slow down the currents, increasing sedimentation; the roots and rhizomes stabilize the substrate of the seabed, so they intervene in the genesis of cumulative forms of relief in it.

The mangroves that develop on the accumulations of sands and clayey silts of the coasts surrounding the different sectors of the shelf and on the coasts of the bays, inlets of the sectors without shelfs and on the keys. They populate 70%

Table 2.1 Types and subtypes of coast according to Decree—Law 212/2000

Type of coast	Subtype
Low terrace	With storm ridges
	In the absence of the ridge
	In the absence of the ridge and in the presence of the cliff on a second level of terrace
	In the absence of the ridge if the area adjoining the low terrace is a coastal lagoon with mangrove
Acantilated	
Beach	With dune
	In the absence of dune
	In the absence of dunes with cliffs
	In the absence of dunes if the area adjacent to the berm, it turns out to be a coastal lagoon with mangroves
Mangrove box	
River mouths	
Antropized	

of the Cuban coasts. They occupy 5321 km^2 (4.8% of the total land area of the Island of Cuba). By their extension, they occupy the ninth place in the world and the third in tropical America. The mangroves grow areal and vertically the cumulative forms on which they develop, because they constitute a source of organic matter (leaves, wood, roots and their associated microbiota) and also to retain between their roots the terrigenous and marine sediments. They dissipate the action of the waves and prevent abrasion and erosion. This allows the growth of the mangrove towards the sea up to several meters per year, changing rapidly the morphological constitution of the media.

Halophilic plants (typical of saline environments), grow in the accumulations of abundant dry sands and other clastic deposits, which allow the development and conservation as a mechanism of attachment of cumulative forms (dunes, bars and storm ridges).

Eolic. The outer morphogenetic zone of the shelfs and the sectors without shelf are under the direct influence of the general directions of the winds of the northeast (trade winds), with predominance of the north in winter and the E-NE in summer. The inner zone of the shelf and its coasts receive only the moderate action of the predominant winds at different times of the year, due to being under the protection of the keys, the mangroves and the coral reefs. The wind process originates the dunes of current and fossilized coastal sands (from the litho-stratigraphic formations of Guanabo, Playa Santa Fe, Cayo Guillermo and Los Pinos). In some cases, the predominant direction of cross-linking of the layers is from zero to 30°, coinciding with the main current directions of the winds, which shows that the variability of the direction of these has remained practically constant at least since the Upper Pleistocene to the present.

The wind process also contributes directly or indirectly to the origin and morphology of another great variety of other cumulative forms of relief (bars, tombs, beaches and banks, among others) or abrasive (terraces, cliffs, tidal niches and others).

Hydrogenic. Lithomorphogenetic-hydrogenic factors and processes are waves, currents and tides. Through its action in the marine-coastal territory three important actions occur: abrasion, transport and deposition, which originate a great diversity of forms in the low coasts (lagoons, estuaries, beaches, conical deltas, tidal deltas and others) and in the raised coasts (linear deltas, tidal niches, abrasive terraces, etc.). That is, they have their particularities in the different regions of the shelf and in the sectors without shelfs.

Paleoclimatic. Under the influence of planetary, regional and local paleoclimatic events, radical transformations have taken place in the composition and relief of the different deposits of the marine-coastal territory of Cuba, through the quaternary regressions and transgressions of great importance to decipher the evolution of their geological constitution. Its main lithomorphogenetic traces are manifested in: (a) the lithification of marine deposits, when exposed to subaerial conditions; (b) relief modelling by karstification and denudation of carbonate rocks (karstic holes tens of meters deep, plains, residual elevations and depressions, with innumerable microforms); (c) deepening of the fluvial channels, particularly those that constitute access to the closed bays; and (d) absence of marine biota, such as corals, at certain intervals of the geological development of the Quaternary.

Chemogenic. The marine-coastal territory of Cuba corresponds to a tropical zone subjected to the action of characteristic chemogenic processes. Its greatest expression is given by the cementing of deposits, dissolution of carbonate rocks and formation of oolites. The oolitic sands have been an important source of contribution for the emergence of cumulative forms in the north central and southwestern shelfs since the Pleistocene (dunes, bars and banks already lithified). Today this process exists only in the east part of the southwestern platform, where in the form of friable deposits they are filling the Batabanó depression.

Lithological. Universally, the deposits of the coastal strip are subject to the onslaught of processes and hydrogenic factors, which physically and chemically abrade them, thus causing the advance of the inland coastline. In Cuba the rhythm of this process occurs quite fast, due to the characteristics of the lithological factor: (a) presence of carbonate rocks of the Quaternary in almost all the rocky coasts; (b) the rocks are found mostly outcropping, so the predominantly karst discovered or covered by storm ridges not lithified, in both cases are fragile before the action of the surf; (d) presence of weakly lithified coastal bars and dunes; (e) presence of friable sediments of different composition; (f) abundant fractures and faults, many of them perpendicular to the coast, facilitating wave action, beating in the same direction; and (g) existence of horizontal lithological contacts, which constitute vulnerable planes before the waves so that the rock mass is separated and slid in blocks.

Hydrographic. Cuba has a poorly developed hydrographic network, due to the narrow and elongated character of its territory, historically with limited access to the sea due to the following factors: (a) tendency to the ascent of an important part of the country, being hung the fluvial valleys or trapped in the bays; (b) presence of coastal karst plains, which capture surface waters, limiting their surface discharge to the sea; (c) existence of fault zones in emerged territory parallel to the coast, which capture the hydrographic network before going out to sea; (d) development of sandy plains in arid zones, where rivers lose their flow in times of drought (endorheic drainage. Example: Imías valley, southeast of Guantanamo province); (e) the existence of dense forest vegetation until only a few hundred years ago; (f) the formation of linear deltas in the mouths of many rivers; and (g) the high level of damming and canalization of rivers.

The mentioned factors have limited the lithomorphogenetic role of the hydrographic network to a scarce contribution of terrigenous sediments and to the poor development of relief forms in front of low coasts (deltas, estuaries, lagoons, lagoons, beaches, marshes and swamps).

During the withdrawal of the sea beyond the current upper edge of the insular slope (regression of Wisconsin), only a few rivers managed to cross that territory, as evidenced by the scarce presence of paleo runway in the territory of the shelf.

Tectonics. The tectonic movements, specifically the neotectonic ones, that in Cuba correspond to the Oligocene-Quaternary, are those that have defined the existing geomorphological regionalization (emerged coastal plain, shelfs, sectors without shelf, insular slope and structures in deep waters). In addition, they have defined the sub-latitudinal lithomorphogenetic zonation of the shelfs (outer and inner zone) and their division into ascending and descending blocks of different orders, separated from each other by faults and flexures.

Anthropogenic. There are innumerable human interventions, which can be located in the emerged island territory, on its coasts and in the open sea, which hinder the natural development of marine-coastal lithomorphogenesis. Among the main ones are the following: (a) civil and engineering constructions; (b) discharges of liquid, gaseous and solid waste; (c) deforestation; and (d) mining, agricultural, livestock, shipping, fishing and diving activities.

2.3 Materials and Methods

These belong to the results of: (a) decades of basic geological-geophysical studies and applied to the prospection of solid minerals, which were synthesized and updated; (b) the projection of engineering works of different types; and (c) the protection of the environment. Many of these materials have been systematized and generalized, mainly through geological and geomorphological cartography, with the aim of contributing to the knowledge of the geological constitution. It takes advantage of the valuable and costly information obtained from the marine territory.

Among the main results processed for the elaboration of this chapter are the following: (a) Geology of the shelf of Cuba [4]; (b) Geological map of Cuba and its marine territory at scale 1: 50,000 [5]; (c) Systematization and generalization of the geology of the shelf of Cuba in relation to the prospection of solid minerals [2]; (d) Contribution to the stratigraphy of the Quaternary deposits of Cuba [6]; (e) Geological map of the neritic seas of the Cuban archipelago at a scale of 1: 10,000 [7]; (f) Geology of the marine territory of Cuba [1]; and (g) Geological nature of the marine-coastal territory of Cuba in the Quaternary [3].

Other more recent investigations are: The Quaternary deposits of the marine territory of Cuba [8]; Neotectonics and the rise of the mean sea level [9]; Geological-geomorphological characterization of the marine and coastal protected areas in the archipelagos of the south of Cuba [10]; and Geology and marine-coastal geomorphology of the Cuban archipelagos and its link with recent tectonic movements [11, 12].

Quaternary deposits for Cuba and its marine territory do not have index fossils, so their age has been determined by the C14 radiometric method [4] or estimated by geomorphological, stratigraphic and paleoclimatic criteria.

2.4 Results

The systematization and updating of the previous investigations allowed to identify in the marine territory of Cuba its morphogenetic complexes and forms of relief, stratigraphy and sedimentology, Neotectonics, the history of geological development and mineral resources.

2.4.1 Morphogenetic Complexes and Relief Forms

They differ according to their location in the shelfs or in the sectors without a shelf. They represent the genesis, development and distribution of geological characteristics in the marine territory, as they govern the development of biotopes in different types of sea bottom. It is, therefore, essential to identify them as a first step to guide marine geological research.

Shelf

Depending on the lithomorphogenetic zones (external and internal, Fig. 2.4a–d), which make up these geomorphological regions, the following morphogenetic complexes can be distinguished: pre-reef terrace, coral reefs, post arrecifal plain, keys and the coasts; as well as the forms of the relief that compose them.

Pre-reef terrace. It is located between the coral reefs and the edge of the insular slope, with a width of hundreds of meters and a surface of smooth slope, that increases gradually until transiting to the insular slope. In those limits are karstic forms of the relief, such as: caverns, tidal niches, funnels, channels, hollows of different shapes and pinnacles. In its transition zone towards the insular slope there are also up to seven spectra of terraces, extending up to 65 m in depth. Two of them are observed in Fig. 2.5 [4, 16, 17].

Coral reefs. In the outer morphogenetic zone they reach their greatest development contiguous to the pre-reef terrace, forming a chain along practically all the shelfs, except in isolated segments due to a greater descent of the rocky substratum. It constitutes an irregular succession of patches hundreds of meters wide, slightly elevated on the pre-reef terrace and the post arrecifal plain, separated from each other by channels and banks of the oldest rocky substratum (Jaimanitas Formation). Sometimes the corals reach their greatest development in front of the keys, where they also have the singularity of surrounding many of them. In such cases, it is where reef crests are closest to the surface of the water. On the southwestern shelf, the reef formations are divided into two large sectors by the Isla de la Juventud. In the west, they develop only in small sections. Towards the east, the patches reach up to 30 km in length and at its easternmost end, they are semi-circular, resembling micro Athlons, with biogenic sands in their interior, their depth does not exceed 0.3–0.7 m.

In the interior morphogenetic zone, coral reefs usually form banks, which occupy considerable areas. The largest of them are found on the northwestern and southeastern shelfs. In the latter, the Gran Banco Buena Esperanza stands out (Fig. 2.4c). Morphologically, in the plane it constitutes a complex labyrinth of curved chains separated by channels, which form a system of annular micro-lagoons. The tops of the reefs, in general, are flat and are close to sea level. Their slopes are very steep (up to 40–50°). According to seism acoustic logging data, the base of these is buried in the sediments up to the depth of 50 m [4], Fig. 2.6), that is, it does not lie on friable sediments as reported by Zlatarski et al. [18].

Post-arrecifal plain. It is a karstic plain with an abrasive surface, with isolated accumulations of sand in the depressions; as well as coral formations, banks and other forms of the underdeveloped relief.

Keys. As a rule, this morphogenetic complex is distinguished by a profuse development of different genetic types of keys and other associated reliefs (channels, tidal deltas, beaches, bars, dunes, different types of coasts, intra-island bays, lagoons, terraces, ridges of storm and salt marshes).

The lowest development is reached on the northwestern shelf. In their outer zone they are scarce and little developed, by their altitude and area. They correspond to the sandbar type, with small and few segments of beaches on the north coast of some of them. In the inner zone they do not abound either, they are small and of the bank type (that grow on a substrate of friable sediments), their sands can form beaches. The keys of this shelf form the Los Colorados archipelago, about 100 km long.

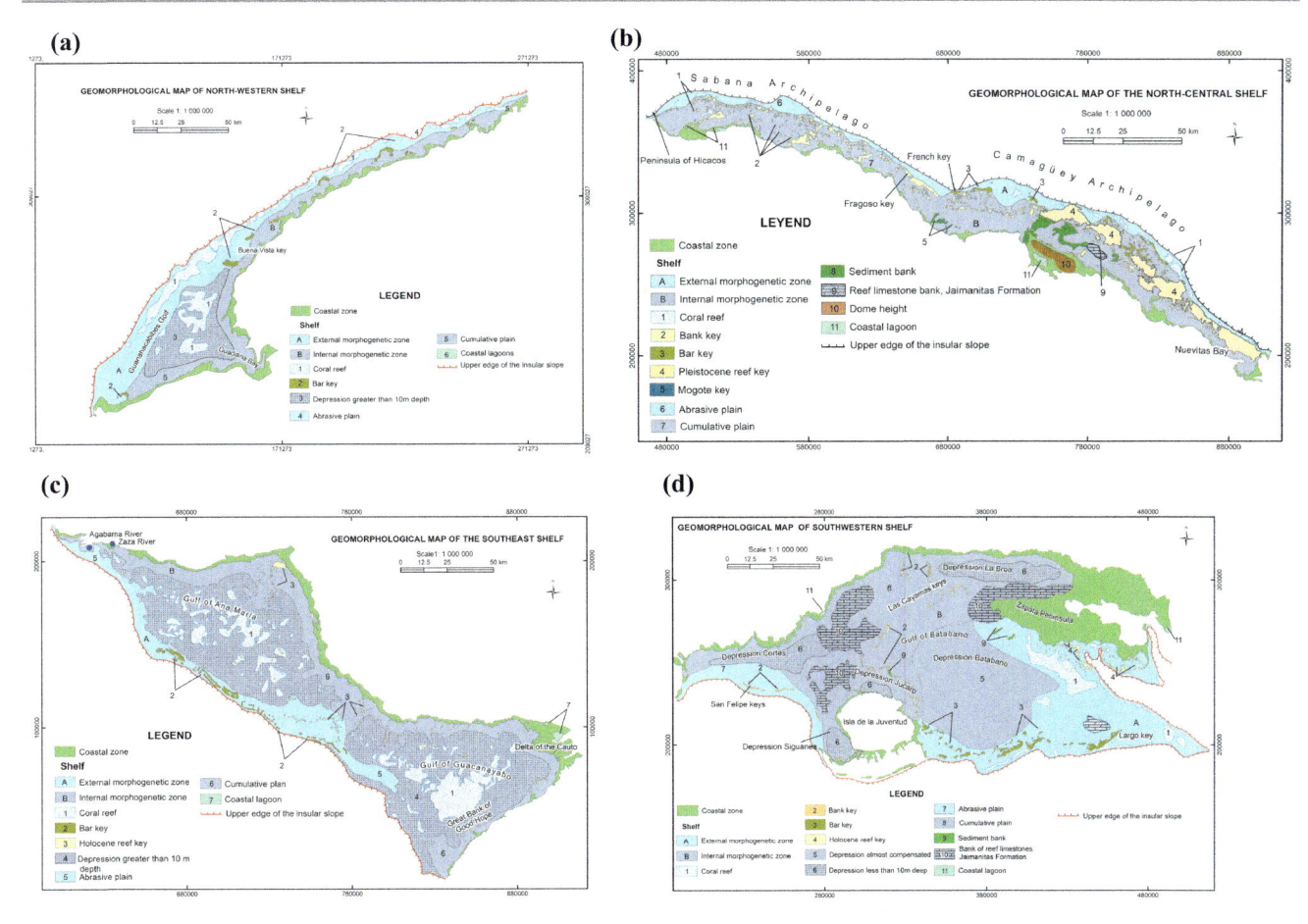

Fig. 2.4 Geomorphological maps of the shelfs: A-northwestern, B-north central, C-southeastern and D-southwestern (based on [3, 13–15])

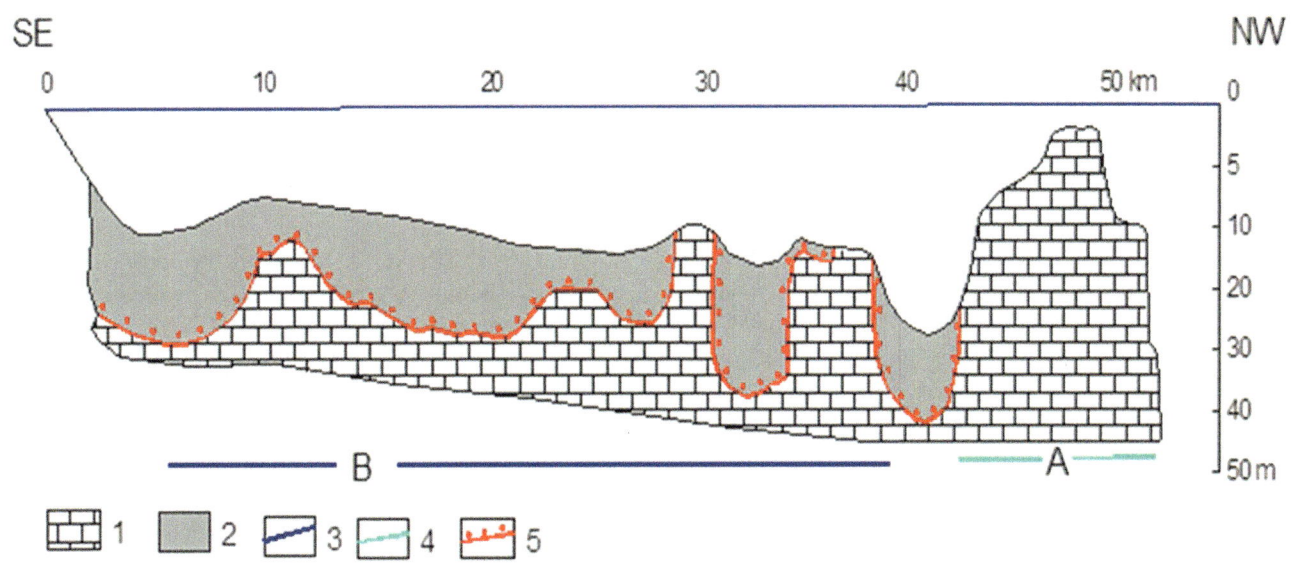

1. Reef limestones, Jaimanitas Formation and probably Vedado; 2. Friable sediments; 3. Internal morphogenetic zone (B); 4. External morphogenetic zone (A); 5. Stratigraphic discordant.

Fig. 2.5 Geological-geomorphological profile, generalized to the west of the northwestern shelf (based on [4, 17]). Note the existence of deep sublatitudinal fluvial channels in the calcareous substrate, in process of clogging and terraces in the vertical wall

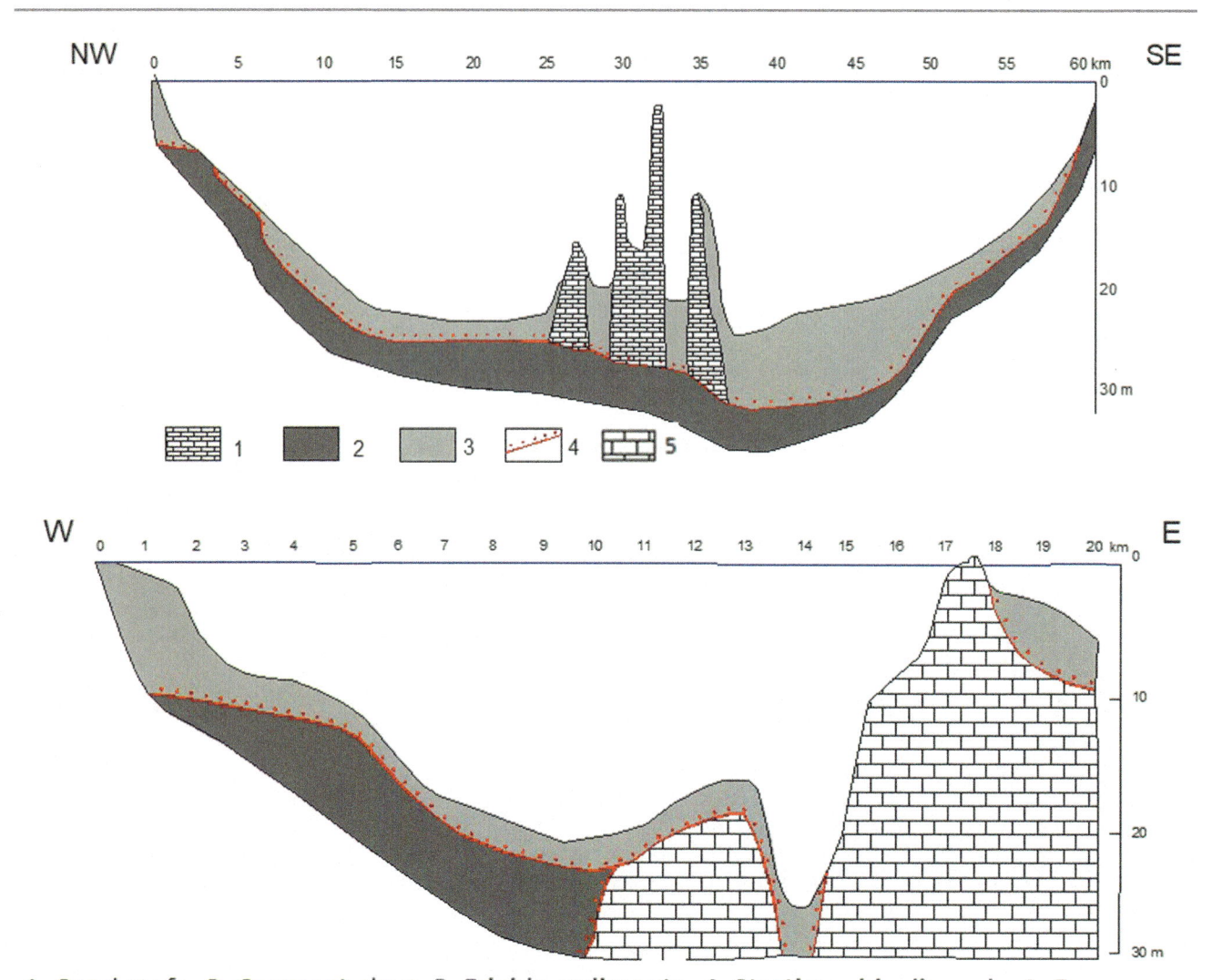

1. Coral reefs; 2. Compact clays; 3. Friable sediments; 4. Stratigraphic discordant; 5. Reef limestones, Jaimanitas Formation.

Fig. 2.6 Morphology of the Gran Banco Buena Esperanza, seen in the upper vertical cut and geomorphological-geomorphological cut across the paleo valley of the Agabama or Manatí River (based on [4]; Avello and Pavlidis [19])

In the north central shelf, the keys are geographically and structurally divided into two groups: (1) Sabana archipelago (Hicacos peninsula–Francés key) and (2) Camagüey archipelago (Francés key-Sabinal key). The former are of the bank type, with less than 1 m average altitude. They are mostly covered by mangroves, small sand bars and narrow (4–20 m wide) and low (1–1.5 m) beaches. They are fragmented by numerous tidal channels, sometimes forming homonymous deltas towards the open sea (Fig. 2.7).

The second ones are of the old reef type (constituted by Pleistocene reefs) and of the bar type, they are distinguished by: (a) their altitude (10–14 m, on average and up to 65 m in the Chair of Romano key) and large territories; (b) presence of spectra of the low Pleistocene terrace (terrace I), with dunes (Fig. 2.8), storm ridges, bars and up to four tidal niches (Fig. 2.8) (c) spectra of the Holocene terrace (terrace Zero); (d) well-structured beaches up to 40 and 50 m wide, which together with the rods form oriented arrows of E-W and northwest-southeast, with the end of the hook towards the northwest (Santa María, Paredón Grande, Guillermo, Cruz and others, Figs. 2.9 and 2.10). They are originated by lateral currents, behind which extends a wide strip of mangroves and lagoons; and (e) tidal channels in the interior of the main keys, with its greater expression in Coco key (Fig. 2.7).

Besides the aforementioned keys, there are the stone keys, in prequaternary rocks, mogotes type, in front of Sancti Spíritus, among which are Judas, Caguanes and Cristo, with levels of caverns raised and superimposed. All the keys mentioned above form the Sabana-Camagüey archipelago or Jardines del Rey, with a length of 481.5 km.

In the outer part of the southeastern shelf, the keys exist only next to the gulf of Ana María (archipelago Jardines de la Reina), the rest of the territory is sunken. In the outer

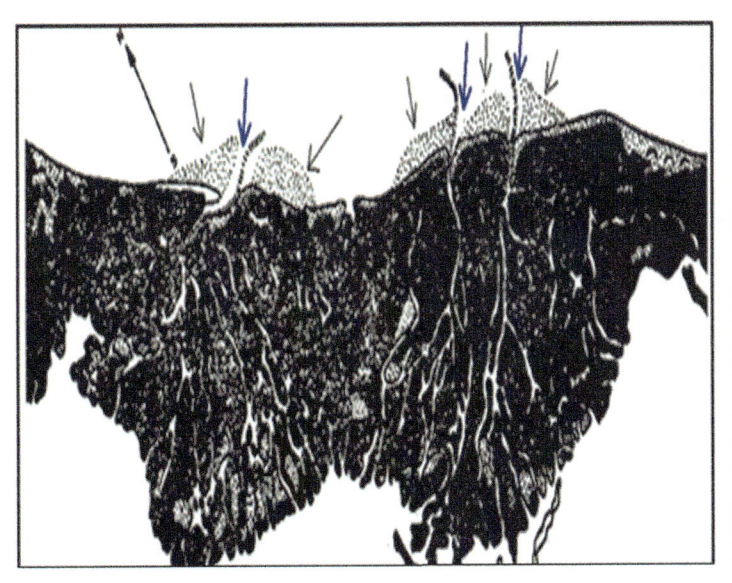

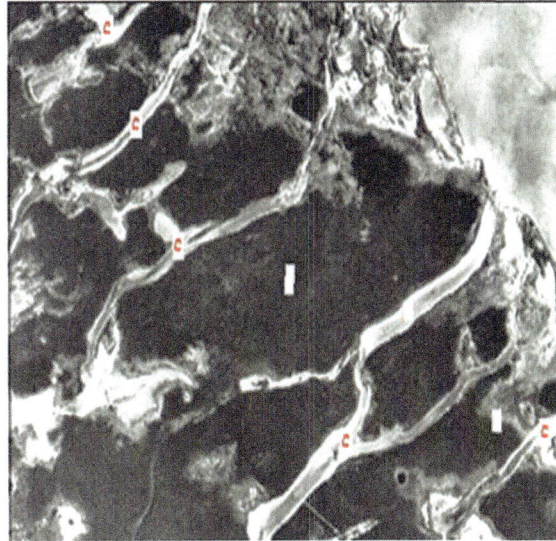

Fig. 2.7 Tidal channels: Left, in the sediments of Fragoso key of bank type (blue arrows), fragmenting homonymous deltas (black arrows); right, on the surface of reef limestones raised above the sea level, on Coco key (c) of the old reef type

Fig. 2.8 Fossil dunes, in Guillermo key, over 12 m altitude (left photo) and Coco key (right photo)

morphogenetic zone, the keys usually have an incipient frontal cliff, in sections of the south coast, where there can also be beaches; as well as bars attached to the coast or in the post beach. On the terrace there are ridges of storm, sometimes far from the coast to tens of meters and with 5–6 m of altitude; as well as large blocks of isolated reef limestones, thrown ashore by the strong waves (hurricane). The rest of the surface of the keys bar is occupied by mangroves and shallow lagoons. Subordinately, there are bank type keys, low, mangroves, sands and large interior lagoons elongated, very shallow and sandy bottom.

In the inner morphogenetic zone, the keys are, mostly, of the Holocene reef type (Fig. 2.11). Subordinately, there are bank and bar type keys, small and very low, located near the coast. Important groups of them form singular alignments, which are oriented N-S (normal to the coast).

In sector west of the outer morphogenetic zone of the southwestern shelf there are two groups of keys (Fig. 2.4d): San Felipe and Los Indios, which are bar-type and on their south coast have beaches. In two of them (Sijú and Real) there are spectra of terrace I, slightly elevated and cliff to windward, covered by sand bars, beaches and storm ridges of 1.5 to 4 m altitude, sometimes far from the coast. Behind these forms of relief, there are mangroves and lagoons.

Fig. 2.9 Fossil bar crowned by storm ridges, Coco key (left photo) and tidal niches, Paredón Grande key (red lines, right photo)

Fig. 2.10 Typical structure of a bar key. Example: Guillermo key. The presence of a rock nucleus is not an obligatory condition for the formation of this type, but it is frequent.1. Reef limestone's, of the Jaimanitas Formation; 2. beach; 3. bar; 4. mangrove; 5. lagoons and 6. terrace

In sector east of this same morphogenetic zone, the following groups of keys are distinguished: (a) bank type, between Isla de la Juventud and Avalos key, they are low, of mangroves and sands; (b) bar type, in Cantiles key-Largo key, constituted by bars of calcarenites of the Cocodrilo Formation, surrounded by mangroves and lagoons, by the south have segments of beaches; and (c) of recent reefs, near the gulf of Cazones, which have a Half moon shaped configuration. They reach less than 1 m in height over the surrounding reefs (Fig. 2.4d).

In the interior morphogenetic zone of this shelf, the keys are of the bank type, with the surface almost at mid-sea level. These form a chain from Isla de la Juventud to the south of the Island of Cuba, in the form of a normal dorsal to the coastline, but individually they are oriented from E-W, coinciding with the main direction of the currents in that part of the shelf.

The Isla de la Juventud and its neighbouring keys form the Canarreos archipelago of 150 km in length.

Karst forms. In both morphogenetic zones of the shelfs abound the carcass forms, surfacing on the seabed, covered totally or partially by friable sediments and in the rocks of the keys and coasts. In addition, of those already mentioned in the morphogenetic complexes of the outer zone of predominantly marine origin (caverns, depressions, channels and holes, among others), others have been identified in the interior zone: channels, depressions, pits and banks and others, which were formed in subaerial conditions by fluvial action during the regression (of Wisconsin).

Only fluvial channels have been discovered crossing the sea floor from north to south on the southeastern shelf, facing the Zaza and Agabama rivers (Fig. 2.10), and other cases could be the deepest parts of the shelfs, which serve as communication with the open sea. On the northwestern and southwestern shelfs, sublatitudinal channels have been identified next to the contact of both lithomorphogenetic zones (Fig. 2.5).

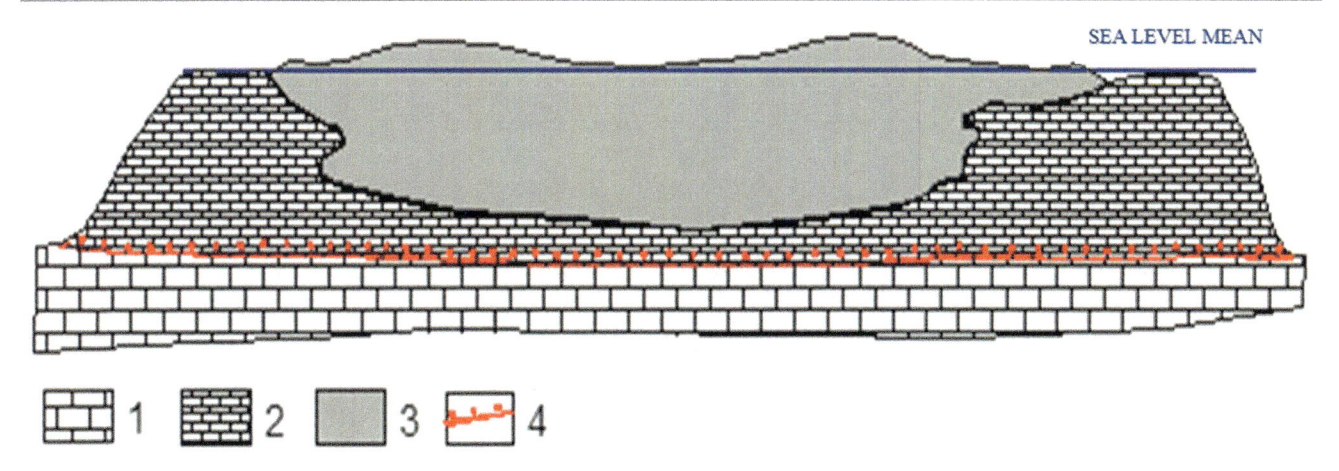

1. Reef edges, Jaimanitas Formation; 2. Present partially dead reefs; 3. Marsh deposits; 4. Stratigraphic discordant.

Fig. 2.11 Schematic representation of the typical structure of the recent reef keys on the southeastern shelf. They constitute a mangrove (swampy bottom), surrounded by a dead reef, which decant, with more than 30° at the bottom of the limestone of the Jaimanitas Formation

The karsts depressions constitute deeper areas on the northwestern (Fig. 2.5) and southwestern (Fig. 2.12) shelfs. In the north central shelf, they are not well identified, but they could be related to the macro-lagoons located between the coasts of the mainland and chains of surrounding keys, which have been denominated as intra-island bays [20] although they do not respond strictly to the definition of this form of relief. For example, Santa Clara and Carahatas (Fig. 2.13).

In the southwestern shelf, the karstic depressions have good expression in the relief, except the call of Batabanó, already compensated, because it is part of the polygon of formation of oolites present in the southwestern shelf, with an abundant supply of oolitic sediments. Due to its extension, it is the fourth world polygon for the formation of this type of deposit, the others are found in the Red Sea, the Caspian Sea and the archipelago of the Bahamas [23].

The karstic holes abound in the abrasive surfaces of the different shelfs. The deepest have been reported in the north central and southwestern, where they reach up to 70–80 m [4, 20, 24]. Similar, but deeper, forms have been reported in neighbouring territories of Cuba: Florida 146 m, Bahamas 198 m and Yucatan 130 m [25].

On the shelfs, in addition to the sediment and coral banks, there are also rock banks (remnants of karstic dissolution of the limestones of the Jaimanitas Formation). They have their greatest expression in the north central and southwestern shelfs (Fig. 3.4b, d).

The coasts and their main forms of relief. The low type of mangrove predominates, poor in forms of coastal relief and incipient deltas in the mouths of some rivers, undeveloped beaches, often with fine thicknesses of sand and shells on a substrate of finer sediments and/or peat. There are cliff segments of low altitude, due to the washing of the coastal sediments. The coastal plain is wide, low, with numerous lagoons, among which is the largest in Cuba (from La Leche, on the north central shelf), estuaries, lagoons and inlets with soft edges at the entrance. In the southwestern and northwestern shelfs, there are small open bays. At the mouth of some rivers there are deltas type cusped or leg of hen little developed. The older ones are found in the southwestern shelf (Cauto, Zaza and Agabama rivers).

There are coastal segments strongly anthropized by different uses: roads, canalization, deforestation, sand extraction and constructions, which have caused the retreat of the coastline to hundreds of meters in the most affected localities, as demonstrated by the cartography compared by aerial photos of different times in the southwestern shelf [26].

On the bank and recent reef keys, the low mangrove coasts predominate, in the former bar and reef coasts the coasts can be beach, low terrace, mangrove and rocky cliffs or washouts.

Sectors without shelf

Unlike shelfs in these geomorphological regions there is no specific lithomorphogenetic zonation. In its narrow surface the following morphogenetic complexes and relief forms can be distinguished: coral reefs, channels, banks, terraces, bays and coasts. There are other forms of lesser extent, less widespread or little studied. For example: cave systems, such as the one identified in Sector south of the Isla de la Juventud, the west of Punto Francés, with thirteen mouths of caves, which along different galleries end in the insular slope, with uneven levels of up to 20–35 m depth [20] in [27].

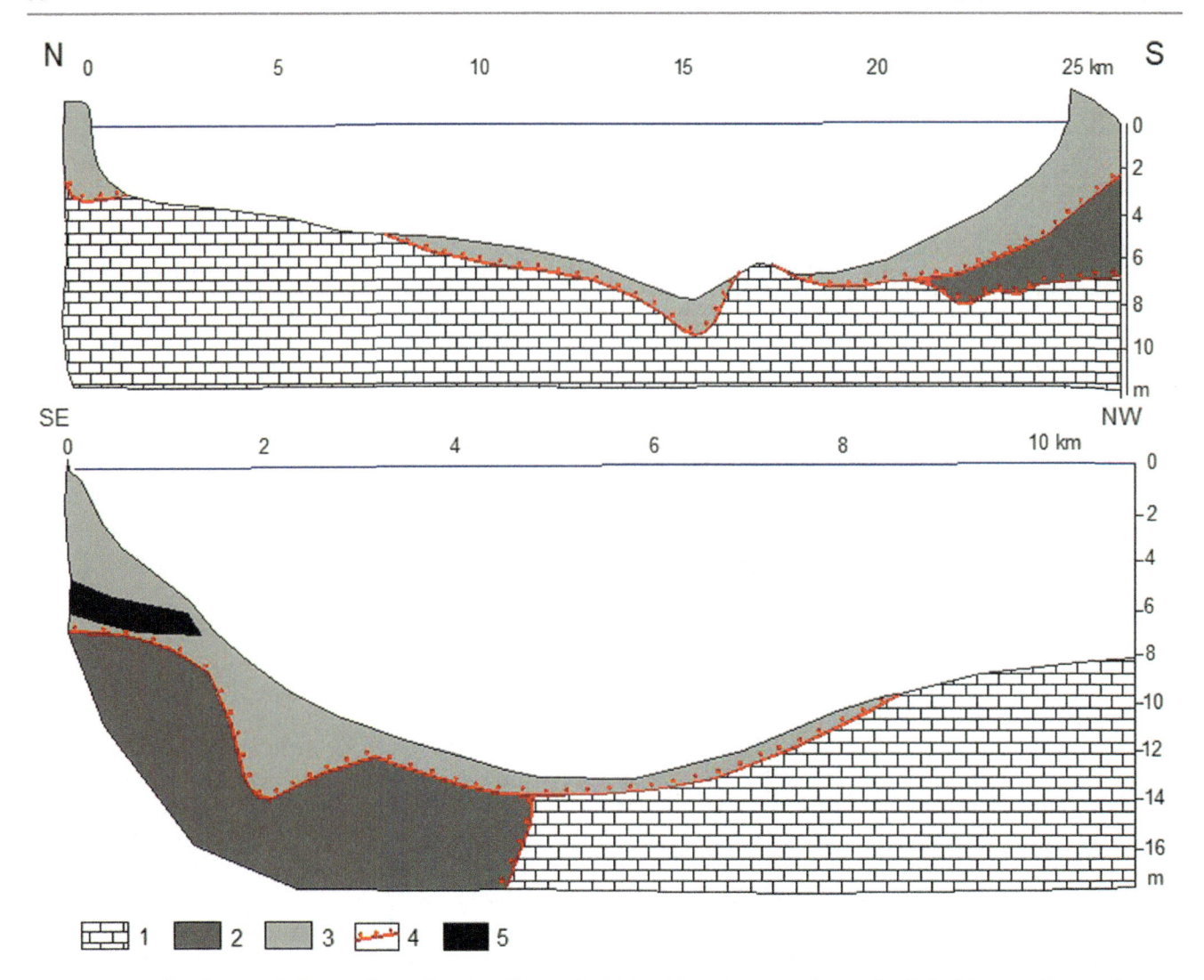

1. Reef edges, Jaimanitas Formation; 2. Weathering spring; 3. Friable sediments; 4. Stratigraphic discordant.; 5. Turf.

Fig. 2.12 Geological-geomorphological cuts through three karst depressions in the southwestern shelf, from top to bottom: La Broa, Siguanea and Batabanó (based on [2, 4, 19–22]

Coral reefs. As a rule, they form scattered patches and, subordinate, reef chains fragmented by narrow channels and banks. For example: in the southeast of the Isla de la Juventud, in La Habana (Santa María-Guanabo) and along the Varadero peninsula. In these last two locations they are constituted by a series of parallel sublatitudinal channels. Between its interior and the coast, they form a kind of lagoon bordered by 9 and 18 km of beaches and dunes. The heights of the coral formations in the sectors without a shelf can reach up to 4 m, similar to the outer zone of the shelfs.

Channels. Regularly they form cannilions of karstic dissolution, shallow and of marine origin. They usually cut in the transverse direction the geomorphological step, which constitute the sectors without a shelf and, consequently, the coral reefs that develop there. They can also be sublatitudinal, like those mentioned in the previous paragraph. In smaller amount there are fluvial channels in front of the main rivers and to the bays of stock market, where they are relatively wide and of tens of meters of depth, sometimes tortuous and deep, due to the influence of the Neotectonics.

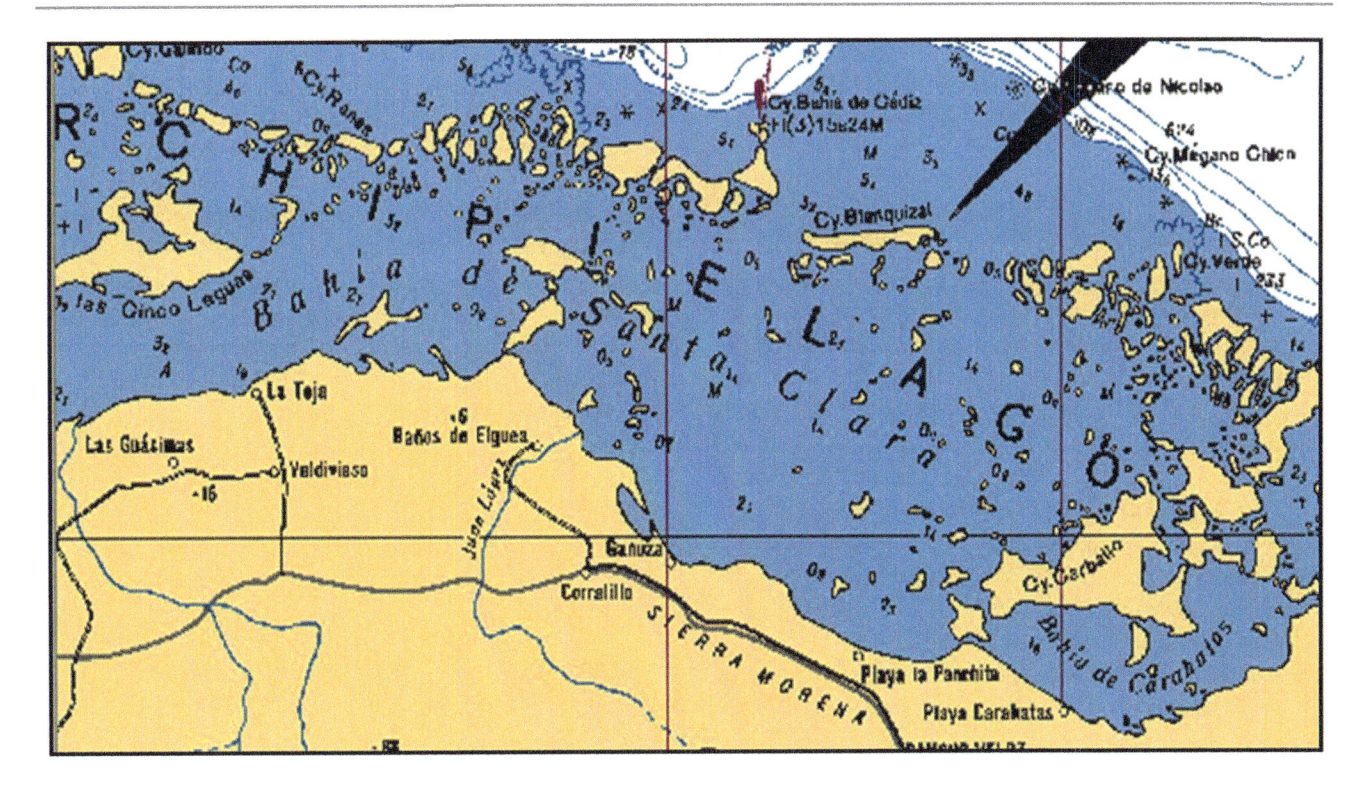

Fig. 2.13 Examples of intra-island bays: Santa Clara and Carahatas

Banks. They exist as remnants of the residual erosion in reef limestones of the Jaimanitas Formation or of modern local coral constructions. Subordinately there are sediments, such as the coral mud from Moa [28, 29] and in the access channels to the main bays.

Terraces. The fragments of the Zero terrace attached to the coastline are frequent. Regularly they seem to be devoid of terraces in their bottom, only in the zone Habana-Matanzas have been identified up to four levels, of tens of meters of width, to the depths 2.6–8 m, 10–16 m, 20–25 m and rarely 40–50 m [16].

Bays. There are three types of bays: (a) open of tectonic origin (from Matanzas and Cochino, with 961 m, the deepest in Cuba, as well as Río Seco and de Jaraguá, a sector without northeastern shelf); (b) intra-island (from Moa key, northeastern sector); and (c) closed or bag. This can be considered a prominent form of relief (or rather a complex of relief forms) and, perhaps is the most polygenic of all forms of relief coastal-marine relief of Cuba, due to the number of factors and processes that intervene in its origin. They are distributed in the northwestern (4), northeastern (18), southeastern (3) and south central (1) sectors. There are several hypotheses about the genesis of this type of bays. In the year 1984 Núñez-Jiménez raised the last of them, based on the following points of view:

(a) When a stream or river flowed down the slope of a limestone slope towards the coast, a narrow valley formed by recessive fluvial erosion; (b) the valley incised in the slope, when deepening, was transformed into a canyon that later allowed the passage to the sea of the waters of the fluvial system developed behind the slope; (c) when sea level rise occurred during the Sangamon, the marine waters penetrated through the fluvial canyon and flooded the upper valley. The fluvial system was converted into a bay of ample sinus (pouch) with a long narrow channel in its mouth. On the new littoral line (mouth and channel of the bay) reefs were built attached to the coast; (d) a decrease in the marine level drained the bay originating in Sangamon during the Wisconsin regression. When the sea level was lower, the river ran again through the canyon, intensified its erosion and therefore deepened it. The old coral reefs, hardened and fossilized, were located above the new level of the sea; and e) when a new ascent of the marine level took place, corresponding to the last transgression, about eleven thousand years ago, the waters of the sea returned to penetrate by the old fluvial canyon and re-emerged the primitive superior valley that now constitutes the stock market bay.

This is the most complete hypothesis that has been based on the genesis of this form of relief, however, suffers from the analysis of the role of tectonic processes, which definitely determined the existence or not of the bays. In the case where the rise of the terrain exceeded the sea level during the last transgression, the bays were "hung" and were definitely fossilized. The most representative examples are those of Imías, Cajobabo and San Antonio del Sur, a sector without a southeastern shelf or the Yumurí valley, a northwestern sector.

Coasts. Two groups of coast types can be identified. The first is related to those located in front of the open sea and the second to those located in the interior of the bays. In the first case, coasts with terraces predominate; cumulative (beaches, bars, storm ridges and dunes); abrasive and cliffs, with block slides. They are cut by deep river valleys where the neotectonic ascents were intense or in the form of rivers where they were weak and did not overcome the rise of the mean sea level. At the mouths of some rivers there is a singular shape of the relief, which are the linear deltas, popularly known as tibaracones (Fig. 2.14). In the second case (inside the bays) there are small segments, practically of all types of coasts.

Figure 2.14 Linear Delta or fluvial-marine bar, known locally in Baracoa, province of Guantánamo, as the Tibaracón. It is typical of some coastal regions of the island of Cuba adjoining mountainous terrain, from which a steep river network is born capable of transporting large volumes of clastic material and depositing it on the coast. In times of drought when the flow decreases, the waves predominate, reworking the deposit and converting it into a bar.

2.4.2 Stratigraphy

From the stratigraphic point of view, the deposits of the marine territory are grouped into lithostratigraphic formations and unnamed deposits. They are also differentiated by: their composition, genesis and distribution, depending on their location on the shelfs or in the sectors without a shelf. The history of the geological development of the Cuban archipelago during the Quaternary period would not be complete unless the knowledge of the deposits of its marine and coastal territory, where, for example, exclusive lithostratigraphic formations exist (Table 2.2).

Table 2.2 Geochronological table of the Pliocene-Quaternary deposits of Cuba. Green: formations located in the coastal territory. Blue: those that appear in keys or in the submarine-coastal territory (Jaimanitas and Vedado). Black: they are exclusive of the mainland. The hiatus in the Lower-Middle Pleistocene appears with an interrogation, because it is only typical of the marine territory, without a clear explanation of its origin.

Lithostratigraphic formations of shelf

The basic stratigraphic information of these geomorphological regions is reflected graphically in the geological map, in their corresponding generalized stratigraphic columns and

Fig. 2.14 Linear Delta or fluvial-marine bar, known locally in Baracoa, province of Guantánamo, as the Tibaracón. It is typical of some coastal regions of the island of Cuba adjoining mountainous terrain, from which a steep river network is born capable of transporting large volumes of clastic material and depositing it on the coast. In times of drought when the flow decreases, the waves predominate, reworking the deposit and converting it into bar

Table 2.2 Geochronological table of the Pliocene-Quaternary deposits of Cuba. Green: formations located in the coastal territory. Blue: those that appear in keys or in the submarine-coastal territory (Jaimanitas and Vedado). Black: they are exclusive of the mainland. The hiatus in the Lower-Middle Pleistocene appears with an interrogation, because it is only typical of the marine territory, without a clear explanation of its origin

LEGEND		CARBONATED FORMATIONS	TERRAIN TREATMENTS
UPPER PLEISTOCENE	LATE	Playa Santa Fe, La Cabaña, Cocodrilo, Cayo Guillermo	Siguanea, El Salado
	EARLY	Jaimanitas	Camacho, Jamaica, Cauto, Cayo Romano
MIDDLE PLEISTOCENE		Versalles, Guanabo	Villaroja
LOWER PLEISTOCENE-MEDIUM		?	Guevara
LOWER BOTTOMEN / UPPER PLYCENE		Río Maya, Vedado, Alegrías	Guane, Bayamo, Dátil

geomorphological-geological sections (Figs. 3.6, 3.8, 3.9, 3.12, 3.15, 3.16 and 3.17).

Additionally, the following descriptions must be added (Figs. 2.15 and 2.16):

Vedado Formation

It was initially described in emerged territories. It has been cut by drilling in the southwestern shelf up to 20 m deep. Between the islands of de la Juventud and of Cuba it is constituted by two facies, the first corresponds to the upper part of the cut formed by organ-detrital limestones, with abundant benthic foraminifera, algae, pseudo oolites, corals, molluscs and microfauna redeposited (miliolids, *Archaias sp, Archais angulatus, peneroplipis proteus, Sorites sp,* and *Sorites magna*). It has basal cement, formed by microcrystalline calcite and clayey cryptocrystalline. Contains sulphurous mineralization as well as iron oxides and hydroxides related to the clay material. The faunistic set corresponds to the Neogene-Quaternary (N-Q) and corresponds to neritic-coastal accumulation conditions. The limestones are of aporcelanated aspect and of cream grey colour.

Pseudo-lithic, organdetritic, lumpy and lumpy-organogenic limestones constitute the second facies, located in the lower part of the cut. Formed by lumps of calcareous clay material or pseudo-oolites up to 60% of 0.2 mm in size, clay impregnations and phosphate in a cement formed by microcrystalline calcite, with some clay material, which predominates on occasions. It contains a scarce micro clastic fraction of quartz and plagioclase, in addition, benthic and redeposited planktonic foraminifera, algae remain, radiolarian moulds, milliolids, molluscs, corals, bivalves, *Archaias angulatus, Peneroplis proteus, Penicillium sp., Sorites sp., Planorbulina sp.* and *Onkolites sp*. The faunistic

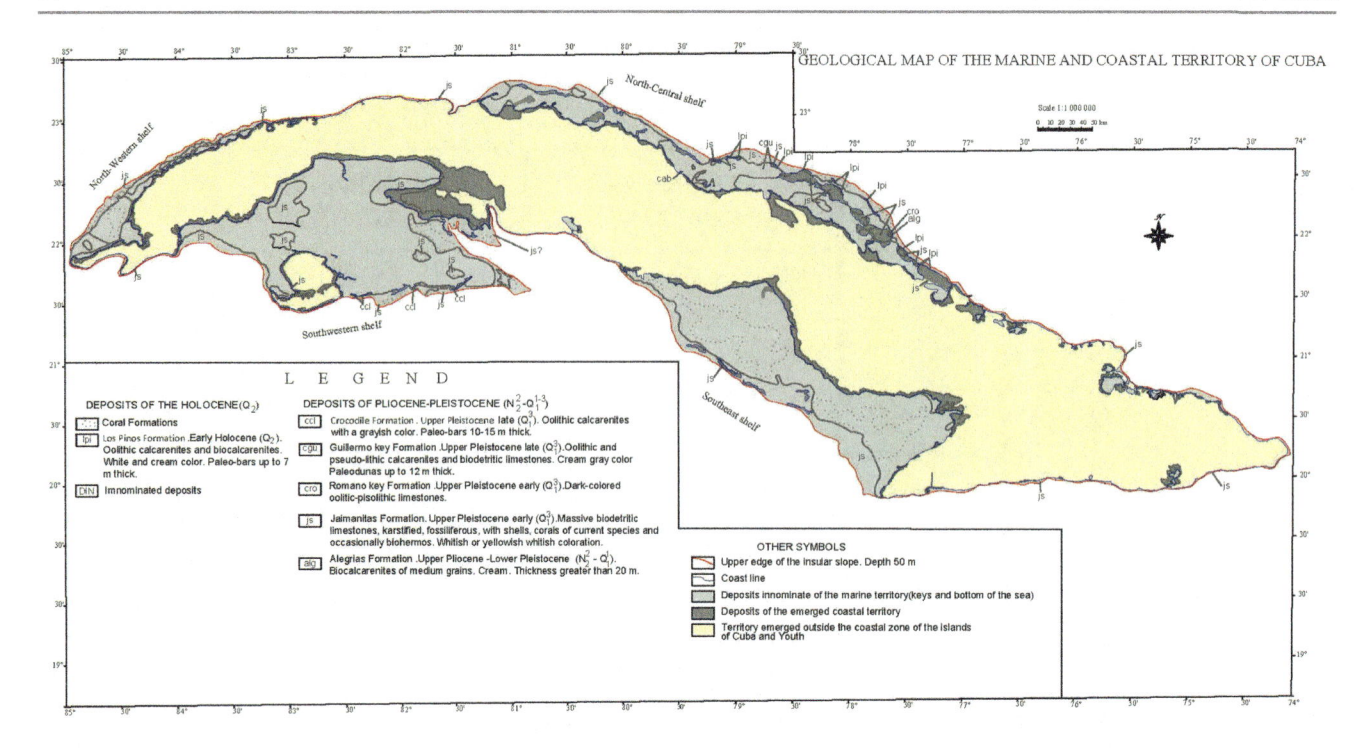

Fig. 2.15 Geological map of the marine and coastal territory of Cuba at scale 1: 1,000,000 (based on [3, 7])

set corresponds to the Neogene-Quaternary and indicates accumulation conditions of an intermediate and open basin, with input from the emerged region.

The limestones of this formation have a well-developed surface on the surface to a depth of 11–13 m, with a yellowish-reddish-variegated evaporite crust and a yellowish clay filling. From that depth there is a second zone, with vertical channels of dissolution filled by secondary calcite. It is probable that this zonation is related to the position of the ancient water table, where the upper part was runoff and the lower one was perennial flood.

Near the southeast coasts of the Isla de la Juventud, the deposits of this formation pass to sandy arenite limestones and hard calcareous sandstones, with abundant quartz grains and subordinate amounts of tourmaline, feldspar and mica grains. Its colour is yellow.

The Vedado Formation could also be found on the northwestern shelf, according to the description of the carbonate rocks of its external lithomorfogenetic zone, offered by Avello et al. [39].

Alegrías Formation

Its type area is located in Sabinal key, a north central shelf, where its holostratotype has the following Lambert coordinates: x- 878,000, and y- 208,000. It is distributed in the higher elevations of the keys of part east of this region.

It is formed by biocalcarenites of medium grains, well cemented, massive, and constituted by rolled remains of calcareous algae, foraminifera and between 2 and 15% of quartz grains, plagioclase and volcanic rocks. They are recrystallized and karstified on the surface, forming a 1–3 m thick cap, because of weathering, which also causes deeper disintegration, turning the rock into an earthy material. No synsedimentary structures are observed. Cream. Its faunistic complex has not been detailed. Its probable Upper Pliocene-Lower Pleistocene age is based on its stratigraphic position and high degree of lithification, similar to that of the Vedado and Río Maya formations. Its sedimentation environment corresponds to a cumulative sandy coast of coastal lowlands, with beaches, bars and dunes, in communication with the territory emerged by means of tidal channels. The thickness exceeds 20 m.

The deposits that underlie it may be composed of black clayey silts, discovered by drilling in a nearby territory.

Jaimanitas Formation

It was initially described in emerged territories. This formation is constituted by massive biogenic limestones and calcarenites of variable cementation, very chalky and with abundant and varied fossils: bivalves, gastropods, echinoderms, algae, corals in position of life, ostracods, *Strombus gigas* and numerous foraminifera of the family *Peneroplidae, Amphisteginidae, Miliolidae, Soritidae, Rotalidae, Homotrematidae and Acervulinidae*. The range of distribution of fauna ranges from the Miocene to the Quaternary. The coloration is whitish, pinkish, light grey or yellowish,

2 Geology of the Marine Territory of Cuba

	STRATIGRAPHIC COLUMN OF THE SHELF					
AGE	NORTH-WESTERN	NORTH-CENTRAL	SOUTHEAST			SOUTH-WESTERN
			GUACANAYABO	ANA MARIA		
HOLOCENE (Q_2)	DIN	DIN / lpi	DIN / arrc	DIN / lpi		DIN
UPPER PLEISTOCENE (Q_3^1) — LATE	H	cgu / lcñ	arc	arc		ccl
UPPER PLEISTOCENE (Q_3^1) — EARLY	js	js / cro	js	js		js
LOWER-MIDDLE PLEISTOCENE (Q_1^1-Q_1^2)	H	H				H
UPPER PLENOCENE-LOWER PLEISTOCENE ($N_2^2 Q_1^1$)	vd	vd / alg	?	?		vd

Legend:
- js — Jaimanitas Formation, reef limestones, sometimes calcareniticas.
- alg — Alegrías Formation, calcarenita.
- vd — Vedado Formation, Reef Limestones.
- cgu — Guillermo key Formation, oolitic calcarenites and biocalcarenites.
- lpi — Los Pinos Formation, calcarenitas oolíticas and biocalcarenitas.
- cro — Romano key Formation, oolito-pisolitic limestones.
- ccl — Crocodile Formation, oolitic calcarenites.
- H — Hiato.
- lcñ — La Cabaña Formation, biocalciruditas.
- DIN — Innominate deposits.
- arrc — Coral reefs.
- arc — Compact red clays.

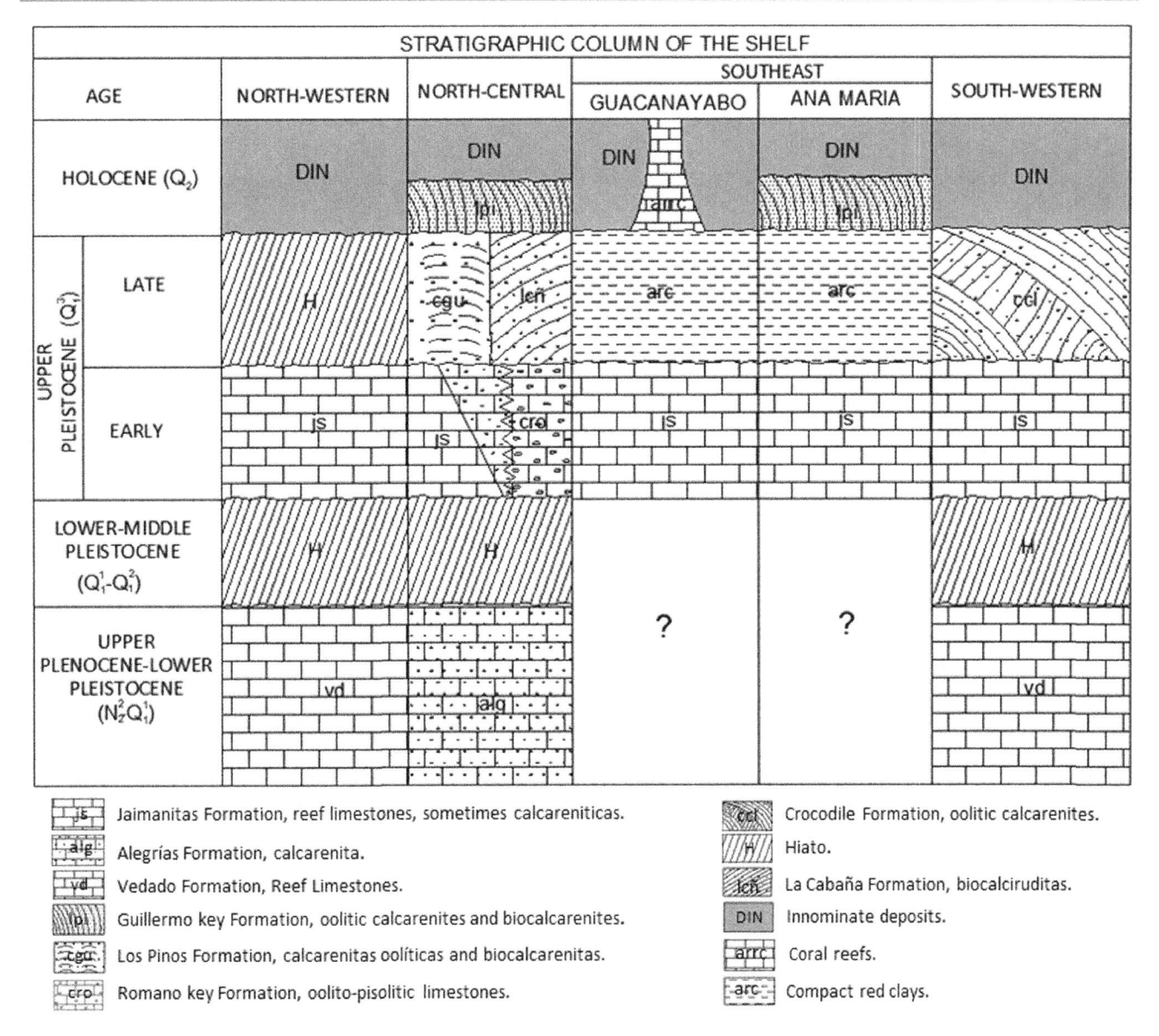

Fig. 2.16 Generalized stratigraphic columns of shelf. **a** northwestern Avello [3, 30–33]; **b** north central (based on Ionin and Pavlidis 1972; Ionin et al. [4, 34]; Cabrera et al. 1997; [35]); **c** southeastern (based on [4, 36, 19]; Cabrera and Batista [3]; and **d** southwestern (based on [37, 38, 21]b; [2, 3]

with a red and white evaporite scab. The sedimentation environment corresponds to reef and beach facies. The rocks of this formation appear in the rocky keys and on the seabed. They constitute, in addition, the paleo bottom, that is to say, the underlying one of the friable deposits, including the substrate of the majority of the present coral formations. Its age according to radiometric determinations by Shantzer et al. [40] and its correlation with deposits of the southeast of the USA, made by Ducloz [41], corresponds to the Early Upper Pleistocene (Sangamon).

This formation has a red and white evaporite crust, which can reach 1–2 m in thickness. The infiltration of the meteoric waters produces a strong weathering under this cover, which disintegrates the rock into an earthy material similar to what occurs in the Alegrías Formation. The relief of the surface is flat and only on the northern coasts of the keys are cliffs, with variable height within the first ten meters, presents up to 2–3 levels of abrasion.

Its roof can be erosive or be covered by younger formations (Cayo Guillermo, La Cabaña and Los Pinos); as well as by mobs and other innominate deposits of the Holocene. It seems to be synchronous with the Cayo Romano Formation. The thickness can reach 20 m and locally in keys, the north central shelf is close to 40 m.

Cayo Romano Formation

It appears at the foot of the slopes, formed in the heights of Romano key and in the Ballenatos keys of the Nuevitas bay. Lambert coordinates: x-847 400, y-244 500.

In its type, section corresponds to oolites pisolitic limestones of dark reddish colour. The oolites pisolitic vary between millimetres and up to 5 cm in diameter. In Romano key, the nucleus can be of rocks or of gastropods filled with oolites and fragments of rocks, cemented by calcium carbonate of dark brown colour. Other nuclei are calcium carbonate of blackish colour, which form a dark layer of 1–2 mm around the mollusc shells. On the outside of this one another layer of brown red colour of 1–2 mm in width is formed, which is the one that contacts with the matrix of the rock. Said layer is formed by thin concentric sheets of calcium carbonate dyed red. The matrix is calcitic and detrital to biodetritic with a reddish brownish colour. In the Ballenatos keys, the nucleus of the pisolites is constituted by breaches of the underlying rocks.

It lies discordantly on the Alegrías Formation in the Romano key or on rocks of the Upper Eocene in the Ballenatos keys. Its roof can be erosive or be covered by Holocene clay sediments. The discordant position on the Alegrías Formation is distinguished by the lithological composition, the presence of terrestrial molluscs and the morphological position, determined by a band around the Romano key hill. Apparently it transitions laterally with the Jaimanitas Formation. Its faunistic set has not been determined. On the basis of the stratigraphic position, it is estimated that it corresponds to the Early Upper Pleistocene. The sedimentation environment was elluvial-colluvial-prolluvial and marine chemogenic, with redeposition of the sediments in a very shallow sea of moderate waves. The thickness is about 10 m.

Cayo Guillermo Formation

Outcrops in the Guillermo, Contrabando, Coco and Hijos de Guillermo keys. Its type area is found in Guillermo key and the holostratotype in the northeast end of it. Its Lambert coordinates are: x- 730 000, y- 311,020.

It corresponds to oolitic and pseudo oolites calcarenites, homogeneous and well-selected biocalcarenites, bioturbated and rounded grains and biodetritic limestones, with micritical matrix, fine to medium grains. You can distinguish up to three horizons of crossed lamination with an inclination of 0 to 30°. They are separated by slightly altered surfaces (diastemes). Cream-grey colour, with a dark grey weathering cap. These deposits form E-W course dunes. They lie discordantly on limestones of the Jaimanitas Formation, their roof is erosive or is covered by fine unconsolidated sands. It is similar in composition to the Late Upper Pleistocene Cocodrilo Formation, described on the southwestern shelf.

Among the biodetrites, there are algae, milliolids, soritiidae, echinoderms and others. Based on the stratigraphic position and the degree of lithification, its age is estimated as possible Late Upper Pleistocene (Wisconsin). The sedimentation environment corresponds to coastal dunes, formed in a tropical climate of dry seasons, with movement of sands and growth of dunes and wet seasons, with proliferation of creeping vegetation and fixation of the dune. This occurred in several cycles as confirmed by the presence of diastemes. It has 12 m thick. Its type area constitutes the highest dune of the Cuban archipelago and is catalogued as a prominent feature of the relief.

La Cabaña Formation

It also appears in the emergent territories of the Island of Cuba, where it was initially described. In the north central shelf only relics of their outcrops remain, which were paleo bars, probably less than 5 m thick, oriented normal to the coastline. Its clear geomorphological differentiation in the terrain and its low degree of lithification turned it into places of easy visual identification for its exploitation and use as material for the construction of roads. Their deposits correspond to limestones and calcarenites with little consolidation and high porosity due to the non-compact packing of the abundant fossiliferous material, mainly shells. White and cream colours. The matrix is contact and filling. The marly content is lower than in other parts of the Cuban archipelago.

In addition to the north central shelf, the deposits of this formation have been identified only in Real key (San Felipe keys, west of the Isla de la Juventud), without a clear stratigraphic position, which is why it was not represented in the generalized lithostratigraphic column. It is formed by fine biocalcirudites of calcarenitic matrix, well cemented, composed of nodules of algae, fragments and well-preserved specimens of *Nerita perolonta* and clasts of corals. They are cream. They constitute deposits of intertidal facies.

Cocodrilo Formation

The deposits of this formation were described in the territories emerged to the southeast of the Isla de la Juventud. In the marine territory, it appears in some keys of the Canarreos archipelago and in Real key (E and west of Isla de la Juventud, respectively). It is constituted by oolites and pseudoolites, with few fragments of corals and shells, cemented by carbonates. They form bars, with oblique laminar stratification. The layers are millimetric and form packages of different thicknesses, oriented indistinctly. Due to discordant lying on the Jaimanitas Formation, it has been considered that they had to be deposited in the Late Upper

Pleistocene, in addition, in a coastal beach environment, under the action of a shallow sea of variable depth, with lateral currents of different directions. Its maximum thickness must not exceed 10 m. Its surface can be abrasive or be covered by marine and marsh deposits.

Los Pinos Formation

It emerges from east to W, in the Sabinal, Cruz, Coco, Guillermo, Paredón Grande and Santa María keys, of the north central shelf. Its type area was established on the north coast of Sabinal key and its holostratotype -beach Los Pinos- to west of Punta Central on this own key, with Lambert coordinates x-889 798, and y-212,119.

It consists of oolitic calcarenites and biocalcarenites with medium grains, well rounded and well selected, sometimes with oblique lamination. White and cream colour. It juts out discordantly to the Jaimanitas Formation, from which it is separated by a paleosoil or an intemperate surface. Sometimes, it is partially covered by current mobs and sands. No similar deposits have been established, with which they can be correlated, nor has fauna been reported. Due to its low degree of lithification and little alteration its age is estimated as possible Holocene. It was deposited in a low cumulative coastal environment, with bars and beaches during the Holocene transgression, when the mean sea level rose above the current one. It is the only lithostratigraphic formation of the Holocene in the territory of the Cuban archipelago. Its thickness can reach up to 7 m.

Innominate deposits of shelfs

These can be found in the keys, mostly classified genetically: (a) marine formed by sands and gravel biogenic-carbonated, cream-coloured. Only in the Los Indios keys (W of Isla de la Juventud) have been reported quartziferous sandstones [4]; (b) swamps constituted by plant remains in different degrees of decomposition, faunal remains and silty-clayey fractions, coloured blackish; (c) ferruginous-carbonated clays and the karst cavities of the Jaimanitas Formation, compact, brick-red, similar to the native residual soils (*terra rossa*) [3, 4] and (d) dead reefs, particularly in the recent reef keys.

On the surface of the seabed, the unnamed deposits are classified lithologically according to their granulometry in: gravels, sands, silts and clays, which often appear mixed. The sands and gravels are biogenic-carbonated (from molluscs, algae and foraminifera), which in the outer morphogenetic zone come to constitute important accumulations in the negative forms of the karstic relief. The said corals are not a source of input for the sediments. In the east part of the southwestern shelf, the sands are of chemical origin (oolitic and pseudo-oolitic).

The sands can also be found in the interior morphogenetic zone, from where they begin to move towards the coast to mud, clays and thicker remains of vegetables, mainly from the mangrove and seagrasses, also decreasing their carbonate content and colours clear. Only near the coast can terrigenous fractions appear.

The silts in the Guadiana cove, northwestern shelf, form two horizons, reported by Ionin et al. [4]. The lower one is viscous and compact, brownish, with few fragments of molluscs and thickness of up to 8 m, corresponding to a lagoon-palustres environment. The upper one is friable, grey clayey silts, with abundant mollusc shells and thickness of 2.0–2.5 m. Similar facies are found in the oolites deposits of the southwestern shelf. Likewise, on the southeast and north central sea shelfs, facial differences have been discovered through the perforation of their sediments [4].

In the terrigenous fractions of the seabed, there are allogenic and autogenous minerals, with very low contents, represented in the northwestern shelf by: pyrite, ilmenite, leucoxene, pyroxenes, amphiboles, olivine, magnetite, limonite, rutile, epidote, andalusite, spheena, staurolite, zircon, barite, brookite, mica, tourmaline, siderite, chromite and native copper [30, 31]. In the detailed study of the content of these fractions in the Cárdenas Bay, north central shelf. Álvarez and Quintana [42] reported low contents of: zircon, quartz, tourmaline, epidote, staurolite and andalusite. Near the coast of Ana María, around Punta Macurije, southeastern shelf, was reported by Alvarez [43] grains of: magnetite, ilmenite, chromite, leucoxene, garnet, epidote, pyroxenes, amphiboles, zircon, spinels, plagioclase and feldspar. Its concentration varies between 3. 3 and 12. 4%. In SW-NW of Isla de la Juventud and south of Pinar del Río, southwest shelf, mineralogical analyses show a wide range of minerals [2, 37, 38, 44, 45]: andalusite, apatite, actinolite, barite, zircon, corundum, distene, epidote, spheena, spinel, staurolite, garnet, hornblende, ilmenite, leucoxene, marcasite, magnetite, monazite, pyrite, pyroxene, tourmaline and tremolite. The most representative for its content and dissemination were distene 19.0%, staurolite 12.0%, zircon 4.0% and garnet 4.0%.

In the terrigenous deposits, coexist chemically stable and unstable minerals as well as a large granulometric heterogeneity. This is due to a low marine hydrodynamic reelaboration existing in its accumulation areas, as coral reefs and the mangrove forest, in addition to the weak hydrographic network, which has little capacity for transportation of material terrigenous carriers of minerals, among other factors, protect it.

Lithostratigraphic formations of the sectors without shelf

The Jaimanitas Formation, in addition to being on the seabed and in many keys of the shelfs, festoons the territory of the Island of Cuba through the sectors without a shelf, both in front of the open sea, as well as in the inlets and interiors of its bays. The limestones of this formation are underlain by the Vedado Formation or its equivalent Río Maya (Table 2.2) and probably in some locations by older formations, for example, south of the Sierra Maestra. Within the bays of Nipe, Puerto Padre and Moa (sector without northeastern shelf), limestones of the Jaimanitas Formation have been drilled up to 10 m thick, overlain by sediments.

Inside some bays, there are keys, which are erosive remains of pre-existing rocks where the lobes of these relief forms were made. They can also be formed by Quaternary deposits, for example, in the Bay of Moa constitute a Holocene reef bank. In the bay of Cienfuegos, deposits of the El Salado and La Cabaña formations appear (Fig. 2.17).

Innominate deposits of the sectors without shelf

In the open sea territory and the entrances of the bays there are isolated accumulations of sands and subordinately biogenic-carbonated gravels, of a light cream colour, several meters thick. There are also modern coral formations, which present their greatest development in front of the east beaches of La Habana and Varadero. There are also patches of bio clastic breccias along the coast (terrace Zero).

Plastic silts form the surface of the bays and in depth, the deposits are heterogeneous for their granulometry and composition, with the presence of mixture or alternation of silts, sands, gravels and peat, which contain marine fauna, mainly molluscs. The colour is variable, predominantly dark, due to the high content of organic matter and anthropogenic contaminants, up to 10 m thick. The best studied case is that of the Puerto Padre bay, in the northeastern sector, where, from the bottom to the top of the geological section, there are indistinctly: (a) silty, brownish-coloured sediments; (b) mounds and remains of plants, with an age of 5–6.5 Ky [4]; (c) silty-sandy sediments; and (d) very compact clays, yellowish grey. More recent greyish brown silts cover them. A similar cut was discovered in the bay of Santiago de Cuba. In Nipe Bay, they have an age of seven Ky at the base of the cut [4]. There are also mobs dated in 1780 years in the bay of Cienfuegos.

2.4.3 Neotectonic

The Cuban archipelago constitutes an elevated mega block, that is, it is reflected as a positive structural element, belonging to the articulation zone of the North American and Caribbean plates. It was formed as a result of an active manifestation of the neotectonic movements and in particular those of the Pleistocene and the Holocene (2.5 My). It extends in the sublatitudinal direction and is limited by the north by the deep Nortecubana fault and probably some flexures. It continues outside the territory of Cuba through the east to La Española. At the S, the mega block is controlled by the deep Surcubana fault that closes its limits by the east and the west by means of the union with the Nortecubana fault, continuing then by the west until the Yucatan Peninsula. Geomorphologically, both faults coincide with the insular slope.

The shelfs have a direct relationship with the general tectonic plane of the emerged territory. These are located in the extension of its most sunken blocks. In general, what is stated by Guilcher [46] for shelfs is true, and is that they are inversely proportional to the force of the emergent relief that surrounds them, being coarse in front of the flat and non-existent regions in front of the mountainous regions,

Fig. 2.17 Quaternary formations in Ocampo key, Cienfuegos bay. Jaimanitas Formation at the base of the cut covered by the La Cabaña Formation (from left to right, photos 1 and 2) and by the El Salado Formation (photo 3)

where the called sectors without shelf. In both cases the neotectonic features are represented in Fig. 2.18.

Neotectonic of the shelf

The northwestern shelf is located in the most sunken part of Pinar macroblock (III1), divided by a fault from north to south in two blocks: Guanahacabibes and Buenavista, which have a marked expression in the geological-geomorphological constitution of the territory. The first is the most sunken, so it has the following characteristics: few keys, the depressed bottom, the greatest sediment thicknesses of this shelf, an important sector without coral reef and marshy coast, slightly cliffed in small segments. In general, the tendency of the macroblock is to the ascent, evidenced by the little development of the coastal plain in width and forms of the relief, the presence of sectors cliffs and deltas in formation.

The north central shelf is located in the most sunken part of the Sabana and Camagüey-Holguín (III3 and III6) macroblocks. The greatest expression in the relief of its faulting and block division of minor orders, is given by the transverse-diagonal direction fault that divides it into two regions, one more sunken (Sabana) and the other more raised (Camagüey). There is a clear geomorphological expression of the current general upward trend in the region, particularly in the more uplifted part (to the E): developing tidal niches and storm ridges far from the coast in the keys (Fig. 2.9), elevation above the mean sea level of deltas and tidal channels (Fig. 2.7), among other indications.

The southeastern shelf occupies the most depressed part south of the macroblocks Guáimaro, Cauto and Escambray (III7, 8, 5), where it is composed of two sunken structures, occupied by the gulfs of Ana María and Guacanayabo; as well as two elevated, one in the centre of both gulfs and another in the Jardines de la Reina archipelago. The macroblocks, in general, undergo neotectonic ascent, which is evidenced by the presence of slightly cliff-like coastal segments, delta formation, slight encasing of the fluvial valleys and emersion of reefs with the formation of keys (Fig. 2.11).

The southwestern shelf corresponds to the most sunken part of the Batabanó macroblock (III3), apparently complicated by faults and blocks of minor orders, which have only been confirmed by geophysical data in the Siguanea depression. There are geomorphological evidences of tectonic ascent in the outer zone of this shelf, which are: a fossil tidal niche, outcrops of Quaternary rocks in some keys (Jaimanitas, Cocodrilo and La Cabaña formations) and, recent coral keys (Holocene reef keys). In addition, of Holocene marine deposits, including those of the terrace Zero in the coasts of the Isla de la Juventud.

Neotectonic of the sectors without shelf

The sectors without shelf are located on the edge of raised macroblocks, which have a general tendency to neotectonic ascent, with an amplitude estimated in the last 5 Ky of up to 5 cm [9]. Faults and tectonic blocks of different minor orders, which have generally undergone differentiated ascents, complicate them. This is evidenced by the presence of: (a) emerged marine terraces, with a variable number of surfaces (for example: in the localities of Vedado and Guanabo, La Habana, reaches four, west of Matanzas bay seven, southeast of Guantanamo province up to 17 and, in other cases only one is seen; (b) tidal niches of different generations located at different altitudes; (c) active and fossil bag bays; (d) open bays of tectonic origin; (e) deep or raised river valleys (fossils); (f) storm ridges of several generations

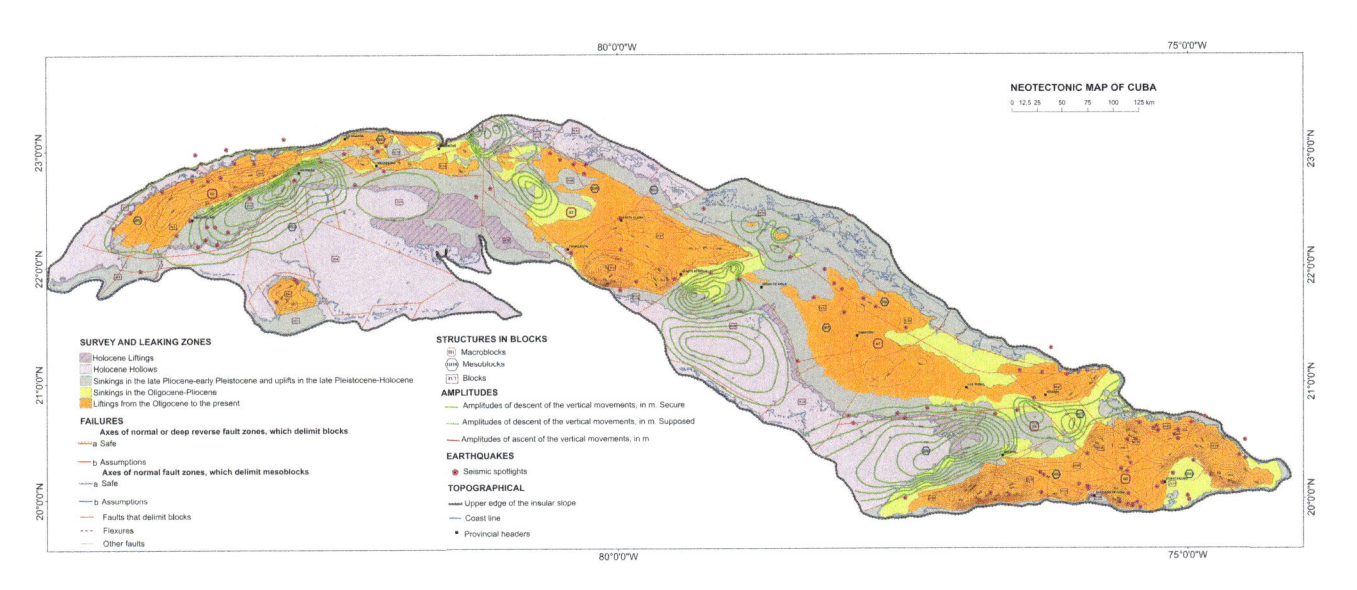

Fig. 2.18 Neotectonic map of Cuba [9]

and at different altitudes; and (g) predominance of raised and abrasive coasts.

2.4.4 Geological Evolution

There is a coincidence in the geomorphological, stratigraphic and tectonic information, which point to the existence of at least five main episodes in the quaternary geological evolution of the marine territory of Cuba. Identified in the following geochronological intervals: (1) Upper Pliocene-Early Lower Pleistocene; (2) Late Lower Pleistocene-Late Middle Pleistocene; (3) Early Upper Pleistocene; (4) Late Upper Pleistocene; and (5) Holocene.

(1) **Upper Pliocene-Early Lower Pleistocene**. The carbonate-biogenic deposits of this interval (Vedado and Alegrías formations) are the result of the occurrence of a transgression. This is evidently associated with the glacio-eustatic ascent known in Cuba, as a Vedado transgression (Fig. 2.19), when a large part of the Cuban mega block was submerged because of the weakening of the intense tectonic ascents produced in this to the Lower Pliocene and the glacio-eustatic ascent. This was related to the first interglacial episode of the Quaternary, dated by stages of oxygen isotopes in other parts of the planet [4, 47].

This transgression had dimensions comparable to the Oligocene-Miocene and consequently Cuba was divided again into several islands [25, 48]. The composition and distribution of its deposits correspond to a carbonated sedimentation environment, in a flat and little dismembered background, with a warm climate and a high hydrodynamic energy. This allowed contemporizing the development of coral reefs and certain contribution of terrigenous material, transported directly through the hydrographic network in sectors without shelfs or through tidal channels to shelfs. This was revealed by the presence of terrigenous components in biogenic-carbonated deposits. The reefs formed in the outer zone of the shelf of the open seas of that time, in which banks grew towards the edge of the insular slopes. The size of this area can be estimated to be similar to contemporary shelfs, based on the presence of its deposits in the northwestern, north central and southwestern shelfs (Vedado and Alegrías formations).

The sectors without shelfs were wider than the current ones, as evidenced by the presence of the Vedado and Río Maya formations in the lands that emerged hundreds of meters beyond the current coastline. Its altitude reaches up to tens of meters, such a hypsometric position is the result of the tectonic acceleration of the terrain after this transgression.

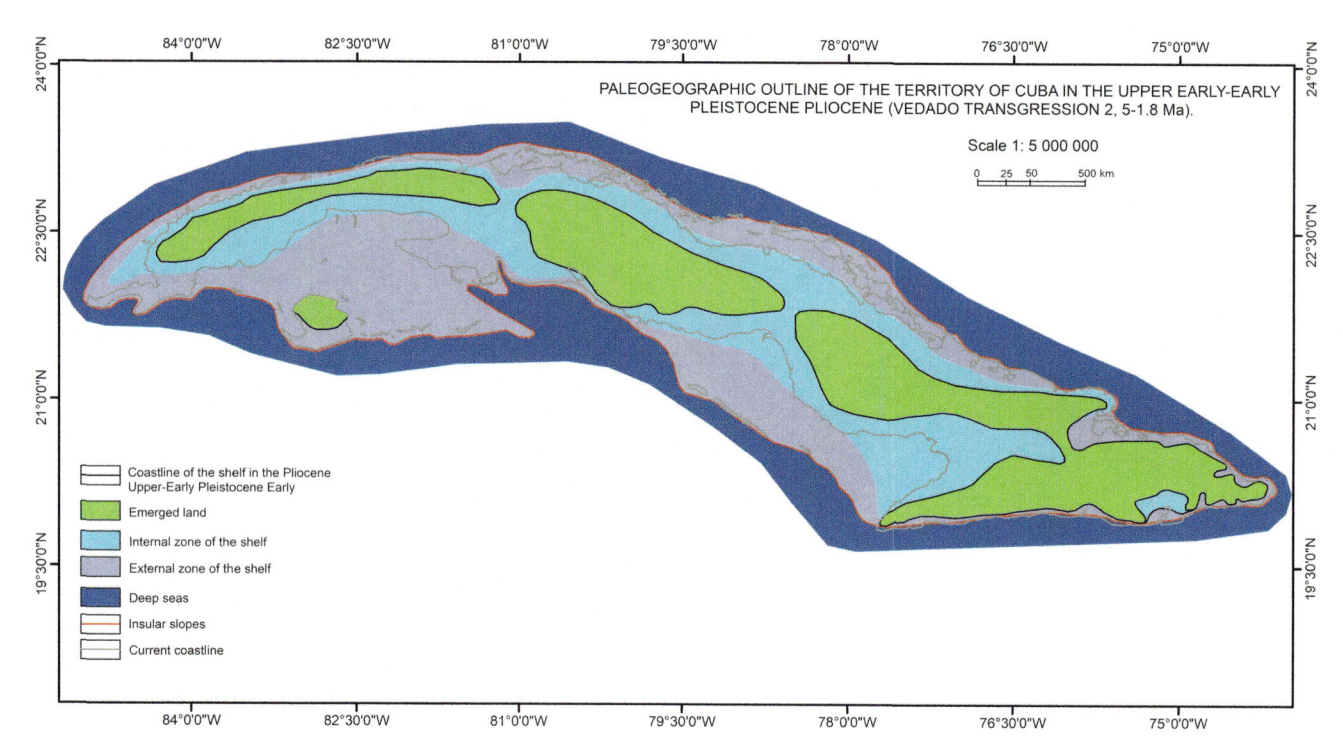

Fig. 2.19 Paleogeographic outline of the territory of Cuba in the Upper Pliocene-Early Lower Pleistocene (Vedado transgression) 2.5–1.8 My)

(2) **Late Lower Pleistocene-Late Middle Pleistocene**. The deposits of this interval (Guevara and Villaroja formations), related by Kartachov et al. [48] with the transgressions Guevara and Villaroja, are absent in the marine territory. This is a big question in the reconstruction of the geological evolution of the marine territory of Cuba during the Quaternary. Other researchers such as Ortega and [49]; Dzulynski et al. [25, 50], does not recognize the marine origin for the mentioned formations. They relate the area of accumulation of this type of lithology, with flat surfaces of denudation located at low altitude above the mean sea level, in marshes and rivers. The aforementioned hypotheses lack quantitative data for their rationale and none refer to the absence of the aforementioned deposits in the current marine territory. On the enigma of the hiatus in the marine territory corresponding to this interval, the following hypotheses could be supported:

(a) If there were marine transgressions they were very weak, with basins of sediment accumulation, due to the revitalization of the ascending neotectonic movements in this stage, which reached up to 600–700 m [9, 48, 51]. This could be the cause of the thinness of the accumulations and of the absence of carbonated environments, except in probable isolated localities, such as the Habana-Matanzas coasts, where carbonate relicts are supposedly of this age (Guanabo and Versalles formations). There was also no biogenic-coastal environment of large wetlands, judging by the absence of ancient mobs.

(b) There was removal and transport to the depths of the marine deposits due to abrasion and erosion during transgressions and regressions, respectively. Such processes were favoured with the presence of friable lithoclastic sediments on the surface of the calcareous floor formed during the Vedado transgression.

There is no evidence of wind transportation, a criterion used by Ortega and Zhuravliova [49] and Ortega [52] to explain the absence of the aforementioned deposits in the offshore shelfs by means of the transportation by southeast winds. In studies carried out at the addresses of the strata of lithified dunes [13, 36, 53] it has been established that the predominant direction of the winds was the NE, which does not validate the hypothesis of transportation by winds of the SE. All the Eolianites and the current dunes are in the north of Cuba, so it can be assumed that at least since the Pleistocene until today the main direction of the winds has not changed.

(3) **Early Upper Pleistocene**. In this interval correspond the limestones of the Jaimanitas Formation product of a great development of reef formations. They festoon the emerged territory of Cuba through its coasts and the marine bottom and paleo bottom, making it clear the transgressive character of this interval, known in Cuba as the transgression of Jaimanitas or Sangamon (Fig. 2.20). She was the elder of the Quaternary sensu lato, again the territory of Cuba became an archipelago, similar to that of the Vedado transgression. Judging by temperature data from different sources on oceanic waters (Pelejero 2003, and Martrat 2004 in [54], it is likely that these were greater by 2 or 3 °C in Jaimanitas times around the Cuban archipelago compared to the current ones.

The limestones of the Jaimanitas and Vedado formations are similar in terms of composition because the paleoclimatic conditions in which they were formed were also. It is likely that something warmer for Jaimanitas, judging by the vigour of this planetary transgression, according to Uriarte [54]. This led to the restoration of corals in the outer area, which occupied most of what are the current shelfs, where they gave rise to banks near the edge of the insular slope, which served as the basis for the subsequent configuration of the contemporary outdoor zone. High hydrodynamic energy prevailed, which allowed contemporizing the development of coral reefs and the contribution of terrigenous sediments through the hydrographic network, particularly in the sectors without southeast and south central shelf.

The territorial scope of this transgression in the sectors without a shelf went beyond the current coastline judging by the presence of reef limestones in the coastal zone, including the interior of the stock bays. However, this range was less than that of the Vedado transgression, due to the tectonic rise of the terrain in the preceding interval.

(4) **Late Upper Pleistocene**. During this interval, there is evidence of a regression of the sea, manifested in: the withdrawal of the corals; lithification of the reefs and the sands formed during the transgression of Jaimanitas, passing to reef limestones and calcarenites of the Jaimanitas Formation. They are correlated with the Pamlico, Anastasia, Limestones Key Largo and Oolita Miami formations, all from the Atlantic coast of the USA [55]. Then its karstification and partial dismemberment occurred, with the formation of plains and erosive remains elevated a few meters above the general

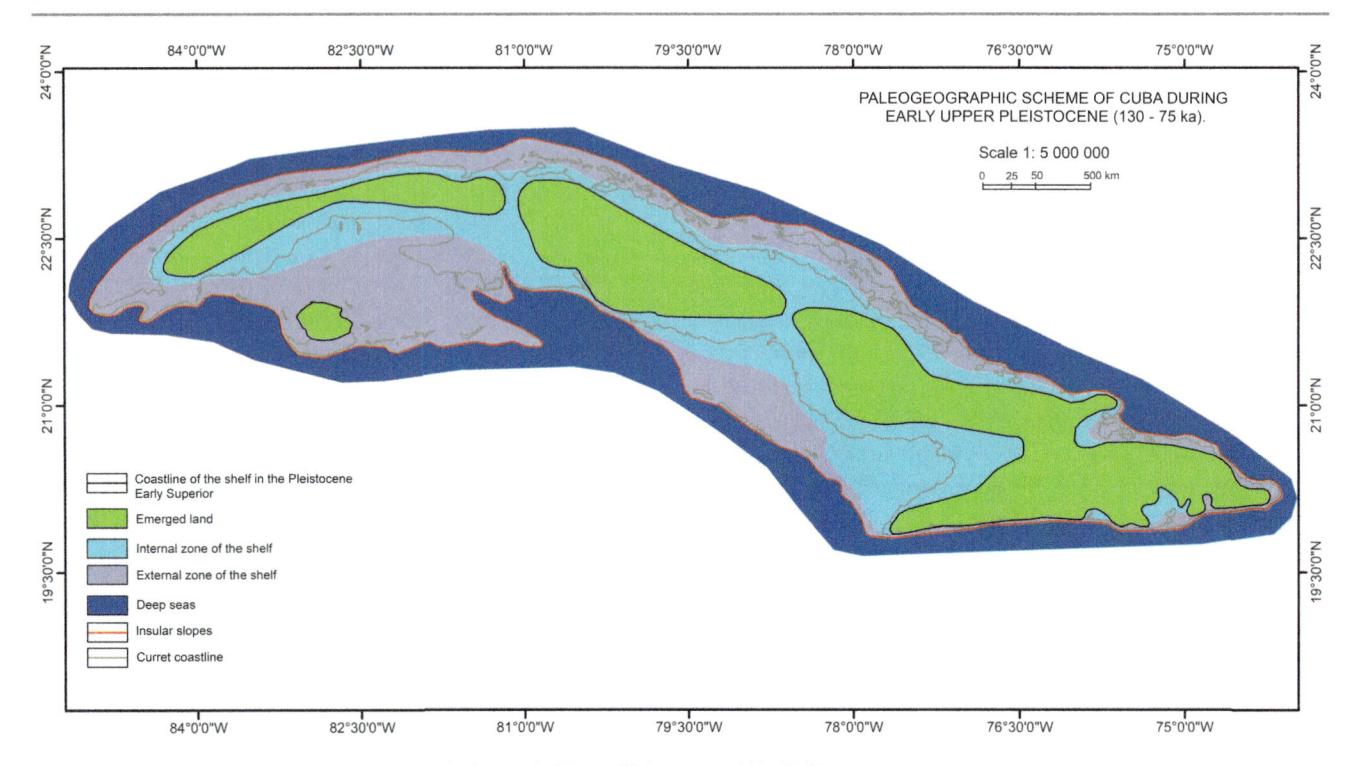

Fig. 2.20 Paleogeographic scheme of Cuba during early Upper Pleistocene (130–75 Ky)

plain (banks) and the extension of the karstic depressions. This event coincides in time with the planetary regression, known as the great Wisconsin regression. It consists of three sub-steps: (a) initial; (b) intermediate; and (c) final. There are traces of them in the current marine-coastal territory of Cuba.

(a) In the initial sub-step there was a planetary descent of the mean sea level of less than 50 m, between 75 and 36

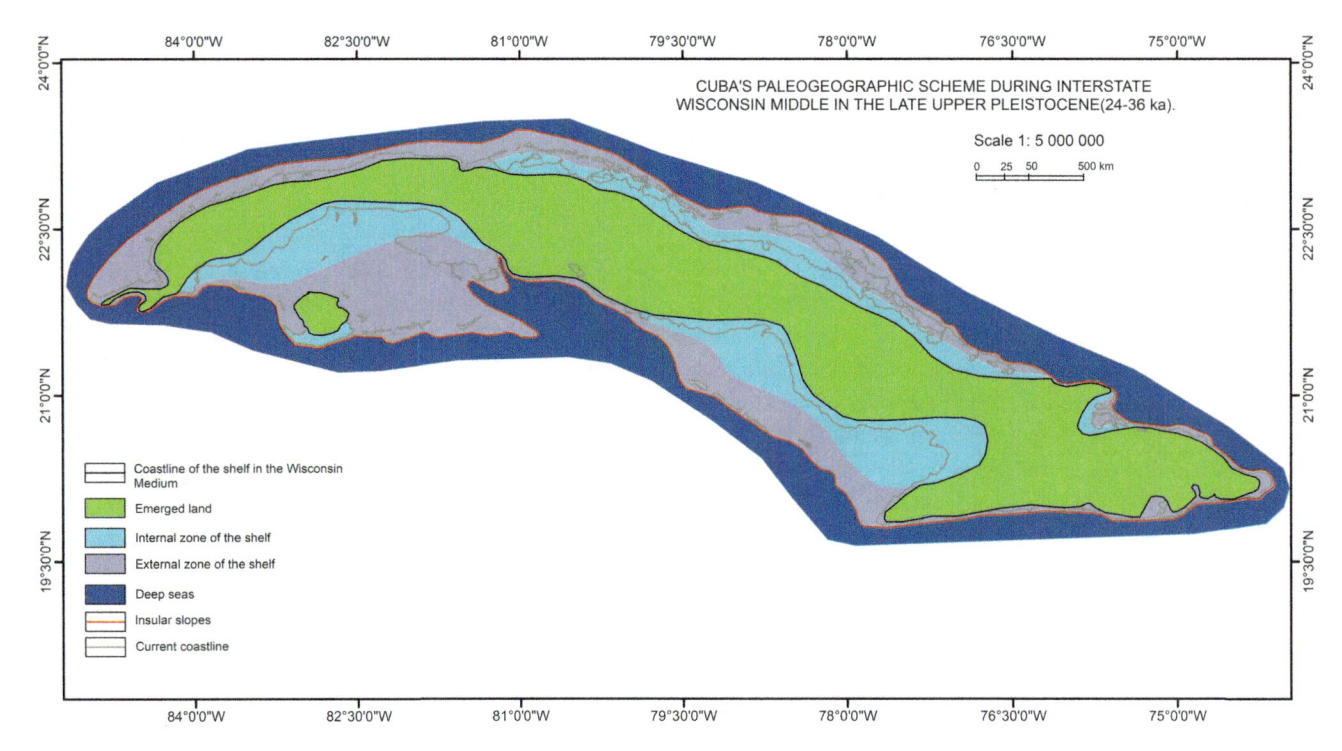

Fig. 2.21 Cuba's paleogeographic scheme during interstate Middle Wisconsin in the late Upper Pleistocene (24–36 Ky)

Ky, which included periods of attenuation [54], with which some of the levels of terraces could be related less deep (up to <40 m), present in the walls of the island slope of Cuba. In addition, other forms of relief such as overlapping caverns and tidal niches.

(b) The intermediate sub-step corresponded to 36–24 Ky, with a high position of the mean sea level in different parts of the planet, known as Interstate of the Middle Wisconsin or transgression of Freeports (Milliman and Emery 1968, Shepard, 1961, and Curray 1961 in [4]. Its possible traces in Cuba are the formations, which overlap discordantly the weathered surface of the Jaimanitas Formation (Table 2.2).

Taking into account the composition and distribution of the deposits of the Middle Wisconsin, it can only be done hypothetically its paleogeographic representation (Fig. 2.21), mainly due to the lack of clarity of the extent of the mean sea level, although it can be affirmed that it came to occupy a part of the current emerged territory. This is corroborated by the presence in the coastal zone and some keys of carbonated deposits of the La Cabaña, Cayo Guillermo, Cocodrilo and El Salado formations. This also shows that the sea was not reduced to a series of lagoons, isolated from the open sea by barriers of islets, bars and the elevated blocks of the outer part of the shelfs, as supposed [4].

The aforementioned deposits are lithomorfogenetic evidence, which corroborates the existence of a relatively warm sea, close to the current, shallow and hydro dynamically active. For example: the thickness reached by some deposits, up to 12 m (Cayo Guillermo Formation); morpholithogenesis represented by dunes and marine bars of some oolitic deposits (Cocodrilo and Cayo Guillermo formations, respectively) and; the abundance of marine macro fauna of snails, molluscs and corals. Although the sea was warm, its average temperature must have been below the minimum necessary for a good development of coral reefs (< + 18 ° C), as only few specimens are appreciated (El Salado and La Cabaña formations). In addition, it was not stable, but with alternation of short periods of aridity (production and accumulation of sand) and humidity (lithification of the sands), as revealed by the diastemes present in the eolianites of the Cayo Guillermo Formation.

The lying of the Jaimanitas Formation on the surface of the Vedado Formation, which is found in the terraces of the Pleistocene sensu lato [40] in localities of the sectors without a shelf; as well as the existence of deposits of the La Cabaña, El Salado and Cocodrilo formations on the surface of the Jaimanitas Formation in different coastal localities, was due to an activation of the neotectonic movements of ascent within this stage and that continued at least in the final sub-step. However, this issue requires more research to do or not correlated regionally, particularly, the dated sensu stricto of the terraces, the aforementioned formations and some attached forms of karst relief, such as tidal niches.

(c) The final sub-stage of the late Upper Pleistocene corresponded to a planetary glacial intensification, known as the last ice age, which reached up to 11 Ky, with a maximum between 17 and 21 Ky (Shepard 1960, Curray 1964, Land 1967 and Karrey 1968 in [4, 54, 56], among others). As a result, the level of the World Ocean dropped below 120 m [57–59], and others).

In the marine-coastal territory of Cuba the traces of maximum descent seem to be related to the formation of karst holes at sea depths of 70–80 m [4, 20, 24]. By then the shelfs and part of the slopes were emerged. The withdrawal from the sea occurred with periods of attenuation, which probably originated the deepest levels of terraces of the insular slope (>40 m).

Other traces of the maximum regressive in the marine-coastal territory is the deepening to dozens of meters of the river valleys in search of their base level, in the sectors without a shelf, reaching to reach the wall of the insular slopes in some cases, between those that are the channels of access to the stock bays. Through the karstic processes emerged depressions and channels, with greater development towards the boundaries between lithomorfogenetic zones exterior and interior. Apparently, the outer zone sheltered the fluvial waters, forcing them to move sublatitudinally, with the consequent formation of channels (paleo valleys) of the same direction. The erosion of a large part of the deposits of the Middle Wisconsin also occurred, leaving only patches.

The limited presence of fluvial valleys on the seafloor, particularly those of normal direction to the coasts, is due to: (a) the great extensions of flat karstic terrains; (b) the weak development of the hydrographic network insufficient to excavate the karstic plains individual; and (c) the scarce rainfall, as a result of the decrease in the evaporation of marine waters, which must have been very intense due to the great magnitude of the glaciation. A particular case is part east of the southeastern shelf (Gulf of Guacanayabo), where the combination of the mighty Cauto River and the low position of the seabed have allowed the valley to remain active, despite the accumulation of large thicknesses of terrigenous deposits in the seabed.

Data from different researchers, obtained in seabed sediments from different parts of the world (Karrey 1968, Hill 1968, Shepard 1969, Kaplin 1973 and Möner 1974 in [4, 54], agree that the end of the maximum Glacial started from the last 17-21 Ky, after a sudden rise in temperatures. In less than 500 years, the average level of the sea rose by 10–

15 m, at a rate of 50 mm/year, then at least two more pulsations occurred until the Pleistocene ended. In Cuba these are still to be identified, for which it is necessary to date the deposits.

(5) **Holocene**. The existence of deposits of different ages between the 11 Ky and the present, the reinstallation of the coral reefs and the formation of a young coastal zone are indications of the presence of the last transgression (Holocene) in the marine territory of the Cuban archipelago. As it is known, this began with the end of the Wisconsin glaciation, when the mean sea level was around −60 or −45 m according to different authors (Mörner 1969 in [60, 61].

The alternation of fine sediment horizons, with different composition and physical-mechanical properties, including peat, in the Holocene deposits means that the Holocene transgression has happened in the marine territory of Cuba, as in other regions of fast way, but with fluctuations [4]; Fairbanks 1989; Edwards 1993; and Hanebuth 2000 in [54]. Among the planetary fluctuations, two intervals stand out [54]: a warmer and more humid (early Holocene), which had its inflection to the second drier 6–5 Ky (late Holocene), with which the desertification of large territories, such as the Sahara.

Six Ky ago significant orbital differences existed with respect to the present: greater eccentricity of the orbit (0.0187 compared to 0.0167), greater inclination of the terrestrial axis (24.1° versus 23.4°) and, above all, different date of perihelion (in the middle of September instead of early January, as now). With the reduction of the summer insolation, the low thermal pressures of the continent, produced by the summer heating, became less deep, which decreased the humidity suction from the Atlantic. The rains that brought from the south the summer monsoons weakened.

According to the radiometric dating of the marine-coastal territory of Cuba, the sediments with 11 and 7–9 Ky should correspond to the early Holocene and the younger ones to the late Holocene. However, there is no palpable limit in light of current information. Only close to the four Ky, there are clearer traces of a change in the mean sea level, due to a decrease in this, fixed by the death of the corals, which had been reinstated when the Wisconsin regression ended, the destruction of its reefs and the formation of Terrace Zero, dated 2–4 Ky [4]. Wetlands also emerged with the formation of mobs.

During the early Holocene the rise in mean sea level was varied, but without a drastic reversal. In this interval, it is probable that shallow terraces and sublatitudinal channel systems were formed in sectors without shelfs, such as those reported to the east of La Habana and the Hicacos peninsula. It is not clear how far the average sea level rose. There is some evidence suggesting a height between 2 and 5 m. Example: sectors where the first Pleistocene terrace is covered by storm ridges, 5–8 m high and at a distance from the coast of 100–200 m, with the presence of soil on its surface. Other indicators of the maximum height reached could be the calcarenites of the Los Pinos Formation and the tidal niches of up to 6.60 m of altitude in deposits of the Jaimanitas and Cocodrilo formations. An accurate diagnosis, in this regard, requires greater geochronological knowledge of the deposits and, consequently, of their geomorphological attributes (tidal niches, caverns, dunes, rods and others), as well as the speed of recent tectonic movements.

It is probable that in the early Holocene the most depressed parts of the southeast shelf formed freshwater lagoons, with accumulation of terrigenous sediments, as evidenced by the presence of deposits of this age reported in the Gulf of Guacanayabo [4] and the foraminiferal complex present. In this locality, the sedimentation regime perhaps passed without interruption until the present, favoured by the depth of its bottom open to the sea.

Probably, less than a thousand years ago the average sea level rose and stabilized in the current general coastline. The corals that had retreated returned to be reinstalled, forming reef constructions on the shelfs and in the sectors without a shelf. This coastal stabilization is a relative and temporary phenomenon, given that: (a) the Holocene transgression at the planetary level has not concluded; (b) it has apparently started to accelerate due to Climate Change [9]; and (c) the general amplitude of recent tectonic movements is lower than the rise of the mean sea level [9], so the glacio-eustatic factor will prevail over the isostatic factor. In the current scenario, the geological development of shelfs and sectors without shelfs under the influence of glacio-eustatic and isostatic factors (primary or founder), with a series of associated (secondary) lithomorfogenetic processes, such as biogenic, hydrogenic, chemogenic, eolic, fluvial and anthropogenic.

2.4.5 Mineral Resources

This work deals with the existence of non-metallic and metallic mineral deposits (Fig. 2.22), some of which have evaluated resources. It also addresses its potential and economic interest estimated from the premises derived from certain processes and lithomorfogenetic factors.

Figure 2.22 Map of mineral deposits and manifestations of the current marine-coastal territory of Cuba.

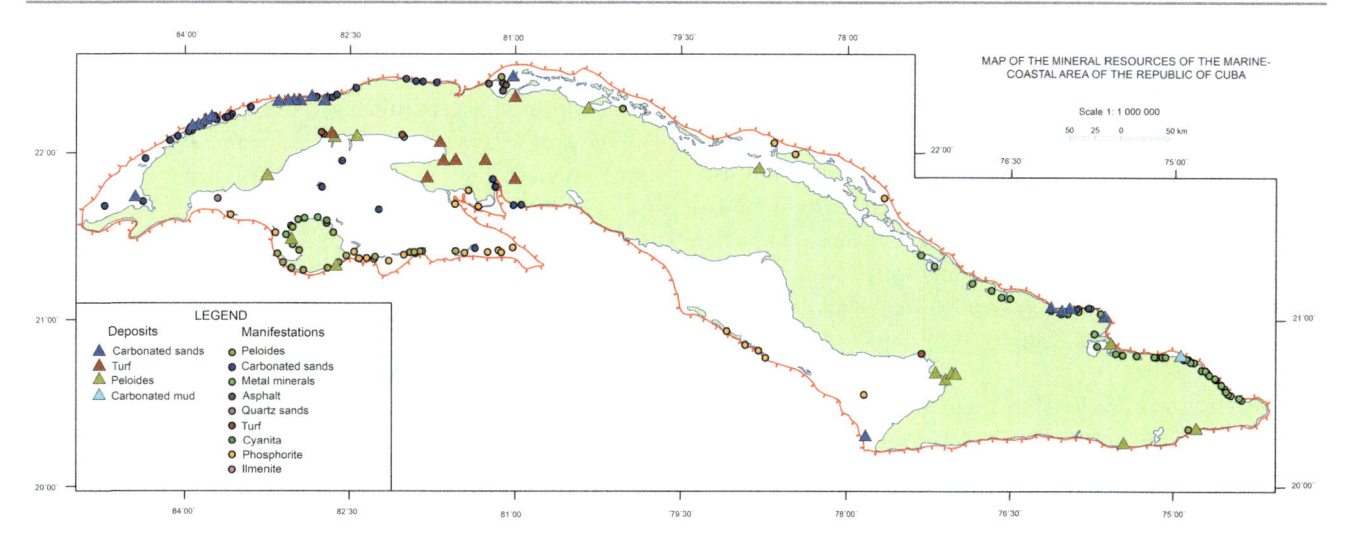

Fig. 2.22 Map of mineral deposits and manifestations of the current marine-coastal territory of Cuba

Non-metallic minerals

The non-metallic mineral deposits are constituted by deposits and manifestations of: sands, peat, peloids, carbonated mud, phosphorite, asphalt and materials for the construction.

Sands. The prospection work has revealed the existence of bio clastic and oolitic carbonated sands [3]. The resources calculated and approved, with quality for fine aggregates and for the rehabilitation and construction of beaches amount to 24×10^6 m^3. They can be increased as suggested by the following premises: (a) presence of negative forms of relief with accumulations of sands not studied and (b) existence of sources of biogenic and chemogenic input.

Turf. The peat deposits of Cuba are associated with the biogenic deposits of the coastal zone of the shelves and have been investigated by numerous projects, which were generalized by González [62]. The resources evaluated amount to $4,926,181 \times 10^6$ m^3 and there are still great potentialities of mobs, according to the extent of unexplored coastal wetlands. However, they do not lend themselves as an energy source, due to their high ash content, high humidity and low degree of decomposition. Until now they are used in a limited way for the reconstruction of soil and garden substrate. On the other hand, its use is highly expensive, related to the impossibility of draining by gravity the areas to exploit from the relative height, causing that its mining is very expensive. To this is added the negative and irreversible impact on biodiversity that would lead to its extraction.

Peloids. The peloids or mineral-medicinal sludge have been studied in numerous localities constituted by marshy deposits, active and inactive salines [3]. The resources evaluated amount to $2,579,157 \times 10^3$ t. The prospects of locating new deposits are favourable in salines, river mouths, estuaries, lagoons and in general, in all the low coasts of the shelfs. Its demand is also favourable, since its use for therapeutic purposes dates back to the times of the Spanish colony. The most studied and rescued uses are the affections of the skin, osteomyoarticular, orthopaedic and in cosmetological treatments, which make it feasible to extend its use.

Carbonated mud. They are mud originated from silty-clayey biodetrites, with high content of CaCO$_3$. They have been studied at the Cayo Moa Grande deposit [28, 29, 63], as a raw material for the nickel sulphide preparation process. Its resources amount to $30,899 \times 10^3$ t. There are favourable premises to locate other deposits of this raw material in the different shelfs, given by the presence of a carbonated sedimentation environment. The Batabanó depression can be particularly interesting, where almost pure carbonated oolites accumulate.

Phosphorite. It was studied in the archipelagos of Los Canarreos, Sabana-Camagüey and Jardines de la Reina; as well as on the south coast of the Zapata peninsula [64, 65]. The results assure that the phosphatization present in deposits of some keys is secondary, through a process of metasomatism, product of the leaching of phosphorus present in the remains of crustaceans, molluscs and guano of birds, which enrich the limestones of the Jaimanitas Formation and Holocene sediments deposited in their karstic depressions. The prospects of locating industrial deposits of this raw material in the territory are null, due to the small size of the areas where they are located and the irregularity of their phosphatization.

Asphalt. Since the nineteenth century the existence of asphalt has been known in the west of the north central shelf, through numerous reports, all inventoried by Cabrera and Batista [3]. It occurs in the form of springs through faults and neotectonic fractures, which have affected deep hydrocarbon deposits, the best-known being those located in Cardenas Bay, where the mines Nueva, Favorita and Casualidad were mined. The latter contributed some 10,800 t between the years 1880 and 1913, being the most productive of the period. The presence of demonstrations in other locations on the northwestern shelf could make the territory favourable to locate new sites.

Construction materials. They come from the rocky part of some keys of the north central shelf, where there are more than 30 quarries (Cabrera et al. 1997). Storm ridges have also been exploited as aggregates for construction in different locations. They are abundant, but their exploitation degrades the environment irreversibly. Another source of construction materials could be the linear deltas (tibaracones), as its regenerative nature allows the part of its deposits that go to the ocean depths during the great floods of the rivers to be extracted rationally.

Metallic minerals. The hypothesis about the existence of marine pleasures of useful metallic minerals, which served as the basis for conducting geological surveys and prospecting of minerals in the marine territory of Cuba, was based on the following indications and premises: (1) location of the territory in a tropical zone and humid, considered the most favourable for the formation of mineral pleasures, for constituting a product of the chemical weathering crusts; (2) probable prolongation to the marine territory in the emergentof rocks that emerge in the emerged territory and that could have been a direct source of contribution of terrigenous material by abrasion at some point of the Quaternary geological development; (3) presence in the emerged territory of auriferous pleasures, accessory minerals and primary deposits; and 4) presence of mineral grains on beaches and the seabed. Under these premises, 15 mineralogical provinces and areas of high prospectivity were identified to locate mineral deposits [4].

So far, the initial thesis has not been proven. Only small concentrations have been found, which are not exploitable pleasures, due to the existence of processes and factors not favourable for their formation, such as: (a) low rate of contribution of terrigenous sediments with useful mineral content towards the sea; (b) low level of fluvial and marine reelaboration of the contribution of terrigenous sediments, for which coexist chemically stable minerals with unstable minerals; and (c) predominance of very fine granulometry (microscopic) of the minerals useful in the deposits, which constitute sources of contribution located in the emergent territories of the country, particularly the auriferous ones.

2.5 Conclusions

The analysis of the existing geological information of the current marine territory of Cuba corresponding to the Quaternary period allows us to reach the following conclusions:

1. Its main geomorphological features are determined by geomorphological regions, constituted by the lithomorphogenetic zones outside and inside of the shelfs and by the sectors without shelf.
2. The geomorphological regions include numerous shapes and complexes of polygenetic relief, such as terraces, reef formations, banks, keys, channels, depressions, bars, dunes, beaches, storm ridges, tidal niches, deltas, coasts of different types and others.
3. The geomorphological regions determine the different sedimentation environments: (a) the external lithomorfogenetic zone corresponds to a biogenic-carbonate environment, sandy sediments, modern reef formations and reef limestones; (b) in the inner lithomorfogenetic zone the following environments can be found: biogenic-carbonate-terrigenous, chemogenic-carbonated and biogenic, of sandy sediments and clay-silt, which can be interspersed with each other and with peat, the paleo bottom is of reef limestones; and (c) in sectors without a shelf, the environments are biogenic-carbonated and alluvial.
4. From a geochronological point of view and composition, the marine territory has deposits represented by lithostratigraphic carbonate units from the Upper Pliocene to the Holocene and innominate, friable, carbonated, terrigenous-carbonated, biogenic, marine and marsh Holocene deposits. Between the Late Lower Pleistocene and Late Middle Pleistocene, there is a hiatus. As in the rest of Cuba, there is a lack of sufficient dating of the Quaternary deposits and, consequently, there are uncertainties in several cases about its strict census age and consequently the relief and the extent of recent tectonic movements.
5. The characteristics of the geological constitution of the marine territory, determined by its composition, distribution, structure and morphology, is a consequence of the singularity of the factors and processes that gave rise to it. They are a historical combination of exogenous and endogenous processes, because of changes in the atmosphere, hydrosphere and lithosphere.
6. It has confirmed the presence of paleoclimatic (transgressions and regressions) and neotectonic (rise of the Cuban mega block and its different morph structures) factors, as the determinants in the definition of the scenarios (founders), which have marked intervals or stages in evolution of the development of the geological

constitution of the marine-coastal territory of Cuba in the Quaternary. For the different scenarios, other secondary exogenous processes and lithomorfogenetic factors were identified, with the following approximate order of relevance: (a) biogenic, (b) hydrogenic, (c) chemogenic, (d) fluvial, (e) eolic and (f) anthropogenic.

7. Regardless of the order of importance of processes and lithomorfogenetic factors, the most significant when evaluating its role lithomorgenesis in each case of study is to establish the interrelation that exists between them, since they tend to converge several in the same time and space, in different scenarios.

8. The traces left by the lithomorfogenetic processes allow reconstructing the following main stages of the geological evolution of the marine-coastal territory of Cuba in the Quaternary: (a) Upper Pliocene-early Lower Pleistocene; (b) Lower Pleistocene—late Middle Pleistocene; (c) early Upper Pleistocene; (d) late Upper Pleistocene; (d) Holocene.

9. The deposits of the shelves have a great potentiality of carbonated sands, peloids, carbonated muds and asphalt, but with limitations for their use due to the fragility of the ecosystems where they are found, so the possibility of their extraction requires studies and environmental assessment, as well as economic feasibility.

10. There have not been able to establish perspectives to locate deposits of metallic minerals with economic interest, because the indispensable condition for the formation of marine pleasures is not fulfilled.

Acknowledgements The authors of this chapter are very grateful to all Cuban and foreign researchers, who have contributed to the geological knowledge of the marine territory of Cuba. We are also grateful for the invitation of Dr M. Pardo Echarte to publish in the prestigious editorial Springer.

References

1. Cabrera M (2006) Geología del territorio marino de Cuba. CDRUM, ISBN: 978-959-7117-15-5, CNDIG, IGP, La Habana
2. Cabrera M, Oro J, Reyes R, Álvarez JL, Franco GF, Rodríguez R y Peñalver LL (1990) Sistematización y generalización de la geología de la plataforma marina de Cuba con relación a la prospección de minerales sólidos, (Inédito) Arch IGP La Habana
3. Cabrera M y Batista R (2009) Naturaleza geológica del territorio marino-costero de Cuba en el Cuaternario. CNDIG, IGP, CDRUM, ISBN: 978-959-7117-17-9, La Habana
4. Ionin AS, Pavlidis YA y Avello O (1977) Geología de la plataforma marina insular de Cuba, AC. URSS, Ed. Naúka, Moscú, 277 p
5. Linares E, Osadchiy PG, Dovbnia VA, Gil S, d. García, García I, Suazo S, González R, Bello V, Brito A, Busch WA, Cabrera M, Capote C, Cobiella JL, Díaz de Villalvilla L, Eguipko OI, Evdoquimov YB, Fonseca E, Hernandez J, Furrazola G, Judoley CM, Kondakov IA, Markovskiy BA, Norman A, Pérez M, Peñalver LL, Tijomirov IN, Trofimov VA, Vtulochkin AI, Vergara F, Zagoskin AM y Zelepuguin VN (1985) Mapa Geológico de la República de Cuba. Esc. 1: 500 000. Cent. Invest. Geolog (CIG). Inst Cient Geol "A. P. Karpinski", Leningrado, URSS
6. Cabrera M y Peñalver LL (2001) Contribución a la estratigrafía de los depósitos cuaternarios de Cuba. Rev. C. & G., 15 (3–4): 37-49, © SEG. AEQUA. GEOFORMA Ediciones
7. Cabrera M, Ugalde C y Pantaleón G (2004) Mapa geológico de los mares neríticos del archipiélago cubano a escala 1: 100 000, (Inédito). Arch IGP La Habana
8. Cabrera M (2011) Los depósitos cuaternarios del territorio marino de Cuba, Minería y Geología. 27(3):1–25. ISSN 1993 8012
9. Cabrera M, Orbera L, Núñez A, Pantaleón G, Núñez K, Triff J, Pérez CM, Santos MR, Chávez ME y González D (2012) Neotectónica y Ascenso del Nivel Medio del Mar. Edit Cent Nac Inf Geol y Publisimex. CDRUM, ISBN 978-959-711732
10. Cabrera M, Denis R, Peñalver LL, Triff J, Núñez A, Batista R, Pantaleón G, Martín D; Díaz S, Rodríguez L; Zúñiga A y García D (2013) Caracterización Geólogo-Geomorfológica. Proyecto GEF-PNUD "Aplicación de un Enfoque Regional al Manejo de las Áreas Protegidas Marinas y Costeras en los Archipiélagos del Sur de Cuba (Archipiélagos del Sur) ", (Inédito). Arch IGP y Cent Nac Áreas Prot (CNAP), La Habana
11. Denis R, Núñez A, Triff J, Peñalver LL, Rodríguez H, Domínguez E, Valdés R, Martín I, Rivada Retejas M, Núñez K, Iglesias E, González Y, Armas A, Nápoles I, Martínez D y Lorenzo S (2015) Geología y Geomorfología Marino-Costera del Archipiélago Cubano y su Vinculación con los Movimientos Tectónicos Recientes, (Inédito). Arch. IGP y Agen. Medio Amb. (AMA)
12. Denis R, Núñez A, Triff J, Peñalver LL, Rodríguez RR, Domínguez E, Valdés R, Martín D, R. Rivada R, Iglesias E y Núñez K (2016) Geología y Geomorfología Marino-Costera del Archipiélago Cubano y su Vinculación con los Movimientos Tectónicos Recientes, (Inédito). Arch. IGP y Agen. Medio Amb. (AMA)
13. Cabrera M (1997) Geología de la plataforma marina insular. Estudios sobre Geología de Cuba. CNDIG, La Habana, pp 179–197
14. Ionin AS y Pavlidis YA (1972a) Relieve de la zona costera y sedimentos del fondo de la costa norte de Cuba en la región del archipiélago Sabana-Camagüey, ser. Oceanol., 12:25
15. Ionin AS y Pavlidis YA (1972b) Depósitos del golfo de Batabanó. En: Procesos de desarrollo y métodos de investigación en la zona somera del mar, Ed. Naúka, Moscú, 62–92 pp
16. Cabrera M (1999) Terrazas marinas sumergidas de Cuba. Bol Sociedad Geológica de Cuba (SCG). 4(3):7–9, La Habana
17. Ionin AS, Medvedev BC y Pavlidis YA (1975) Terrazas submarinas del de la plataforma marina de Cuba, Revista Estructura del Shelf zonas tropicales del océano, Moscú. 53-125 pp
18. Zlatarsky V y Martínez Estalella north (1982) Les Scléractiniaires de Cuba, Ed. de l´Academiebulgare des Sciences, Sofía, 471 p
19. Avello O y Pavlidis A (1986a) Sedimentos de la plataforma cubana. II. Golfos Ana María y Guacanayabo, Rep Invest Inst Geol Paleont (IGP), ACC 7:27
20. Núñez Jiménez A (1982) El Archipiélago, t.1, Ed. Letras Cubanas, La Habana, 669 p
21. Avello O y Pavlidis YA (1986b) Sedimentos de la plataforma cubana. III. Golfo de Batabanó, IGP, ACC Rep Invest 6:42
22. Estrada V (1992) Informe del levantamiento geólogo-geofísico de la ensenada la Broa, (Inédito). Arch, IGP, La Habana
23. Pavlidis YA, Ionin SA, Ignatov EI, Lluís M y Avello O (1973) Condiciones de formación de la oolita en las regiones someras de los mares tropicales, Rev Serie Oceanol, 18:18

24. Núñez Jiménez A (1990) Medio Siglo Explorando a Cuba. Imprenta Central de las FAR, La Habana, t. II, p 458–485
25. Iturralde Vinent M (2011) Geología del Cuaternario de Cuba. En: Iturralde-Vinent, M (2011) ISBN 959-7117-II-8: Compendio de Geología de Cuba y del Caribe. DVD ROM, ISBN 978-9592-372-863
26. Hernández Zanuy AC, Alcolado PM y Caballero H (2007) Evaluación de las posibles afectaciones del Cambio Climático a la Biodiversidad Marina y Costera de Cuba. Integración de los resultados, (Inédito). Arch Inst Oceanología, La Habana
27. Borrego R (2011) Propuesta de zonificación para los ecosistemas marinos costeros del área protegida "Punta Francés". Tesis presentada en opción al grado académico de Master en Manejo Integrado de Zonas Costeras
28. Nápoles E, Álvarez M, Ortega F, Reyes R, Corrada R, Ayra CM y Ramos V (1992) Exploración Orientativa de Cienos Carbonatados al sur de Cayo Moa Grande. Inventario, (Inédito). Arch. ONRM, La Habana
29. Nedved B (1966) Búsqueda de Fango Coralino en el yacimiento Cayo Grande de Moa, (Inédito). Arch, ONRM, La Habana
30. Estrada V, Corrada R, Ramos V, Oviedo A, Ayra C y Sánchez M (1989) Informe sobre la prospección de arenas marinas para la construcción en el tramo costero Santa Fe-bahía de Santa Lucía (plataforma noroccidental). Escala 1: 50 000, (Inédito). Arch IGP, La Habana
31. Suyí C, Ramos V, Hernández A y González R (1981) Informe sobre los resultados de los trabajos de reconocimiento pronóstico-evaluativo para minerales sólidos en el shelf noroccidental, (Inédito). Arch. ONRM, La Habana
32. Ionin AS y Avello O (1975) Sedimentos de la plataforma cubana. I. Golfo de Guanahacabibes, Rev Oceanol. 30:23
33. Peñalver LL (1982) Correlación estratigráfica entre los depósitos cuaternarios de la plataforma noroccidental de Pinar del Río y las zonas emergidas próximas. Rev Ciencias de la Tierra y del Espacio 5:63–84
34. Gil A (1988) Características de los sedimentos en dos bahías al sur de los cayos Coco y Romano, Inst. Oceanol. ACC Rep Invest 18:36
35. Iturralde Vinent M y Cabrera M (1998) Estratigrafía de los cayos del archipiélago Sabana-Camagüey entre Ciego de Ávila y Las Tunas, t. 1, Memorias III Cong. Cub. Geología y Minería, La Habana, 319–322 pp
36. Iturralde Vinent M (1981) Depósitos Cuaternarios, En: Belmustakov E, Dimitrova E, Ganev M, (1981) Texto explicativo al Mapa Geológico a escala 1: 250 000 de las provincias de Camagüey y Ciego de Ávila, (Inédito, Arch. IGP, La Habana
37. Hernández CE, Ramos V, Corrada R y Álvarez JL (1985) Informe sobre los trabajos regionales de apoyo a la geología de la plataforma marina suroccidental de la República de Cuba para minerales sólidos, (Inédito). Arch ONRM, La Habana
38. Hernández CE, Ramos V, Corrada R, Álvarez JL, Sánchez M y Rodríguez R (1988) Informe sobre los trabajos de levantamiento geológico y búsqueda de minerales sólidos de la Isla de la Juventud, (Inédito). Arch ONRM
39. Avello O, Ionin AS, Pavlidis YA, Dunaev NN (1975) Particularidades de la constitución de la plataforma marina según los datos del perfilaje sismoacústico, ed. Inst. Ocean AC. URSS., Moscú. Soc America 23:75.
40. Shantzer EV, Petrov OM y Franco GL (1976) Sobre las terrazas marinas costeras de Cuba y los sedimentos vinculados con ellas, En: Kartachov IP. ed., Sedimentación y formación del relieve de Cuba en el Cuaternario, Ed. Naúka, Moscú, 34–80 pp
41. Ducloz C (1963) Etude geomorphologique de la region de Matanzas, Cuba (avec une contribution á l'étude des dépots quaternairés de la zona Habana-Matanzas).Sep de Archives des Sciences, Géneve, vol. 15, 2 tex. figs.1-20, carte
42. Álvarez M, Quintana MC (1988) Caracterización sedimentológica preliminar de la bahía de Cárdenas. Inst Oceanol ACC, Rep Invest 3:16
43. Álvarez M (1992) Algunas consideraciones acerca de la composición mineralógicas en una franja costera del golfo de Ana María, Cuba, Invest Oceanol Acad Cien Cuba (ACC) Rep Invest 14:14
44. Estrada V, Hernández CE, Álvarez JL y Corrada R (1985) Sobre los trabajos regionales de apoyo de geología marina suroccidental de la República de Cuba para minerales sólidos, (Inédito). Arch Ofic Nac Rec Miner (ONRM), La Habana
45. Rodríguez R, Cabrera M, Hernández A y Guerra R (1984) Distribución mineralógica en el shelf occidental de la Isla de la Juventud, ser Geol CIG, 1: 102-121
46. Guilcher (1957) Morfología litoral y submarina, Ed. Omega S. A, España, 261 p
47. Emiliani C (1970) Pleistocene Paleotemperatures. Science 168 (3933)
48. Kartachov IP, Cherniajovski AG y Peñalver LL (1981) El Cuaternario de Cuba, vol. 356, Ed. Naúka, Moscú, 145 pp
49. Ortega F y Zhuravliova I (1983) Critica a la hipótesis de los "dos" pleistocenos, a la luz de la información edafológica. Cien Tierra Espacio 6:63–85
50. Dzulynski S, Pszczółkowski A y Rudnicki J (1984) Observaciones sobre la génesis de algunos sedimentos terrígenos cuaternarios del occidente de Cuba Rev Ciencias de la Tierra y del Espacio, 9:75–90
51. Bielousov TP, Orbera L y Nikolaev L (1978) Mapa de fallas y Superficies de Cuba Central, escala 1: 250 000, (Inédito). Arch IGP La Habana
52. Ortega F (1994) La sedimentación en la plataforma insular cubana en relación con los cambios climáticos pleistocénicos. Rev Cien Tierra Espacio 23(24):21–31
53. Franco GL (1975) Las eoleanitas del Occidente de Cuba. ACC. Serie Geolog. 7:1–12
54. Uriarte A (2003) Historia del Clima de la Tierra, Servicio Central de Publicaciones del Gobierno Vasco, 306 pp. ISBN 84-457-2079-1
55. Léxico Estratigráfico de Cuba (2012) Inédito. IGP, La Habana
56. Wyrwoll KH, Dong B, Valdés P (2000) On the position of southern hemisphere westerlies at the Last Glacial Maximum: an outline of AGCM simulation results and evaluation of their implications. Quatern Sci Rev 19:881–898
57. Cutler KB, Edwards RL, Taylor FW, Cheng F, Adkins J, Gallup CD, Cutler PM, Burr GS, Bloom AL (2003) Rapid sea-level fall and deep-ocean temperature change since the last interglacial period. Earth Planet Sci Lett 206:253–271
58. Mitrovica JX (2003) Recent controversies in predicting post-glacial sea-level change. Quatern Sci Rev 22:127–133
59. Wehmillera JF, Simmons KR, Edwards HRL, Martin McNaughton J, Yorka II, Krantza ED, Chuan Chou S (2004) Uranium-series coral ages from the US atlantic coastal plain–the "80 ka problem" revisited. Quatern Int 120:3–14
60. Dunaev NN (1977) Análisis Cuantitativo de los movimientos tectónicos verticales en la plataforma marina insular de Cuba. Rev. Oceanología. T. XVII, No 6, Moscú, 1050–1054 pp

61. Kevin F, Johnston P, Zwartz D, Yokoyama Y, Lambeck K and Chappell J (1998) Refining the eustatic sea-level curve since the Last Glacial Maximum using far- and intermediate-field sites. Earth and Planetary Science Letters163 (1-4): 327-342
62. González V (1997) TTP. Evaluación y actualización del potencial de Turba en Cuba y generalización de la información sobre sus usos y procesamiento tecnológico, (Inédito) Arch ONRM, La Habana
63. García AE, Palaú R, Díaz F, Fernández L, Pérez Almaguer E, Azcuy Rodríguez L, Olivera J, Collante A, DeHuelves J, Ayra CM y Campos M (2001) Informe Geológico de Exploración Detallada y de Explotación del Yacimiento de Cienos Carbonatados al sur de Cayo Moa Grande, (Inédito). Arch. ONRM, La Habana
64. Pokrishkiin V (1966) Trabajos de búsqueda y reconocimiento de las fosforitas en Cuba, 1964–65, (Inédito). Arch. ONRM, La Habana
65. Pokrishkiin V y Shirokov VN (1964) Informe preliminar sobre la búsqueda de fosforitas en Los cayos del Archipiélago de Los Canarreos y la costa sur de la Península de Zapata, (Inéd. Arch. ONRM, La Habana

Synthesis of Fossil Record of Cuba—A Bibliographic Compilation

Reinaldo Rojas Consuegra

Abstract

In general, the "Cuban Fossil Record", which covers approximately the last 200 million years of life on Earth, is rich in very varied fossils, witnessing a wide diversity of organisms, both animals and plants, that inhabited the Antillean and Caribbean region and constitute the inheritance of the biological diversity exhibited by the current Cuban Archipelago. The most characteristic fossils of Jurassic Period in Cuba are petrified bones of marine reptiles, shells and molds of ammonites, petrified skeletons and molds of ganoid fish, bivalve mollusc shells, fronds and carbonized plant stems, mainly ferns, and very abundant pollen and spores. The fossil record of the Cretaceous Period is characterized by shell varieties of rudist molluscs, ammonites, and *aptychus*, endoskeletons and radioles of echinoids, gastropod shells such as the acteonellids, nerineids and naticids, ostreids, and others. Other fossils such as corals, ichnofossils, the very diverse planktic and benthic foraminifera, and radiolaria are also common. The fossils that characterize the Paleogene are abundant echinoderms, shark teeth, ichnofossils, shells and molds of turritellid and naticid gastropods, ostreids, and various foraminiferal genera, especially large orbitoids. Neogene rocks contain abundant shells and molds of bivalve and gastropod molluscs, and mineralized endoskeletons of various sea urchins are also common. Corals and frequent skeletons and molds of marine crustaceans can also be commonly found. Among vertebrates, fish are very common, mainly teeth of sharks, rays, and skeletons of bony fish, and a single whale tooth have also been found. The fossil remains of sirenians are relatively common. Very important is the finding of terrestrial mammal vertebrate remains, among them are monkey, rodents, and sloths. The greatest feature exhibited by the fossil record of the Quaternary in the Cuban Archipelago is perhaps the peculiar fossil material produced by the diverse megafauna of terrestrial vertebrates, which inhabited it in the last hundreds of thousands and thousands of years ago. It also highlights the bones and teeth of large sloths, various rodents, the giant predatory birds and gunboats, small and giant insectivorous, numerous bats, reptiles, and amphibians, among other animals that disappeared in the recent past. Fossils are part of the Cuban natural heritage, and as such, they deserve to be studied, conserved and protected, as a legacy to future generations, to contribute to a better understanding of our origins and to the full enjoyment of our island nature. The numerous literature about Cuban fossils allows us to know the varied degree of study exhibited by the different fossil groups reported to this day. The irregular development of the investigations carried out so far reveals the possibilities of study that the Cuban fossil record still needs, and which points out the future path for new researchers interested in different topics on this paleontological richness, where there are many questions to be solved, even waiting to assess correctly, from modern work basis.

Keywords

Cuba • Fossil record • Jurassic • Cretaceous • Paleogene • Neogene • Quaternary

R. Rojas Consuegra (✉)
Centro de Investigación del Petróleo (Ceinpet),
Churruca, No. 481, e/Vía Blanca y Washington,
El Cerro, C.P. 12000 La Habana, Cuba
e-mail: rojas@ceinpet.cupet.cu

Abbreviations

CFR	Cuban Fossil Record
IGP	Instituto de Geología y Paleontología (Cuba)
MNHNC	Museo Nacional de Historia Natural de Cuba
My	Millions of years
SCG	Sociedad Cubana de Geología

3.1 Introduction

This chapter is dedicated to fossils from Cuba, essentially to mega- or macrofossils, although some groups of microfossils are mentioned. It aims to offer the general information known to date about the "Cuban Fossil Record" (CFR).

The fossils (conserved elements or entities) are essential pieces for reconstructing the past of our planet, in recognizing the diverse organisms that inhabited it, the varied communities, ecosystems and environments, which have existed on it until the present.

The oldest known macrofossils of the Cuban territory are from the Lower to Middle Jurassic Period (ferns and molluscs), which are less than 200 million years old. Comparing this data with the age of the Earth, it is stated, geologically speaking, that Cuba is very young! The CFR represents approximately 4.6% of the time of existence of our planet.

A significant utility of this synthesis is revealing the degree of study that has been achieved on the different fossils groups that it contains, which can serve as a guide for those interested in the investigations related to Cuban fossils, trying to move forward the knowledge already accumulated about them.

3.2 Theoretical Framework

The fossil concept that is used is summarized as follows: "Fossils are the remains, traces or signs of the organisms that lived in the geological past and have been preserved in the rocks".

Thus, a fossil can be the skeleton or parts of it, belonging to any organism, such as bones, shells, corals, algae, or spicules of sponges. Also, they may be the dental pieces of various marine or terrestrial animals, or other organic parts, resistant to mechanical degradation or that have been in relative equilibrium with the taphonomic processes traversed. The charred plant remains, the trunks of wood replaced by silica, or the impressions of leaves and branches in films of iron hydroxides are common fossil elements left by the plants.

In addition, molds, replicas, or impressions left by organisms on rocks are considered fossils, along with the traces, perforations, prints, burrows, marks of bites, coprolites, etc., produced by the vital activity of different organisms. This group is studied within the ichnofossils. In this case, the presence of organisms in the past is deduced through the substances generated from them and that today form some types of rocks or minerals; these are the chemofossils (Fernández López 1989; López Martínez y Truyols Santonja 1994).

From the latter, it is inferred that the fossils can be material (bone remains, molds, teeth, etc.) or not (traces, tracks, traces, etc.), where the direct passage of substance from the biosphere was not necessarily the lithosphere, but where it is always possible to recover information about life in the Earth's past from the rocks (Fernández López 1989, 2000).

As it is known, fossils are useful to know the age of the rocky sequences where they are contained, although in some cases they may have been brought (by the action of erosive and transport processes) from layers of other ages, and they would provide distorted information, but at the same time of significant utility. To know the processes that have acted during the formation of the fossils, their taphonomic characteristics, and even their absence in the rocks, among other objectives, an important paleontological subdiscipline is occupied, the Taphonomy (Efremov 1940). Taphonomy has reached a deep conceptual and methodological development, and reinforces the application of modern Paleontology (López Martínez y Truyols Santonja 1994; Fernández López 1989, 2000).

From it, the term *taxoregister* is taken in the present work, used to count those evidence on any fossil (at any taxonomic level: class, order, family, genus, or species), which has been recognized in the rocks of the Cuban geological substrate.

Paleogeography and paleobiogeography are essential to understand the diversity and meaning of fossils. When talking about Cuba or Cuban fossils, "Cuba" should be understood as a current geographic unit of reference, but originated through a complex evolution of independent geological elements, now spatially related. That is, for millions of years, other geographies have taken place in the region, where rock series were accumulated in different spaces and today are located hundreds or thousands of kilometers away from those original positions, forming part of the Cuban territory, known to sometimes only of this complex geological substrate (Iturralde Vinent 2012).

3.2.1 General Characteristics

In general, the accumulation of sedimentary deposits has occurred over thousands, millions, or tens of millions of years, with the conformation of Cuba's stratigraphic record. The accumulation, transport, and re-accumulation of the fossils, processes parallel to the previous ones, gave rise to the fossil record (paleontological). This is integrated then, by the set of fossil entities or elements that have been conserved in these rock series.

The CFR collects the fossil groups, recorded or reported, that have been found in the rocks that today constitute the geological substrate of the Cuban Archipelago (Rojas Consuegra 2014).

The geological record of Cuba, including all existing rock complexes (magmatic, metamorphic, and sedimentary) in its current substrate, which originated at different times, in spaces of thousands of kilometers in width (millions of square kilometers), while today is stacked-imbricated in a space of the Earth's crust (transitional type), only about 200 km wide by something over a thousand in length. This material and spatial reduction have also led to the inexorable loss of significant information about its natural history, including its geological evolution. This is about reconstructing through the scientific data that is being systematically recovered, thanks to the work of the different disciplines of geosciences and those related.

The general characterization of the CFR can partially shed a light on the diversity of organisms that inhabited the past of a large region, with which the origin and geological evolution of the current Cuban territory are related. It is thus that the CFR reflects information elements on the paleogeography, the paleobiogeography, and the paleobiotas that have existed along the last 200 million years approximately, in the dynamic geographic space that would have covered: the confluence of Tethys sea (Early Jurassic–Early Cretaceous) with the Pacific Ocean (Early Cretaceous), the Atlantic Ocean (Upper Cretaceous), and all of them with the proto-Caribbean (Cretaceous–Paleogene) and up to the present-day Caribbean (Paleogene–Quaternary) (see Paleogeography of Cuba in Iturralde Vinent 2012).

3.2.2 Conserved Fossil Entities

By its nature, each group of organisms produces biogenic materials in the course of their life activity. They would be related to the intrinsic characteristics of their own skeletons, ontogeny (modifications of growth), ethology (life behavior), their death, and subsequent evolution, until their relative stabilization as fossil material; even in constant interaction with the environmental conditions of the present.

It is expected that fossil materials, even if they originate from the same group of organisms, may have different structures and compositions; this is how different types of fossils could also be produced. In this sense, vegetable fossils are represented by wood in the form of trunks or branches, leaves appear as molds or impressions oxidized, coals, spores, and pollen, pellets, etc. In addition, in other countries, there is amber (fossil resine), flowers, fruits, and seeds.

Included in the invertebrates are a wide diversity of fossil elements, complete or partial, such as exoskeletons or endoskeletons, shells, tests, valves, opercula, fixed or loose tubes, plates, radioles, spicules, spines, discs, radules, rods, elytra, etc. Their molds and their replicas can also be preserved; these variants are also valid for bone skeletal elements under certain fossilization conditions, that is, in specific taphonomic environments.

The group of vertebrates contains a high abundance of fossil elements, such as the dissimilar bones (and other organs) that make up their complex articulated skeletons, such as diversity of bones in general, dental plates or individual and morphologically multiple dental pieces, plates, scales, coals, and many other elements, sometimes of microcoscopic dimensions.

The ichnofossils complete the large groups of preserved materials (with recoverable palaeontological information). They include all kinds of traces, prints, perforations, excavations, burrows, galleries and tunnels, nests, room pipes, bites, scrapes, and varied marks, as well as their molds or fillings (replicas); biogenic depositions such as coprolites (fecal pellets) and microcoprolites (*Favreina*) are also included.

3.2.3 Sizes of the Fossil Entities

A striking feature of fossils can be their size. In the CFR, the largest conserved elements or entities, among the invertebrates, are the shells of rudists, such as hippuritids and antillocaprinids (60–150 cm); followed by the Miocene giant ostreid (*Crassostrea vaughani insularis*) or the

Cretaceous leaflets (*Arctostrea aguileae*); rolled shells of ammonites (40–70 cm); shells and molds of gastropods (*Strombus gigas*) (20–30 cm); the exoskeletons of the very diverse echinoderms, such as clypeasteroids (10–25 cm). Fossils smaller than 10 cm, belonging to different groups, are very common.

In the registry of fossil vertebrates, stands out the carapace (shell) of Miocene turtles (up to approximately one meter). It is followed by impressions and skeletons of Jurassic fishes (30–80 cm); the skeletal remains of Jurassic marine reptiles (30–60 cm); long bones of Quaternary sloths and jaws of crocodiles (20–30 cm). The most abundant are the small bone elements of different types of mammals (capromids, insectivores); in addition, there are thousands of bone remains and very small dental pieces of the most varied groups (birds, chiroptera, shrews, reptiles, amphibians, fishes, etc.).

The ichnofossils, represented by traces, prints, or burrows, have been recovered from a few centimeters in length to the first meters, with the longest ones appearing as molds for tunnels and galleries systems in the paleogenic turbiditic rocks.

Among the most significant plant remains, due to their size, stand out the silicified Quaternary trunks (30–90 cm or greater); wood fragments (20–40 cm) and fronds of Jurassic ferns (15–30 cm); other gravel and coals (less than 5–10 cm) can be relatively abundant.

Finally, there are the incalculable millions of microfossils (less than one millimeter), extremely diverse, belonging to the most varied fossil groups.

3.2.4 Coloration of Materials

The general composition of fossils is relatively homogeneous, mainly calcareous, siliceous, and rarely others. However, the action of the dissimilar processes of fossilization through which the original biogenic materials become conserved entities, leads to compositional or substantial variations that can also be diverse. Even when certain compositions and general colorations predominate, local variations can not be ruled out, and very heterogeneous conserved elements can appear.

Cuban Jurassic fossils (including those that are contained in the rocks of that age that make up the geological substrate of present-day Cuba), tend to be carbonated, with the influence of carbonization (reducing conditions). Entities of this age have dark colorations, from black to dark gray in the fresh state (without strong action of weathering—hypergenesis). On the other hand, the vegetable remains embedded in alluvial-marine sediments can be black (manganese or iron oxide), dark gray-greenish, reddish-orange to cream-yellowish to white-grayish; in different weathering states.

The predominant Cretaceous fossils are calcareous or carbonate, with clear colorations of creams to yellowish, sometimes aporcelanated. In sequences related to paleovolcanic environments take darker colorations, from gray to black. The dark to light red and orange colors stand out in the marble limestones rich in ferrous oxides, widely used as marble in wall coverings or veneer throughout the country.

Among the Paleogene fossils of invertebrates predominate the calcareous ones, very recrystallized, sometimes marmorized, with clear, pink, cream, and yellowish colorations, also can get from grayish to white. It is also common to observe them on the walls and floors of different buildings in the country. Locally recovered remains of invertebrate gastropods and teeth of cartilaginous fish impregnated in manganese oxide are partially dark gray to black colors.

Neogene fossil entities and elements are predominantly calcareous, of light colors, mainly yellowish to whitish, sometimes grayish and rarely creamy. There are vertebrate elements that can show dark cream colorations, with reddish tones due to weathering. In some levels, they can be dark by phosphatization, and acquire reddish to orange tones, produced by oxidation, dominant in the weathered limestones, which originate residual red clays or red ferralitic soils, common in large regions of Cuba.

The most recent, Quaternary fossils are quite similar to the Miocene fossils, wherein many of them some of the original composition of the biogenic entity produced can be conserved. The colorations, in general, tend to be clear, yellowish to whitish or grayish; in some cases, there are original religious tones and colors. In the case of skeletal remains, mainly of terrestrial vertebrates, they usually maintain light colors, but often have different degrees of impregnation with clays, ferrous, manganese, and rarely other oxides.

3.3 Materials and Methods

This work is a compilation of the Cuban geological and paleontological literature, produced by foreign and criollo geoscientists since the eighteenth century (Parra 1787) to date. It also gathers data obtained from the formation and study of the "Paleontological Collection" (Fig. 3.1) of the National Museum of Natural History of Cuba (MNHNC, acronym in Spanish), consolidated during the last 30 years (Rojas Consuegra 2007, 2009, 2014). Its content also reflects the field experience and the author's opinion, which exempts any other social entity from any responsibility.

Fig. 3.1 View of a piece of furniture from the fossil invertebrate collection of the MNHNC; a true scientific—cultural treasure

3.4 Results and Discussion

3.4.1 Composition of the Fossil Record of Cuba

In the future, a synthetic information is offered about the main groups of fossils found in the rocks that make up the geological record of the Cuban Archipelago.

In the Mesozoic era began to originate the first rocks that appear today as part of the complex geological substrate of the Cuban territory, thus initiating the geological history of Cuba, and therefore, the main natural materials that make up the CFR.

3.4.1.1 Jurassic Period

Cuba did not exist in the Jurassic period, only began to accumulate different types of sediments, which then gave rise to the Jurassic rocks that today are part of the complex geological substrate of the Cuban territory (see Paleogeography, in Iturralde Vinent 2012) (Table 3.1).

Sedimentary rocks, predominantly siliciclastic, formed during the Jurassic are alluvial-marine, marine, and partly alluvial (San Cayetano Formation). The Jurassic series are distributed in Cordillera de Guaniguanico (Sierra de Los Órganos and Sierra del Rosario) and Isla de la Juventud, in the western region of Cuba; Sierra de Guamuaya in the central region; and very limited in the eastern region (Asunción). They also exist in several locations along the northern part of the island. These rocks contain different horizons rich in fossils, witnesses of the life in that geological past.

Other geological formations originated in different settings of the marine environment, from shallow to deep, within the sea of Tethys, already in expansion between the supercontinents Laurasia (North America and Europe) and Gondwana (South America, Africa, Australia, and Antarctica); when even the South Atlantic Ocean had not opened (see Tectonic Plates, in other sources).

The oldest stratigraphic record of Cuba, consolidated during the opening of the sea of American Tethys, shows the influence possibly of a Mediterranean type climate, warm and humid. From the Late Jurassic, there is evidence of the continuity of a warm climate and the maximum increase in sea level, in a general transgressive cycle. It manifests an underwater volcanism in an oceanic rift (submarine magmatic ridges), with direct influence on the deepest media. Anoxic conditions, possibly generalized to the oceanic environment from basal to abyssal, are deduced, determined by the division of the basins into tectonic blocks, discontinuously dislocated.

Jurassic Plants, Pollen, and Spores

There are abundant plant reports in the Jurassic fossil record, represented by molds and oxidized films, sometimes charred, corresponding to parts of ferns, fronds, and even trunks of several centimeters in size; in addition, a high variety of palinomorphs (Vachrameev 1965, 1966; Wierzbowski 1976; Haczewski in: Pszczolkowski 1978; Areces Mallea 1990,

Table 3.1 Some Cuban lithostratigraphic units with Jurassic—Early Cretaceous fossil record

Ages	Formations	Invertebrates									
		Corals and Briozoans	Brachiopods	Gastropods	Cephalopods	Nautiloids	Aptychus	Bivalves	Echinoids	Crinoids	Decapods
		Mainly fossil groups									
Middler-Lower Jurassic	Punta Alegre										
	San Cayetanos		r	c[1]	a			a	r	r	r
Oxfordian Upper Jurassic	Jagua	r	r	r	a	r	c	c	r	r	r
Oxfordian Upper Jurassic—Valanginian Lower Cretaceous	Guasasa				a		a	c		c	
	Artemisa				a		a	c		c	
Tithonian Upper Jurassic—Neocomin Lower Cretaceous	Cayo Coco									c	c
Tithonian Upper Jurassic—Barremian Lower Cretaceous	Trocha		r		a		c	c			
	Constancia			c	a		c	c			
	Veloz				a		c	a			

Ages	Formations	Vegetables		Marine vertebrates		Microfossils					
		Ferns, plants	Palynomorphs	Reptils	Bony fishes	Calpionellids	Tintinids	Foraminifera	Nanoplankton	Radiolarians	Favreina s.l
		Mainly fossil groups				Mainly fossil groups				
Middler-Lower Jurassic	Punta Alegre		c								
	San Cayetanos	a[2]	a								a
Oxfordian Upper Jurassic	Jagua	c		a	a			c	c		
Oxfordian Upper Jurassic—Valanginian Lower Cretaceous	Guasasa					c		c	c		
	Artemisa					c	c	c			
Tithonian Upper Jurassic—Neocomin Lower Cretaceous	Cayo Coco						c	c			c
Tithonian Upper Jurassic—Barremian Lower Cretaceous	Trocha					a			c		
	Constancia	c	c			a		c	c	c	
	Veloz				r	c		a	a	c	

Keys a—abundant, c—comun, r—rare, c[1]—from continental lakes, a[2]—continental

1991, 1992; Dueñas Jiménez y Linares Cala 2001; Flores Nieves y Sánchez Arango 2005; López Almirall 2005; Rojas Consuegra y Alabarreta Pérez 2009; Flores Nieves 2015, 2016; Diez et al. 2016).

- **Ferns**

In Cuban Jurassic rocks, of alluvium-marine type (rift stage), fossil plant remains are common (San Cayetano and Constancia Formations); among them, *Piazopteris branneri* is very abundant. In addition, a new fern is also known from the Early to Middle Jurassic (Fig. 3.2). The fossil record of the family to which it belongs (Hymenophyllaceaea) is very scarce; only one taxon (*Hopetedia*) has been reported from the Upper Triassic of North America, and this new fossil resembles it. Also, fragments of stems and other parts of ferns from the Late Jurassic, still unidentified, are conserved in collections.

- **Gymnosperms**

An extinct gymnosperm (*Araucarioxylon* sp.), known only by its fossil remains and reported for the North American continent, has been identified. Several fragments of Jurassic fossil wood without identification exist in the Cuban paleontological collections (Jagua and Guasasa formations).

At least curious is the fact that among the current Cuban vegetation is known the "cork palm" (*Microcycas calocoma*), considered a primitive species, descendant from a group of pre-Jurassic plants (Cycadales, Zamiaceae), and that could have inhabited the American region since that time, where it apparently survived to this day.

- **Palynomorphs**

From Early to Middle Jurassic, very diverse and abundant palynomorphs have been recognized by fossil spores and possibly even the Triassic boundary (San Cayetano Formation). Carbonated rocks of the Late Jurassic, containing many ammonites, fish, and marine reptiles, are also filled with carbonatized fragments of branches and remains of unidentified gymnosperms (Paleontological Collection of the MNHNC). Palynomophs have also been determined in evaporites (Punta Alegre Formation).

Jurassic Invertebrates

The fossil invertebrates comprise a large set of remnants or conserved elements related to different organisms, where most of them had an external skeleton in the form of shells,

Fig. 3.2 Fossil fern attributed to the Hymenophyllaceaea family of Lower–Middle Jurassic. San Cayetano Formation

tests, plates, spicules, etc.; generally of calcareous or corneous composition, but there are also some siliceous ones, and exhibit a wide variety of shapes and structures (Torre y Huerta 1909, 1910; De Golyer 1918; Brown and O'Connell 1919, 1920, 1922; Sánchez Roig 1920, 1951; Lewis 1932; Dickerson and Butt 1935; Vermunt 1937; Imlay 1942; Palmer R 1945; Schevill 1950; Kromemlbeinn 1956; Torre 1960; Judoley y Furrazola Bermúdez 1964, 1968; Gutiérrez Domech 1968; Housa 1969; Khudoley and Meyerhoff 1971; Pszczolkowski et al. 1978; Núñez Jiménez 1988; Pszczolkowski 1989; Furrazola Bermúdez 1997; Myczyński et al. 1998; López Martínez et al. 2013, 2014).

- **Cephalopods**

Among the Jurassic marine invertebrates, the group of ammonites (cephalopod molluscs) are one of the best-known fossils in Cuba (Fig. 3.3). They show a high diversity, made up of hundreds of species. This criterion makes them very useful as fossils guides; particularly in the dating of the rocks of this age and in the stratigraphic correlation (Constancia, Trocha, Veloz, Jagua, Guasasa, Artemisa formations, etc.).

The ammonites are contained in different Jurassic rock series, which accumulated in regions located at different latitudes (see Paleogeography), separated hundreds or thousands of kilometers from each other, which is confirmed by the study of their fossil associations. While some species appear in various locations (recorded in all rock series), others are unique for each rock set, thus acquiring a decisive role in the division of the different stratigraphic units. Some are even considered "endemic" to those regions of the Jurassic past, since they have not appeared anywhere else in the world.

Among the genera of ammonites that appear associated with the marine rock sequences of the continental margins of

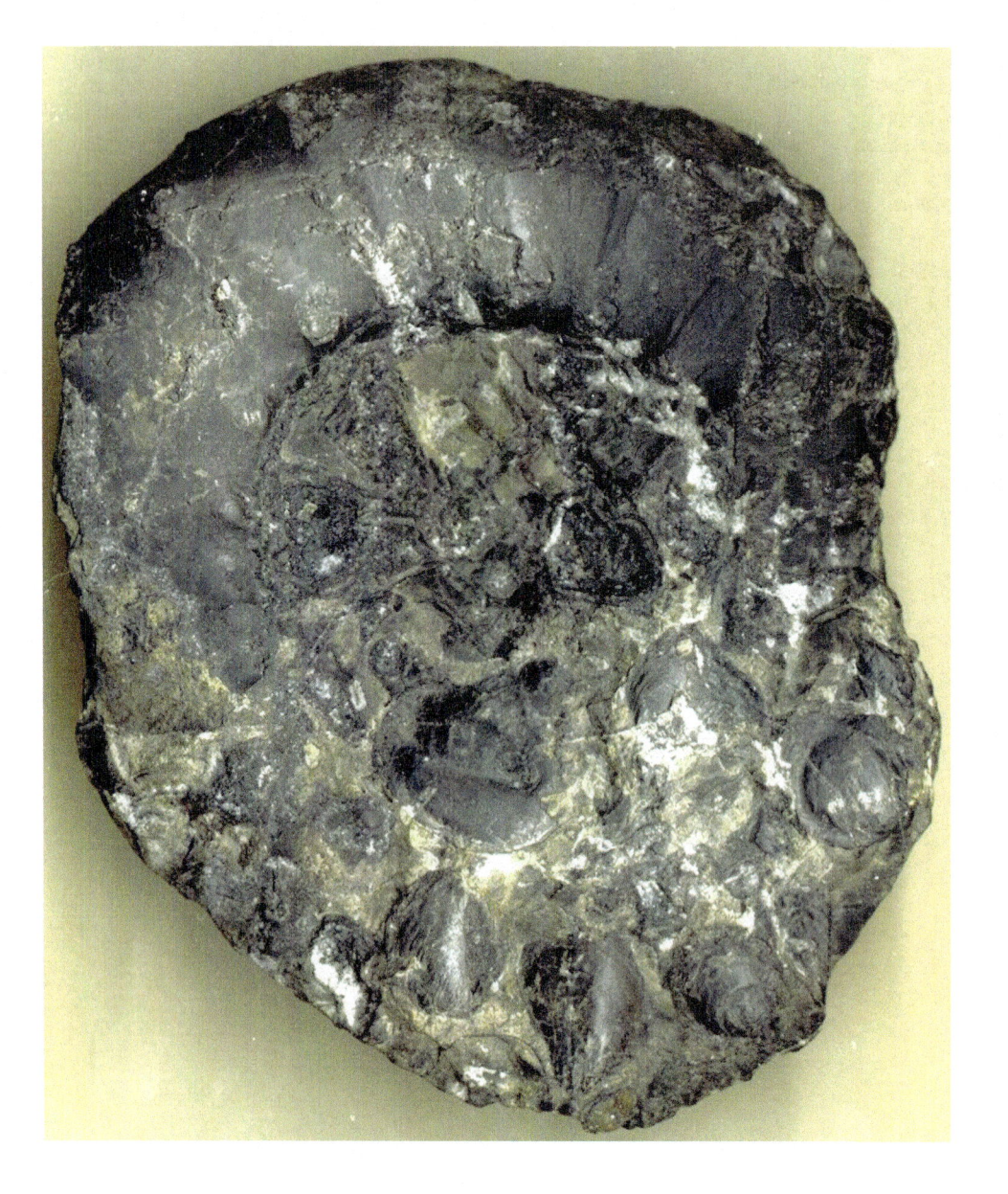

Fig. 3.3 Ostreidos on ammonite shell, a fossil association of the Upper Jurassic. Jagua Formation

the western region, are *Perisphinctes*, *Glochiceras*, *Ochotoceras*, *Vinalesphinctes*, *Discosphinctes*, *Euaspidoceras*, *Cubaochetoceras*; in the northern part of Central Cuba, *Hamites*, *Karakaschiceras*, *Oostrella*, *Melchiorites*, *Astieridiscus*, *Corongoceras*, *Substreblites*, and others appear. Shared by both territories emerge *Micracanthoceras*, *Protancyloceras*, *Paralytohoplites*, *Pseudolissoceras*, *Himalayites*, *Butticeras*, among others.

Other fossil elements, of parataxonomic value, related to cephalopods have been reported from Jurassic rocks abundant in ammonites. Among them, it is common the appearance of the aptics (*Aptychus* or *Lamellaptychus*), which are the fossil opercula of ammonites.

Less well-known are the reports of other much more rare cephalopods. A "cephalopod beak" has been identified as *Hadrocheilus* Tall. Fossil element apparently similar to a ricolites (Rhycholites), which are calcareous tip-type pieces of arrows, a few centimeters in size, attributed to the jaws of cephalopods; and the majority has been related to the nautiloids.

Another rare cephalopod is a single sepoid (*Voltzia palmeri*) reported from the Late Jurassic. A particular mention for the Jurassic belemnites in Cuban rocks, which are still without a specific taxonomic study, and rare pieces in collection.

- **Pelecypods**

From the Jurassic, the marine bivalve molluscs (pelecypods) called trigonias (*Trigonia krommelbeini*) are very well known; this species is the oldest fossil macroinvertebrate in Cuba (San Cayetano Formation). In addition, several genera and species of molluscs have been reported, among them are ostreids, and are mentioned (San Cayetano, Jagua, Guasasa formations, among others): *Buchia*, *Catenula*, *Corbula*, *Cuspidaria*, *Eocallista*, *Exogyra*, *Gervillaria*, *Gryphaea*, *Inoceramus*, *Liostrea*, *Modiolus*, *Neocrassina*, *Ostrea*, *Plicatula*, *Posidonomya*, *Quenstedtia*, and *Vaughonia*. Curiously, there are also molluscs typically reported from continental lakes (*Utschamiella* cf. *asiatica* and *Utschamiella lacelata*).

- **Brachiopods and gastropods**

Along with the previous macrofossils, there are few gastropod molluscs (Gastropoda) and small brachiopods (Brachipoda).

- **Echinoderms**

There are few equinoids (sea urchins) in the Jurassic rocks, and they are known mainly for their loose radiolas. On the other hand, the disintegrated remains of the pelagic crinoid *Saccocoma* are very common and abundant on a microscopic scale.

- **Crustaceans**

The finding of an unidentified Jurassic decapod is reported; but, above all, it is deduced from the presence of crustaceans through the abundant fecal pellet (*Favreina* sp., *F. joukovky*, *F. salevensi*, *Parafavreina* sp.), contained in rocks of that age, of considerable thickness and decametric (Jagua y Guasasa formations).

- **Microfossils**

Microfossils in some Jurassic rock sequences are scarce. The most useful group in this age belong to the nanoplankton, such as calpionellids, tintinnids, and rare calcispherids. Algae of different types (*Codiaceae*, *Corallinaceae*, *Solenoporaceae*, *Dasycladaceae*; *Clypeina* sp., ?*Salpingoporella*, ?*Likanella* sp., ?*Cimopalia* sp.) are also present.

Interesting results, from the taphonomic point of view are the report of foraminifera (*Fusulinacea* and *Tetrataxacea*) together with bryozoans, belonging to the Permian period, contained in clasts redeposited in Upper Jurassic sandstones.

It is common to see, as microscopic detritus, echinoid radioles, and sponge spicules.

Jurassic Vertebrates

In general, they are represented by organisms with an internal bone skeleton and a developed vertebral column, as well as with manifest skeletal appendages of multiple aspects. Various types of fossil elements, such as those mentioned, appear in the Cuban Jurassic fossil record, mostly belonging to marine reptiles or saurians and to the diversity of ganoid fish (Gregory 1923; De La Torre y Cuervo 1939; White 1942; Torre y Rojas 1949; Torres y Callejas 1949; Colbert 1969; Arratia y Schultze 1985; Thies 1989; Iturralde Vinent y Norell 1996; Brito 1997, 1999; Arratia 2000; Fernández e Iturralde Vinent 2000; De la Fuente e Iturralde Vinent 2001; Gasparini e Iturralde Vinent 2001; Gasparini et al. 2002, 2004; Iturralde Vinent 2004; Gasparini e Iturralde Vinent 2006; Gasparini 2009; Iturralde Vinent y Gasparini 2013; Iturralde Vinent y Ceballos 2015).

- **Marine reptiles**

Among the remains of vertebrate fossils, the most known and striking in Cuban Jurassic rocks, are those related to several types of marine reptiles (Jagua Formation), such as plesiosaurs (*Vinialesaurus caroli*), pliosaurs (*Gallardosaurus iturraldei*), ichthyosaurs (*Ophthalmosaurus*), turtles (*Caribemys oxfordiensis*), and primitive metriorinquid crocodiles (*Geosaurus* or *Cricosaurus*). López Conde et al. (2016), denote that *C. oxfordiensis* is the oldest member of the clade of the platychelyid turtles.

- **Flying reptiles**

In addition, two species of pterosaurs (flying reptiles) (*Nesodactylus hesperius* and *Cacibupteryx caribensis*) are known. A unique rest of dinosaur (camarasaurid sauropod), although contained in marine limestone, has been identified.

This association of fossils, due to its stratigraphic and biogeographical position, represents a valuable link in the knowledge, not only of the distribution of that fauna of the past, in particular, for the relations with the fauna of the Mediterranean Tethys, but also in the understanding of its global evolution, in the connections with the North American interior sea and the Niuquen basin of South America.

- **Fishes**

The fossils of ganoid fish (Fig. 3.4) are very abundant toward the Late Jurassic (Guasasa and Francisco formations), represented by species of several genera: *Aspidorhynchus*, *Gyrodus*, *Leptolepis*, *Luisichthys*, *Caturus*, *Sauropsis*, *Eugnathides*. In addition, there are remains identified at family level (Caturidae, Pachycormidae, Pycnodontiformes, Varasichthyidae). Some of these taxa are known only from this region of the world. It is evident, according to the diversity and abundance of the fossil material existing in the collections, that this group deserves even deeper studies.

Jurassic Ichnofossils

The Cuban ichnological record referring to the Jurassic period is relatively scarce, but includes interesting structures, mainly bioerosion and biodeposition (Pszczólkowski et al. 1987; Arai y Cuevas 2007; Pszczółkowski and Myczyński 2009; Villegas Martín et al. 2012; Martínez López et al. 2014).

- **Bioerosion**

Among the Jurassic ichnofossils, the *Chondrites*, *Planolites*, and bioturbations, such as channels made by invertebrates

Fig. 3.4 Concretion of carbonate containing a gonoid fish of the Upper Jurassic, Oxfordian. Jagua and Francisco formations

Fig. 3.5 Perforations in wood (lower half of the image) and ammonite section (above) of the Upper Jurassic, Oxfordian. Jagua Formation

(Jagua and Guasasa formations), have been mentioned or deduced. In addition, perforations in fossil wood have been described, belonging to the ichnogenus *Teredolites* (*T. clavatus* and *T. longisimus*) produced by xylophagous organisms (Fig. 3.5).

- **Microbioerosion**

Very curious are the microperforations in fossil pollen (*Diqueiropollis* sp.), with submicrometric pustules resulting from the bacterial action, given to the Jurassic of Cuba. There are also very little studied manifestations of microbioerosion: shells of gastropods embedded by bryozoans and microperforations in echinoderms.

- **Biodeposition**

Another type of ichnofossils are the biogenic depositions, represented by different *Favreina* sp., referred to microscopic coprolites of crustaceans and other organisms, which produced powerful accumulations during the Late Jurassic.

3.4.1.2 Cretaceous Period

Cuba still did not exist in the Cretaceous Period. At this time, there were large volumes of volcanic, volcano-sedimentary, and sedimentary accumulations, mainly in the marine environment, within the so-called Tethys Sea. The first proto-Antillean land emerges, formed by an archipelago of volcanic islands located between the North and South American continents, but still possibly in the Pacific Ocean.

The activity of this system of volcanic arcs ceases toward the Late Cretaceous, nailed in the primitive Caribbean. These stratigraphic series currently appear in large territories of the Cuban geological complex substrate, commonly rising toward the central part of the main island.

Toward the northern part of that sea eventually emerged a cord of calcareous islands associated with the southern margin of the North American continent. In the marine platforms and their slopes, sediments were accumulated from the erosion of the limited temporally emerged lands (evanescent islands); and also, as products of the vital activity of a diverse marine fauna (Paleogeography of Cuba, in Iturralde Vinent 2012).

During the Cretaceous, the influence of igneous activity associated with arcs of volcanic islands prevailed, with eventual impact on regional climatic conditions, due to the expulsion of enormous volumes of materials, including greenhouse gases. Toward the middle and end of the period, episodes of normal marine conditions are reported, in a warm climate, with shallow production of biogenic carbonates. Oceanic anoxic events are denoted in some stratigraphic units.

The cease of magmatism benefited the intense tropical weathering of the silicates in the landmasses and the subtraction of the atmospheric CO_2. Terrestrial conditions of rapid erosion, the formation of kaolinitic crusts, and the establishment of local anoxic shallow conditions are denoted.

With the transgressive advance and the expansion of the shallow seas toward the end of the period, a diverse benthic and planktonic fauna was developed, with a possibly exuberant vegetation. This is revealed as an important time in the fixation and accumulation of carbon in the marine environment, in the form of skeletogenesis. The formation of soils rich in organic matter could be favored. The final stage, with a very hot and humid climate, is known as the "Cretaceous greenhouse world". Today the heating and cooling events that took place at smaller time scales are discussed.

The Cretaceous series in Cuba are rich mainly in invertebrate fossils, ichnofossils, and microfossils. The remains of identifiable vertebrates are very rare.

Cretaceous Plants

The fossil record of Cretaceous plants in Cuba is relatively poor (Judoley y Furrazola Bermúdez 1964; Kantchev et al. 1978; Radocz y Nagy 1983; Pszczólkowski et al. 1987; Areces Mallea 1990; Smiley 2002; Iturralde Vinent 2004, 2013; López Almirall 2005; Rojas Consuegra 2009).

Geography, climate, and sea level in this period were changeable. During the deposition of marine evaporites (halite, anhydrite, gypsum) in the Late Jurassic and Early Cretaceous (Punta Alegre Formation), according to the floristic composition that shows pollen and fossil spores, the prevailing climate was warm and dry.

In the sedimentary basin formed behind the volcanic arc (Provincial Formation), toward the continent, plant remains have appeared (Fig. 3.6), such as fragmentary remains of plants, and isolated charcoal, where a leaf with an angiosperm's own rib is recognized, which lived near the boundary Early–Late Cretaceous (Albian–Cenomanian).

In some stratigraphic sequences of the Cretaceous in Cuba, carbonized fragmentary remains of plants appear, or their impressions in films of iron hydroxide. Also, and more rarely, molds of branches or stems are found on the surfaces of the strata, in the volcano-sedimentary series (Piragua Formation).

Fig. 3.6 Stem of wood in sandstones of the Lower Albian Cretaceous. Provincial Formation

In sedimentary deposits at the end of the period, coals have been discovered in well-defined layers, and small fragments (Picota Formation). According to the pollen and spores preserved in these rocks, palm trees and ferns grew in the eastern region, among them the arborescent ones and other gymnosperms.

It is to be expected that with the appearance of the first volcanic islands in the primitive Caribbean, they will also begin conquest by plants, some of which have reached the present, such as *Microcycas* (*M. calocoma*) and *Pinus* (*P. tropicalis*).

Cretaceous Invertebrates

Bivalve molluscs (rudists), gastropods, and foraminifera predominate among the Cretaceous invertebrates. Also common are other pelecypods, cephalopods, and echinoderms, among others macrofossils, which have been preserved in a variety of marine, carbonated, and siliciclastic sedimentary rocks (Fernández de Castro 1884; Spencer 1896; De La Torre 1915; Golyer 1918; O'Connell 1921; Sánchez Roig 1926a, 1948, 1949; Valette 1926; Douvillé 1926, 1927; Lewis 1932; Palmer R 1933, 1948; Rutten 1936; Thiadens 1936; Trauth 1936; Vermunt 1937; MacGillavry 1937; Palmer D 1941; Wells 1941; Imlay 1942; Hermes 1945; Álvarez Conde 1957; Seiglie 1960; Dommelen 1971; Kuffman and Shol 1974; Lupu 1975; Tshunev et al. 1976; Kantchev et al. 1978; Kier 1984; Torre e Iturralde Vinent 1990; Rojas et al. 1995; Iturralde Vinent 1996; Furrazola Bermúdez et al. 2000; Varela y Rojas Consuegra 2009, 2011c; Barragán, Rojas Consuegra y Szives 2011; Aguilar Pérez et al. 2015; Rojas Consuegra 2005, 2015; Arano Ruiz et al. 2018) (Tables 3.2 and 3.3).

- **Bivalves**

Among invertebrate organisms, Cretaceous marine molluscs are represented by several fossil groups. The bivalves stand out for their abundance, and among them, particularly, the "rudists". Rudists were a group of organisms, sometimes gigantic shells, whose fossils appear contained in rock series of different ages, mainly in those associated with the Cuban and Antillean volcanic terrain, where they colonized the shallow seabed, in the stages of recess or low volcanic activity.

The shells of the rudists serve as fossils guides to date and to correlate the rocks of this period, in particular between the volcano-sedimentary sequences (Guáimaro, Provincial, and Piragua formations). They also appear in rocks belonging to the continental margin of North America (Remedios, Palenque, and Purio formations). In the Cuban rocks, there is a wide diversity of shells belonging to practically all known families of the Western Hemisphere (Los Negros, Vía Blanca, Cantabria, Isabel, Presa Jimaguayú, Tianjita, etc.): requienids (*Apricardia*), caprinids (*Caprina*, *Pachitraga*, *Offneria*, *Amphitriscoelus*, *Caprinuloidea*, *Coalcomana*, *Kimbleia*), policonitids (*Tepeyacia*), radiolitids (*Macgillavryia*, *Biradiolites*, *Bournonia*, *Durania*, *Radiolites*, *Tampsia*, *Eoradiolites*, *Thyrastylon*, *Sauvagesia*, and *Chiapasella*), hippuritids (*Praebarrettia*, *Barrettia*, *Parastroma*, *Torreites*, *Vaccinites*, *Hippurites*), plagioptichids (*Plagioptychus* and *Mitrocaprina*), and antillocaprinids (*Antillocaprina* and *Titanosarcolites*) (Fig. 3.7).

In particular, recently from Loma Rioja in Cruces, Cienfuegos (Provincial Formation), new rudist taxa have been provided for Cuba: *Guzzyella bisulcata*, *Ichthyosarcolites* (*Mexicaprina*) *alata*, *Mexicaprina* cf. *cuadrata*, *M. cornuta*, and *Requienia* sp., along with others already known in the country.

Also in Cretaceous marine rocks (Piragua, Durán, Vía Blanca, Presa Jimaguayú, Isabel, Cantabria and others formations), together with the rudists, appear other bivalve molluscs, such as ostreids (*Arctostrea aguilerae*, *Exogira costata*; *Ostrea*, *Alectrinia*, *Arca*, *Chama*, *Cyprimera*, *Cyrena*, *Depanocheilus*, *Eriptycha*, *Fusas*, *Granocardium*?, *Idonearca.*, *Lima*, *Neithea*, *Pseudamussium*, *Pterotrigonia*, *Pecten*, *Trigonarca*, *Trigonia*, *Vaniella*, *Verenicardia*, *Voluta*, *Volutoderma*), and also *Condrodontos*, *Cardium* and others.

- **Cephalopods**

Cephalopod molluscs decrease throughout the Cretaceous. Ammonites are recognized in the series of the North American continent margin of the Lower Cretaceous (Veloz and Cifuentes formations) and in the volcanic terrain of the Lower Cretaceous (Provincial Formation) and Upper Cretaceous (Arimao, Moscas, Piragua, and Contramaestre formations). At the beginning of the period are present: *Acanthoceras*, *Barrioisiceras*, *Crioceras*, *Hystericeras*, *Mortoniceras*, *Ostlingoceras*, *Turrilites*, and others; while toward the second half (Monos and Cantabria formations), *Eupachydiscus*, *Paralenticeras*, *Paratexanites*, *Schloenbachia*, and other genera are reported; but at the end of the period only *Bostrychoceras*, *Hamites*, *Pachydiscus*, and *Sphaerodiscus* are known (Fig. 3.8).

Near Loma Las Nueces (Provincial Formation) several species were identified: *Protetragonites* cf. *aeolus*, *Desmoceras* cf. *latidorsatum*, ?*Discohoplites* sp., *Mortoniceras* sp., *Cantabrigites spinosum*, *C. wenoensis*, *Stoliczkaia* cf. *clavigera*, and *Algerites* sp., which correspond to a set of Tethys, typical of the later Upper Albian, *Stoliczkaia* (*Stoliczkaia*) *dispar* subzone.

It is striking that none of the species related to the Cretaceous volcanic arc is present in the sedimentary series

Table 3.2 Some lithostratigraphic units with Lower Cretaceous to Early Upper Cretaceous fossil record

Cretaceous age	Formations	Corals and Briozoans	Gastropods	Cephalopods	Aptychus	Bivalves	Echinoids	Plants, Palynomorphs	Bony fishes
		Mainly fossil groups							
Barremian–Valanginian	Margarita			c					
Valanginian–Albian	Polier			c	r				
Hauterivian–Turonian	Pons				r				c
Aptian–Cenomanian	Palenque	c	c	a		a			
	Guajaibón		r	a					
	Santa Teresa								
	Guáimaro	r	r			r			c[1]
	Mata				r				
	Provincial	c	a	r		a*	c	r	u[2]
Cenomanian–Santonian?	Mata								
Coniacian–Santonian	Orozco								
Santonian	Arimao	r		c		c			
	Mbr. Moscas	r		c		c			
Turonian–Campanian	Crucero Contramaestre			c		r		r	
Santonian–Campanian	Piragua	r	a	r		a	r		

Cretaceous age	Formations	Sponges	Algae	Calpionellids	Foraminifera	Nanoplankton	Radiolarians	Ostracods	Ichnofossils
		Mainly fossil groups							
Barremian–Valanginian	Margarita			a	a	a			
Valanginian–Albian	Polier				a	a			
Hauterivian–Turonian	Pons				a				
Aptian–Cenomanian	Palenque			a	a				
	Guajaibón				a				
	Santa Teresa				a		a		
	Guáimaro				c				
	Mata				a				
	Provincial	c	c		a		r		c
Cenomanian–Santonian?	Mata				a		a		
Coniacian–Santonian	Orozco								
Santonian	Arimao		r		r		r	r	
	Mbr. Moscas		r		c				
Turonian–Campanian	Crucero Contramaestre				a		r	r	
Santonian–Campanian	Piragua				a			a	

Keys See Table 3.1, *—mixed fossil association, c[1]—unidentified bony remains, u[2]—*Ptychodus cyclodontis*

Table 3.3 Some lithostratigraphic units with Late Upper Cretaceous fossil record

Cretaceous age	Formations	Corals and Briozoans	Gastropods	Cephalopods	Bivalves	Echinoids	Crinoids	Decapods	Vermes
		Mainly fossil groups							
Campanian	Martí	r							
Campanian—Maastrichtian	Vía Blanca	c	c		c	c			
	Durán	r	r		c	c			
	Monos	c	c	r	c	r		r	a
	Yáquimo			c		c*			
	La Jiquima								
	La Picota	c							
Maastrichtian	Tinajita	r	r		a	c			
	Los Negros	c	r		a	r			
	Presa Jimaguayú	c	a		a	a	c	r	c
	Isabel	c	c		a	c	r		
	Cantabria	c	c		a	c	c	r	a
	Mícara								
Cretaceous-Paleogene (Basal Danian)	Peñalver	c	r	c	c	c			

Cretaceous age	Formations	Plants, Palynomorphs	Vertebrates	Algae	Calpionellids	Foraminifera	Nanoplankton	Radiolarians	Ichnofossils
		Mainly fossil groups							
Campanian	Martí					r			
Campanian—Maastrichtian	Vía Blanca			r		a		r	c
	Durán	r				a			r
	Monos	a	u[1]	r		a			c
	Yáquimo					a			
	La Jiquima			r		a	c		
	La Picota			r		c			
Maastrichtian	Tinajita			r		a			r
	Los Negros			c		a			c
	Presa Jimaguayú	r	u[2]	c		a			c
	Isabel			c		a			c
	Cantabria	r		c		a			c
	Mícara	r[1]	u			c	c		r
Cretaceous-Paleogene (Basal Danian)	Peñalver	r		r	r	a	r	r	

Keys See Table 3.1, u[1]—Possible saurus, u[2]—unidentified mineralized remains, u[3]—mosasaurus, r[1]—coal, c*—incliude a unidentified starfish

Fig. 3.7 Shell of the rudists *Titanosarcolites giganteous* in marmorized limestones of the Upper Cretaceous, Maastrichtian. Cantabria Formation

belonging to the continental margin of North America; a fact that suggests the probable existence in that past, of some biogeographical barrier, that prevented the interchange between both peleogeográficos domains.

It is interesting to know the abundant presence of the *aptychus*: *Lamellaptychus angulocostatus* (Peters), *L. longa* (Trauth), *Lamellaptychus beyrichi* (Oppel), *Lamellaptychus cubensis* O'Connell, *Lamellaptychus didayi* (Coquand), *Lamellaptychus excavatus* Trauth, *Lamellaptychus expansus* Sánchez Roig, *Lamellaptychus rectecostatus* (Peters) enm. Trauth, *Lamellaptychus seranonis* (Coquand), fossil opercula of the shells of the ammonites (Constancia, Trocha, Veloz, and Artemisa formations). These parataxonomic elements, in Cuba, have a certain value in biostratigraphy, since they appear with significant abundance in series of rocks where the shells of the ammonites are scarce or absent, and therefore, are propitious to be used as guide fossils (Fig. 3.9).

There is a large specimen (more than 15 cm long) of unidentified belemnites in the collection of the Instituto de Geología y Paleontología (IGP), which claims to come from Soroa.

This group, due to its diversity and importance, deserves a deepening in its research.

- **Inoceramus**

The inoceramus are more scarce, which have a high bio-chronostratigraphic value, useful in dates within the volcano-sedimentary sequences (Piragua and Abreus formations). Some of the mentioned taxa are *Inoceramus barabini*, *I.* aff. *fibrous*, *I.* ex. gr. *labiatus* (Fig. 3.10).

- **Brachiopods**

Only one species of brachiopod (*Orthothyris radiata*) of the Cretaceous has been described, perhaps due to the very low degree of study of this group.

- **Gastropods**

Gastropod molluscs are mainly composed of acteonelids, nerineids, naticids, and turritellids (*Acteonella grussduvre*, *A.* cf. *crassa*, *A. cubensis*, *A.* cf. *coquiensis*, *A. bornensis*, *Angoria aculata*, *Ampullina semiglobosa*, *Astralium densiporatum*, *Cypraea* sp., *Haustator rigidus*, *H. fittonianus*, *Helicaulax sinuata*, *Nerinea* cf. *gigantea*, *N. forojulensis*, *N. parvula*, *N.* cf. *requieni*, *N. epelys*, *N. buchardi*, *Tylostoma cossoni*, *Trochacteon renauxi*, *T. rossi*, *T. palmeri*), which

Fig. 3.8 Ammonite of the Upper Cretaceous, Maastrichtian. Monos Formation. Courtesy of Alberto Arano

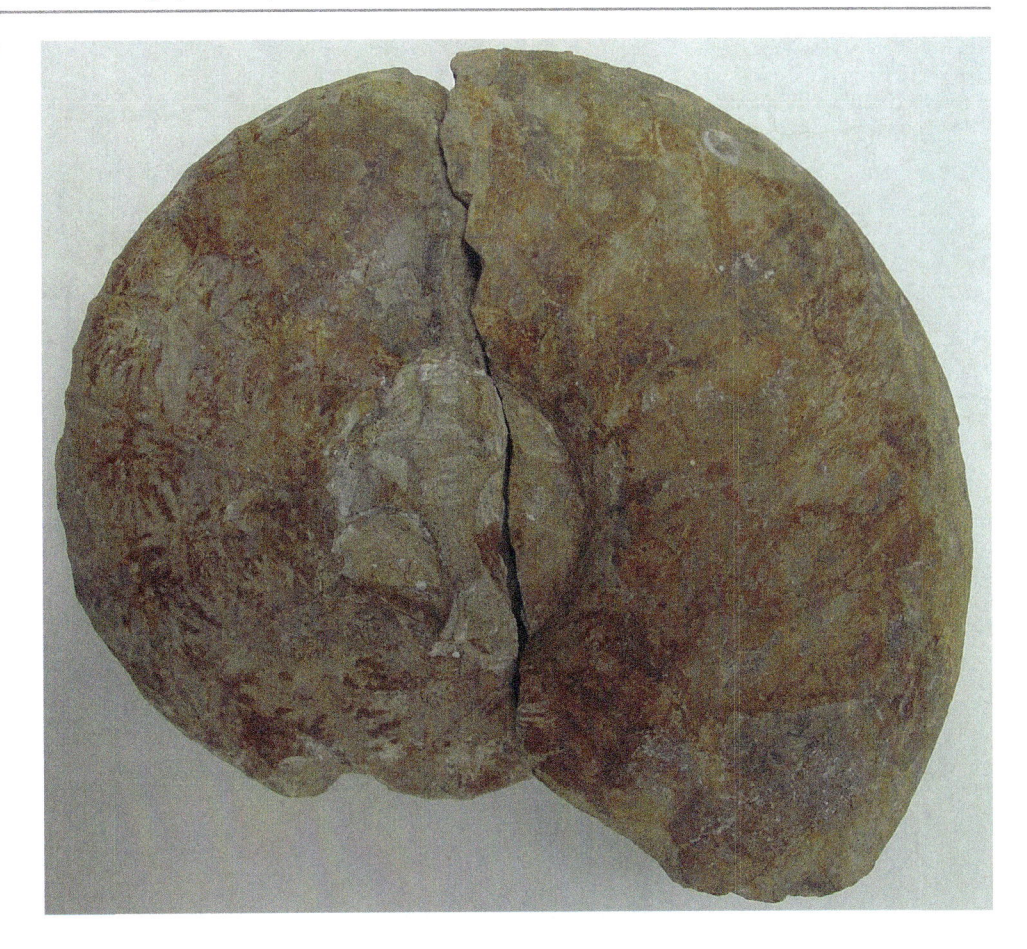

become abundant at the end of the Cretaceous (Piragua, Durán, Jimaguayú, Cantabria and Isabel formations).

- **Echinoids**

The echinoids, from this period and until today, began to be common in Cuban rocks (several formations). Some of the reported genera are *Anorthopygus, Anorthopygus, Austerobrissus, Cardiaster, Catopygus, Cidaris, Clypeolampas, Clypeopygus, Codiopsis, Conoclypeus, Conulus, Cylindrolampas, Discoides, Douvillaster, Echinokorys, Echinolampas Echinometra, Globator, Goniopygus, Hemiaster, Hemiaster, Holectypos, Holeptypus, Isolampas, Lanieria, Lithia, Micraster, Pedina, Phyllacanthu, Phymosoma, Polydiadema, Procassidulus, Pseudorthopsis, Rhynchopygus Salenia, Schisaster, Trachyaster*. There are common and abundant loose radioles of several species (Vía Blanca, Durán, Cantabria, and Tinajita formations, among others).

- **Crinoids**

They were reported by M Sánchez Roig which were the first identifiable fossil pieces of a benthic crinoid (sea lily) in Cuban rocks, named *Austinocrinus cubensis* (Presa Jimaguayú Formation); of them, several fossil elements (articular discs) have been collected, even for studying in detail (Fig. 3.11).

- **Asteroids**

Recently, a new fossil asteroid (starfish), in excellent state of preservation, has been collected in the region of Rodas (Monos Formation), Cienfuegos (Granma 2018).

- **Crustaceans**

In siliciclastic, polymictic rocks, the oldest crustacean of Cuba was found (*Lophoranina precocious*) possibly of Campanian—Maastrichtian (Upper Cretaceous) age (Monos Formation). This species has passed to a new taxonomic arrangement (Fig. 3.12): *Vegaranina rivae* (Arano Ruiz et al. 2018). Also, some decapod chelae have appeared, loose in other Maastrichtian locations. In addition, new pieces have been collected, already under study.

- **Corals**

Corals are present in Cretaceous carbonate rocks (Provincial and Isabel formations, among others), and the genera are known: *Ellamophylla, Enalloelia, Astrocoenia, Trochoseris,*

Fig. 3.9 The *aptychus* can be very abundant in some facies of the Lower Cretaceous, Barremian. Veloz Formation

Trochocyathus, *Haplaraea*, *Leptophyllia*, *Paracycloseris*, *Goniopora* y *Montastrea*, along with algae (Corallinaceae, Dasycladaceae; *Salpingoporella annulata Cimopalia* sp. *Clypeina*? sp. *Likanella* sp.), but in general, they demand a better study (Fig. 3.13).

- **Hydrozoans**

It is reported for this age, the first hydrozoan in Cuban rocks: *Stomatopora veneszuelensis*. In addition, some specimens have been recovered in the same locality (Loma La Rioja, Provincial Formation), where the aforementioned species was found; currently under study.

- **Bryozoans**

The bryozoans (phylum Bryozoa) are revealed as small loose elements or adhered on the shells of several calcareous benthic organisms of different ages. Despite this, there are no particular studies in the country dedicated specifically to them.

- **Sponges**

Large sponges (phylum Porifera) have also been found from the Uppermost Cretaceous (Provincial Formation), still unidentified and under study.

- **Vermes**

Despite being common, only two species of sedentary fossil polychaete worms (Vermes) have been reported: *Pyrgopolon onyx* and *Serpulia* sp. The latter consists of small calcareous tubes, very abundant, that appear along with some equinoids (*Salenia salieri*), ostreids, and large orbitoidal benthic foraminifera of the Maastrichtian (Mono and Cantabria formations). Very thin vermiform tubes of sedentary polychaetes (possibly *Serpulia*) are also observed on shells

Fig. 3.10 Internal mold and remains of the shell of an *Inoceramus* of the Upper Cretaceous, Turonian–Coniacian. Abreus Formation

Fig. 3.11 Elements of the crinoid *A. cubensis*, collected in Ciego de Ávila, west of the type locality. Presa Jimaguayú Formation

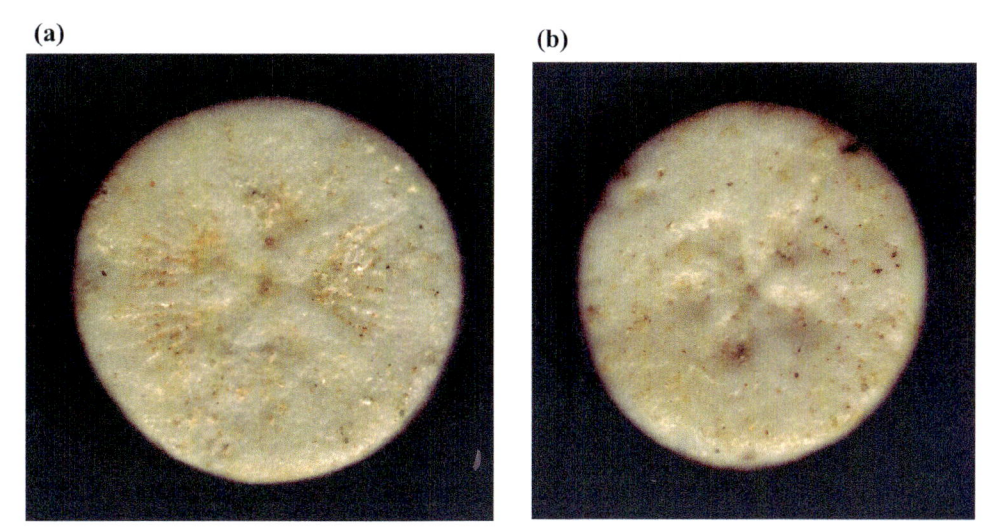

of different molluscs (Fig. 3.14), with profusion over the rudists (Jimaguayú and Isabel formations).

- **Microfossils**

Particularly striking are the varied and abundant microfossils that contain the sedimentary rocks of the Cretaceous period, where the foraminifera stand out, both benthic (Fig. 3.15), sometimes large (millimeters to centimeters) and planktonic (49 genera and 162 species; IGP 2018), along with ostracodes and other microfossils. The radiolarians, with siliceous shells, sometimes appear substituted in carbonate, and are common in rocks of deep marine environment of all represented ages.

Fig. 3.12 Carapace of crutacean *Vegaranina rivae* of the Upper Cretaceous. Monos Formation. Photo courtesy of Alberto Arano

Fig. 3.13 Corals are abundant in the carbonate rocks of the Upper Cretaceous, Maastrichtian. Cantabria Formation. Photo courtesy of Carlos Borges and Alberto Arano

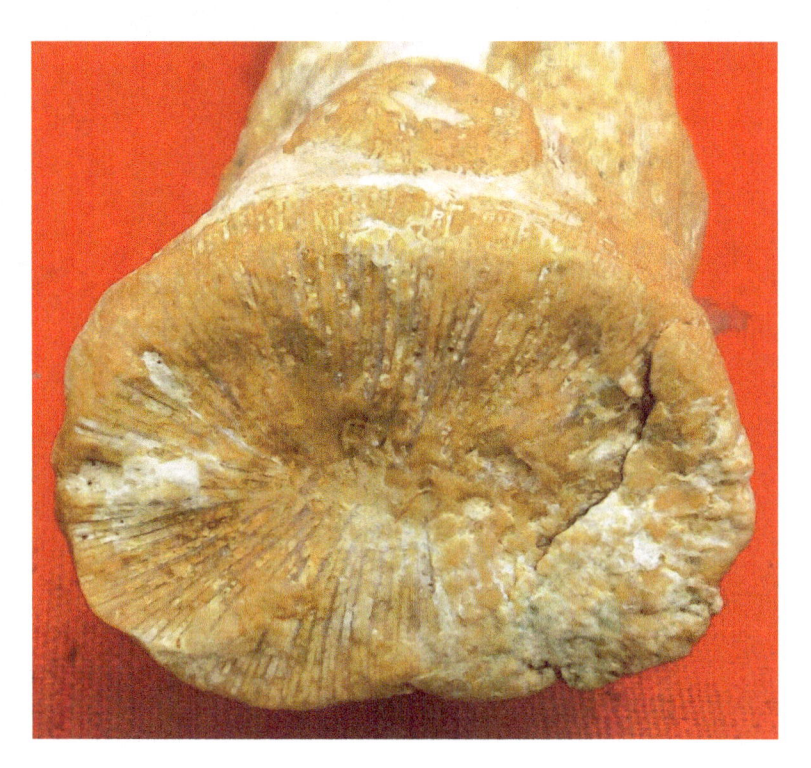

Fig. 3.14 Calcareous tubes of vermes on rudist shell of the Upper Cretaceous. Presa Jimaguayú Formation

Fig. 3.15 Macroforaminifera, abundant in the rocks of the Upper Cretaceous, Maastrichtian

Cretaceous Vertebrates

The remains of Cretaceous vertebrates in Cuban rocks have been scarcely described. The fish are related, consisting of small skeletal parts, scales, and teeth; and rare remains of tetrapods and a marine reptile (Kantchev et al. 1978; Mutter et al. 2005; Granma 2014, 2015; Rojas Consuegra 2015, 2018; Borges Sellén et al. 2016a, b; Linares Cala Pers. Comun. 2019; Martínez Molina 2019). There is an opinion

Fig. 3.16 Coprolite found in clay-sandy facies of the Upper Cretaceous, Maastrichtian. Cantabria Formation

that amphibians (e.g., the *Eleutherodactylus* frog) arrived in the Antilles from the Late Cretaceous or Early Paleogene (Hedges 2006).

- **Fishes**

Belonging to marine vertebrates (Mata Formation), only fossil teeth of a shark (*Ptychodus cyclodontis*) have been reported (Mutter et al. 2005). However, in some unpublished research reports, the presence of remains (mainly teeth and little bones) of unidentified bony fish, contained in Cretaceous rocks, is mentioned (Kantchev et al. 1978). There had been news of other findings (Rojas Consuegra 2018). In addition, in collection, there are some pieces not yet studied (Fig. 3.16).

- **Marine reptile**

The news is very striking (Granma 2015), on the discovery of a single fossil tooth corresponding to a marine reptile, possibly a mosasaur, recovered from Maastrichtian limestones (Cantabria Formation), rich in shallow water fossil invertebrates, in the territory of central Cuba. Significant bone remains from this region, along with other diverse fossils, are under study.

- **Flying reptile**

Just now, the first fossil remains of a Pterosaur in marine rocks of the Upper Cretaceous, Campanian-Maastrictian, in the region of Rhodes, province of Cienfuegos in central Cuba (Martínez Molina 2019), have been discovered.

- **Tetrapoda indet**

A site with fossilized bone remains of vertebrates possibly belonging to the Cretaceous Period, has been revealed in the central region, Ciego de Avila Province (Jimaguayú Formation). The few fossil elements, strongly transformed, were recovered from a sequence of breccia—conglomerates, calcareous, from the end of the Cretaceous or the Early Paleogene, but it is not ruled out that they are in particular deposits of the K/Pg boundary (Rojas Consuegra 2015). These entities, due to their intense taphonomic transformation, have been able to be identified taxonomically until now, only as belonging to some tetrapod animal (amphibians, reptiles, birds, or mammals). Complementary material is needed for more accurate identification.

Cretaceous Ichnofossils

In general, in the Cuban geological literature, the ichnofossils are reported under various terms, such as bioglyphs, hieroglyphs, worms, coprolites (Fig. 3.17), perforations, traces, burrows, etc. (Broniman and Rigassi 1963; Kantschev et al. 1978; Albear e Iturralde Vinent 1985; Pszczólkowski et al. 1987; Pszczołkowski 2002). In recent years, a significant contribution has been made to the knowledge of this fossil group in the country (Villega Martín y Rojas Consuegra 2009, 2010, 2011, 2012; Menéndez Peñate et al. 2011; Villega Martín, Rojas Consuegra and Klompmaker 2016).

In the Cretaceous fossil record, various ichnofossils frequently appear, which generally correspond to traces of life activity and behavior (ethology) of benthic marine invertebrate organisms, both infaunal and epifaunal, conserved on

Fig. 3.17 Coprolite found in clay-sandy facies of the Upper Cretaceous, Maastrichtian. Cantabria Formation

the surface or inside sedimentary strata accumulated in the seabed of that stage.

- **Bioerosion**

Among the Cretaceous ichnotaxon on a solid substrate are listed *Oichnus simplex* and *Gastrochaenolites* isp. (Presa Jimaguayú, Isabel and Cantabria formations) on shells of rudists, ostreids, and coral skeletons. Interesting are the microperforations on the calcareous tubes of the *P. onix*, considered as the ichnospecies *Oichnus ovalis*.

A fragment of fern, and other remains of wood, recovered together with several invertebrates (inoceramus, cephalopods, gastropods, bivalves, echinoderms, crustaceans), in sandstones of the Upper Cretaceous (Mono Formation), reflects a high density of perforations belonging to the ichnogenus *Teredolites* (Fig. 3.18).

Also, as a result of the activity of the xylophagous organism, at least two ichnospecies (*Teredolites clavatus*, *T. longissimus*) have been described in the megablocks associated with deposits of the K/Pg boundary) (Peñalver Formation). In addition, there is a wealth of material under study.

- **Bioturbation**

Some traces and molds of bioturbations (*Chondrites*, *Thalassinoides*, *Planolites*) have been collected in Cretaceous rocks (Vía Blanca Formation). As it is known, crustaceans are active producers of ichnofossils, and in general, of various types deformations of sediments (bioturbations); thus, the ichnogenus *Thalassinoides* has been reported in Late Cretaceous sediments (Presa Jimaguayú Formation).

Mass Extinction of the Cretaceous–Paleogene

The end of the Mesozoic era was marked by the impact of a giant meteorite on Earth, today marked by the well-known Chicxulub crater, located in the Yucatan peninsula (Mexico). That cataclysm caused by such a global event produced drastic environmental changes throughout the planet, which put an end to one of the greatest mass extinctions that the development of life on the planet has suffered.

In several localities of Cuba, a wide registry of those particular processes is contained in powerful successions of rocks of 66 million years ago; well documented as the result of intensive research conducted during the last 20 years in the western and central regions of our country (Takayama et al. 1999; Kiyokawa et al. 2002; Tada et al. 2004; Alegret et al. 2005; Rojas Consuegra et al. 2007; Goto et al. 2008; Yamamtomo et al. 2010; Arenillas et al. 2011; Arz et al. 2012; Meléndez Hevia et al. 2013; Rojas Consuegra y Núñez Cambra 2017); also, studies have already begun in the eastern region.

The Cuban fossil record shows the extinction of several taxa related to the Cretaceous–Paleogene boundary (K/Pg). Among these bioevents stand out the disappearance of the rudists at their time of maximum diversity; the ammonites already in decline; various taxa of bivalves (Inoceramids), gastropods (actaeonellids), echinoderms, and several

Fig. 3.18 Endo-skeleton well preserved from an equinoid of the Paleogene

microorganisms, including many foraminiferal taxa, mainly planktonic.

Such a global catastrophe was well marked in the fossil record preserved in the rocks of the K/Pg boundary in Cuba (Moncada, Cacarajícara, Peñalver, and Amaro formations) and the chaotic non-nominated clastic units between formations: Isabel/Fomento, Santa Clara/Ochoa, Babiney, Mícara, and others.

3.4.1.3 Paleogene Period

The Cenozoic era begins. An archipelago of islands, unstable and changing, was bordering the northern region of the Caribbean Sea to the south of the southern margin of the North American continent. The intense geotectonic processes, which occurred in the first half of this period, are conforming the complex geological substrate of most of Cuba, in their current geographical position. Toward the southeast, volcanism is reactivated, which ceases toward the middle of this period. A widespread production and accumulation of sediments of varied compositions nourishes the marine basins, as a result of erosion combined with intense tectonic processes (see Iturralde Vinent 2004, 2012).

At the end of the Paleogene, a temporal terrestrial connection between the Greater Antilles and South America is established, forming a long peninsula toward the interior of the Caribbean, named GAARlandia (MacPhee and Iturralde Vinent 1999), which seems to have played an important role in the Antillean Biogeography (Iturralde Vinent and MacPhee 2004; Iturralde Vinent 2009).

The asteroid impact of the Cretaceous–Paleogene (K/Pg) boundary induced a widespread extinction of life on Earth (Schulte et al. 2010). The geological record of Cuba contains important evidence of that global environmental cataclysm. The marine sedimentary stuffing of the Paleogene is witness to an extremely hot and humid climate, tempered by intense rains and deep continental erosion. Abundant plant organic matter was buried by profuse currents of turbidity in the marine basins, with significant subtraction of atmospheric CO_2.

The global temperature declined toward the middle part of the period, leading to a marked speciation and biological irradiation in all environments, very noticeable in the open ocean and in deep environments (Zachos 2001). Here too, siliceous organisms of relatively cold water proliferated. Toward the end of the stage, rapid eustatic changes occurred, infringing marked spatial variations of the emerged lands, which led to the colonization of new West Indian island territories.

Paleogene Plants, Pollen, and Spores

The vegetation underwent significant variations that responded to its adaptations to the climatic changes experienced during this period.

Among the enormous volumes of sediments carried by the rivers, which eroded the lands that emerged in at least the central archipelago (the land that is today the Macizo de Guamuhaya was already rising); the central part of the Camagüey territory and the southeast end are very proliferous in plant remains, carbonized or replaced by iron hydroxides (Albear and Iturralde Vinent 1985; Rojas Consuegra y Núñez Cambra 1998; Rojas Consuegra y Denis 2011; Villegas Martín et al. 2016). Also, these rocks are very rich in pollen and spores that bear witness to the existence of a vegetative cover, but are still scarcely studied (Areces Mallea 1985, 1987, 1988; Areces Mallea y García 1990). From the Cuban stratigraphic record of this period, severe stages of high rainfall (with the production of turbiditic sediments), probably alternated with others of low humidity, are deduced.

It is inferred that toward the end of this period, many of the current plants had been established in the Caribbean region. A fact that justifies such data: 75% of the genera of identified plants, by means of analysis of the fossil pollen in Puerto Rico, are today among the Antillean flora (Grahan and Jarzen 1969): *Eugenia*, *Casearia*, *Bombax*, *Guarea*, *Faramea*, and several others.

Apparently, *Pinus* (*Pinus* cf. *silvestris*) was also present in the primitive Cuban flora, also recorded in sedimentary deposits of this age (Areces Mallea 1978). The most effective route for such plant colonization has been the bridge established through GAARlandia (MacPhee and Iturralde Vinent 1999). In this period, many of the plant species that make up the current Cuban flora were established (López Almirall 2005, 2015).

Paleogene Invertebrates

In Cuba, the profuse siliciclastic and carbonatic sequences of the Paleogene period are rich in echinoderms, other macroinvertebrates, abundant ichnofossils, diverse foraminifera, radiolarians, and nanofossils (Valette 1926; Sánchez Roig 1926b, 1949; Bermúdez 1950; Downs 1950; Cooper 1955; Miller and Furnish 1956; Juno 1966; Franco Álvarez 1983; Galacz 1988; Iturralde Vinent et al. 1998, 2001). Vertebrate remains are more rare, excluding shark's fossil teeth (Table 3.4).

- **Echinoids**

The echinoids (Fig. 3.19) are abundant and dozens of species are counted (Nuevitas, Jatibonico, Camazán, Maquey formations). Other genera of fossil sea urchins are listed: *Agassizia*, *Aguayoaster*, *Amblypygus*, *Anomalanthus*, *Antillaster*, *Asterostoma*, *Brissoides*, *Brissopatagus*, *Brissus*, *Bunactis*, *Caribbaster*, *Cidaris*, *Clypeaster*, *Cubanaster*, *Cylindrolampas*, *Cypholampas*, *Deakia*, *Echinolampas*, *Echinometra*, *Echinoneus*, *Encope*, *Eupatagus*, *Gauthieria*, *Goniocidaris*, *Habanaster*, *Habanaster*, *Haimea*, *Hemiaster*, *Hernandezaster*, *Jacksonaster*, *Laganidea*, *Laganum*, *Lajanaster*, *Lambertona*, *Linthia*, *Macropneustes*, *Meoma*, *Neorumphia*, *Oligopygus*, *Paraster*, *Pauropygus*, *Pedina*, *Periarchus*, *Pericosmus*, *Peronella*, *Phymosoma*, *Prenaster*, *Procassidulus*, *Progonolampas*, *Pseudorthopsis*, *Pygorhynchus*, *Rhynchopygus*, *Rojasaster*, *Sanchezaster*, *Schizaster*, *Scutella*, *Tarphypygus*, *Toxaster*. Some taxa, already existed from the end of the Cretaceous, and others, that appear in this period, will extend to the recent periods and up to the present.

There is also a crinoid (sea lily): *Balanocrinus cubensis*. Particularly, interesting is that two species of fossil asteroids (starfishes) are known from the Paleogene: *Nymphaster miocenicus* and *Stauranderaster sanchezi*.

- **Molluscs**

Marine fossil invertebrates are composed of gastropod molluscs (*Cypraea semen*, *Lerifusus angelicus*, *Hemisinus costatus*, *Margarites naticoide*, *Bursa* (*Bufonaria*) *ricardi*, *Tonna jamaicensis*, *Elmira naticoides*, *Terebellum subdistortum*, *Turritella* spp., *Natica* spp.) and some bivalve molluscs (*Chama engonia*, *Pseudomiltha haitiensis*, *Myrtaea asphaltica*, *Unio bitumen*; vesicomids, solemids). These organisms were abundant in the marine bottoms of the shallow water platforms of the Middle Paleogene (Eocene), where the biogenic limestones that contain them were originated (Charco Redondo Formation) (Fig. 3.20).

Among the cephalopod molluscs, at least one species of nautilus has been described in Paleogene rocks of Cuba, *Aturia aturi*. Some pieces have also been recovered, and are currently under study.

- **Brachiopods**

Interestingly, several species of fossil brachiopod (*Orthothyris radiata*, *Phragmothyris subplana*, *Ph. cubensis*, *Ph. costellata*, *Ph. rotunda*, *Ph. palmeri*) of the Paleogene that were only known in Cuban rocks were also described. A low level of study is still maintained.

- **Corals**

Fossil corals, mainly of Oligocene age, do not show a wide distribution, nevertheless, below them are known several

Table 3.4 Some lithostratigraphic units with Paleogene to Neogene fossil record

Age	Formations	Corals, Briozoans, algae	Bivalves and Gastropods	Decapods	Balanids	Echinoids	Nautiloids	Plants, Palynomorphs	Cartilaginous fishes	Bony fishes	Xenarthra
		Mainly fossil groups									
Paleocene—Middle Eocene	Group El Cobre								c	r	
Lower Paleocene	Gran Tierra										
Paleocene	Apolo										
Low. Eocene	Capdevila					r		c			
Low.-Mid. Eocene	Group Universidad					r			r		
Mid. Eocene	Lesca										
	Charco Redondo	c	a								
Upper Oligocene—Low. Miocene	Colón	c	a	a	c	a			a	a	c
	Lagunitas	r	a	c	c	c		r	c		c
Oligocene—Low. Miocene	Paso Real	r	a			r					
Low.-Mid. Mioceno	Vázquez	c	a			c	c				
	Arabos	r	a			c					
Low.-Upp. Miocene	Güines	r	a	c		a	r			c	
Pliocene	El Abra	c	c						a	c	

Age	Formations	Primates	Sirenids	Rodents	Crocodiles	Turtles	Ostracods	Foraminifera	Nanoplankton	Radiolarians	Icnofossils
		Mainly fossil groups									
Paleocene—Middle Eocene	Group El Cobre							a			c
Lower Paleocene	Gran Tierra						a	a			a
Paleocene	Apolo							a			a
Low. Eocene	Capdevila							a	c	a	a
Low.-Mid. Eocene	Group Universidad						a	a	a	c	c
Mid. Eocene	Lesca							a		c	
	Charco Redondo							a			c
Upper Oligocene—Low. Miocene	Colón				r		a	a	c		c
	Lagunitas		a	r	c	c	c	c			c
Oligocene—Low. Miocene	Paso Real						a	a			c
Low.-Mid. Miocene	Vázquez						a	a			c
	Arabos							a			r
Low.-Upp. Miocene	Güines						a	a			c
Pliocene	El Abra							a			

Fig. 3.19 Wood intensely bioturbed by a xylophagous, which produced a high concentration of perforations (inchnogenus Teredolites sp.). Monos Formation. Photo courtesy of Carlos Borges and Alberto Arano

Fig. 3.20 Abundant shells of gastropods in marmorized limestones of the Paleogene, Middle Eocene. Charco Redondo Formation

genera and species: *Acropora panamensis, A. salutensis, Agathiphyllia angullensis, A. antiguensis, A. splendens, A. tenus, Antiguastrea cellulosa, A. guantanamensis, A. meineri, Diploastrea crassolamellata, Diploria antiguensis, Goniopora decaturensis, Manicina willoubiensis, Montastrea costata, M. imperatoris, Pironastrea antiguensis, Pocillopora arnoldoi, Pocillopora guantanamensis, Porites panamensis, Siderastrea conferta, Stylophora canalis, Trochoseris meiazeri, Stylophora affinis, Stylophora granulata*; mainly in the eastern region of the country.

- **Vermes**

It is also mentioned, along with the *Lanieria* echinod, as resedimented material, calcareous tubes of the polychaete worm *P. onix*, typical of the Late Cretaceous.

- **Algae**

Calcareous algae are frequently observed and relatively abundant in shallow marine limestones, in high-energy and well-oxygenated media, where they formed oncolites and carpets or covers, laminar, and irregular, on hard substrates.

Some taxa of paleogenic algae are recognized: *Distichoplax biserialis, Terquemella* sp., *Mesophyllum* sp., *Jania* sp., *Melobesia* sp., *Lithothamnium* sp.

- **Microfossils**

Among the invertebrates, microfossils are the most varied and diverse group during the Paleogene, when also appear the large orbitoidal foraminifera (Fig. 3.21), which reach true giant macroscopic sizes (from several millimeters to more than one centimeter), and formed coquines or "lumaquelas", a kind of rock constituted almost exclusively of its shells.

The radiolarians are a very abundant group in marine sequences of deep waters, many times where there are no other microfossils. The knowledge about nanofossils is in clear increase.

The use of microfossils is still weak in Cuba, taking into account its usefulness in paleogeographic reconstructions, including paleobatimetric and paleoclimate reconstructions, as well as global geo-events, with ample attention on the part of the scientific community at present at an international level.

Paleogene Vertebrates

There are no known fossils of terrestrial vertebrates of the Paleogene in Cuba. Marine vertebrates are represented by cartilaginous fish. However, it is possible that a fossil turtle found in the central region (Sagua La Grande), might be of this age (Fernández de Castro 1871; Iturralde Vinent et al. 1996, 1998, 2001; Rojas Consuegra 2009; Iturralde Vinent 2012).

- **Fishes**

Several species of cartilaginous fish of this age are known by their fossil teeth, at least three genera of fossil sharks (*Carcharodon landenensis, C. auriculatus, Striatolamia*

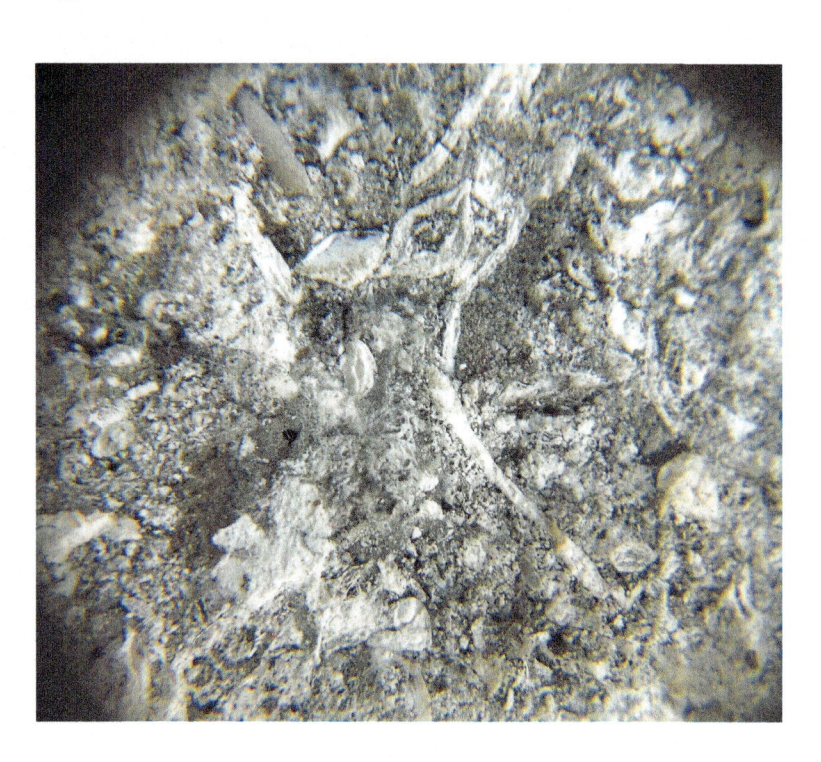

Fig. 3.21 Macroforaminifera in asphalt matrix of the Paleogene, Middle Eocene. Peñón Formation

Fig. 3.22 Traces fossils (*Scolicia* isp.) produced in sandy-clayey sediments of the Paleogene, Lower Eocene. Capdevila Formation

macrota, *Striatolamia* sp., *Isurus* sp.), and possibly a ray (*Aetobatus poeyi*).

There also have been collected for years at least two fossil skeletons of bony fish, with very good conservation, found in the volcano-sedimentary series of the eastern region, but they still remain unidentified.

Paleogene Ichnofossils

The ichnofossils are especially abundant in the Paleogene rocks, especially among the clay-sandy and clay-carbonated turbidites, accumulated in the wide marine basins of this period, throughout the Cuban Archipelago in consolidation, and on its current geographical position or very close to it (Bröniman and Rigassi 1963; Villegas Martín y Rojas Consuegra 2010, 2012; Menéndez Peñate et al. 2011; Villegas Martín et al. 2016; Rojas Consuegra et al. 2018).

- **Bioturbation**

The most common known fossil traces (Fig. 3.22) of this age have been species belonging to the genera *Scolicia*, *Chondrites*, *Planolites*, *Psammichnites*, *Taenidium*, *Helminthorhaphe*, *Cosmorhaphe*, *Ophiomorpha*, *Entobia*, *Zoophycos*, and *Paleodyction* (Gran Tierra, Apolo, Capdevila, Lesca, Ranchuelo, Vertientes, Nazareno, among the other formations).

- **Prints of plants**

In a section of turbidites, common in the Paleogene of the western region of Cuba, fossil plant tracks have been reported: rhizobioturbation, which reflect a shallow and temporarily exposed coastal marine paleoenvironment (Villegas Martín et al. 2014).

3.4.1.4 Neogene Period

In this period, the land connection of GAARlandia disappears. An extensive archipelago is consolidated, formed by three groups of large islands and numerous keys, joined by a wide marine platform, covered by the shallow sea with warm, clean, and well-oxygenated water, crossed at the beginning by deeper channels, which were gradually being filled with marine sediments (Iturralde Vinent 2004).

In some regions of the marine basins, clay-sandy sediments were predominantly deposited, under the influence of the first rivers in formation. In others, large volumes of carbonate materials accumulated, produced by the rich and diverse marine benthic fauna, widely established. Toward the end of the period, an extensive marine regression occurs (about 11–10 My ago), with the general emersion of the archipelago. It gives rise to the formation of large terrestrial areas, more extensive than the current ones, since they included the marine platform, which is today submerged. The modeling of the current relief of karstic plains on the

Fig. 3.23 Oxidized impressions of plants of the Upper Neogene, Pliocene. El Abra Formation

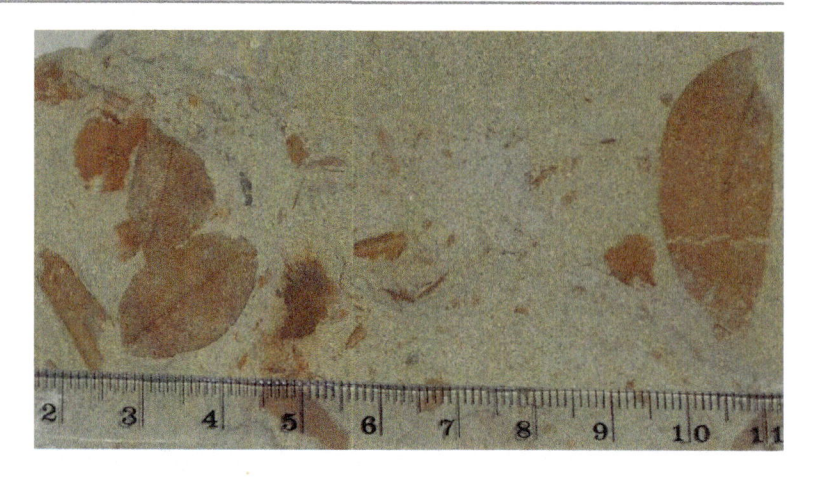

ancient shallow sea beds and the low mountain system would begin (Iturralde Vinent 2012).

Neogene Plants, Pollen, and Spores

In Cuba, there are few fossil remains of plants from Neogene Period, although the rocks formed in that stage commonly contain carbonized remains or films of iron and manganese oxides and hydroxides, but still not identified (Fig. 3.23). In addition, they are rich in fossil pollen, still not sufficiently studied to this day. Among the carbonate rocks abundant marine algae, typical of the phytoplankton of that past, are reported.

The Neogene terrestrial flora was very similar to the present one. In Cuban sediments, it has been possible to identify fossil remains of plants belonging to several families (Leguminosae, Meliaceae, and Arecaceae), and also *Calophyllum* (Berry 1939). These taxa, as well as others known from the Caribbean, mostly show a direct relationship with the vegetation of the South American continent.

After the disappearance of GAARlandia, several taxa of plants may have arrived in Cuba, such as the Austral Pines (*Pinus caribaea*, *P. cubensis*, Western *Pinus*, *P. elliotti* and *P. maestrensis*); probably, from the North American continent (López Almirall 2005, 2015).

From the stratigraphic recording of the period, the conformation of a very complex lithological and geomorphological substrate throughout that primitive Cuban Archipelago is deduced, with the formation of varied types of insipient soils, and therefore, a clear ecological conditioning for a wide diversification of Neogene flora.

Neogene Invertebrates

The rocks originated during the Neogene are generally very rich in fossils of marine macroinvertebrates and diverse microfossils. An extensive list of fossil marine organisms is known, where echinoderms stand out for their high diversity; perhaps followed by molluscs, both bivalve and gastropod. The first news about fossil invertebrates is due to the work of the Portuguese personality Ramón de la Sagra y Huerta (1787).

Numerous studies have contributed to the knowledge of the fossil record of the Neogene in the Cuban territory (Lea 1841; Cortázar 1880; Coteau 1881; Sánchez Roig 1920b, 1949; Jaume y Pérez Farfante 1942; Bermúdez 1950; Palmer D 1950; Bronnimann and Rigassi 1963; Nagy et al. 1976; Franco, Nagy y Radocz 1983; Albear e Iturralde Vinent 1985; Sánchez Arango 1985; Franco Álvarez et al. 1992; Franco Álvarez y Delgado Lamas 1997; Schweitzer and Iturralde Vinent 2004; Schweitzer et al. 2006; López Martínez 2007; Díaz Franco et al. 2008; López Martínez et al. 2008; López Martínez y Rojas Consuegra 2008; Varela y Rojas Consuegra 2009, 2011a, b; Varela and Schweitzer 2010; Varela 2013). These fossils are common in numerous formations, this can be found in the Léxico Estratigráfico de Cuba (Franco Álvarez et al. 1992; IGP 2013) (Table 3.4).

- **Echinoids**

In general, fossil equinoids show a high diversity; represented by some 40 families, 109 genera and subgeneras, and more than 400 species and subspecies. Among the Miocene sea urchin are present: *Agassizia*, *Antillaster*, *Amblypneustes*, *Brissopsis*, *Cassidulus*, *Coronanthus*, *Cyclaster*, *Echinopedina*, *Encope*, *Eupatagus*, *Fernandezaster*, *Gomphechinus*, *Goniocidaris*, *Habanaster*, *Hemicidaris*, *Hernandezaster*, *Homeopetalus*, *Jacksonaster*, *Lajanaster*, *Macrolampas*, *Macropneustes*, *Meoma*, *Migliorinia*, *Nucleolites*, *Nucleopygus*, *Paraster*, *Pliolampas*, *Plistophyma*, *Prosostoma*, *Rhyncholampas*, *Sanchezaster*, *Schizobrissus*, *Trochalosoma*.

Fig. 3.24 Internal mold of a gastropod of the Neogene, Miocene. Güines Formation

- **Crinoids**

Some unidentified crinoid specimens have also appeared; this is a group that needs a better study. There is material in collections.

- **Bivalves**

The marine bivalve molluscs within the Miocene series are very common and diverse, and numerous genera are known: *Aequipecten*, *Amusium*, *Anomia*, *Argopecten*, *Cardia*, *Chione*, *Chlamys*, *Glycymeris*, *Crassostrea*, *Hyotissa*, *Kuphus*, *Malea*, *Ostrea*, *Orthualax*, *Pecten*, *Spondylus* and others; and very diverse species. The fossil oysters (*Ostrea*) are one of the most abundant groups in the early and middle Neogene.

- **Gastropods**

Gastropod molluscs also show a very significant diversity (Fig. 3.24). This group has been revised, proving the presence of several genera and species in the Cuban Neogene rocks: *Conus*, *Oliva*, *Siphocypraea*, *Turritella*, *Polyneices*, *Xenophora*, *Scalina*, *Mitra*, *Strombus*, *Melongena*, *Olivella*, *Turritella*, *Sconcia*, and others; and there are some taxa to be identified.

- **Nautiloids**

It corresponds to a cephalopod mollusc one of the oldest fossil descriptions for Cuba, *Nautilus cubaensis*. They have been found in Miocene rocks, along with several internal molds, belonging to this nautiloid (Arabos and Güines formations).

- **Corals**

The fossil corals are very common in the marine limestones of Miocene age (Paso Real, Arabo, Güines and other formations), where several species of different genera appear: *Acropora*, *Agathiphyllia*, *Antiguastrea*, *Diploastrea*, *Diploria*, *Goniopora*, *Manicina*, *Montastrea*, *Pironastrea*, *Pocillopora*, *Porites*, *Siderastrea*, *Stylophora*, *Trochoseris*, *Solenastrea*, *Stephanocoenia*, *Saidcoenia*, and *Thysanus*.

- **Briozoans**

Frequently they are also, in the rocks of this period, bryozoans and algae (*Mesophyllum* sp., *Jania* sp., *Melobesia* sp., *Lithothamnium* sp., *Arqueolithothamnium* sp., *Lithophyllum* sp., *Amphiroa* sp.), but they have been scarcely studied.

- **Crustaceans**

For the Neogene of Cuba, there is a high diversity of fossil crustaceans, including decapods (crabs and crabs). Several genera are known for the Caribbean region, and more than 20 taxa are listed in the country, some species only known in this territory. The genera have been replaced: *Arenaeus*, *Callinectes*, *Eriosachila*, *Eriphia*, *Euphylax* (Fig. 3.25), *Eurytium*, *Hepathus*, *Iliacantha*, *Libinia*, *Necronectes*,

Fig. 3.25 Decapod species *Euphylax* sp. from the Lower Miocene. Colón Formation

Panopeus, *Paraeuphylax*, *Persephona*, *Portunus*, *Spinolambrus*, and *Mithrax*.

- **Cirripeds**

Also, the cirriped crustaceans are common (Fig. 3.26), represented by the balanids (*Balanus* sp.). In addition, there are other microscopic crustaceans, very abundant and diverse, the ostracods. Along with the balanid crustaceans, attached to the shells of molluscs, the fossil footprints of the activity of the polychaete worms are frequent, and similar associations can be observed in the current seas.

- **Microfossils**

The shells of the foraminifera, both microscopic and macroscopic, are very abundant and diverse, since these organisms occupied varied means in the marine environment, from the shallow bottoms to the great depths; and, in addition, they floated in the whole column of seawater. In

Fig. 3.26 Numerous specimens of cirripeds (*Balanus* sp.), adhered on an oyster shell. Lagunitas Formation

the Cuban geological literature, numerous taxa are listed, in all formations of this age (see Léxico Estratigráfico de Cuba: Franco Álvarez et al. 1992; IGP 2013). The algae of this age are still to be studied in detail.

Neogene Vertebrates

In the locality "Domo Zaza", in Sancti Spíritus Province, fossil remains of marine vertebrates have been found, such as dugongs (sirenids), crocodiles, turtles, sharks, rays, and whales. This fossil association is a witness of the paleofauna that inhabited the seas and coasts of the Cuban central archipelago in the Neogene past, and that, due to its uniqueness, acquires an exceptional value for the biogeography, regionally and globally.

Neogene Cuban marine rocks are characterized by containing various fossil elements (vertebrae, dental plates, teeth, and caudal spines) of sharks and rays; in addition, there are teeth, scales, and skeletal parts of some bone fish.

The fossil teeth of different shark species (some exclusively known to science for their fossil remains) are some of the most characteristic paleontological elements of the Miocene rocks in Cuba. His finding is frequent in the La Habana–Matanzas region, where he draws the attention of a wide public, being frequently published several news (Valdés 1855; MacPhee and Iturralde Vinent 1994; Iturralde Vinent et al. 1996, 1998, 2001; Franco Álvarez y Delgado Damas 1997; Iturralde Vinent and Case 1998; MacPhee et al. 2003; Gaffney 2003; Iturralde Vinent 2004; Domning and Aguilera 2008; Díaz Franco y Rojas Consuegra 2009; Aranda Manteca et al. 2011; Rojas Consuegra y Viñola López 2013, 2016; Brochu and Jiménez Vázquez 2014; Viñola López et al. 2019).

- **Terrestrial mammals**

In rocks of the Neogene (Lagunitas Formation) of the Domo Zaza deposit, the oldest fossils of terrestrial mammals of our country have been discovered, related to typical animals of South America. Among them are known: a primate (*Paralouatta marianae*), a primitive sloth (*Imagocnus zazae*), and a rodent (*Zazamys veronicae*) (MacPhee and Iturralde Vinent 1994; MacPhee et al. 2003).

- **Aquatic mammals**

In the Miocene rocks (Colón, Jaruco, and Güines formations), rest of sirenids of the *Metaxytherium* group are relatively common (Fig. 3.27). But the opinion that in the Domo Zaza (Outstanding Natural Monument) are skeletal remains of this and another dugong of the genus *Nanosiren*, is very attractive.

Here, too, it is interesting that the only whale tooth (Cetaceus) has been discovered for Cuba and the Antilles.

- **Cartilaginous fishes**

The first published report of fossil shark teeth in Cuban territory dates back a century and a half ago, and corresponded precisely to the well-known *Carcharodon megalodon* (today belonging to the genus *Carcharocles*, according to some authors); found in the excavation of a

Fig. 3.27 Mold of the endocrane of a sirenium found in limestones of the Neogene. Colón Formation. Courtesy of Lázaro Viñola López

Fig. 3.28 Fossil teeth of Neogene sharks, Miocene. Güines Formation

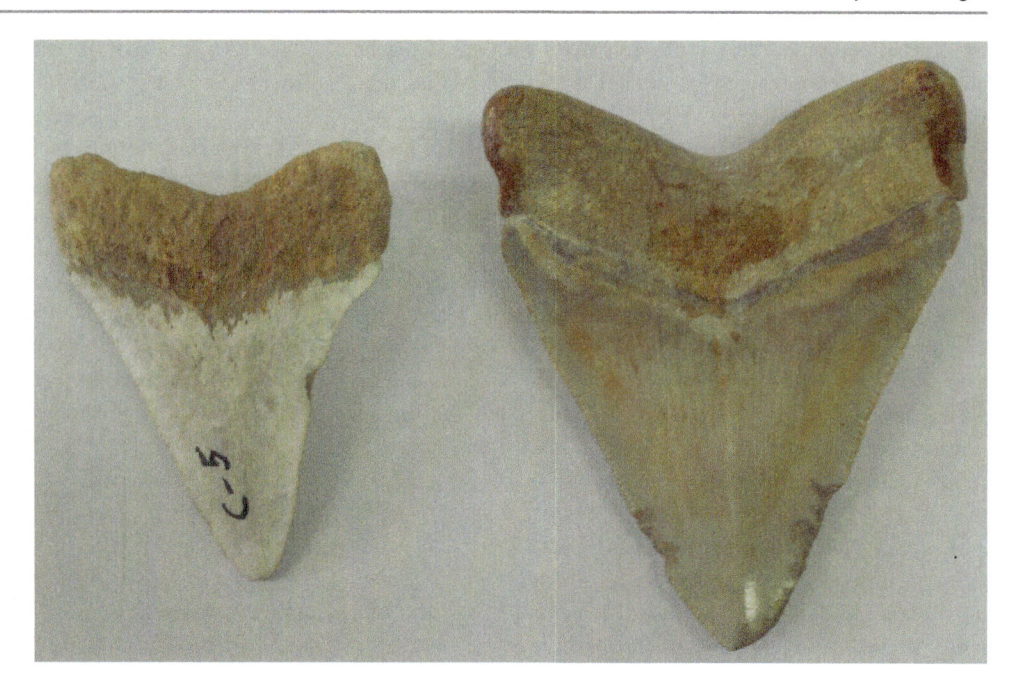

creole well at a depth of 12 ft. (about 4 m), in the Cárdenas region (Matanzas). This finding was described by Fernándo Valdés y Aguirre in 1855, in the Revista de La Habana, and for this, he had as a consultant the renowned Cuban ichthyologist Don Felipe Poey Aloy. In addition, the existence of several examples of fossil teeth in paleontological collections of the colonial era is already mentioned.

Currently, there is a wide diversity of Miocene fishes (Fig. 3.28), mainly cartilaginous (Colón, Cojímar, Güines formations), such as sharks (*Carcharhinus altimus*, *C. amblyrhynchos*, *C. limbatus*, *C. longimanus*, *C. obscurus*, *C. perezii*, *C. plumbeus*, *Carcharias taurus*, *Carcharocles megalodon*, *Carcharodon carcharias*, *C. subauriculatus*, *C. chubutensis*, *Galeocerdo contortus*, *Hemipristis serra*, *Isurus hastalis*, *I.* cf. *desori*, *Negaprion brevirostris*, *Sphyrna mokarran*, *S. prisca*) and also rays (*Aetobatus arcuatus*, *Aetomylaeus cubensis*, *A. cojimarensis*, *Myliobatis* sp.). There is a new material under study.

- **Bony fishes**

There are two species of bony fish *Diodon* or Parrot Fish (*Diodon scillae*, *D. circumflexus*, *D.* cf. *circumflexus*), several fossil teeth of "picua" (*Sphyraena* cf. *barracuda*), two new species of *Sparus* (*S. cinctus* and *S. neogenus*), and probably another of the genus *Opleognatus*. In addition, two species of *Balistes* (*B. vegai* nov. sp. and *B. crassidens*) are added.

We have news of the finding of some more molds in the province of Matanzas, where there is the greatest wealth of

Fig. 3.29 Fragment of bone skeleton of fish, from the Lower Miocene, collected in the region of Jaguey Grande, during the extraction of limestone blocks for construction. Colón Formation. Courtesy of Lázaro William Viñola López

this type of pieces in the country (Colón Formation). There is also fossil bone material from fish under study (Fig. 3.29).

- **Reptiles**

In the Domo Zaza site (Lagunitas Formation), numerous teeth and bony remains of crocodiles have also been recovered, possibly corresponding to two different lineages (sebecid and planocranid), since they resemble those of South America, and are not related to the Quaternary and current crocodiles.

In this site, they have also collected, bone remains and carapace of turtles (pelumedúcides), apparently corresponding to at least two taxa (flattened large carapaces and other small concaves). In several other locations, belonging to various formations, remains of turtles (Colón, Güines and in other formations) have been reported.

Neogene Ichnofossils

In Neogene rocks it is possible to recognize various ichnofossils, such as root prints, different types of perforations, bioerosion structures, traces of epibiot organisms, fossil burrows, etc.; but there are still few studies devoted to this subject (López Martínez 2006; López Martínez et al. 2008; Villegas Martín 2009; Villegas Martín y Rojas Consuegra 2012).

- **Bioerosion**

Bioerosion is common in the rocks of the wide shallow marine platforms of the Neogene. It is common to find corals (*Montastrea*) with traces of perforations (*Gastrochaenolites* isp.) of molluscs, sometimes particularly large. It can be considered that the first period of ichnofossils in Cuba was due to Ramón de la Sagra y Huerta (1855), when it illustrates, among the gastropod molluscs, at least, the current *Gastrochaenolites* ichnotaxons (Paleontology, Volume 3, Tab. Figs. 1–4, Tab VI, Figs. 11 and 12, and *Oichnus* isp., Tab VIII, Fig. 3).

A significant finding is the presence of the *Entobia*, *Oichnus*, and *Centrichnus* ichnogenera, on mollusc bioclasts, in lateritic deposits (Fig. 3.30).

- **Bioconstruction**

Some of the most common fossils of this period are the calcareous tubes made by the perforation molluscs of the

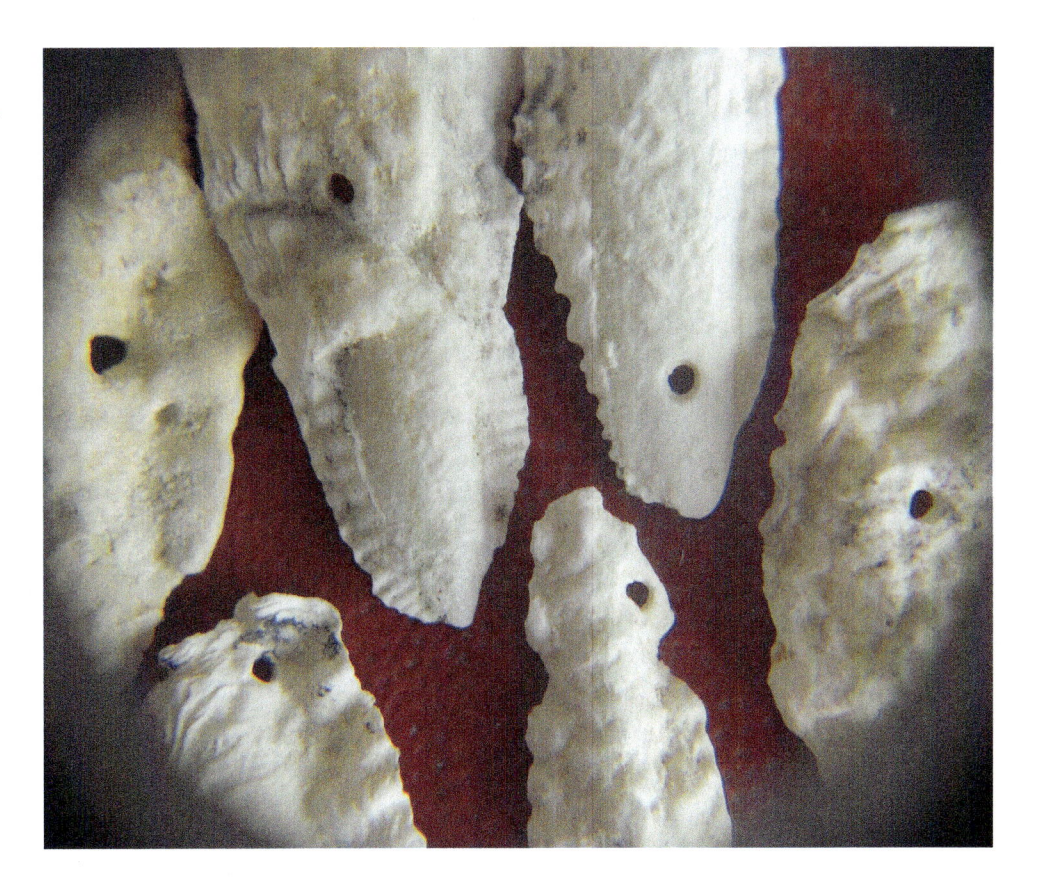

Fig. 3.30 Very small perforations of a predator on bivalve shells *Lima* sp. Lagunitas Formation

group of teredos, called *Kuphus incrassatus* or "Teredo tubes", which are molds of their housing cavities. These calcareous tubes are very abundant in some Miocene outcrops of loams and clay limestone, even preserved in production position.

In the Domo Zaza site, among the shallow marine rocks or keys, a possible bioconstruction appears, apparently originated by the filling of a burrow of some animal, possibly terrestrial, or at least large benthic, which inhabited the deltaic environment.

- **Bioturbation**

In levels of paleosoils may appear the marks left by the plants that inhabited an area in the past. Traces of roots are known in lithified lagoon clays (Lagunitas Formation); they begin from the surface of a paleosoil (purple color), under a layer of yellowish sandy clay, which buried it later (Domo Zaza, Sancti Spíritus).

3.4.1.5 Quaternary Period

During the Quaternary period, the Cuban Archipelago acquires its current physiographic features, going through the changing climatic conditions that characterized the most recent stage of the evolution of life and our planet; which extends to the present.

The geography of our region experienced significant oscillations in the marine level, with the consequent intense dynamics in the relationship between the areas of emerged lands and those eventually flooded by the sea, taken temporarily by swamps and marshes (see Iturralde Vinent 2004, 2013).

In the emerged territories, the processes of tropical karstification took place, giving rise to the so-called sequential formations (speleothems), of a peculiar subterranean environment, with a wide extension in present-day Cuba (with 65% karstifiable surface).

The drastic climatic changes of the Quaternary marked the physiography of the current Cuban Archipelago, with varied biotopes, exclusive ecosystems, and unique natural landscapes. In the Early Quaternary, there is generally a global terrestrial cooling. The climate is changing, initially warm and humid. The seas contain warm shallow waters, well oxygenated and rich in nutrients. In the hemerited lands, weathering and physical–chemical erosion are intense.

For the Middle Quaternary, the climate continues unstable, it is warm and humid, with cold and dry stages. There are rapid changes in sea level. In general, tropical weathering is intense, conditioning rapid karstification and increased pedogenesis. The vegetation cover is diversified profusely.

Toward the Late Quaternary, the climate is still changing. Warm and humid stages stand out, with a wide transgression and the formation of a shallow carbonate platform. There are marked episodic lapses of cold and dry weather. The increase of the carbon sinks in the generalized mangroves and in the consolidated marine vegetation, also by the burying of organic matter in the extended wetlands and swamps. They are inferred episode of paleolluvias, with clay deposits, among fossilized eolianites accumulated possibly in cold and dry stages. Karstification and the formation of diverse soils are intense. The vegetation becomes even more abundant and heterogeneous, similar to the current one.

The Quaternary marine rocks are arranged around almost all the Cuban territory, in the form of strips generally associated with the current coasts, a few meters above sea level. They contain a fossil record made up of fossil entities produced by organisms very similar to those that inhabit the current marine platform; with some living species in our underwater territory.

Quaternary Plants, Pollen, and Spores

The climatic variations, and the particular geological and edaphic mosaic (types of soils), which characterize the Cuban Quaternary, among other factors, had a definite influence on the conformation of the current flora of our archipelago.

The fossil record of Cuban flora is relatively scarce. It highlights the presence of some fossil plants, silicified, preserved due to the substitution of wood by silica (silicon oxide-xylopalo). In some sandy-clay alluvium-marine rocks, there are few vegetable molds (leaves and stems), generally replaced by iron hydroxide.

An interesting case is the excellent conservation of woods, leaves, and fruits, impregnated in hydrocarbons associated with asphalt or natural "chapapote" springs, which formed deposits of asphaltite and grahamite, in several places in Cuba. This was the consequence of the natural oil spill to the earth's surface, thousands or millions of years ago. Among the main fossil species reported, in this special type of paleontological site, savanna taxa are recognized (*Pinus caribaea*, *Coccothrinax* sp., *Cordia galeottina* and *Thrinax radiata*); others of freshwater lagoons (*Caraphyllum* sp., *Spondia lutea*, *Picus* sp.); and also, those of coastal areas (*Cordia sebestena* and *Chrysobalanus icaco*), together with marine molluscs (Berry, 1934, 1939; Iturralde Vinent et al. 1996, 1999, 2000; Díaz Franco 2005; Suárez 2006).

In some coastal marshes and swamps, peat deposits have formed during the last thousands of years, with the preservation of potentially identifiable parts of the native vegetation. The Cuban mobs have thrown an age of about 5–11,000 years, which indicates that the process of formation of the cienagas that contain them is relatively recent. In the argillaceous rocks, it is common that microscopic vegetal elements (pollen and spores) have been preserved, contributed by the herbaceous vegetation, cortadera, mangroves, yanas, etc.; but they still await particular paleobotanical and paleopaulological research (Cosculluela 1918; Olenin 1985).

In several parts of the country, related to springs and streams of saturated carbonate water, currently (under normal natural conditions) excellent molds of different species of the flora of the areas surrounding these sources are formed, which are embedded in layers successive travertine.

Among the sedimentary fill of the extended karstic cavities (caves, caverns, lapels, casimbas, etc.) the appearance of charcoal is not rare, which in many cases, witness the occurrence of fires produced by natural causes in the prehistoric past.

From the Pliocene, most of the components of the rainforests arrived in Cuba, probably from Central America and through the Cayman crest; and in addition, savannahs and xeromorphic bushes are enriched. It seems to have arrived at this time, among others: Asteraceae, Rubiaceae (*Psychotria*), and Orchidaceae (*Lepanthes* and *Pleuothallis*). They also followed this path, numerous taxa originating from the South American continent, taking advantage of the closure of the ism of Panama, such as Dillenaceae (*Davilla*), Eriocaulaceae (*Paepalanthus*), Cyperaceae (López Almirall 2005, 2015).

Likewise, species originated in the continent of Laurasia arrived to the Cuban territory, such as *Quercus* (*Q. cubena*) and *Salix* (*S. caroliniana*). And more recently, others entered from the peninsula of Florida, *Fraxinus* (*F. carolina* var *cubensis*) (López Almirall 2005, 2015). The paleobotanical studies that could confirm these considerations are still to be faced in Cuba.

Quaternary Invertebrates

Abundant shells, calcareous skeletons, and other elements (teeth, scales, weavings, coprolites, etc.) were contributed by the rich fauna of marine invertebrates that swarmed in those waters in the last thousands of years; many of which still live in the varied mediums of the shallow seas that bathe the islands, their thousands of cays, bays, lagoons, and coastal marshes, characteristic of our archipelago (Sánchez Roig 1920b, 1949; Jaume y Pérez Farfante 1942; Richards 1935; Brönnimann and Rigassi 1963; Franco Álvarez, Nagy y Radocz 1983; Kier 1984; Albear e Iturralde Vinent 1985; De la Torre Callejas y González Guillén 1998; Espinosa y Ortea 1999; Díaz Franco y Rojas Consuegra 2001, 2009; Díaz Franco y Jiménez Vázquez 2008; Fernández Millera 1997; Franco Álvarez y Delgado Lamas 1997; Torre y Kojumdgieva 1985; Iturralde Vinent et al. 1999, 2000; Iturralde Vinent 2004, 2012; Perera Montero y Rojas Consuegra 2005).

In general, studies of Cuban fossil invertebrates require even more attention, and although some groups have already been treated in detail, almost all of them suffer from a rigorous taxonomic update.

- **Bivalves**

Among the most well-known marine invertebrates of this period are bivalve molluscs, represented by hundreds of species of different genera (*Arca, Argopecten, Barbatia, Hyotissa, Cardium, Chama, Chione, Chlamys, Lima, Lopha, Lucina, Mytilus, Ostrea, Pecten, Spondylus, Tellina, Xenophora*, and others), preserved in the most recent marine rocks (Jaimanitas Formation).

- **Marine gastropods**

Something similar reflect marine gastropod molluscs, contained in the younger limestones, where numerous genera have been recognized (*Architectonica, Bulla, Cassis, Cancellaria, Cerion, Ceritium, Citharium, Conus, Cypraea, Diodora, Fissurella, Melongena, Natica, Nassarius, Nerita, Oliva, Polinices, Rissoina, Strombus, Regula, Tricolia, Turbo* and others); most of which present several species each, which inhabit today in the seas surrounding Cuban Archipelago (Fig. 3.31).

- **Scaphopods**

More rarely mentioned, among the Quaternary marine molluscs, the presence of scaphopods (*Dentalium antillarum, Dentalium* sp.), a group of fossils practically untreated in specialized bibliography.

- **Echinoids**

Another fossil group common in the Quaternary rocks is one of the equinoids, with several sorts (*Brissus, Clypeaster, Echinoneus, Schizaster, Tarphypygus, Mellita*, and others), to which several living species belong.

- **Corals**

In the carbonate rocks formed in the last hundred to tens of thousands of years ago, which are almost completely distributed around the Cuban coastlines, only a few meters high from the current normal level of the sea, there are particular associations of corals (*Acropora, Agaricia, Cladocora, Colpophyllia, Davia, Diploria, Eusmilia, Isophyllia,*

Fig. 3.31 *Strombus giga*, a large gastropod common in the marine rocks of the Quaternary. It is easily seen on the colonial walls of Habana Vieja. Jaimanitas Formation. This species has current representatives (known as Cobo), which inhabit the shallow seas of Cuba

Fig. 3.32 The corals are diverse and abundant in the limestone rocks of the Quaternary. Jaimanitas Formation

Madracis, Meandrina, Montastrea, Mycetophyllia, Porites, Siderastrea, Solenastrea, Stephanocoenia, Stylophora and others), very well preserved, fossilized in situ (Fig. 3.32). Many of these species live today in the shallow seabed of the Cuban marine platform.

- **Hydrozoans**

Also in the literature, a species of hydrozoans (*Nellia oculata*) is mentioned; a fossil group quite common, but practically not studied until now in Cuba.

Fig. 3.33 True coquina of diverse shells of terrestrial gastropods, formed in a Quaternary cave

- **Crustaceans**

Among the few fossil crustaceans of the Quaternary, apparently only two genera of decapod crabs (*Calianassa* sp. and *Mithrax hispidus*) have been reported.

- **Terrestrial gastropods**

Of this group, more than 1400 species live in Cuban territory. In the clay deposits, recent or present, formed in the karst cavities, the accumulations of shells of the rich populations of terrestrial pulmonate molluscs are very common and abundant (Fig. 3.33).

In the subterranean Cuban hypogeum environment, due to the pluvial drag, true coquines of hundreds and thousands of copies of these terrestrial molluscs are formed, which could be potentially useful in ecological or other studies.

More rarely, coastal lung mollusc shells, of living species (*Cerion*, *Ligus*, *Planorbis*), conserved in the lithified Quaternary paleodunes, also appear.

Striking are the shells, which, inside the caves, are covered by secondary carbonate, due to the precipitation of the salts (carbonate) contributed by the saturated waters that infiltrate and accumulate in the cave interior. In this case, the calcareous crust or envelovement of the snail (shell) gets to turn them into real concretions (Fernández López 2000), very similar to the own "pearls of cave"—giant ooncolites; often, the shell of the gastropods is unrecognizable externally, and its existence can only be verified by mechanically breaking the carbonate coverture.

- **Insects**

In the Quaternary of Cuba, it is rare to find insects preserved in asphaltite or fossil tar in the San Felipe site (Matanzas), which can be considered the only ones known to date in the country (Valdés 1999). On the other hand, a curious case is the insects trapped in stalagmite in a cave, also in Matanzas (Díaz Franco y Rojas Consuegra 2001).

Quaternary Terrestrial Vertebrates

Most of the terrestrial Cuban vertebrates, extinct and living, have their ancestors in the southern part of South America. The insectivores, in changes, seem to have arrived from North America, where they are distinctive. The other groups of vertebrates show varied origins, depending, among other factors, on the dispersion mechanisms of each one (MacPhee and Iturralde Vinent 2003; Silva et al. 2008). On some

Fig. 3.34 Large bony remains of sloths on a sandy asphalt matrix of the Quaternary Las Breas deposit in San Felipe, Matanzas

groups of fossil vertebrates, the current knowledge is still limited, and deserve in general, a greater attention, such as saurians and amphibians.

The greatest wealth exhibited by the Cuban Quaternary fossil record is provided by the peculiar and diverse fossil material produced by the megafauna of terrestrial vertebrates, which inhabited the territory in the last hundreds of thousands and thousands of years ago. Likewise, the skeletal remains of the large sloths, the various rodents, the gigantic birds (predators and gunboats), the small and giant insectivores, numerous chiroptera, reptiles, and amphibians stand out; among other animals disappeared in the recent past.

The most widespread Quaternary fossiliferous deposits in Cuba are those accumulated in the extensive karstic cavities that occupy large regions of the country (De la Torre y Huerta and Matthew 1915; De la Torre y Huerta 1917; Wetmore 1928; Aguayo 1950; Silva Taboada 1979; Acevedo González y Arredondo de la Mata 1982; Kartachov et al. 1981; Suárez y Díaz Franco 2003; Suárez 2004; Silva Taboada et al. 2008; Suárez and Olson, 2015; Aranda Pedroso et al. 2015, 2017).

- **Primates**

Among the most suggestive fossils of the Cuban Quaternary, primates stand out, which had already appeared in the Miocene fossil record (Domo Zaza site), and which are represented, also in the Pleistocene, by a species of howler monkey (*Paralouatta varonai*).

- **Xenarthra**

Among the Pleistocene megalonichids fossil species, with Miocenic ancestors, are several different types of sloths (*Megalocnus rodens*, *Parocnus browni*, *Neocnus gliriformis*, the major *Neocnus* and *Acratocnus antillensis*) (Silva Taboada et al. 2008); all now extinct, even though they apparently cohabit with the primitive man who populated the Antillean islands (Fig. 3.34).

- **Rodents**

Another group with Miocene relatives in the Cuban fossil record are the rodents (MacPhee and Iturralde Vinent 1995 and other citations). A very interesting aspect of this group is the high diversity of fossil species that shows the Cuban Quaternary. Abundant, are the remains of large rodents called "Jutías", with several genera and fossil species: the capromids (*Capromys*, *Mesocapromys*, and *Geocapromys*), which surpass in quantity the living species in Cuba. In addition, fossils of the echyimid (*Brotomys* sp.) or spiny rats appear.

- **Insectivores**

In the Cuban fossiliferous deposits have been recognized two types of insectivores, some of them considered giants, such as the "Almiquí" (*Solenodon arredondoi*), and others, very small, called shrews (*Nesophontes*). These mammals, of primitive characteristics, have a living descendant (*Solenodon cubanus*); the other species, previously mentioned, are only known extinct, through their fossil remains.

The distribution in the Pleistocene past of the various groups of vertebrates, as shown by their fossil record, was very broad, occupying all of the Cuban territory (Silva Taboada et al. 2008); in contrast to the current impoverished diversity, it is an undoubted sign of the recent extinctions suffered in the region, and in particular in the Cuban archipelago.

- **Chiroptera**

Bats are a very diverse group of flying mammals today in Cuban territory, but also in the Quaternary past, a fact that attests to their well-known fossil record, since several genera and species are known (*Pteronotus pristinus*, *Mormoops magna*, *Cubanycteris silvai*, *Desmodus puntajudensis*, *Phyllops vetus* and *Phyllops silvai*—vampire bat). Most fossil species have no living representatives (Silva Taboada 1979; Suárez y Díaz Franco 2003); apparently as a result of the drastic ecological changes that occurred during the Pleistocene–Holocene period.

- **Birds**

Although the fossil record of birds in Cuba is relatively scarce, it does exhibit a wide diversity; despite the fact that the delicate nature of the bone material of this group tended not to favor fossilization in Quaternary karst taphonomic environments. A large group of fossil birds is represented in the country's karst deposits, and many of the species still inhabit the regions where their fossils have been recorded. Thus, a relatively high diversity of genera and species have been described in the fossil state, among which are the giant predatory and carrion birds, such as *Bubo*, *Pulsatrix*, *Ornimegalonyx oteroi*; *Buteo*, *Geranoaetus*, *Amplibuteo*, *Titanohierax*, *Gigantohierax*; *Caracaora*, *Milvago*, *Falco*; the condors (New World vultures, *Gymnogyps*); and the owls (*Tyto alba*, *T. noeli*, *T. pollens* and *T. cravesae*). In addition, fossils of other types of birds appear, like *Ciconia*, *Mycteria*, *Teratornis*, *Grus*, *Nesotrochis*, *Burhinus*, *Capella*, *Ara*, *Siphonorhis*, and *Scytalopus* (Suárez 2004, Suárez and Olson 2015).

Fig. 3.35 Quaternary turtle bones preserved in karstic deposit. Courtesy of Lázaro Viñola López

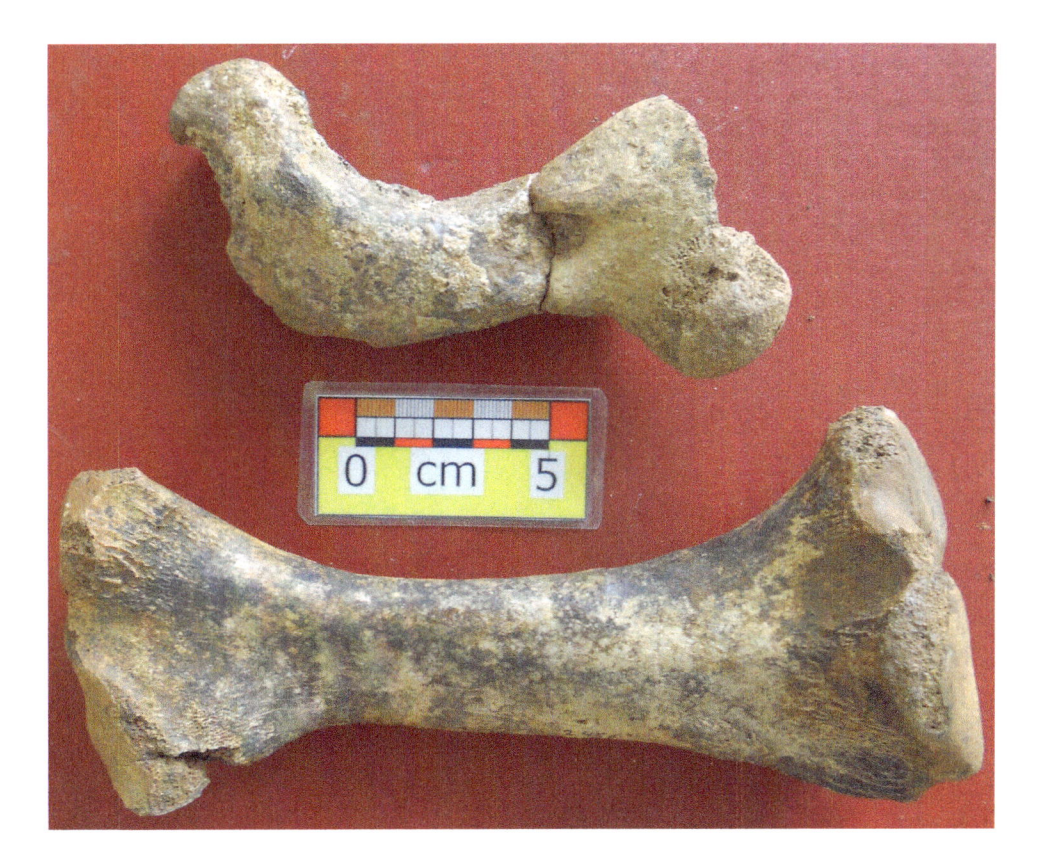

- **Reptiles**

The remains of fossil reptiles are common mainly in the cave sediments of the Cuban Quaternary. Often appear the turtle *Geochelone cubensis*, known only for its fossils (Fig. 3.35), and the Antillean crocodile (*Crocodylus antillensis*). In addition, the remains of iguanas (*Cyclura nubila* and the subspecies *C. nubila nubila*), boas (*Epicrates angulifer*), snakes and "jubos" (*Cubophis* sp., *Chilabothrus* sp., *Tropidophis* sp.) and lizards are still insufficiently studied.

- **Amphibians**

Toads and frogs (*Peltophryne empusus*, *Osteopilus septentrionalis*, *Lithobates* sp.), and other small vertebrates, are also recorded (Aranda Pedoso et al. 2017), and are even less known. In general, these are groups with few systematic studies that have been conducted in the country.

Quaternary Marine Vertebrates

Quaternary marine vertebrate fossils are rare, fragments of bones are mentioned among conglomerates of limestone, coral, shell of marine and terrestrial molluscs, oolites and pseudoolites (Kartachov et al. 1981); apparently accumulated as products of the eventual alluvial contribution to the shallow marine zone of the coast.

Human Bone Remains

Finally, in Cuban cave deposits (Fig. 3.36), skeletal remains of the aborigines (Antillean natives) have been preserved, sometimes mixed (mainly because of anthropic action) with older vertebrate fossils, and with the relicts of their own tools and diets.

This set of elements (artificially buried together) are objects of study of Archeology, Anthropology, and Archaeozoology, also historical disciplines and life, but with different methods and purposes, which are proposed to know the most recent History. However, often neophytes confuse Paleontology with the aforementioned sciences.

Quaternary Ichnofossils

Conserved in Quaternary rocks and recent sediments, ichnofossils are also recognized, mainly represented by perforations and galleries, due to the activity of the *Lithophaga* mollusc (Franco Álvarez y Delgado Lamas 1997).

The biodegradation marks on skeletons of marine invertebrates in general are common and diverse, but they remain without studying.

The products of the biodeposition of vertebrates (coprolites) are common in the cave sediments, the icnites themselves are rare, the traces of plant roots are frequent; also some impressions of trunks among eolianites from paleodunes (Núñez Jiménez 1988; Iturralde Vinent 2012; Villegas Martín y Rojas Consuegra 2012; Abreu Vega 2015; López Valdés y Ochoa Hernández 2016; Ochoa Hernández et al. 2017; Rojas Consuegra et al. 2017).

- **Ichnites—fossilized footprints**

There are only a few messages to confirm, in the Cuban literature, about possible vertebrate footprints—icnitas sensu stricto (Villegas Martín y Rojas Consuegra 2012): Cretaceous (Rojas Consuegra 1999); Quaternary (Fernández de Castro 1864; Torres 1910).

- **Bioerosion**

In general, the ichnofossils product of the bioerosional activity is common in many Quaternary rocks in the current coasts, in particular in ridges of storm and rare in coastal

Fig. 3.36 Various coprolites or pellets of rodents, mainly of "Jutias", preserved by the impregnation in asphaltite. Las Breas de San Felipe, Matanzas

Fig. 3.37 The vast majority of the fossiliferous vertebrate sites of the Quaternary in Cuba appear in the diverse manifestations of the extensive karst of their territory. Cave in Cayo Paredones, Sancti Spíritus

dunes. Many of them have already been described in the Antillean area, but still without attention in our country.

- **Biodeposition**

It is peculiar preservation of fossil excrements (*pellets*), called coprolites (Fig. 3.37). These fossils have been preserved in the cave clay sediments. They present special preservation in the natural asphalte deposits, due to the rapid impregnation suffered by these in contact with the hydrocarbons, which can reach its complete compositional substitution and even a certain mineral recrystallization.

- **Plant footprints**

Even without studying and describing taxonomically, the traces of the vegetation in the rocks, mainly Quaternary (Jaimanitas Formation and eolianites), are not rare to observe, as much in the plating of the walls in constructions of the city, as in outcrops (Fig. 3.38). Particularly, interesting are the fossil footprints of trees (external mold o cast), preserved among the Quaternary eolianites (Santa Fe Formation).

Molds of roots (rhizolites) are known in the Miocene or Quaternary laterites of eastern Cuba.

Fig. 3.38 Traces of plants in the Quaternary eolianits

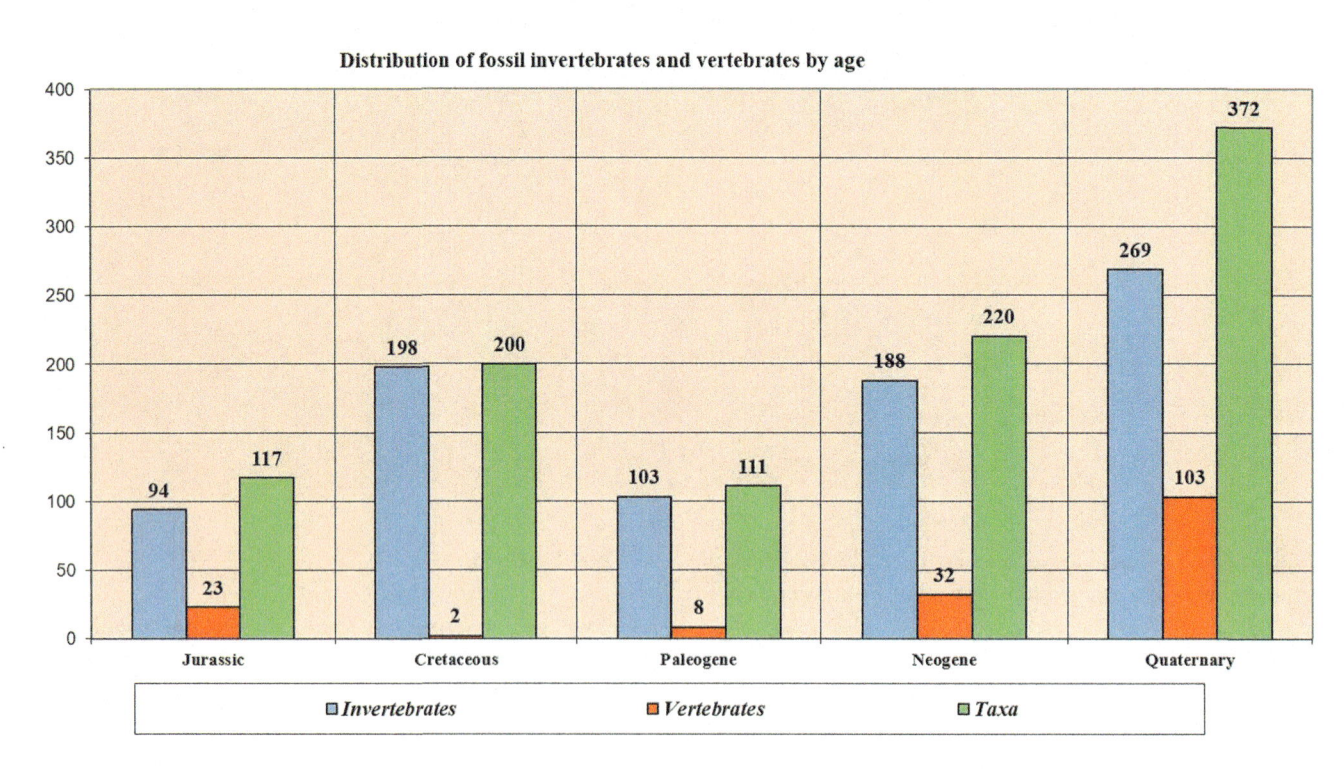

Fig. 3.39 Wealth in the composition of the fossil record in Cuba for the five geological periods represented by its rocks

In summary, the fossil record of Cuba, up to now, is reflected as follows (Fig. 3.39).

For all geological periods, represented by the sequences of rocks studied, the fossil record of invertebrates predominates. This picture is in line with the wide presence of marine sedimentary rocks in the current Cuban territory.

In the Quaternary, the fossils of vertebrates correspond to essentially terrestrial taxa, and are the best represented, qualitatively, with 103 taxa. This responds to the prevailing environmental conditions in the Cuban Archipelago, already emerged before the Quaternary. Most of these materials come from fossil deposits associated with profuse karst.

3.4.2 Major Contributions to the Paleontology of Cuba

In this section, the bibliography compiled on Paleontology of Cuba is presented. The scientific articles of the references found, by means of an extensive review of the Cuban and foreign geological and paleontological heritage, are chronologically ordered from the oldest contributions to the most recent ones.

There are brief comments on some milestones in the development of paleontological studies, on the contributions of some very prominent authors for their pioneering contributions, or that mark important stages in Cuban fossil research.

Contributions are divided, from the most varied sources and content levels, according to the large fossil groups (plants, invertebrates, vertebrates, and ichnofossils) and ages (Jurassic to Quaternary).

This section can be useful to those persons interested in the study of some of the main groups of mega-fossils reported in Cuban territory up to the present.

3.4.2.1 Contributions on the Jurassic Period

Paleontological studies on the Jurassic of Cuba have more than 95 contributions, divided into invertebrates (57), vertebrates (27), paleobotanical and palynological (10), and ichnofossils (2).

Works on Jurassic Plants, Pollen, and Spores

In the early 60s of the twentieth century, an important discovery about Jurassic pteraphytes was due to Vakhrameev (1965–1966), which shed light on the confirmation of the age of the lower and middle part of the San Cayetanos Formation. To this same aim Areces Mallea (1990–1992) also contributed, with the correct identification of the fern previously reported. More recently, from 2003 to 2015 (Dueñas Jiménez, Flores Nieves, Weber, and others) the knowledge of the palynology of this Jurassic age in Cuba has deepened.

1965 Vakhrameev, V.A., 1965. First discovery of Jurassic flora in Cuba. Paleont. Zhur., 3: 123–126.

1966 Vakhrameev, V.A., 1966. Primer descubrimiento de flora del Jurásico en Cuba. Tecnológica, 4 (2): 22–25.

1990 Areces-Mallea, A., 1990. *Piazopteris branneri* (White) Lorch, helecho del Jurásico Inferior-Medio de Cuba. Revista Sociedad Mexicana de Paleontología, 3 (1): 25–40.

1991 Areces-Mallea, A., 1991. Consideraciones paleobiogeográficas sobre la presencia de *Piazopteris branneri* (Pterophyta) en el Jurásico de Cuba. Revista Española de Paleontología, 6 (2): 126–134.

1992 Areces-Mallea, A., 1992. Alcance paleobiogeográfico de de *Piazopteris branneri* (White) Lorch en Cuba (II). Ciencias de la Tierra y el Espacio, 20: 99–105.

2003 Dueñas Jiménez, H., Linares-Cala, E., y García-Sánchez, R., 2003. Palinomorfos en las rocas de la Formación San Cayetano, Pinar del Río, Cuba. Minería y Geología, 19 (1–2): 59–70.

2005 Flores-Nieves, A., y Sánchez-Arango, J.R., 2005. La Formación San Cayetano en el contexto de nuevos enfoques en la exploración de petróleo y gas. Memoria I Convención Cubana de Ciencias de la Tierra. Sociedad Cubana de Geología, La Habana, CD-Rom.

2008 Weber, R., 2008. *Phlebopteris* (Matoniaceae) en el Triásico y Jurásico de México. En: Weber, R., (Editor), Plantas triásicas y jurásicas de México: Universidad Nacional Autónoma de México, Instituto de Geología. Boletín, 115 (2): 85–115.

2013 Flores-Nieves, A., 2013. Estudio palinológico de la Formación San Cayetano y su vinculación con la exploración de hidrocarburos. Memorias V Convención Cubana de Ciencias de la Tierra (Geociencias' 2013). Sociedad Cubana de Geología, La Habana, CD-Rom.

2015 Flores-Nieves, A., 2015. Palinología de la Formación San Cayetano, provincia Pinar del Río, Cuba. Descripciones sistemáticas. Memorias VI Convención Cubana de Ciencias de la Tierra. Sociedad Cubana de Geología, La Habana, CD-Rom.

Works on Jurassic Invertebrates

The first reports of Jurassic fossil invertebrates date from the 20s of the twentieth century, referring to cephalopod molluscs (ammonites, *aptychus*, and other elements), with important contributions to the stratigraphy of Western Cuba and its regional significance.

The first communication was due to Carlos de la Torre y Huerta (1910–1915), who was followed by Brown and O'Connell, with important contributions to the Jurassic biostratigraphy; in addition, Trauth, Wells, Imlay stood out, mainly in the 40s.

Krommelbein and Torre (1956–1960) are untied, which indicated the Lower-Middle Jurassic Age of the San

Cayetano Formation by identifying the *Trigonia* bivalve, contained therein.

A very significant leap in knowledge about the paleontology of this age, occurred between the end of the 60s and the beginning of the 2000s, with numerous contributions from several authors from Eastern Europe, among which stand out Myczyński, Pszczółkowski, Wierzbowski, Housa, Pugaczewska, and others.

1909 De la Torre y Huerta, C., 1909. Excursión científica a Viñales, descubrimiento de ammonites del período Jurásico en Cuba. Anales de la Academia de Ciencias Médicas, Físicas y Naturales de la Habana, pág. 99–103.

1910 De la Torre y Huerta, C., 1910a. Comprobation de l'existence d'un horizon Jurassique dans la region occidental de Cuba: Compte Rendu du XI Congress Geologique Internationnel, Stockholm, 1021–1022.

1910 De la Torre y Huerta, C., 1910b. Investigaciones paleontológicas en las Sierras de Viñales y Jatibonico: Anales de la Academia de Ciencias Médicas, Físicas y Naturales de la Habana, 33 pp.

1912 De la Torre y Huerta, C., 1912. Comprobation de l'existence d'un horizon Jurassique dans la region occidental de Cuba (Con discusión). Compte Rendu, XI Congress Geologique Internationel, Stockholm, p. 1021–1022.

1915 De la Torre y Huerta, C., 1915a. Revisión del catálogo de la fauna cubana (Introducción). Memorias de la Sociedad Cubana de Historia Natural "Felipe Poey", 1: 31–36.

1919 Brown, B., O'Connell, M., 1919. Discovery of the Oxfordian in Western Cuba: Geological Society of America Bulletin, 30, p. 152.

1920 O'Connell, M., 1920. The Jurassic ammonite fauna of Cuba. Bulletin American Museum of Natural History, 52: 643–692.

1920d 1920 Sánchez-Roig, M., 1920d. La fauna jurásica de Viñales: Secretaría de Agricultura, Comercio y Trabajo, Boletín Especial, 61 p.

1921 O'Connell, M., 1921. New species of ammonite opercula from the Mesozoic rocks of Cuba. American Museum Novitates, 28: 1–20.

1922 Brown, B., O'Connell, M., 1922. Correlation of the Jurassic formations of western Cuba: Geological Society of America Bulletin, 33: 639–664.

1922 O'Connell, M., 1922. Phylogeny of Ammonite genus *Ochetoceras* (Cuba). Bulletin of the American Museum of Natural History, 46: 387–411.

1935 Dickerson, R.E., y Butt, W.H., 1935. Cuban Jurassic. American Association of Petroleum Geologists Bulletin, 19: 116–118.

1936 Trauth, F., 1936. Uber Aptychenfunde auf Cuba. Proc. Kon. Akad. Wet. Amsterdam, 39 (1): 66–76.

1940 Jaworski, E., 1940. Oxford-Ammoniten von Cuba. N. Jb. Min. Geol. B. B. LXXXIII, p. 87.

1941 Wells, J.W., 1941. Upper Cretaceous Corals from Cuba. Bulletin of American Paleontology, 26 (97): 1–16; 282–298.

1942 Imlay, R.W. 1942. Late Jurassic fossils from Cuba and their economic significance. Geological Society of America Bulletin, 53: 1417–1478.

1950 Schevill, W.E., 1950. An Upper Jurassic Sepiod from Cuba. Journal of Paleontology, 94 (1): 99–101.

1951b Sánchez-Roig, M., 1951b. La fauna jurásica de Viñales. Anales de la Academia de Ciencias Médicas, Físicas y Naturales de La Habana, tomo 89, fas. 2: 46–94.

1953 De la Torre y Capablanca, C., 1953. Dos casos de impresiones de las partes blandas de dos Ammonoideos del Oxfordiense superior de Viñales (Cuba). Estudios geológicos, 10: 407–414.

1956 Krommelbein, K., 1956. Die ersten marinen fossilien (Trigoniidae lamellibr) aus Cayetano-formation West-Kubas. Senckenbergiana Lethaia, 37 (3–4): 331–335.

1960 De la Torre y Callejas, A., 1960a. Fauna de la Formación Cayetano, del Jurásico Medio de Pinar del Río. Memorias de la Sociedad Cubana de Historia Natural "Felipe Poey", 25 (1): 65–72.

1960 Krommelbein, K., 1960. Los primeros fósiles marinos (Trigoniidae, Lamellibr.) procedentes de la formación Cayetano del Oeste de Cuba. Memorias de la Sociedad Cubana de Historia Natural "Felipe Poey", 25 (1): 43–47.

1969 Housa, V., 1969. Neocomian rhyncholites from Cuba. Journal of Paleontology, 43 (1): 119–124.

1973 Myczyński, R., y Pszczółkowski, A., 1973. La fauna ammonoidea pretithoniana en la Sierra del Rosario, provincia de Pinar del Río. Actas, Academia Ciencia de Cuba, 3: 31–34.

1973 Wierzbowski, A., 1973. Sistemática de los ammonites del Oxfordiano de la Sierra de los Órganos, provincia de Pinar del Río. Actas, Instituto de Geología, Academia de Ciencias de Cuba, 3: 26–31.

1974 Housa, V., 1974. Los ápticos de Cuba I. Serie Geológica, 14.

1975 Housa, V., y M.L. de la Nuez, 1975. Ammonite fauna of the Tithonian and Lowermost Cretaceous of Cuba. Colloque Sur la lim. Jur-Cret. Lyon, Neuchatel, Mem. du B.R. G.M. No. 86.

1976 Kutek, J., Pszczółkowski, A., y Wierzbowski, A., 1976. The Francisco Formation and an Oxfordian ammonite faunule from Artemisa Formation, Sierra del Rosario, Western Cuba. Acta Geologica Polonica, 26 (2): 299–319.

1976 Myczyński, R., y Pszczółkowski, A., 1976. The ammonites and age of the San Cayetano Formation from the Sierra del Rosario, Western Cuba. Acta Geologica Polonica,

26: 321–330. Versión en Español: En: Pszczółkowski, A., (Ed.) 1987. Contribución a la Geología de la Provincia Pinar del Río. Editorial Científico-Técnica, La Habana, p. 221–227.

1976 Wierzbowski, A., 1976. Oxfordian ammonites of Pinar del Rio province (western Cuba), their revision and stratigraphic significance: Acta Geologica Polonica, 26(2), 137–260.

1976a Myczyński, R., 1976a. A new ammonite fauna from the Oxfordian of the Pinar del Río province, western Cuba. Acta Geologica Polonica, 26: 261–298.

1976b Myczyński, R., 1976b. *Organoceras* gen.n. (ammonoidea) from the Oxfordian of Cuba. Acta Paleontologica Polonica, 21 (4): 391–396.

1977 Myczyński, R., 1977. Lower Cretaceous ammonites from Sierra del Rosario (western Cuba). Acta Paleontologica Polonica, 22 (2): 139–179.

1978 Millán-Trujillo, G., y Myczyński, R., 1978. Fauna jurásica y consideraciones sobre la edad de las secuencias metamórficas del Escambray. Academia de Ciencias de Cuba, Informe Científico Técnico, 80: 1–14.

1978 Pugaczewska, H., 1978. Jurassic pelecypods from Cuba. Acta Paleontologica Polonica, 23 (2).

1979 Millán-Trujillo, G., y Myczyński, R., 1979. Jurassic ammonite fauna and age of metamorphic sequences of Escambray. Bulletin of the Polish Academy of Sciences. Earth Science, 27 (1–2).

1980 Myczyński, R., y Brochwicz-Lewiński, W., 1980. Cuban Oxfordian Aspidoceratids: Their relation to the European Ones and their stratigraphic values. Bulletin de L'Academie Polonaise des Sciences, Serie de Sciences des la Terre, 28 (4).

1984 De Albear, J.F., y Myczyński, R., 1984. Ammonites en el conglomerado Río Piedras, Cuba. Ciencias de la Tierra y el Espacio, 9: 117–123.

1987 Myczyński, R., y Pszczółkowski, A., 1987. Tithonian stratigraphy in the Sierra de los Órganos, western Cuba: Correlation of the ammonite and microfossil zones. Fossili, Evoluzione, Ambiente: Atti del Secondo Convegno Internazionale, Pergola, p. 405–415.

1987a Myczyński, R., 1987a. *Simocosmoceras* Spath (Perisphinctidae, Ammonitina) in the Lower Tithonian of Sierra del Rosario (western Cuba), Fossili, Evoluzione, Ambiente: Atti del Secondo Convegno Internazionale, Pergola, p. 401–403.

1987b Myczyński, R., 1987b. Correlaciones paleobiogeográficas de los ammonites del Jurásico Superior y Cretácico Inferior de Cuba occidental. En: Pszczółkowski, A., (Ed.) 1987. Contribución a la Geología de la Provincia Pinar del Río. Editorial Científico-Técnica, La Habana, p. 248–255.

1989 Myczyński, R., 1989. Ammonite biostratigraphy of the Tithonian of western Cuba. Annales Societatis Geologorum Poloniae, 59: 43–125.

1989 Pszczółkowski, A., 1989. Late Paleozoic fossils in cobbles collected from the San Cayetano formation, Sierra del Rosario, Pinar del Río province. Annales Societatis Geologorum Poloniae, 59: 27–40.

1990 Myczyński, R., y Meléndez, G., 1990. On the current state of progress of the studies on Oxfordian ammonites from Western Cuba. 1st. Oxfordian Meeting, Zaragoza, p. 185–189.

1990 Pszczółkowski, A., 1990. Fósiles del Paleozoico tardío de los guijarros colectados de la Fm. San Cayetano en la Sierra del Rosario, provincia de Pinar del Río. Boletín de Geociencias, 4 (1): 1–17.

1994 Myczyński, R., 1994. Caribbean ammonite assemblages from the Upper Jurassic-Lower Cretaceous sequences of Cuba. Studia Geologica Polonica. Geology of Western Cuba, 105: 91–109.

1994 Myczyński, R., y Pszczółkowski, A., 1994. Tithonian stratigraphy and microfacies in the Sierra del Rosario, western Cuba. Studia Geologica Polonica. Geology of Western Cuba, 105: 7–38.

1998 Myczyński, R., Olóriz, F., y Villaseñor, A.B., 1998. Revised biostratigraphy and correlations of the Middle-Upper Oxfordian in the Americas (southern USA, Mexico, Cuba, and northern Chile). Neues Jahrbuch Geol. Paläont., 207 (2): 185–206.

1999 Pszczółkowski, A., Myczynski, R. 1999. "Nannoconid assemblages in Upper Hauterivian Lower-Aptian limenstones of Cuba: their correlation with ammonites and some planktonic foraminifers" in Studia Geologica Polonica, 114: 35–75.

1999a Myczyński, R., 1999a. Inoceramids and buchiids in Tithonian deposits of western Cuba: a possible faunistic link with South-Eastern Pacific. Studia Geologica Polonica, 114: 77–92.

1999b Myczyński, R., 1999b. Some ammonite genera from the Tithonian of western Cuba and their paleobiogeographic importance. Studia Geologica Polonica, 114: 93–112.

2003 Pszczółkowski, A., y Myczyński, R., 2003. Stratigraphic constraints on the Late Jurassic-Cretaceous paleotectonic interpretations of the Placetas Belt in Cuba. En: Bartolini, C., Buffler, R.T., y Blickwede, J.F., (Editores) The circum-Gulf of Mexico and the Caribbean; hydrocarbon habitats, basin formation, and plate tectonics. American Association of Petroleum Geologists Memoir, 79: 545–581.

2010 Pszczółkowski, A., y Myczyński, R., 2010. Tithonian-early Valanginian evolution of deposition along the proto-Caribbean margin of North America recorded in Guaniguanico successions (western Cuba). Journal of South American Earth Sciences, 29: 225–253.

2013 López-Martínez, R.A., Barragán, R., Reháková, D., y Cobiella-Reguera, J.L., 2013. Calpionellid distribution and microfacies across the Jurassic/Cretaceous boundary in

western Cuba (Sierra de los Órganos). Geologica Carpathica, 64 (3): 195–208.
2014 López-Martínez, R.A., Barragán, R., Buitrón-Sánchez, B.E., y Rojas-Consuegra, R., 2014. Sea level changes through the Jurassic/Cretaceous boundary in western Cuba indicated by taphonomic analysis. Boletín de la Sociedad Geológica Mexicana, 66 (3): 431–440.
2016a Zell, P., y Stinnesbeck, W., 2016a. *Salinites grossicostatum* (Imlay, 1939) and *S. finicostatum* sp. nov. from the latest Tithonian (Late Jurassic) of northeastern Mexico. Boletín de la Sociedad Geológica Mexicana, 68 (2): 305–311.
2016b Zell, P., y Stinnesbeck, W., 2016b. Paleobiology of the Latest Tithonian (Late Jurassic) Ammonite *Salinites grossicostatum* Inferred from Internal and External Shell Parameters. PLoS ONE 11(1): e0145865. https://doi.org/10.1371/journal.pone.0145865.

Works on Jurassic Vertebrates

The Jurassic vertebrates of Western Cuba, mainly represented by ganoid fish and marine saurians, began to be studied from the 20s to the 40s with the first findings in the Sierra de los Órganos (Gregory, De la Torre y Madrazo, White, Torre y Callejas), new reports are made in the late 60s and 80s (Gutiérrez Domech, Colbert, Arratia, Schultze, Thies), and from the 90s to the date (2017) the greatest contribution is made to these groups, including pliosaurs and pterosaurs, also turtles and crocodyliformes (Gasparini, Brito, Norell, Iturralde Vinent, Fernández, De La Fuente, Bardet and others).

1923 Gregory, W.K., 1923. A Jurassic fysh fauna from western Cuba with an arrangement of the families of holostean ganoid fyshes: Bulletin of the American Museum of Natural History, 48: 223–242.
1939 De la Torre y Madrazo, R., Cuervo, A., 1939. Dos nuevas especies de Ichthyosauria del Jurásico de Viñales: Universidad de La Habana, Departamento de Geología y Paleontología, 9 pp.
1942 White, T.B., 1942. A new leptolepid fish from the Jurassic of Cuba: New England Zoological Club Proceedings, 21: 97–100.
1942 White, T.B., 1942. A new leptolepid fish from the Jurassic of Cuba: New England Zoological Club Proceedings, 21: 97–100.
1949 De la Torre y Callejas, A., 1949. Hallazgo de un hueso de dinosaurio terrestre en el Jurásico de Viñales, Pinar del Río. Universidad de La Habana, Departamento de Geología y Paleontología, 19 pp.
1949 De la Torre y Madrazo, R., Rojas, L., 1949. Una nueva especie y dos subespecies de Ichthyosauria del Jurásico de Viñales, Cuba: Memorias de la Sociedad Cubana de Historia Natural "Felipe Poey", 19(2): 197–202.

1968 Gutiérrez-Domech, R., 1968. Breve reseña sobre el período Jurásico en la Provincia de Pinar del Río. Instituto Nacional de Recursos Hidráulicos, La Habana. Publicación Especial Número 5: 3–23.
1969 Colbert, E., 1969. A Jurassic pterosaur from Cuba. American Museum Novitates, 2370: 1–26.
1981 Gutiérrez, R., 1981. Hallazgo de restos de reptil en rocas jurásicas en Punta de la Sierra, provincia Pinar del Rio: Voluntad Hidráulica, 18(56): 4–11.
1985 Arratia, G., Schultze, H-P., 1985. Late Jurassic teleosts (Actinopterygii, Pisces) from Northern Chile and Cuba: Palaeontographica A, 189: 29–61.
1989 Thies, D., 1989. *Lepidotes gloriae*, sp. nov. (Actinopterygii: Semionotiformes) from the Late Jurassic of Cuba: Journal of Vertebrate Paleontology, 9(1): 18–40.
1996 Iturralde-Vinent, M., y Norell, M., 1996. Synopsis of Late Jurassic marine reptiles from Cuba. American Museum Novitates, 3164: 1–17.
1997 Brito, P.M., 1997. Révision des Aspidorhynchidae (Pisces—Actinopterygii) du Mésozoïque: ostéologie et relations phylogénétiques: Geodiversitas, 19: 681–772.
1999 Brito, P.M., 1999. Description of *Aspidorhynchus arawaki* sp. nov. from the Late Jurassic of Cuba, with comments on the phylogeny of aspidorhynchid (Actinopterygii: Aspidorhynchiformes), in G. Arratia, Schultze, H-P., (Eds.), Mesozoic Fishes 2: Systematics and Fossil Record: Verlag Dr. F. Pfeil, München, 239–248.
2000 Fernández, M., Iturralde-Vinent, M., 2000. An Oxfordian Ichthyosauria (Reptilia) from Viñales, Western Cuba: Paleobiogeographic signiphycance: Journal of Vertebrate Paleontology, 20, 191–193.
2001 De la Fuente, M., Iturralde-Vinent, M., 2001. A new pleurodiran turtle from the Jagua Formation (Oxfordian) of western Cuba: Journal of Vertebrate Paleontology, 75: 860–869.
2001 Gasparini, Z., 2001. El corredor caribeño: evidencias aportadas por la biogeografía de los reptiles marinos jurásicos de América del Sur y Cuba. En: IV Congreso Cubano de Geología y Minería, Memorias. La Habana.
2001 Gasparini, Z., e Iturralde-Vinent, M., 2001. Metriorhynchid crocodiles (Crocodyliformes) from the Oxfordian of western Cuba. Neues Jahrbuch für Geologie und Paläontologie Monatshefte, 9: 534–542.
2002 Gasparini, Z., Bardet, N., Iturralde-Vinent, M., 2002a. A new cryptoclidid Plesiosaur from the Oxfordian (Late Jurassic) of Cuba: Geobios, 35: 201–211.
2002 Gasparini, Z., Bardet, N., e Iturralde-Vinent, M., 2002b. A new cryptoclidid plesiosaur from the upper Jurassic of Cuba. Geobios, 35 (201): 211–217.
2003 Gasparini, Z., e Iturralde-Vinent, M., 2003. Oxfordian reptiles in the Caribbean Corridor. En: V Congreso Cubano de Geología y Minería, Memorias, La Habana.

2004 Gasparini, Z., Fernández, M., De la Fuente, M., 2004. A new pterosaur from the Oxfordian of Cuba: Palaeontology, 47: 919–927.

2006 Gasparini, Z., Iturralde-Vinent, M., 2006. The Cuban Oxfordian herpetofauna in the Caribbean Seaway: Neues Jahrbuch für Geologie und Palaöntologie, 240: 343–371.

2009 Gasparini, Z., 2009. A new oxfordian pliosaurid (Plesiosauria, Pliosauridae) in the Caribbean Seaway. Palaeontology, 52: 661–669.

2013 Iturralde-Vinent, M., y Gasparini, Z., 2013. Animales del Caribe primitivo y sus costas. Editorial Oriente, Santiago de Cuba. 61 pp.

2015 Iturralde-Vinent, M., y Ceballos Izquierdo, Y., 2015. Catalogue of Late Jurassic Vertebrate (Pisces, Reptilian) specimens from Western Cuba. Paleontología Mexicana, 3 (65): 24–39; 4: 24–39 (*versión electrónica*).

2016 López-Conde, O.A., Sterli, J., Alvarado-Ortega, J., y Chavarría-Arellano, M.L., 2016. A new platychelyid turtle (Pan-Pleurodira) from the Late Jurassic (Kimmeridgian) of Oaxaca, Mexico. Papers in Palaeontology, 1–14.

Works on Jurassic Ichnofossils

Bioturbation structures in the Jurassic rocks are mentioned in some contributions to the geology and stratigraphy of Western Cuba (Judoley y Furrazola Bermúdez, Pszczółkowski y others). Only one work has been dedicated in particular to ichnotaxonomy until now (Villegas Martín, de Gibert, Rojas Consuegra, and Belaústegui).

2007. Arai, M., Cuevas de Azevedo-Soares, H. L. 2007. Palinoicnofóssets: macas de beocorrosao em Palinomorfos. Capítulo 11. En: Cravalho, Ismar de Souza y Fernades, A. C. S. (Eds): Icnologia. Sociedad Brasileera de Geologgia, Serie Textox N. 3, 2007: 177 p. Sao Paulo. pp. 118–121.

2012 Villegas Martín, J., de Gibert, J.M, Rojas-Consuegra, R., y Belaústegui, Z., 2012. Jurassic Teredolites from Cuba: New trace fossil evidence of early wood-boring behavior in bivalves. Journal of South American Earth Sciences, 38: 123–128.

3.4.2.2 Contributions on the Cretaceous Period

In the Paleontology of the Cretaceous in Cuba, there are more than 87 contributions, mainly distributed in invertebrates (70), vertebrates (11), inchnofossils (4), and plants (3).

Works on Cretaceous Plants, Pollen, and Spores

Mentions are scarce to the study of plants, pollen, and spores of the Cretaceous in Cuba. Most of the reports correspond to the reports of the regional geological surveys.

1983 Radocz, G.Y., y Nagy, E., 1983a. Manifestaciones carboníferas en la molasa del Cretácico Superior de Cuba oriental. En: Contribución a la Geología de Cuba Oriental. Editorial Científico-Técnica, La Habana, pág. 186–192.

2004 Iturralde-Vinent, M. 2004. Catálogo de localidades de plantas fósiles de Cuba. En: Iturralde-Vinent, M. A. (Editor), 2012. Compendio de Geología de Cuba y del Caribe. Segunda Edición. DVD-ROM. Editorial CITMATEL, La Habana, Cuba.

2019 Rojas-Consuegra, R. 2019. Hoja fósil cretácica. Boletín de la SCG (*in press*).

Works on Cretaceous Invertebrates

The first news about ammonites in rocks of the volcanic arc of the Cretaceous in Cuba was due to Don Carlos De La Torre y Huerta (1892), discovered in the central region; subsequently, several other authors have contributed (Lewis, Myczyński and others).

Also, to De La Torre y Huerta, we owe the first report on Cretaceous rudists (1915) in Cuba. To the taxonomic and biostratigraphic study of the rudists, perhaps the group best known today among the Cretaceous invertebrates, have contributed numerous foreign and Cuban authors (Sanchez Roig, Douvillé, Palmer R, Rutten, Thiadens, MacGillavry, Vermunt, Hermes, Albear, De la Torre y Callejas, Jung, Dommelen, Iturralde Vinent, Lupu, Jakus, Skelton, Rojas Consuegra, Aguilar, Arano and others).

In addition, other fossil invertebrates (gastropods, echinoderms, other bivalves, etc.) have been given attention by several researchers (Sánchez Roig, Lambert, Bermudez, Brönnimann, Rigassi, Palmer DK, Weisbord, Seiglie, Ayala Castañares, Brodermann, Knipscheer, Dilla, Díaz de Villalvilla, Furrazola Bermúdez, Otero, Kier, Arz, Arenilla, Alegret, among many others). Thus, in general, the Cretaceous sequences of Cuba exhibit a high degree of paleontological study; nevertheless, an update on dissimilar aspects is necessary.

1892 De la Torre y Huerta, C., 1892. Observaciones geológicas y paleontológicas en la región central de la Isla (de Cuba). Anales de la Academia de Ciencias Médicas, Físicas y Naturales de la Habana, 29: 121–124.

1883 Felix, J., 1883. Die fossilen Hölzer West-Indies. 11 International Geological Congress, Stockholm, The Iron Resources of the World, 2: 793–797.

1915 De la Torre y Huerta, C., 1915a. Descubrimiento de interesantes fósiles del género Barrettia y otros rudistas, característicos del período Cretáceo en Camagüey. Anales de la Academia de Ciencias Médicas, Físicas y Naturales de la Habana, 52: 824–827.

1916 Mestre, A., 1916. La Vida de la Sociedad Poey de 1915 a 1916. Presentación de un segundo ejemplar de ammonites del Cretácico de Santa Clara por Carlos de la Torre. Memorias de la Sociedad Cubana de Historia Natural "Felipe Poey", 2: 90.

1925 Raymond, P.E., 1925. A new oyster from Cretaceous of Cuba. Boston Soc. Nat. Hist., Occ. Papers, 5, p. 183–185.

1926 Douvillé, H., 1926a. Quelques fossiles de Crétacé supérieur de Cuba. Bulletin of the Geological Society of France, 4 (26): 127–138.

1926 Douvillé, H., 1926b. Quelques fossiles de Crétacé supérieur de Cuba, Planche VII et VIII. Bull. Soc. Géol. France 26 (3-4-5): 127–138.

1926 Sánchez-Roig, M., 1926a. Contribución a la paleontología cubana. Los equinodermos fósiles de Cuba. Boletín de Minas, 10: 1–179.

1926 Sánchez-Roig, M., 1926b. La fauna cretácica de la región central de Cuba. Memorias de la Sociedad Cubana de Historia Natural "Felipe Poey", 7 (1, 2): 83–102.

1927 Douvillé, H., 1927. Nouveaux rudistes du Crétacé de Cuba. Bulletin of the Geological Society of France, 4 (27): 49–56.

1930 Lambert, J., y Sánchez-Roig, M., 1930. Nuevas especies de equinodermos fósiles cubanos. Inst. Invest. Cien. y Mus. Hist. Nat., Mem. 1, p. 143–150.

1931 Lambert, J., 1931. Notes sur le groupe des Oligopygus, la nouvelle famille des Haimeidae et sur quelques Echinides fossiles de Cuba. Bulletin of the Geological Society of France, 5 (1): 289–304.

1932 Lewis, J.W., 1932. Probable age of Aptychus-bearing formation of Cuba. American Association of Petroleum Geologists Bulletin, 16 (9): 943–944.

1933 Palmer, R.H., 1933. Nuevos rudistas de Cuba. Revista de Agricultura, 14: 95–125.

1934 Palmer, D.K., 1934. Some large foraminifera from Cuba. Memorias de la Sociedad Cubana de Historia Natural "Felipe Poey", 8 (4): 235–269.

1934 Weisbord, N.E., 1934. Some Cretaceous and Tertiary echinoids from Cuba. Bulletin of American Paleontology, 20, (70): 165–268.

1936 Rutten, M.G., 1936. Rudistids from the Cretaceous of nortern Santa Clara, Las Villas Province, Cuba. Journal of Paleontology, 10 (2): 134–142.

1936 Thiadens, A. 1936a. On some Caprinids and a Monopleurid from Southern Santa Clara, Cuba. Konin. Akad. Van Wetensch. Te Amsterdam, 39(9): 1132–1141.

1936 Thiadens, A., 1936b. Rudistids from Southern Santa Clara, Cuba. Konin. Akad. Van Wetenschap. Te Amsterdam, 39 (8): 1010–1019.

1937 MacGillavry, H.J., 1937. Geology of the province of Camagüey, Cuba, with revisional studies in rudist paleontology. Geogr. en Geol. Meded., Physiog. Geol. reeks, 2 (14): 1–168.

1937 Vermunt, L.W.J., 1937. Cretaceous rudistids of Pinar del Río province, Cuba. Journal of Paleontology, 11 (4): 261–275.

1940 De la Torre y Madrazo, R., 1940. Estado actual de los conocimientos de la era Mesozoica en Cuba. En: Proceedings of the 8th American Scientific Congress, Volumen 4, Washington.

1943 Wessem, A. Van, 1943. Geology and paleontology of central Camagüey, Cuba. Geogr. Geol. Mededeel (Utrecht), Phys. Geol. Reeks, 5:1–91.

1945 Hermes, J., 1945. Geology and paleontology of East Camagüey and West Oriente, Cuba. Geographische en Geologische Mededeelingen, Physiographisch-Geologische Reeks, 2 (7): 1–75.

1947 De Albear, J.F., 1947. Stratigraphic paleontology of Camagüey district, Cuba. American Association of Petroleum Geologists Bulletin, 31 (1): 71–91.

1951 Mullerried, F. K. G., 1951. Paquiodontos nuevos del Cretácico Superior de Cuba. Revista Sociedad Malacológica Carlos de la Torre 8 (2): 83–92 ilustr.

1956 Chubb, L. J., 1956. Rudist assemblages of the Antillean Upper Cretaceous. Bull. Amer. Pal. 37 (161): 1–23.

1960 De la Torre y Callejas, A., 1960b. Notas sobre rudistas. Memorias Sociedad Cubana de Historia Natural, 25 (1): 51–64.

1961 Chubb, L. J., 1961. Rudist assemblages in Cuba. Bull. Amer. Paleont., 43 (198): 413–422.

1963 Brönnimann, P., y Rigassi, D., 1963. Contribution to the geology and paleontology of the area of the city of La Habana, Cuba, and its surroundings. Eclogae Geologicae Helvetiae, 56: 193–480.

1963 Seiglie, G.A., y Ayala-Castañares, A., 1963. Sistemática y bioestratigrafía de los foraminíferos grandes del Cretácico Superior (Campaniano y Maestrichtiano) de Cuba. Paleontología Mexicana, 13: 1–56.

1970 Jung, P., 1970. Torreites sanchezi (Douvillé) from Jamaica. Palaeontographica Americana, 7 (42).

1971 Alencáster, G., 1971. Rudistas del Cretácico Superior de Chiapas (Parte I). UNAM Inst. Geol. Paleontol., México, (34): 1–91.

1971 Dommelen, H. Van. 1971. Ontogenetic, Phylogenetic and Taxonomic Studies of the American Species of Pseudovaccinites and of Torreites and the Multiple Hippuritids. Amsterdam, University of Amsterdam. Doctoral thesis, 125 p.

1974 Lübimova, P.S., y Sánchez-Arango, J.R., 1974. Los ostrácodos del Cretácico Superior y Terciario de Cuba. Instituto Cubano del Libro, La Habana, 171 p.

1974 Myczyński, R., 1974. Los ammonites del Cretácico Inferior de la Sierra del Rosario, provincia de Pinar del Río. Archivo del Instituto Geología y Paleontología, Academia Ciencia de Cuba.

1974 Kauffman, E. G. y N. F. Sohl, 1974. Structure and Evolution of Antillean Cretaceous Rudist Frameworks. Verhandl. Naturf. Gas. Basel. 84 (1): 399–467.

1975 Lupu, D., 1975. Faune sénonenne á rudistes de la province de Pinar del Río (Cuba). Instiutul de Geologie si Geofizicá. Bucuresti. Dári de Seamá ale sedintelor, 61 (1973–1974): 223–254.

1978 De la Torre y Callejas, A., Jakus, P., y De Albear, J.F., 1978. Nuevos datos sobre las asociaciones de rudistas en Cuba. Geologie en Mijnbow, 57 (2): 143–150.

1983 De la Torre y Callejas, A., Jakus, P., y De Albear, J.F., 1983. Nuevos datos sobre las asociaciones de rudistas en Cuba. En: Contribución a la Geología de Cuba Oriental. Editorial Científico-Técnica, La Habana, p. 206–216.

1983 Jakus, P., 1983. Formaciones vulcanogeno-sedimentarias de Cuba oriental. En: Contribución a la Geología de Cuba Oriental. Editorial Científico-Técnica, La Habana, pág. 17–85.

1984 Knipscheer, H., 1938. On Cretaceous Nerineas from Cuba (Nerineas del Cretácico de Cuba). Procc. Kon. Neederl. Aka. Wesemchap., 41 (6): 673–676.

1984 Kier, P.M., 1984. Fossil spatangoid echinoids of Cuba. Smithsonian Contributions to Paleobiology, 55, 348 p.

1986 Dilla, M., y Díaz de Villalvilla, L., 1986. Sobre la edad de algunas vulcanitas de las provincias camagüeyanas. Serie Geológica, 2: 91–103.

1990 Iturralde-Vinent, M., y de la Torre y Callejas, A., 1990. Posición estratigráfica de los rudistas de Camagüey, Cuba. Transactions 12th Caribbean Geological Conference, Miami Geological Society, p. 59–67.

1993 Luperto Sinni, E., e Iturralde-Vinent, M., 1993. Lower Cretaceous algae from a Cuban carbonate platform sequence: studies on fossil benthic Algae. Boll. Soc. Paleont. Ital., Spec. Vol. 1, Mucchi, Modena, 1993: 281–285.

1995 Rojas-Consuegra, R., e Iturralde-Vinent, M., 1995. Checklist of Cuban rudistids taxa. Revista Mexicana de Ciencias Geológicas, 12 (2): 292–293.

1995 Rojas-Consuegra, R., Iturralde-Vinent, M., y Skelton, P.W., 1995. Stratigraphy, composition and age of Cuban rudist-bearing deposits. Revista Mexicana de Ciencias Geológicas, 12 (2): 272–291.

1996 Johnson, C. C. y E. G. Kauffman, 1996. Maastrichtian extinction patterns of Caribbean province rudistids. En: MacLeod, N., G. Keller. Cretaceous-tertiary mass extinctions: biotic and environmental changes. New York: W.W. Norton, 1996. p. 231–273.

1998 Skelton, P.W., y Rojas-Consuegra, R., 1998. Overview of rudist biostratigraphy in the Volcanic Arc Sequences of the Greater Antilles. 15th Caribben Geological Conference, Kingston, Jamaica, p. 10–11.

2000 Rojas-Consuegra, R., 2000a. El límite Cretácico/Terciario en la Formación Peñalver (Cuba Occidental): observaciones tafonómicas, paleoecológicas y paleogeográficas. Rev. Geotemas (2) 355–358.

2000 Rojas-Consuegra, R., 2000b. Taphonomic and palaecological observations on the Peñalver formation, western Cuba. En: Abstract of the Intern. Confer. on Catastrophic Events and Mass Extintions: Impacts and Beyond. University of Viena, Austria. LPD Contribution No. 1053: 183–184.

2003 Furrazola-Bermúdez, G., Díaz-Otero, C., Rojas-Consuegra, R., y García-Delgado, D., 2003. Generalización bioestratigráfica de las formaciones volcanosedimentarias del arco volcánico cretácico y su cobertura, en Cuba central. En: Memorías IV Congreso Cubano de Geología y Minería, La Habana.

2004 Rojas-Consuegra, R., 2004. Los rudistas de Cuba: Estratigrafía, tafonomía, paleoecología y paleogeografía. Tesis doctoral. Biblioteca Nacional de Ciencia y Técnica, La Habana. 180 p.

2005 Alegret, L., Arenillas, I., Arz, J.A., Díaz-Otero, C., Grajales-Nishimura, J.M., Meléndez, A., Molina, E., Rojas-Consuegra, R., y Soria, A.R., 2005. Cretaceous-Paleogene boundary deposits at Loma Capiro, central Cuba: evidence for the Chicxulub impact. Geology Boulder, 33 (9): 721–724.

2007 López-Martínez, R. y R. Rojas-Consuegra, 2007. Taxonomía y Tafonomía de los Gasterópodos del Mioceno en Cuba. En: Memorias II Convención sobre Ciencias de La Tierra (GEOCIENCIA'2007). Centro Nacional de Información Geológica. IGP. La Habana. CD ROM. 2007. GEO2-08: 1–18. ISBN 978-959-7117-16-2.

2009 López-Martínez, R. A. y R. Rojas-Consuegra, 2009. Análisis paleontológico complejo del sector Punta Gorda—Camarioca Sur, Moa, Cuba oriental. En: Resúmenes VIII CONGRESO DE GEOLOGÍA (GEOLOGIA'2009), GEO2-O6. CD-ROM, La Habana. ISBN 978-959-7117-19-3.

2011 Barragán, R., Rojas-Consuegra, R., y Szives, O., 2011. Late Albian (Early Cretaceous) ammonites from the Provincial Formation of central Cuba. Cretaceous Research, 32: 447–455.

2011 Villegas-Martín, J., R. Rojas-Consuegra, R. Barragán-Manzo y R. López-Martínez, 2011. Presencia del icnogénero Chondrites en formaciones del Cretácico temprano y del Paleógeno medio de México y Cuba. En: Resúmenes VI Congreso Cubano de Geología (GEOCIENCIA'2011), GEO2-P7. CD-ROM, La Habana. ISBN 978-959-7117-30-8.

2011 Menéndez-Peñate, L., R. Rojas-Consuegra, J. Villegas-Martín y R. A. López-Martínez, 2011. Taphonomy, Cronostratigraphy and paleoceanographic implications at turbidite of Early Paleogene (Vertientes Formation), Cuba. Revista Geológica de América Central, 45: 87–94. ISSN 0256-7024.

2012 Arz, J.A., Arenillas, I., Menéndez-Peñate, L., Rojas-Consuegra, R., Meléndez, A., Grajales-Nishimura, J.

M., Rosales-Domínguez, M.C. y Ceballo-Melendres, O., 2012. Resultados preliminares sobre la edad y emplazamiento de una unidad clástica relacionada con el impacto de Chicxulub, en Fomento (Cuba central). En: XXVIII Jornadas de la Sociedad Española de Paleontología, 19–22. España. ISBN: 978-84-370-8993-5.

2015 Aguilar-Pérez, J., Rojas-Consuegra, R., Pichardo-Barrón, Y., Arano-Ruiz, A., 2015. Nuevos rudistas del Albiano temprano en Loma Rioja, perteneciente a la formación Provincial, en Cuba centrooccidental. En: Memorias XI Congreso Cubano de Geología. Sociedad Cubana de Geología. La Habana. CD ROM, Geo2-01: 1 p. ISSN 2307-499X.

2016 Javier Aguilar Pérez, J., Rojas-Consuegra, R., Pichardo Barrón, Y. y Arano Ruiz, A. 2016. Nuevos registros de rudistas del Albiano temprano en loma Rioja, perteneciente a la Formación Provincial, en Cienfuegos, Cuba. En: Programa IX Congreso Latinoamericano de Paleontología. Lima, Perú. Museo de la Nación. 1 p.

2016 Arenillas, I., Arz, J.A., Grajales-Nishimura, J.M., Meléndez, A., y Rojas-Consuegra, R., 2016. The Chicxulub impact is synchronous with the planktonic foraminifera mass extinction at the Cretaceous/Paleogene boundary: new evidence from the Moncada section, Cuba. Geologica Acta, 14 (1).

2017 Arano-Ruiz, A., Rojas-Consuegra, R. 2017. Valores paleontológicos en la zona de La Rioja, municipio de Cruces, Cienfuegos, Cuba. I Conferencia Científica Internacional. Cienfuegos: Editorial Universo Sur. Recuperado de 8 pp. ISBN: 978-959-257-454-0. https://biblioteca.ucf.edu.cu/biblioteca/eventos.

2017 Oliva-Martín, A., Rojas-Consuegra, R. 2017. Nueva especie de Sogdianella (allogastropoda: itieriidae) del Maastrichtiano en Las Pozas, provincia de Sancti Spíritus, Cuba central. En: Trabajos del XII Congreso de Geología, VII Convención de Ciencias de la Tierra. La Habana. Soc. Cub. Geología. CD ROM, Geo2-03. 13 pp. ISSN 2307-499X.

2018 Martínez Molina, J. 2018. Hallan en Rodas estrella de mar fósil más antigua de Cuba. Granma, 28 de mayo.

2018 Martínez Molina, J. 2018. Hallan en Rodas cangrejo fósil del Cretácico superior. Granma, 1 de octubre.

2018 Arano-Ruiz, A., Viñola-López, L. W., Rojas-Consuegra, R., Borges Sellen, C. R. 2018. Reevaluation of the taxonomic status of Vegaranina (Decapoda: Raninidae) from the Late Cretaceous of description of a new species. Zootaxa 4527(4): 588–594. https://doi.org/10.11646/zootaxa.4527.4.10.

Works on Cretaceous Vertebrates

The remains of vertebrate Cretaceous fossils identified taxonomically in Cuba until today are really rare, as can be seen from the scarce references existing on the subject in the country. Given the material recently found, new contributions can be glimpsed in the near future.

2005 Mutter, R.J., Iturralde-Vinent, M., Fernández-Carmona, J., 2005. The first Mesozoic Caribbean shark is from the Turonian of Cuba: Ptychodus cyclodontis sp. nov. (? Neoselachii): Journal of Vertebrate Paleontology, 25(4): 976–978.

2014 Peláez, O., 2014. Notable hallazgo de vertebrados fósiles. (entrevista a R.Rojas-Consuegra). Granma, Julio 18, p. 2.

2014. Noticiero Nacional de la Televisión Cubana (NTV-ICRT), 2014. Reportaje de la peridista Gladys Rubio, sobre hallazgo de restos óseos cretácicos. Emisión del Medio Día, viernes 13-07-2014.

2014 Rojas-Consuegra, R. 2014. PROFCA: comentario sobre vertebrados fósiles pioneros. El Explorador, Periódico Digital Espeleológico, No. 123. ISSN 1819-3765.

2015 Rojas-Consuegra, R. 2015. San Vicente (Ciego de Ávila), primera localidad de vertebrados cretácicos para Cuba y Las Antillas. En: VI Convención Cubana de Geociencias (GEOLOGÍA'2015). La Habana. CD ROM. Edit. Sociedad Cubana de Geología. Geo2-05: 22 pp. ISSN 2307-499X.

2015 ANC, 2015. En Cienfuegos hallan pieza dental de reptil marino del período Cretácico. 17 de junio de 2015.

2015 Borges García, Y. 2015. Confirman autenticidad de un diente de dinosaurio del periodo Cretácico Superior en Cienfuegos, único conocido en Cuba y el Caribe insular. https://kuentasklaras.wordpress.com/2015/06/18/.

2016 Borges-Sellén, C.R. 2016. El diente fósil de Rodas. Juventud Técnica. 10 de octubre.

2016 Borges-Sellén, C.R., Arano-Ruiz, A.F., y Ceballos Izquierdo, Y., 2016a. The mistery tooth of Rodas, Cuba: A Mosasaur in the Cretaceous Caribbean? Fossil News, 19 (3): 39–44.

2016 Borges-Sellén, C.R., Arano-Ruiz, A.F., y Ceballos Izquierdo, Y., 2016b. Monstruos marinos: El diente fósil de Rodas. Juventud Técnica, 392: 10–15.

2018 Rojas-Consuegra, R. 2018. Pececillo fósil "Veloz". Boletín Ceinpetillazo (Cienpet), 1(3): 13.

2019. Martínez Molina, J. 2019. Hallan restos fósiles de Pterosaurio. Granma, 26 de marzo, 2019.

Works on Cretaceous Ichnofossils

Only after 2000 have appeared the first works directed specifically to the study of existing ichnofossils in the Cuban territory (Pszczółkowski, Villegas Martín, Klompmaker, Rojas Consuegra). In general reports, and in some publications, collateral data appear on this group, although it presents a wide variety; and it demands a sustained systematic approach.

2002 Pszczółkowski, A., 2002. Crustacean burrows from Upper Maastrichtian deposits of South-Central Cuba. Bulletin of the Polish Academy of Sciences. Earth Science, 50 (2): 147–163.

2008 Villegas Martín, J., y Rojas-Consuegra, R., 2008. Algunas icnitas presentes en el registro estratigráfico cubano. II Simposio de Museos y Salas de Historia Natural, Museo Nacional de Historia Natural (CITMA), La Habana. ISBN: 978-959-282-072-6.

2011 Villegas Martín, J., y Rojas-Consuegra, R., 2011. Presencia del icnogénero Teredolites en un megabloque de la Formación Peñalver, límite Cretácico-Paleógeno (K/Pg), Cuba occidental. Revista Española de Paleontología, 26 (1): 45–52.

2016 Villegas-Martín, J., Rojas-Consuegra, R., y Klompmaker, A.A., 2016. Drill hole predation on tubes of serpulid polychaetes from the Upper Cretaceous of Cuba. Palaeogeography, Palaeoclimatology, Palaeoecology, 455: 44–52.

3.4.2.3 Contributions on the Paleogene Period

About paleontology of Paleogene in Cuba, there are more than 42 contributions, grouped into invertebrates (27), plants, pollen, and spores (6), vertebrates (4), and ichnofossils (5).

Works on Paleogene Spore, Pollen, and Spore

Although the identifiable macroscopic remains of Paleogene plants are very rare, there are already several contributions to the palynological identification of the sequences of that age; and the most outstanding contribution to its knowledge corresponds to Areces Mallea (1985–2000). The wide distribution of the Paleogene rocks in Cuba, generally rich in vegetal organic matter, deserve deep investigations on its palynological registry, still pending to face.

1969 Grahan, A. y Jarzen, D. M. 1969. Studies in neotropical paleobotany. The Oligocene communities of Puerto Rico. Ann. Missouri Bot. Gard. 56: 308–357.

1985 Areces-Mallea, A., 1985. Una nueva especie de Bombacacidites Couper emmend. Krutzsch del Eoceno Medio de Cuba. Revista Tecnológica. Serie Geológica, 15 (1): 3–7.

1987 Areces-Mallea, A., 1987. Consideraciones sobre la supuesta presencia de *Pinus sylvestris* L. en el Oligoceno de Cuba. Centro de Investigaciones y Desarrollo del Petróleo e Instituto de Geología y Paleontología. Serie Geológica, 2: 27–40.

1988 Areces-Mallea, A., 1988. Palinomorfos de la Costa del Golfo de Norteamérica en el Eoceno medio de Cuba. Tecnológica, 18: 15–25.

1990 Areces-Mallea, A., y García-Rodríguez, F., 1990. Nuevo género-forma de palinomorfo para el complejo 3-colporado-3-pseudocolporado del Eoceno Medio de Cuba. Ciencias de la Tierra y el Espacio, 17: 33–40.

2000 Grahan, A., D. Cozadd, A. Areces-Mallea y N. O. Frederiksen, 2000. Studies in neotropical paleobotany. XIV. A palynoflora from the middle Eocene Saramaguacan Formation of Cuba. Amer. J. Bot. 87: 1526–1539.

Works on Paleogene Invertebrates

The study on the Paleogene invertebrates began very early in the nineteenth century in Cuba, thanks to the work of Cotteau (1881–1887), dedicated mainly to the equinoids, a very diverse group in the Cuban territory. However, the most important to the knowledge of the group was achieved during the decades of 30 to 50 of the twentieth century, and was undoubtedly due to Mario Sánchez Roig (1923–195).

Other fossil groups (bivalve and gastropod molluscs, brachiopods, corals, cirripeds, nautiloids, crinoids, asteroids, and dissimilar microfossils) have also been reported by several authors, mainly foreigners (Egozcue, Withers, Wells, Miller, Downs, Woodring, Cooper, Miller, Furnish, Kojumdjieva, Franco Álvarez, Nagy, Radocz, De la Torre y Callejas, Galacz, among others).

It is suggestive that, after the end of the 1980s, contributions to invertebrate paleontology of this age have been almost nil.

1881 Cotteau, G.H., 1881. Description des Echinides fossiles de l'ile de Cuba. Soc. Geol. Belgique, 9: 3–49.

1882 Cotteau, G.H., 1882. Sur les Echinides fossiles de l'ile de Cuba. Comptes Rendus de l'Academie des Sciences, Paris, 94: 461–463.

1897 Cotteau, G.H., y Egozcue y Cía., J., 1897. Descripción de los echinoides fósiles de la isla de Cuba. Boletín de la Comisión del Mapa Geológico de España, 22 (2): 1–99.

1923 Sánchez-Roig, M., 1923a. Nuevas especies de equínidos fósiles cubanos. Memorias de la Sociedad Cubana de Historia Natural "Felipe Poey", 7: 83–192.

1923 Sánchez-Roig, M., 1923b. Revisión de los equínidos fósiles cubanos. Memorias de la Sociedad Cubana de Historia Natural "Felipe Poey", 6: 6–42.

1926 Withers, T.H., 1926. *Scalpellum sanchezi*, sp. n., a Cirripede from the Lower Miocene? (Eocene) of Cuba. Ann. and Mag. Nat. Hist. 9th. ser 18, pág. 616–621.

1934 Wells, F.G., 1934. Eocene Corals, pt. 1, from Cuba; pt. 2, A new Species of Madracis from Texas. Bull. Amer. Pal. 20, 20 pág.

1950 Bermúdez, P.J., 1950. Contribución al estudio del Cenozoico cubano. Memorias de la Sociedad Cubana de Historia Natural "Felipe Poey", 19 (3): 204–375.

1950 Miller, A.K., y Downs, H.R., 1950. Tertiary nautiloids of the Americas: Supplement. Journal of Paleontology, 24 (1).

1951 Sánchez-Roig, M., 1951a. Faunula de equinodermos fósiles del Terciario, del término municipal de Morón, Provincia de Camagüey. Memorias de la Sociedad Cubana de Historia Natural "Felipe Poey", 20 (2): 37–64.
1951 Sánchez-Roig, M., 1951a. Faunula de equinodermos fósiles del Terciario, del término municipal de Morón, Provincia de Camagüey. Memorias de la Sociedad Cubana de Historia Natural "Felipe Poey", 20 (2): 37–64.
1952 Sánchez-Roig, M., 1952c. Nuevos géneros y especies de equinoideos fósiles cubanos. Torreia, 17:1–18.
1952 Sánchez-Roig, M., 1952a. El género Cubanaster (Equinidos fósiles irregulares). Torreia, 16: 3–8.
1952 Sánchez-Roig, M., 1952b. Nuevos géneros y especies de equinodermos fósiles cubanos. Memorias de la Sociedad Cubana de Historia Natural "Felipe Poey", 21 (1): 4–30.
1952 Woodring, W.P., 1952. A Nerinea from southwestern Oriente Province, Cuba. Journal of Paleontology, 26 (1): 60–62.
1955 Cooper, G.A., 1955. New brachiopods from Cuba. Journal of Paleontology, 29 (1): 64–70.
1956 Miller, A.K., y Furnish, W.M., 1956. An Aturia from eastern Cuba. Journal of Paleontology, 30 (5): 1154.
1960 Seiglie, G.A., 1960. Notas sobre el límite Oligogeno-Mioceno. Memorias de la Sociedad Cubana de Historia Natural "Felipe Poey", 25 (1): 21–31.
1979 Cooper, G.A., 1979. Tertiary and Cretaceous brachiopods from Cuba and the Caribbean. Smithsonian Contributions to Paleobiology, 37, 56 pp.
1982 Kojumdjieva, E., y De la Torre y Callejas, A., 1982. Paleogene mollusk from Central Cuba. Bulgarian Academy of Science, 17: 3–23.
1982 Kojumdjieva, E., y Popov, N., 1982. Distribución geográfica de los moluscos y equinodermos del Terciario de la antigua provincia de Las Villas, Cuba. Ciencias de la Tierra y el Espacio, 4: 71–79.
1983 Franco-Álvarez, G.L., Nagy, E., y Radocz, G., 1983. Desarrollo de facies coralinas desde el Oligomioceno hasta el reciente en Cuba Oriental. En: Contribución a la Geología de Cuba Oriental. Editorial Científico-Técnica, La Habana, p. 217–238.
1983 Radocz, G.Y., y Nagy, E., 1983b. Algunas novedades paleontológicas de Cuba Oriental. En: Contribución a la Geología de Cuba Oriental. Editorial Científico-Técnica, La Habana, p. 193–205.
1983 De la Torre y Callejas, A., 1983. Nueva especie de molusco gastrópodo del género *Bursa*, del Eoceno de Cuba Oriental. En: Contribución a la Geología de Cuba Oriental. Editorial Científico-Técnica, La Habana, p. 263–265.
1986 Kojumdjieva, E., y De la Torre y Callejas, A., 1986. Paleogene mollusks from Central Cuba and their paleobiographic significance. Contrib. Bulgarian Geol. 27th Intern. Geol. Congress., Moscú.

1988 Galacz, A., 1988. First record of Paleocene nautiloids from Cuba. Paläont. Z., 62: 265–269.
2007 Medina Batista, A., 2007. Nuevos datos estratigráficos del desarrollo de arrecifes coralinos a partir del Eoceno superior y su extensión hasta el Mioceno inferior. En: Memorias II Convención Cubana de Ciencias de la Tierra. Centr. Nac. Inform. Geológ., Instituto de Geología y Paleontología de Cuba, La Habana, CD-Rom.

Works on Paleogene Vertebrates

Among the vertebrates of the Paleogene only fishes, essentially cartilaginous, make up the fossil record so far, some remains under study could expand it in the future.

1872 Fernández de Castro, M., 1872. Nota sobre un diente placoide fósil de la isla de Cuba, el *Aetobatis poeyi*. Anales de la Academia de Ciencias de la Habana, 8: 643–645.
1873 Fernández de Castro, M., 1873. *Aetobatis poeyi*, nueva especie fósil procedente de la isla de Cuba. Anales de la Sociedad Española de Historia Natural, 2: 193–212.
1874 Fernández de Castro, M., 1874. *Aetobatis poeyi*, nueva especie fósil procedente de la isla de Cuba. Anales de la Academia de Ciencias de la Habana, 10: 368–374, Anales 11: 61–70, 93–109.
1996 Iturralde-Vinent, M., Hubbell, G., y Rojas-Consuegra, R., 1996. Catalogue of Cuban fossil Eslamobranchii (Paleocene-Pliocene) and paleogeographic implications of their Lower to Middle Miocene occurrence. The Journal of the Geological Society of Jamaica, 31: 7–21.

Works on Paleogene Ichnofossils

The paleogenic sequences are very rich and diverse in fossil traces, on which the first specific contributions (Villegas Martín, Rojas Consuegra, and others) have begun in 2000.

2009 Rojas-Consuegra, R. y J. Villegas-Martín, 2009. Icnofósiles e Icnofacies en algunas formaciones geológicas cubanas. III Convención sobre Ciencias de La Tierra. GEOCIENCIAS'2009. Memorias, Trabajos y Resúmenes en CD-ROM, La Habana, GEO2-P14: 397–424.
2011 Menéndez Peñate, L., Rojas-Consuegra, R., Villegas-Martín, J., y López, R., 2011. Taphonomy, cronostratigraphy and paleoceanographic implications at turbidite of Early Paleocene (Vertientes Formation), Cuba. Revista Geológica de América Central, 45: 87–94.
2014 Villegas Martín, J., Netto, R.G., Correa-Lavina, E.L., y Rojas-Consuegra, R., 2014. Ichnofabrics of the Capdevila Formation (Early Eocene) in the los Palacios Basin (western Cuba): Paleoenvironmental and paleoecological implications. Journal of South American Earth Sciences. 56: 214–227.

2016 Rojas-Consuegra, R. y Arano-Ruiz, A., Borges, C. y Villegas-Martín, J. 2016. Fósiles difíciles. Boletín de la Sociedad Cubana de Geología, 16(1): 15–16. ISSN 0864-3646.

2018 Rojas-Consuegra, R., Villegas-Martín, J., García-Sánchez, R. 2018. Presencia de los icnogéneros *Zoophycos* Massalongo 1855 y *Paleodictyon* Meneghini 1850 en el Paleógeno de Cuba. Libro de Resúmenes: IX Jornadas Científico Técnicas Ceinpet 2018. p. 29.

3.4.2.4 Contributions on the Neogene Period

Studies of paleontology of the Neogene of Cuba have, in general, more than 45 contributions, divided into vertebrates (23), invertebrates (18), and plants, pollen, and spores (4).

Works on Neogene Plants, Pollen, and Spores

In the first half of the twentieth century, several works were devoted to the macroscopic remains of plants found in Matanzas, in particular by Roca (1919) and years later by Berry (1934–1940). This site, the only one known to date in Cuba with such an abundance of well-preserved material, needs modern studies of its fossil plant association, where new contributions are foreseeable.

1919 Roca, M., 1919. Nota acerca de un yacimiento de fósiles vegetales del Abra del Yumurí, Matanzas, Cuba. Memorias de la Sociedad Cubana de Historia Natural "Felipe Poey", (4): 120–124.

1934 Berry, E.W., 1934. Pleistocene plants from Cuba. Bulletin of the Torrey Botanical Club, 61 (5): 237.

1936 Berry, E.W., 1936. A Miocene flora from the gorge of the Yumuri river, Matanzas, Cuba. John Hopkins Studies in Geology, 13: 95–134.

1940 Berry, E.W. 1940. Pleistocene plants from Cuba. Bull Torrey Bot. Club 61(5): 237–240.

Works on Neogene Invertebrates

Among the invertebrates of the Neogene, the study of echinoids, corals, molluscs, and the very diverse planktonic and benthic foraminifera stands out. The works go from the decade of the 20 to the 60 mainly, with the contributions of important authors (Sanchez Roig, Vaughan, Palmer, De la Torre y Callejas and others). Since the 2000s, decapods have gained attention (Shweitzer, Iturralde Vinent, Hetler, Velez Juarbe, Varela, Rojas Consuegra) and, in addition, the gastropods (López Martínez).

1920 Sánchez-Roig, M., 1920. Fósiles del Mioceno de la Habana. Boletín de Minas, 6.

1924 Vaughan, T.W., 1924. Fossil corals from Central America, Cuba and Puerto Rico, with an account of the American Tertiary, Pleistocene and Recent European Tertiary larger foraminifera. Geological Society of America Bulletin, 35: 785–822.

1941 Palmer, D.K., 1941a. Foraminifera of the Upper Oligocene Cojimar formation of Cuba (part 4). Memorias de la Sociedad Cubana de Historia Natural "Felipe Poey", 15 (3): 181–197.

1941 Palmer, D.K., 1941b. Foraminifera of the Upper Oligocene Cojimar formation of Cuba (part 5). Memorias de la Sociedad Cubana de Historia Natural "Felipe Poey", 15 (3): 281–310.

1953 Sánchez-Roig, M., 1953a. Dos nuevos géneros de equinoideos cubanos: Lambertona y Neopatagus. Memorias de la Sociedad Cubana de Historia Natural "Felipe Poey", 21 (3): 257–262.

1966 De la Torre y Callejas, A., 1966. El Terciario Superior y Cuaternario en los alrededores de Matanzas. Academia de Ciencias de Cuba, 51 pp.

1983 Franco-Álvarez, G.L., 1983. Fauna redepositada del Eoceno en el Aquitaniano de Cuba Oriental. En: Contribución a la Geología de Cuba Oriental. Editorial Científico-Técnica, La Habana, p. 261–262.

2006 Shweitzer, C.E., Iturralde-Vinent, M., Hetler, J.L. & Velez-Juarbe, J. 2006. Oligocene and Miocene decapods (Thalassinidae and Brachyura) from the Caribbean. Annals of Carnegie Museum, 75: 111–136.

2008 López-Martínez, R.A., y Rojas-Consuegra, R., 2008. Análisis tafonómico de los gasterópodos miocénicos de Cuba. Implicaciones paleobiogeográficas. Rev. Minería y Geología, 24(2):1–21. ISSN 1993 8012.

2008 López-Martínez, R.A., y Rojas-Consuegra, R., Urra-Abraira, J.L., y Martínez-Vargas, A., 2008. Análisis paleoambiental de sedimentos lateríticos del depósito Camarioca, Moa, Cuba. Rev. Minería y Geología, 24(1):1–14. ISSN 1993 8012.

2009 Varela, C., y Rojas-Consuegra, R., 2009. Crustáceos (Decapoda: Brachyura) fósiles de Cuba. Solenodon, 8: 118–123.

2011 Varela, C., y Rojas-Consuegra, R., 2011. El registro fósil de los crustáceos decápodos (Arthropoda, Crustacea) marinos de Cuba. En: Memorias IV Convención Cubana de Ciencias de la Tierra (Geociencias'2011). Centro Nacional de Información Geológica, Instituto de Geología y Paleontología de Cuba, CD-Rom. GEO2-P7, 10 p. La Habana.

2011 Varela, C., y Rojas-Consuegra, R., 2011a. Especie nueva de Eriosachila Blow y Manning, 1996, (Crustacea: Decapoda), de la formación Colón, Cuba. Novitates Caribaea, 4: 17–20.

2011 Varela, C., y Rojas-Consuegra, R., 2011b. Crustáceos fósiles (Decapoda: Brachyura) de la Formación Colón, Matanzas, Cuba. Solenodon, 9: 66–70.

2011 Varela, C., y Schweitzer, C.E., 2011c. A new genus and new species of Portunidae Rafinesque, 1815 (Decapoda,

Brachyura) from the Colón Formation, Cuba. Bulletin of the Mizunami Fossil Museum, 37: 13–16.
2013 Varela, C., 2013. Nuevos datos sobre los crustáceos fósiles (Decapoda: Brachyura) de Cuba. Solenodon, 11: 1–5.
2015 Kiel, S., y Hansen, B.T., 2015. Cenozoic Methane-Seep Faunas of the Caribbean Region. PLoS ONE 10 (10): e0140788. https://doi.org/10.1371/journal.pone.0140788.
2017 Frias Cárdenas, A. 2017. Balánidos del Neogéno en las Formaciones Colón y Lagunitas, Cuba cetroccidental. Resumen en: XII Congreso Cubano de Geología (Geología'2017).

Works on Neogene Vertebrates

The first report (1855) of the discovery of a tooth of *Carcharodon megalodon* is due to Valdés Aguirre, with the opinion of the prominent Cuban ichthyologist Felipe Poey y Aloy. Fishes were the first marine vertebrates of the Neogene studied by Sanchez Roig in the 20s of the twentieth century. More than 70 years later they were studied again in relation to the stratigraphy and paleogeography of the stage (Iturralde Vinent et al. 1998, 2001) and more recent contributions (2011–2016). The other widely studied group has been the sirenids, since the 30s and 70s (Trelles Duelo, Varona), attended in greater detail after 2000 (Domning and others). But a transcendental contribution to the study of Neogene terrestrial vertebrates in Cuba is due to MacPhee and Iturralde Vinent (1993–2003), which has a far-reaching impact on the knowledge of the Antillean-Caribbean fauna.

1855 Valdés-Aguirre, F. 1855. Fósiles cubanos. Revista de la Habana, 4.
1920 Sánchez-Roig, M., 1920a. Escuálidos del Mioceno y Plioceno de La Habana, Cuba. Boletín de Minas, 6: 1–16.
1936 Trelles Duelo, L., 1936. Restos fosilizados de un manatí extinguido del Oligoceno Inferior. Memorias de la Sociedad Cubana de Historia Natural "Felipe Poey", 9 (4): 269–270.
1972 Varona, L.S., 1972. Un dugóngido del Mioceno de Cuba (Mammalia: Sirenia). Memoria de la Sociedad de Ciencias Naturales La Salle, 32 (91): 5–20.
1993 MacPhee, R.D.E., 1993. From Cuba: A mandibule of *Paralouatta*. Evolutionary Anthropology, 2 (2).
1994 MacPhee, R.D.E., e Iturralde-Vinent, M., 1994. First Tertiary land mammal from Greater Antilles: an early Miocene sloth (Xenarthra, Megalonychidae) from Cuba. American Museum Novitates, 3094: 1–13.
1995 MacPhee, R.D.E., e Iturralde-Vinent, M., 1995a. Origin of the Greater Antillean land mammal fauna, 1: New Tertiary fossils from Cuba and Puerto Rico. American Museum Novitates, 3141: 30 pág.
1995 MacPhee, R.D.E., e Iturralde-Vinent, M., 1995b. Earliest monkey from Greater Antilles. Journal of Human Evolution, 28: 197–200.
1996 Iturralde-Vinent, M., G. Hubbell y R. Rojas, 1996. Catalogue of Cuban fossil Eslamobranchii (Paleocene—Pliocene) and Paleogeographic implications of their Lower to Middle Miocene occurrence. The journal of the Geological Society of Jamaica, 31: 7–21.
1998 Iturralde-Vinent, M., Mora, C.L., Rojas-Consuegra, R., y Gutiérrez-Domech, R., 1998. Myliobatidae (Elasmobranchii: Batomorphii) del Terciario de Cuba. Revista de la Sociedad Mexicana de Paleontología 8 (2): 135–145.
1998 Iturralde-Vinent, M., y Case, G.R., 1998. First report of the fossil fish *Diodon* (family Diodontidae) from the Miocene of Cuba. Revista de la Sociedad Mexicana de Paleontología, 8 (2): 123–126.
2003 MacPhee, R.D.E., Iturralde-Vinent, M., y Gaffney, E. S., 2003. Domo de Zaza, an early Miocene vertebrate locality in south-central Cuba: with notes on the tectonic evolution of Puerto Rico and the Mona Passage. American Museum Novitates, (3394): 1–42.
2004 Vázquez de la Torre, I. y E. Grau González-Quevedo, 2004. Descubren pez fósil del mioceno en Matanzas. In Noti-Cem. Rev. Electrónica (1861: Revista de Espeleología y Arqueología), 1:37.
2008 Díaz-Franco, S., Rojas-Consuegra, R., Hernández, I., y Chávez Marrero, M.E., 2008. Nueva localidad de sirenio (Mammalia) fósil, para el occidente de Cuba. En: Memorias II Simposio de Museos y Salas de Historia Natural, Museos, símbolos de comunicación cultural. Museo Nacional de Historia Natural.
2008. López-Martínez, R. A. y R. Rojas-Consuegra, 2008. Análisis tafonómico de los gasterópodos miocénicos de Cuba. Implicaciones paleobiogeográficas. Rev. Minería y Geología, 24(2):1–21. ISSN 1993 8012.
2008 Domning, D. and Orangel, A. 2008. Fossil sirenia of the West Atlantic and Caribbean region. VIII. *Nanosiren garciae*, gen. et sp. nov. and *Nanosiren sanchezi*, sp. nov. Journal of Vertebrate Paleontology 28(2): 479–500.
2011 Aranda-Manteca F.J., Rojas-Consuegra, R., y Jiménez-Vázquez, O., 2011. Carcharhinidae de Cuba y Haití, en la Colección del Museo Nacional de Historia Natural de Cuba. En: XII Congreso Nacional de Paleontología (SOMEXPAL), Benemérita Universidad de Puebla. 1 p.
2014 Vélez-Juarbe, J., y Domning, D.P., 2014. Fossil Sirenia of the West Atlantic and Caribbean region. Ix. Metaxytherium albifontanum, sp. nov. Journal of Vertebrate Paleontology, 34 (2): 444–464.
2016 Viñola-López, L. W. y Rojas-Consuegra, R. 2016. Distribución del género *Sparus* (Perciforme: Sparidae) en el Terciario de Cuba Occidental. Revista Geológica de América Central, 54: 57–66.

2017 Viñola López, L. W., R. Rojas Consuegra, O. Jiménez Vázquez. 2017. Nuevos registros de Sphyraena (Perciformes: Sphyraenidae) para el Neógeno de Cuba y La Española. Novitates Caribaea 11: 89–94.
2018 Rojas-Consuegra, R. 2018. Pez fósil matancero. Boletín Ceinpetillazo (Cienpet), 1(2): 19.
2019 Viñola-López, L. W., Carr, R., Lorenzo, L. 2019. First occurrence of fossil *Balistes* (Tetradontiformes: Balistidae) from the Miocene of Cuba with the description of a new species and a revision of fossil Balistes. Historical Biology (*in press*).

3.4.2.5 Contributions on the Quaternary Period

The Quaternary paleontology in Cuba has a wide literature, essentially dedicated to the different groups of fossil vertebrates, where more than 130 contributions are counted, and grouped into vertebrates (207), invertebrates (26), and plants, pollen, and spores (2).

Works on Quaternary Plants, Pollen, and Spores

It is evidently the needess of the particular studies on these fossil groups in Cuba till date.

1937 Bonazzi, A., 1937. Estudio sobre las turbas de Cuba. Memorias de la Sociedad Cubana de Historia Natural "Felipe Poey", 11: 5–30.
1990 Ferrera, M. M., E. Hernández Fuentes y M. Cabrera Castellanos. 1990–91. Análisis polínico de sedimentos marinos del occidente de la Isla de la Juventud (Cuba). Acta Bot. Hung. 36: 145–161.

Works on Quaternary Invertebrates

Among the first studied invertebrates of the Cuban territory are the corals, already from the nineteenth century and the beginning of the twentieth century (Crosby, Agassz, Vaughan). In the decades of 30 and 40 of the twentieth century, the study of molluscs, marine, and terrestrial of Cuba (Aguayo, Clench, Richards, Rehder, Jaume, Borro) had a wide development. A new impulse received the attention of marine molluscs, mainly, between the 70s and 80s (De la Torre y Callejas y Kojumdgieva). In the 1990s and 2000s, some other contributions appear. The first fossil insects in the country are reported (Valdéz, Díaz Franco, Iturralde Vinent, and others).

1882 Crosby, W.O., 1882. On the elevated coral reefs of Cuba. Boston Soc. Nat. Hist., 22: 124–130.
1893 Agassiz, A., 1893. Investigation of the Coral Reefs of the West Indies. American Journal of Science, 78 p.
1893 Agassiz, A., 1893. A reconnaissance of the Bahamas and the elevated reefs of Cuba, in the Steam-yacht Wild Duck. January to April 1893. Bulletin of the Museum of Comparative Zoology at Harvard College, 26: 1–203.
1919 Vaughan, T.W., 1919. Fossil corals from Central America, Cuba and Puerto Rico, with an account of the American Tertiary, Pleistocene and Recent coral. United States National Museum, Bulletin, 103: 189–524.
1933 Aguayo, C.G., y Clench, W.J., 1933. A new fossil Cepolis from Cuba. The Nautilus, 47: 129–138.
1935 Aguayo, C.G., 1935. Especilegio de moluscos cubanos. Memorias de la Sociedad Cubana de Historia Natural "Felipe Poey", 9 (4): 107–128.
1935 Aguayo, C.G., y Clench, W.J., 1935. A new Pleistocene Mecolitia from Cuba. The Nautilus, 49: 91–93.
1935 Richards, H.G., 1935. Pleistocene mollusks from western Cuba. Journal of Paleontology, 9 (3): 253–258.
1936 Aguayo, C.G., y Rehder, H., 1936. New marine molluscs from Cuba. Memorias de la Sociedad Cubana de Historia Natural "Felipe Poey", 9 (4): 263–268.
1938 Aguayo, C.G., 1938. Moluscos pleistocénicos de Guantánamo, Cuba. Memorias de la Sociedad Cubana de Historia Natural "Felipe Poey", 12 (2): 97–118.
1942 Jaume, M.L., y Pérez Farfante, I., 1942. Moluscos pleistocénicos de la zona franca de Matanzas. Memorias de la Sociedad Cubana de Historia Natural "Felipe Poey", 16 (1): 37–44.
1946 Aguayo, C.G., y Borro, P., 1946a. Nuevos moluscos del Terciario Superior de Cuba. Revista de la Facultad de Malacología Carlos de la Torre, 4 (2): 9–12.
1946 Aguayo, C.G., y Borro, P., 1946b. Algunos moluscos terciarios de Cuba. Revista de la Facultad de Malacología Carlos de la Torre, 4 (2): 43–49.
1948 Aguayo, C.G., 1948. Moluscos fósiles de la provincia de Oriente, Cuba. Revista Sociedad Malacológica, 6 (2): 55–63. (Se describen también moluscos de la provincia de Matanzas).
1949 Aguayo, C.G., 1949. Nuevos moluscos fósiles de Cuba y Panamá. Revista de la Facultad de Malacología Carlos de la Torre, 7 (1): 11–14.
1971 De la Torre y Callejas, A., 1971. Nueva especie de Pectinidae y notas sobre la distribución estratigráfica de Chlamys (Nodipecten) pittiery (Dall) en Cuba. Serie 4, Ciencias Biológicas, 13.
1985 De la Torre y Callejas, A., y Kojumdgieva, E., 1985. Asociaciones y niveles faunales de moluscos del Plioceno-Cuaternario del occidente de Cuba, y sus implicaciones estratigráficas. Reporte de investigación del Instituto de Geología y Paleontología, #5.
1988 De Huelbes Alonso, J., 1988. *Chlamys cruciana* (Bivalvia: Pectinidae) en los sedimentos de la Formación la Cruz de los alrededores de la ciudad de Santiago de Cuba. Tecnológica, 18 (3): 30–39.
1998 De la Torre-Callejas, A., y A. González-Guillen, 1998. Una segunda nueva especie de molusco terrestre del genero

Liguus Montford (Mollusca: Pulmonata) del Pleistoceno temprano de Cuba. Mem. III Congr. Cub. Geol. Min.'98, Tomo1: 700–701.

1999 Espinosa, J. y Ortea, J. 1999. Moluscos terrestres de Cuba. Avicennia, 1: 111–124.

1999 Valdéz, P., 1999. Primeros insectos fósiles cubanos (Coleoptera: Scarabaeidae: Dytiscidae). Cocuyo, 9: 17–18.

2001 Díaz-Franco, S., y Rojas-Consuegra, R., 2001. A new type of Holocene deposit in Cuba: "Trapped" insects within stalagmitic calcium carbonate. Caribbean Journal of Earth Science, 35: 37–38.

2005 Perera Montero, Y., y Rojas-Consuegra, R., 2005. Distribución facial de los corales de la Formación Jaimanitas en un área al oeste de Cójimar, Ciudad de la Habana. Memorias de Geomin 2005. La Habana. CD ROM. 2005. GEO3-4: 1–16. ISBN 959-7117-03-7.

2006 Schweitzer, C.E., Iturralde-Vinent, M., Hetler, J.L., y Velez-Juarbe, J., 2006. Oligocene and Miocene decapods (Thalassinidea and Brachyura) from the Caribbean. Annals of Carnegie Museum, 75 (2): 111–136.

2008 Portell, R.W., McCleskey, T. & Toomey, J.K. 2008. Pleistocene marine Mollusca. Fossil Invertebrates of the U.S. Naval Station Guantanamo Bay, Cuba, 1: 1–32.

2015 Suárez-Torres, A., 2015. Nueva especie fósil de *Cerion* Röding, 1798 (Mollusca: Pulmonata: Cerionidae) de Cuba oriental. Novitates Caribaea, 8: 120–127.

Works on Quaternary Vertebrates

During the second half of the nineteenth century (1865–1884), the discovery of fossil skeletal remains of terrestrial vertebrates on the island of Cuba aroused enormous interest in the world scientific community at the time, given the hypotheses that were open about the Antillean biogeography. In these pioneering investigations are the work of Fernández de Castro, Leidy, Pomel, Poey y Aloy, Egozcue y Cía. Throughout the twentieth century and so far in the twenty-first century, the studies of the various groups of vertebrates that integrate Cuba's Quaternary fossil megafauna have not ceased. Perhaps the vertebrate Quaternary fossils are the groups of taxonomically better-known fossils of the Cuban territory, nevertheless, their biochronology, for example, as well as other aspects, still remain without being specified. A broad field of studies remains clearly open to the talent of young scientists.

1864 Fernández de Castro, M., 1864. De la existencia de grandes mamíferos fósiles en la Isla de Cuba. Anales de la Academia de Ciencias de la Habana, 1: 17–21, 54–60, 66–107.

1865 Fernández de Castro, M., 1865. De la existencia de grandes mamíferos fósiles en la Isla de Cuba. Revista Minera, Madrid, 16: 161–178. 193–210.

1868 Leidy, J., 1868. Notice of some vertebrate remains from the West Indian Islands. Proceedings of the Academy of Natural Sciences of Philadelphia.

1868 Pomel, A., 1868. Sur le Myomorphus cubensis, Sousgenre Nouveau du Megalonyx. Acad. Sci. Paris, C. R. 67, pág. 665–668.

1871 Fernández de Castro, M., 1871a. El *Myomorphus cubensis*, nuevo sub-género del Megalonyx. Anales de la Academia de Ciencias de la Habana, 7: 463–476.

1871 Fernández de Castro, M., 1871b. El *Myomorphus cubensis*, nuevo sub-género del Megalonyx. Revista Minera, Madrid, 22: 165–178, 190–205.

1871 Poey y Aloy, F., 1871a. Indagación acerca de ciertos fósiles de Cuba. En: Anales de la Real Academia de Ciencias Médicas, Físicas y Naturales de La Habana T. 7. La Habana, pág. 656 y 698.

1871 Poey y Aloy, F., 1871b. Mamíferos fósiles de la Isla de Cuba. En: Anales de la Real Academia de Ciencias Médicas, Físicas y Naturales de La Habana T. 8. La Habana, pág. 124 y 163.

1872 Egozcue y Cía., J., 1872. Descripción de algunas piezas fósiles correspondientes a grandes mamíferos de América. Anales de la Academia de Ciencias de la Habana, 8: 627–634.

1881 Fernández de Castro, M., 1881. Pruebas paleontológicas de que la isla de Cuba ha estado unida al continente americano y breve idea de su constitución geológica. Boletín de la Comisión del Mapa Geológico de España, 8: 357–372.

1884 Fernández de Castro, M., 1884. Pruebas paleontológicas de que la isla de Cuba ha estado unida al continente americano y breve idea de su constitución geológica. Anales de la Academia de Ciencias de la Habana, 21: 146–165.

1902 Vaughan, T.W., 1902. Notes on Cuban fossils mammals. Science, 15 (369): 148–149.

1910 De la Torre y Huerta, C., 1910c. Excursión a la Sierra de Jatibonico: osamentas fósiles de *Megalocnus rodens* o *Myomorphus cubensis*: comprobación de la naturaleza continental de Cuba a principios de la época Cuaternaria. Anales de la Academia de Ciencias Médicas, Físicas y Naturales de la Habana, 47: 204–217.

1910 Spencer, J.W., 1910. The discovery of fossil mammals in Cuba and their great geographical importance. Science, 32 (825).

1911 De la Torre y Huerta, C., 1911. Comunicación sobre dos fósiles nuevos cubanos. Anales de la Academia de Ciencias Médicas, Físicas y Naturales de la Habana, 48: 599–602.

1912 Vanatta, E.G., 1912. Pleistocene fossils from eastern Cuba. The Nautilus, 26: 69.

1913 Matthew, W.D., 1913. Cuban fossils mammals, preliminary notes. Geological Society of America Bulletin, 24: 118–119.

1913 Brown, B., 1913. Some Cuban fossils: a hot spring yields up the bones animals that lived before the Advent of Man. The Journal of the American Museum 13: 221–228.

1915 De la Torre y Huerta, C., y Matthew, W.D., 1915. *Megalocnus* and other Cuban ground sloths. Geological Society of America Bulletin, 26: 152.

1916 Miller, G.S., 1916. Bones of mammals from Indian Sites in Cuba and Santo Domingo. Smithsonian Misc. Colls., 66, 10 pág.

1916 Miller, G.S., 1916. The teeth of a monkey in Cuba. Smithsonian Misc. Colls., 66 (13): 1–3.

1917 De la Torre y Huerta, C., 1917a. Nuevas especies de mamíferos fósiles de Cuba y otras antillas. Memorias de la Sociedad Cubana de Historia Natural "Felipe Poey", 2: 234–251.

1917 De la Torre y Huerta, C., 1917b. Presentación -del esqueleto restaurado del Myomorphus o Megalocnus rodens. Memorias de la Sociedad Cubana de Historia Natural "Felipe Poey", 2: 94–101.

1917 Peterson, O.A., 1917. Report upon the fossil material collected in 1913 by the Messrs. Link in a Cave in the Isle of Pines. Annals of Carnegie Museum, 11 (24): 359–361.

1918 Matthew, W.D., 1918. Affinities and origins of the Antillean mammals. Geological Society of America Bulletin, 29: 657–666.

1918 Matthew, W.D., 1918. Skeletons of the Ground Sloths in Habana and American Museums. The Journal of the American Museum, 18, p. 303.

1919 Matthew, W.D., 1919. Recent discoveries of fossil vertebrates in the West Indies and their bearing on the origin of Antillean Fauna. Proceedings of the American Philosophical Society, 58: 161–181.

1922 Kraglievich, L., 1922. *Amphiocnus paranense*, n gen., n. sp., un probable precursor del *Megalocnus* de la isla de Cuba en la Formación Entrerriana. Physys. Revista de la Sociedad Argentina de Ciencias Naturales, 6: 73–87.

1928 Wetmore, A., 1928. Bones of birds from the Ciego Montero deposits of Cuba. American Museum Novitates 301: 1–5. En: 1928. On Fossil Birds from Nebraska and Cuba.

1929 Miller, G.S., 1929. The characters of the genus *Geocapromys* Chapman. Smithsonian Misc. Colls., 82 (4): 1–3.

1931 Matthew, W.D., 1931. Genera and new species of ground sloths from the Pleistocene of Cuba. American Museum Novitates, 511: 1–6.

1948 Arredondo de la Mata, O., y Rivero de la Calle, M., 1948. Sensacional descubrimiento en las cuevas de Bellamar. Nuevos Rumbos, IV (I) 14–15.

1950 Aguayo, C.G., 1950. Observaciones sobre algunos mamíferos cubanos extinguidos. Boletín de la Sociedad Cubana de Historia Natural, 1 (3): 121–134.

1950 Williams, E.W., 1950. *Testudo cubensis* and the evolution of Western Hemisphere tortoises. Bulletin of American Museum of Natural History, 95: 1–36.

1951 Arredondo de la Mata, O., 1951. De prehistoria cubana: Sorprendente hallazgo del almiquí en Pinar del Río. Lux, La Habana, 15–17.

1951 Arredondo de la Mata, O., 1951. Manifestaciones paleontológicas descubiertas por el ICA en un abrigo rocoso de Santa Fe. En: Minero de la fauna extinguida de Cuba en Santa Fe, La Habana, Boletín Informativo del ICA, III (247): 46–59.

1952 Williams, E.E., y Koopman, K.F., 1952. West Indians fossil monkeys. American Museum Novitates, 1546: 1–16.

1954 Arredondo de la Mata, O., 1954. Informe a la Sociedad Espeleológica de Cuba del resultado, hasta el presente, de las investigaciones paleontológicas llevadas a cabo por esta institución en la Cueva de Pío Domingo, ubicada la Sierra de Sumidero, frente al Valle de Pica-Pica y en la Cueva del Salón, Quemado de Pineda, en Pinar del Río (Inédito).

1955 Arredondo de la Mata, O., 1955. Contribución a la Paleontología de la Sociedad Espeleológica de Cuba. Resumen de las actividades paleontológicas realizadas por la Sección de Geología y Paleontología. Boletín de la Sociedad Espeleológica de Cuba, I (2): 3–31. Anexo.

1955 Koopman, K.F., y Ruibal, R., 1955. Cave-fossil vertebrates from Camagüey, Cuba. Breviora, 46: 1–8.

1956 Arredondo de la Mata, O., 1956. Como identificar los restos óseos de algunos mamíferos cubanos extintos que resultan de suma importancia en las investigaciones arqueológicas y antropológicas. Boletín de la Sociedad Espeleológica de Cuba, 3: 1–10. [Parte I].

1956 Arredondo de la Mata, O., 1956. Descripción de los maxilares de la posible nueva especie de roedor hallada en Cueva de "Pio Domingo" Sierra de Sumidero en Pinar del Río. Apéndice de: Cómo identificar los restos de algunos mamíferos cubanos extinguidos que resultan de suma importancia en las investigaciones arqueológicas y antropológicas. Boletín de la Sociedad Espeleológica de Cuba, 5: 6–10.

1956 Arredondo de la Mata, O., 1956. Una posible especie nueva más dentro del grupo de los Boromys y "Geoboromys" (FM. Echimyidae). Apéndice de: Cómo identificar los restos de algunos mamíferos cubanos extinguidos que resultan de suma importancia en las investigaciones arqueológicas y antropológicas. Boletín de la Sociedad Espeleológica de Cuba, 6: 4–6.

1956 Paula Couto, C. de, 1956. On two mounted skeletons of Megalocnus rodens. Journal of Mammalogy, 37 (3): 423–427.

1958 Arredondo de la Mata, O., 1958. Aves gigantes de nuestro pasado prehistórico. El Cartero Cubano, 17 (7):10–12.

1958 Arredondo de la Mata, O., 1958. El Mesocnus torrei: un mamífero extinguido del Pleistoceno cubano. Scout, La Habana, 2–3, 14.

1958 Arredondo de la Mata, O., 1958. Los roedores cubanos extintos. El Cartero Cubano, 17 (12): 8–11 y 48.

1958 Koopman, K.F., 1958. A fossil vampire bat from Cuba. Breviora, 90: 1–5.

1959 Matthew, W.D., y Paula Couto, Carlos de, 1959. The Cuban Edentates. Bulletin American Museum of Natural History, New York, 117 (I): 56 pág.

1960 Arredondo de la Mata, O., 1960. Origen, evolución y extinción del Megalocnus rodens. El Cartero Cubano, Enero, 9–12.

1961 Arredondo de la Mata, O., 1961. Descripciones preliminares de dos nuevos géneros y especies de edentados del Pleistoceno cubano. Boletín del Grupo Exploraciones Científicas, 1: 19–40.

1966 Varona, L.S., 1966. Notas sobre los Crocodílidos de Cuba y descripición de una nueva especie del Pleistoceno. Poeyana, Serie A., 1–34.

1967 Paula Couto, C. de, 1967. Pleistocene edentates of the West Indies. American Museum Novitates, 2304: 1–47.

1969 Mayo, N.A., 1969. Nueva especie de Megalonychidae y descripción de los depósitos cuaternarios de la Cueva del Vaho, Boca de Jaruco, La Habana. Mem. Fac. Cien. Ser. Cien. Biol., 3:1–58.

1970 Arredondo de la Mata, O., 1970. Dos nuevas especies subfósiles de mamíferos (Insectívora: Nesophontidae) del Holoceno Precolombino de Cuba. Memoria de la Sociedad de Ciencias Naturales La Salle, 30 (86): 122–152.

1970 Arredondo de la Mata, O., 1970. Nueva especie de ave pleistocénica del Orden Accipitriformes (Accipitridae) y nuevo género para las Antillas. Ciencias 4, Ciencias biológicas 8, Universidad de la Habana, 19 pp.

1970 Mayo, N.A., 1970. Depósitos pleistocénicos de los cauces subterráneos abandonados de la Sierra de los Órganos: evidencias de periodos pluviales. Academia de Ciencias de Cuba, Actas 2: 57–62.

1970 Mayo, N.A., 1970. La fauna vertebrada de Punta Judas. Academia de Ciencias de Cuba, Serie Espeleológica y Carsológica, 30: 1–45.

1971 Arredondo de la Mata, O., 1971. Nuevo género y especie de ave fósil (Accipitriformes: Vulturidae) del Pleistoceno de Cuba. Memoria de la Sociedad de Ciencias Naturales La Salle.

1971 Fischer, K.H., 1971. Riesenfaultiere (Megalonychidae, Edentata, Mammalia) aus dem Plesitozänr der Pio-Domingo-Höhle in Kuba. Wissenschafliche Zeitschrift der Humboldt—Universität zu Berlin, Math., -Nat., R., XX (1971) 4/5.

1972 Arredondo de la Mata, O., 1972a. Especie nueva de lechuza gigante (Strigiformes: Tytonidae) del Pleistoceno cubano. Boletín de la Sociedad Venezolana de Ciencias Naturales, 30 (124/125): 129–140.

1972 Arredondo de la Mata, O., 1972b. Nueva especie de ave fósil (Strigiformes: Tytonidae) del Pleistoceno superior de Cuba. Boletín de la Sociedad Venezolana de Ciencias Naturales, 29 (122/123): 415–431.

1972 Martin, R.A., 1972. Synopsis of Late Pliocene and Pleistocene bats of North America and the Antilles. The American Midland Naturalist, 87 (2): 326–335.

1973 Kurochkin, E., y Mayo, N.A., 1973. Las lechuzas gigantes del Pleistoceno Superior de Cuba. Actas, resúmenes, comunicaciones y notas del V consejo científico, La Habana, 3: 56–60.

1973 Mayo, N.A., y Peñalver-Hernández, L.L., 1973. Los problemas básicos del Pleistoceno de Cuba. Actas, Resúmenes, comunicaciones y notas del V consejo científico. La Habana, 3: 61–64.

1973 Woloszyn, B.W., 1973. El vampiro fósil de Cuba. Actas, Instituto de Geología, Academia de Ciencias de Cuba, 3: 54–55.

1974 Arredondo de la Mata, O., y Varona, L.S., 1974. Nuevos género y especie de mamífero (Carnivora: Canidae) del Cuaternario de Cuba. Poeyana, 131: 1–12.

1974 Silva-Taboada, G., 1974. Fósil Chiroptera from cave deposits in central Cuba, with descriptions of two new species (genera *Pteronotus* and *Mormoops*) and the first West Indian record of *Mormoops megalophylla*. Acta Zool. Cracoviencia.

1974 Varona, L.S., 1974. Catálogo de los mamíferos vivientes y extinguidos de Las Antillas. Academia de Ciencias de Cuba, La Habana, 139 p.

1974 Woloszyn, B.W., 1974. Informe final del asesor polaco, Dr. Bronislav Woloszyn, sobre investigaciones de los mamíferos cuaternarios de Cuba (1972–1974). Instituto de Geología y Paleontología, Academia de Ciencias de Cuba, 23 p.

1974 Woloszyn, B.W., y Mayo, N.A., 1974. Postglacial remains of a vampire bat (Chiroptera: Desmodus) from Cuba. Acta Zool. Cracoviensia, 19 (13): 253–265.

1975 Arredondo de la Mata, O., 1975. Distribución geográfica y descripción de algunos huesos de Ornimegalonyx oteroi, Arredondo 1958 (Strigiformes: Strigidae) del Pleistoceno superior de Cuba. Memoria de la Sociedad de Ciencias Naturales La Salle, 101 p.

1976 Arredondo de la Mata, O., 1976. Contribuciones del Dr. Alexander Wetmore a la Paleornitología del Continente Americano. Memorias de la Sociedad de Ciencias Naturales La Salle, 105: 229–248.

1976 Arredondo de la Mata, O., y Olson, S.L., 1976. The great predatory birds of the Pleistocene of Cuba. En: Olson, S.L., (Editor) Smithsonian Contributions to Paleobiology, 27: 169–187.

1977 Arredondo de la Mata, O., 1977. Nueva especie de Mesocnus (Edentata: Megalonychidae) del Pleistoceno de Cuba. Poeyana, 172:1–10.

1977 Fischer, K.H., 1977. Quartäre Mikromammalia Cubas, vorwiegend aus der Höhle San José de la Lamas, Sante Fé, Provinz Habana. Z. Geol. Wiss., Berlin, 5: 213–255.

1977 Woloszyn, B.W., y Silva-Taboada, G., 1977. Nueva especie fósil de *Artibeus* (Mammalia: Chiroptera) de Cuba, y tipificación preliminar de los depósitos fosilíferos cubanos contentivos de mamíferos terrestres. Poeyana, 161: 1–17.

1978 Mayo, N.A., 1978. Ein neves genus and zwei neve arten fossiler Riensenfaultiere von der familier der Megalonychidae (Edentata: Mammalia) aus dem Pleistozan Kubas. Eclogae Geologicae Helvetiae, 71 (3): 687–697.

1978 Mayo, N.A., 1978. Revision of *Neocnus minor* Arredondo, 1951 (Edentata: Megalonychidae) from the Pleistocene of Cuba. Koninklijke Nederlandse Akademie, Amsterdan, series B, 81 (3): 322–338.

1979 Silva-Taboada, G., 1979. Los murciélagos de Cuba. Editorial Academia, La Habana, 423 pp.

1979 Varona, L.S., 1979. Subgénero y especie nuevo de *Capromys* (Rodentia: Caviomorpha) para Cuba. Poeyana, 194, 33 pág.

1979 Varona, L.S., y Arredondo de la Mata, O., 1979. Nuevos taxones fósiles de Capromyidae (Rodentia: Caviomorpha). Poeyana, 195: 1–51.

1980 Mayo, N.A., 1980. Nueva especie de *Neocnus* (Edentata: Megalonychidae de Cuba) y consideraciones sobre la evolución, edad y paleoecología de las especies de este género. Actas II Congreso Argentino de Paleontología y Bioestratigrafía y I Congreso Latinoamericano de Paleontología. Buenos Aires. T. III: 223–236.

1980 Mayo, N.A., 1980. Revision de *Neocnus major* Arredondo, 1961 (Mammalia: Edentata del Pleistoceno de Cuba), con descripción de un cráneo y algunos huesos postcraneales. Estudios geol., 36: 427–440.

1980 Morgan, G.S., Ray, C.E., y Arredondo de la Mata, O., 1980. A giant extinct insectivore from Cuba (Mammalia: Insectivora: Solenodontidae). Proceding of Biological Society of Washington, 93 (3): 597–608.

1980 Varona, L.S., 1980. Sobre las jutías fósiles de Cuba (Rodentia: Capromyidae). Memoria de la Sociedad de Ciencias Naturales La Salle, 40 (113): 129–137.

1981 Arredondo de la Mata, O., 1981. Nuevos género y especie de mamífero (Carnivora: Canidae) del Holoceno de Cuba. Poeyana, 218.

1981 Arredondo de la Mata, O., 1981. Reemplazo de *Paracyon* por *Indocyon* (Carnivora: Canidae). Miscelánea Zoológica, 39.

1982 Acevedo-González, M., y Arredondo de la Mata, O., 1982. Paleozoogeografía y geología del Cuaternario de Cuba, características y distribución geográfica de los depósitos con restos de vertebrados. IX Jornada Científica del Instituto de Geología y Paleontología, Academia de Ciencias de Cuba, p. 59–84.

1982 Arredondo de la Mata, O., 1982. Los Strigiformes fósiles del Pleistoceno cubano. Boletín de la Sociedad Venezolana de Ciencias Naturales, 140.

1982 Arredondo de la Mata, O., y de la Cruz, J., 1982. Hallazgo del desdentado fósil *Neocnus gliriformis* (Mattew, 1931) (Edentata: Megalonychidae) en la provincia de Holguín, Cuba. Miscelánea Zoológica, 15: 2–4.

1983 Arredondo de la Mata, O., y Varona, L.S., 1983. Sobre la validez de *Montaneia anthropomorpha* Ameghino, 1910 (Primates: Cebidae). Poeyana, 255: 1–21.

1984 Arredondo de la Mata, O., 1984. Sinopsis de las aves halladas en depósitos fosilíferos pleisto-holocénicos de Cuba. Reporte de Investigación del Instituto de Zoología, 17, 33 pp. La Habana.

1984 Rodríguez, R., et al., 1984. La convivencia de la fauna de desdentados extinguidos con el aborígen de Cuba. Kobie. Diputación floral de Vizcaya, Bilbao, 14: 561–566.

1984 Rosenberger, A.L., 1984. Fossil New World Monkeys Dispute the Molecular Clock. Journal of Human Evolution, 13: 737–742.

1984 Varona, L.S., 1984. Los cocodrilos fósiles de Cuba. Caribbean Journal of Science, 20 (1–2): 13–17.

1984 Varona, L.S., 1984. Nueva especie fósil de *Capromys* (Rodentia: Capromyidae) del Pleistoceno Superior de Cuba. Poeyana, 285: 1–6.

1984 Varona, L.S., 1984. Otra especie fósil de *Capromys* (Rodentia: Capromyidae). Poeyana, 286: 1–5.

1985 Olson, S.L., 1985. A new species of *Siphonorhis* from Quaternary cave deposits in Cuba (Aves: Caprimulgidae). Proceedings of the Biological Society of Washington, 98 (2): 526–532.

1985 Pino, M., y Castellanos, N., 1985. Acerca de la asociación de perezosos cubanos extinguidos con evidencias culturales de aborígenes cubanos. Reporte de Investigación del Instituto de Ciencias Sociales, 4:1–29.

1985 Pino, M., y Febles, J., 1985. Restos de Nesophontes major Arredondo, en la región central de Cuba. Miscelánea Zoológica, 25: 3–4.

1985 Wetherbee, D.K., 1985. The extinct Cuban and Hispaniolan Macaws (Ara: Psittacidae) and description of a new species *Ara cubensis*. Caribbean Journal of Science, 21 (3–4): 169–175.

1986 Morgan, G.S., y Woods, C.A., 1986. Extinction and zoogeography of West Indian land mammals. Biological Journal of the Linnean Society, 28: 167–203.

1987 Olson, S.L., y Kurochkin, E.N., 1987. Fossil evidence of a *Tapaculo* in the Quaternary of Cuba (Aves:

Passeriformes: Scytalopodidae). Proceedings of the Biological Society of Washington, 100 (2): 353–357.
1988 Arredondo de la Mata, O., 1988. Nueva subfamilia de Megalonychidae (Mammalia: Edentata). Miscelánea Zoológica, 39, 2 p.
1989 MacPhee, R. D. E. et al. 1989. Pre-Wisconsinan land mammals from Jamaica and models of late Quaternary extinction in the Greater Antilles. Quat. Res. (31): 94–106.
1989 Woods, C.A., 1989. The biogeography of West Indian Rodents. En: Woods, C.A., (Editor) Biogeography of the west Indies: past, present and future, pág. 741–797. Sandhill Crane Press, Gainesville, Florida.
1990 Woods, C.A., 1990. The fossil and recent land mamals of West Indies: an analysis of the origin, evolution and extinction of an insular fauna. En: Azzaroli, A., (Editor) Biogeographical aspects of insularity, Atti Convegni Lincei., 85: 641–680. Rome, Accad. Naz. Lincei, Rome.
1991 Rivero de la Calle, M., y Arredondo de la Mata, O., 1991. *Paralouatta varonai*, a new Quaternary platyrrhine from Cuba. Journal of Human Evolution, 21: 1–11.
1993 Jiménez-Vázquez, O., y Crespo, R., 1993. Las alturas del Cacahual y el registro fósil de sus cuevas. Jornada Científica XVI Aniversario del Grupo Borrás (SEC), Resúmenes, pág. 12–13.
1993 Morgan, G.S., Franz, R., y Crombie, R.I., 1993. The Cuban crocodile, *Crocodylus rhombifer*, from Late Quaternary fossil deposits on Grand Cayman. Caribbean Journal of Science, 29 (3–4): 153–164.
1993 Morgan, G.S., y Ottenwalder, J.A., 1993. A new extinct species of Solenodon (Mammals: Insectivora: Solenodontidae) from the Late Quaternary of Cuba. Annals of Carnegie Museum, 62 (2): 151–164.
1993 Ramos, J., 1993. Hallazgos de fauna vertebrada fósil en cuevas de Sierra Las Damas, Sancti Spíritus. Grupo Caonao, Sociedad Espeleológica de Cuba, Cabaiguán (Inédito).
1993 Jiménez-Vázquez, O., y Crespo, R., 1993. Las alturas del Cacahual y el registro fósil de sus cuevas. Jornada Científica XVI Aniversario del Grupo Borrás (SEC), Resúmenes, pág. 12–13.
1993 Woods, C.A., 1993. Rodentia: Hystricognath: Capromyidae (Capromyinae). En: Wilson, D.E., y Reeder, D.M., (Editores) Mammals species of the world, a taxonomic and geographic reference, Segunda edición, pág. 800–803, Smithsonian Inst. Press, Washington.
1994 Arredondo de la Mata, O., y Olson, S.L., 1994. A new species of owl of the genus *Bubo* from the Pleistocene of Cuba (Aves: Strigiformes). Proceedings of the Biological Society of Washington, 107: 436–444.
1994 Arredondo-Antúnez, C., 1994. Distribución geográfica de los restos de mamíferos extintos (Edentata: Megalonychidae) del Terciario y Cuaternario de Cuba. II Congreso Latinoamericano de Terología, La Habana, p. 162–163.

1995 Jiménez-Vázquez, O., y Valdés, P., 1995. Los vertebrados fósiles de la Cueva del Indio, San José la Lajas, Habana, Cuba. Congreso Internacional 55 Aniversario de la Sociedad Espeleológica de Cuba, La Habana, Resúmenes, p. 62–63.
1995 MacPhee, R. D. E, Horovitz, I., Arredondo, O., Jiménez, O. 1995. A new genus for the extinct Hispaniolan monkey *Saimiri bernensis* Rimoli, 1977, with notes on its systematic position. American Museum Novitates, 3134:1–21.
1996 Arredondo de la Mata, O., 1996. Lista de especies extinguidas de vertebrados halladas en las provincias orientales de Cuba. Garciana, 24–25: 1–2.
1996 MacPhee, R.D.E., y Rivero de la Calle, M., 1996. Accelerator mass spectrometry 14C age determination for the alleged "Cuban spider monkey", *Ateles* (=*Montaneia*) *anthropomorphus*. Journal of Human Evolution, 30: 89–94.
1997 Arredondo de la Mata, O., 1997. Aportes al conocimiento de los caprómidos fósiles del Cuaternario de Cuba. La Habana, 76 pp.
1997 Arredondo de la Mata, O., y Rivero de la Calle, M., 1997. Nuevo género y especie de Megalonychidae del Cuaternario cubano. Revista de Biología, 11: 105–112.
1997 Arredondo-Antúnez, C., 1997. Composición de la fauna de vertebrados terrestres extintos del Cuaternario de Cuba. Revista electrónica Órbita Científica, 8 (2). Instituto Superior Politécnico Enrique José Varona. La Habana.
1997 Jiménez, O. 1997. Seis nuevos registros de aves fósiles en Cuba. El Pitirre, 10(2):49.
1997 Suárez, W., y Arredondo de la Mata, O., 1997. Nuevas adiciones a la paleornitología cubana. El Pitirre, 10 (3): 100–102.
1999 Arredondo. O., y Arredondo-Antúnez, C., 1999. Nuevos géneros y especie de ave fósil (Falconiforme) del Cuaternario de Cuba. Poeyana, 470–475.
1999 Arredondo-Antúnez, C., 1999. Los edentados extintos del Cuaternario de Cuba. Tesis doctoral. Facultad de Biología, Universidad de la Habana, Cuba, 97 pp.
1999 Díaz-Franco, S., 1999. Dos registros nuevos de aves endémicas en depósitos fosilíferos de Cuba. El Pitirre, 12 (1): 12–13.
1999 Iturralde-Vinent, M., MacPhee, R.D.E., Díaz-Franco, S., y Rojas-Consuegra, R., 1999. A small "Rancho La Brea" site discovered in Cuba. The Journal of the Geological Society of Jamaica, (33): 20.
1999 MacPhee, R.D.E., Flemming, C., y Lunde, D.P., 1999. "Last occurrence" of the Antillean insectivoran *Nesophontes*: new radiometric dates and their interpretation. American Museum Novitates, 3261: 1–20.
1999 Suárez, W., 1999. Dos registros nuevos de aves endémicas en depósitos fosilíferos de Cuba. El Pitirre, 12 (1): 12–13.

2000 Arredondo-Antúnez, C., y Arredondo de la Mata, O., 2000. Nuevo género y especie de perezoso (Edentata: Megalonychidae) del Pleistoceno de Cuba. Revista de Biología, 14 (1): 1–7.

2000 Iturralde-Vinent, M., MacPhee, R.D.E., Díaz-Franco, S., Rojas-Consuegra, R., Suárez, W., y Lomba, A., 2000. Las Breas de San Felipe, a Quaternary Fossiliferous Asphalt Seep near Martí (Matanzas Province, Cuba). Caribbean Journal of Science, 36 (3–4): 300–313.

2000 MacPhee, R.D.E., Jennifer, L., White, y Woods, C.A., 2000. New Megalonychid Sloths (Phyllophaga, Xenarthra) from the Quaternary of Hispaniola. American Museum Novitates, 3303, 32 pp.

2000 Suárez, W., 2000a. Contribución al conocimiento del estatus genérico del cóndor extinto (Ciconiiformes: Vulturidae) del Cuaternario cubano. Ornitología Neotropical, 11: 109–122.

2000 Suárez, W., 2000b. Fossil evidence for the occurrence of Cuban Poorwill Siphonorhis daiquiri in western Cuba. Cotinga, 14: 66–68.

2001 Condis-Fernández, M.M., 2001. Revisión taxonómica del género *Nesophontes* (Insectívora: Nesophontidae) en Cuba. Tesis de maestría, Facultad de Biología de la Universidad de La Habana, 61 pp.

2001 Díaz-Franco, S., 2001a. Situación taxonómica de *Geocapromys megas* (Rodentia: Capromyidae). Caribbean Journal of Science, 37 (1–2): 72–80.

2001 Díaz-Franco, S., 2001b. Estructura dental interna y modificación del diseño oclusal inferior en Boromys offella (Rodentia: Echimyidae). Revista de Biología, 15 (2): 152–157.

2001 Ottenwalder, J.A., 2001. Systematic and biogeography of the West Indian genus *Solenodon*. En: Woods, C.A., y Sergile, F.E., (Editores) Biogeography of the West Indies: Patterns and perspectives, 2da Edición, CRC Press, pág. 253–329.

2001 Suárez, W., 2001a. Deletion of the flightless *Ibis xenicibis* from the fossil record of Cuba. Caribbean Journal of Science, 37 (1–2): 109–110.

2001 Suárez, W., 2001b. A reevaluation of some fossils identified as vultures (Aves: Vulturidae) from Quaternary cave deposits of Cuba. Caribbean Journal of Science, 37 (1–2): 110–111.

2001 Suárez, W., y Olson, S.L., 2001a. A remarkable new species of small falcon from the Quaternary of Cuba (Aves: Falconidae: Falco). Proceedings of the Biological Society of Washington, 114 (1): 34–41.

2001 Suárez, W., y Olson, S.L., 2001b. Further characterization of *Caracara creightoni* Brodkorb based on fossils from the Quaternary of Cuba (Aves: Falconidae). Proceedings of the Biological Society of Washington, 114 (2): 501–508.

2001 Woods, C.A., 2001. Introduction and historical overview of patterns of West Indian biogeography. En: Woods, C.A., y Sergile, F.E., (Editores) Biogeography of the West Indies: patterns and perspectives, Segunda edición, pág. 1–14, CRC. Press, Boca Raton.

2001 Woods, C.A., Borroto, R., y Kilpatrick, C.W., 2001. Insular pattern and radiations of West Indian rodents. En: Woods, C.A., y Sergile, F.E., (Editores) Biogeography of the West Indies: patterns and perspectives, Segunda edición, pág. 335–353, CRC. Press, Washington.

2001 Woods, C.A., y Sergile, F.E., 2001. Biogeography of the West Indies: patterns and perspectives, Segunda edición, 582 pág, CRC. Press, Boca Raton.

2001 Williams, M.I., y Steadman, D.W., 2001. The historic and prehistoric distribution of parrots (Psittacidae) in the West Indies. En: Woods, C.A., y Sergile, F.E., (Editores) Biogeography of the West Indies Patterns and Perspectives. Ed. 2. CRC Press: Baton Rouge.

2002 Arredondo de la Mata, O., y Arredondo-Antúnez, C., 2002a. Nueva especie de ave (Falconiforme: Teratornithidae) del Pleistoceno de Cuba. Poeyana, 470–475: 15–21.

2002 Arredondo de la Mata, O., y Arredondo-Antúnez, C., 2002b. Nueva especie de perezoso arborícola (Edentata: Megalonychidae) del Pleistoceno de Cuba. Poeyana, 470–475: 36–40.

2002 Arredondo de la Mata, O., y Arredondo-Antúnez, C., 2002c. Nuevos género y especie de ave fósil (Falconiforme: Accipitridae) del Cuaternario de Cuba. Poeyana, 470–475: 9–14.

2002 Díaz-Franco, S., 2002. La variación del diseño oclusal inferior en *Boromys torrei* (Rodentia: Echimyidae). Revista de Biología, 16 (1): 60–65.

2002 Jiménez-Vázquez, O., y Fernández Milera, J., 2002. Cánidos precolombinos de las Antillas: mitos y verdades. Boletín del Gabinete de Arqueología, 2 (2): 78–87.

2003 Jiménez, O.; M. M. Condis, 2003. Génesis de los yacimientos fosilíferos cavernarios del Cuaternario de Cuba. Proceedings of the International Symposium "Insular Vertebrates Evolution: The Palaeontological Approach". Palma de Mallorca: 18 pp.

2003 Jiménez, O., Milera, J. F. 2003. Cánidos precolombinos de Las Antillas: Mitos y verdades. Gabinete de Arqueología, 2(2):78–87.

2003 Suárez, W., y Díaz-Franco, S., 2003. A new fossil bat (Chiroptera: Phyllostomidae) from a Quaternary cave deposit in Cuba. Caribbean Journal of Science, 39 (3): 371–377.

2003 Suárez, W., y Emslie, S.D., 2003. New fossil material with a redescription of the extinct condor *Gymnogyps varonai* (Arredondo, 1971) from the Quaternary of Cuba (Aves: Vulturidae). Proceedings of the Biological Society of Washington, 116 (1): 29–37.

2003 Suárez, W., y Olson, S.L., 2003. A new species of *Caracara* (Milvago) from the Quaternary asphalt deposits in Cuba, with notes on new material of *Caracara creightoni* Brodkorb (Aves: Falconidae).

2003 Suárez, W., y Olson, S.L., 2003. New records of storks (Ciconiidae) from Quaternary asphalt deposits in Cuba. Condor, 105: 150–154.

2003 Suárez, W., y Olson, S.L., 2003. Red-shouldered hawk and aplomado falcon from Quaternary asphalt deposits in Cuba. Journal of Raptor Research, 37 (1): 71–75.

2004 Díaz-Franco, S., 2004a. Análisis de la extinción de algunos mamíferos cubanos, sobre la base de evidencias paleontológicas y arqueológicas. Revista de Biología, 18 (2): 147–154.

2004 Díaz-Franco, S., 2004b. Type specimens of fossil vertebrates housed in the Museo Nacional de Historia Natural de Cuba. Revista de Biología, 18 (2): 155–159.

2004 Jull, A.J.T, Iturralde-Vinent M, O'Malley, J.M., et al. 2004. Radiocarbon dating of extinct fauna in the Americas recovered from tar pits. Nucl Instrum Methods Phys Res Sect B 223–224, 668–671.

2004 Suárez, W., 2004a. The identity of the fossil raptor of the genus *Amplibuteo* (Aves: Accipitridae) from the Quaternary of Cuba. Caribbean Journal of Science, 40: 120–125.

2004 Suárez, W., 2004b. The enigmatic snipe *Capella* sp. (Aves: Scolopacidae) in the fossil record of Cuba. Caribbean Journal of Science, 40 (1): 156–157.

2004 Suárez, W., 2004c. Biogeografía de las aves fósiles de Cuba. En: Iturralde-Vinent, M., (Editor) 2004. Origen y Evolución del Caribe y sus biotas Marinas y Terrestres. Centro Nacional de Información Geológica, La Habana, Cuba.

2005 Condis-Fernández, M.M., Jiménez-Vázquez, O., y Arredondo-Antúnez, C., 2005. Revisión taxonómica del género *Nesophontes* (Insectivora: Nesophontidae) en Cuba. Análisis de los caracteres diagnósticos. Procedings of the Internacional Symposium "Insular Vertebrate Evolution: the Palaeontological Approach" (Alcocer, J.A., y Bover, P., Editores), Monografies de la Societat d'Historia Natural de les Balear, 12: 95–100.

2005 Jiménez Vázquez, O., Marjorie M. Condis, Elvis García Cancio. 2005. Vertebrados postglaciales en un residuario fósil de *Tyto alba* Scopoli (Aves) del occidente de Cuba. Revista Mexicana de Mastozoología, 9: 85–111.

2005 Mancina, C.A., y García-Rivera, L., 2005. New genus and species of fossil bat (Chiroptera: Phyllostomidae) from Cuba. Caribbean Journal of Science, 41 (1): 22–27.

2005 Núñez-Torres, L., 2005. Caracterización osteológica del género *Neocnus* del Cuaternario de Cuba (Edentata: Megalonychidae). Tesis de Maestría, Facultad de Biología, Universidad de La Habana, 114 pág. (Inédito).

2005 Suárez, W., 2005. Taxonomic status of the Cuban vampire bat (Chiroptera: Phyllostomidae: Desmodontinae: Desmodous). Cariebean Journal Science, 41 (4): 761–767.

2006 Díaz-Franco, S., 2006. Los mamíferos fósiles del yacimiento "Las Breas de San Felipe", Martí, Matanzas, Cuba. Tesis de Maestría, Universidad de La Habana, 64 pp.

2006 Fleischer, R.C., Kirchman, J.J., Dumbacher, J.P., Bevier, L., Dove, C., Rotzel, N.C., Edwards, S.V., Lammertink, M., Miglia, K.J., y Moore, W.S., 2006. Mid-Pleistocene divergence of Cuban and North American ivory-billed woodpeckers. Biology Letters, 2: 466–469.

2006 Hedges, S.B., 2006. Paleogeography of the antilles and origin of west indian terrestria vertebrates. Ann. Missouri Botanical Garden, 93: 231–244.

2006 MacPhee, R.D.E., y Meldrum, J., 2006. Postcranial remains of the extinct monkeys of the Greater Antilles, with evidence for semiterrestriality in *Paralouatta*. American Museum Novitates, 3516, 65 pp.

2006 Suárez, W., 2006. La avifauna fósil de Cuba. Tesis de maestría, Museo Nacional de Historia Natural, La Habana.

2007 MacPhee, R. D., et al. 2007. Prehistoric sloth extinction in Cuba: Implications of a new last appearance date. Caribbean Journal of Science (43) (1): 94–98.

2007 MacPhee, R.D.E., Iturralde-Vinent, M., y Jiménez-Vázquez, O., 2007. Prehistoric sloth extinctions in Cuba: implications of a new "last" occurrence date. Caribbean Journal of Science, 43 (1): 94–98.

2007 Silva-Taboada, G., Suárez, W., y Díaz-Franco, S., 2007. Compendio de los mamíferos terrestres autóctonos de Cuba: vivientes y extinguidos. Ediciones Boloña, La Habana, 465 pp.

2007 Suárez, W., y Olson, S.L., 2007. The Cuban fossil eagle *Aguila borrasi* Arredondo: a scaled up version of the great Black-Hawk *Buteogallus urubitinga* (Gmelin). Journal of Raptor Research, 41 (4): 288–298.

2008 Díaz-Franco, S., y Jiménez-Vázquez, O., 2008. *Geocapromys browni* (Rodentia: Capromyidae: Capromyinae) en Cuba. Solenodon, 7: 41–47.

2008 Olson, S.L., y Suárez, W., 2008a. A fossil cranium of the Cuban Macaw *Ara tricolor* (Aves: Psittacidae) from Villa Clara Province, Cuba. Caribbean Journal of Science, 44: 287–290.

2008 Olson, S.L., y Suárez, W., 2008b. Bare-throated Tiger-Heron (*Tigrisoma mexicanum*) from the Pleistocene of Cuba: a new subfamily for the West Indies. Waterbirds, 31 (2): 285–288.

2008 Olson, S.L., y Suárez, W., 2008c. A new generic name for the Cuban Bare-legged Owl *Gymnoglaux lawrencii* Sclater and Salvin. Zootaxa, 1960: 67–68.

2009 Díaz-Franco, S., y Rojas-Consuegra, R., 2009. Dientes fósiles de *Sphyraena* (Perciformes: Sphyraenidae) en el Terciario de Cuba occidental. Solenodon, 8: 124–129.

2009 Jiménez Vázquez, O. 2011. Los monos extintos. 44–49. En: Mamíferos en Cuba (Eds. R. Borroto Páez y C. A. Mancipa). UPC Print, Vaasa, Finlandia. 271 pp.

2009 MacPhee, R.D.E., 2009. Insulae infortunatae: establishing a chronology for Late Quaternary mammal extinctions in the West Indies. En: Haynes, G., (Editor), American megafaunal extinctions at the end of the Pleistocene, 169–193.

2009 Suárez, W., y Olson, S.L., 2009. A new genus for the Cuban teratorn (Aves: Teratornithidae). Proceedings of the Biological Society of Washington, 122 (1): 103–116.

2010 Condis-Fernández, M.M., 2010. Inferencias paleontológicas sobre especies de la mastofauna cuaternaria cubana, conservada en el depósito superficial de la Caverna Geda, Pinar del Río, Cuba. Tesis doctoral, Universidad de Pinar del Río y Universidad de Alicante, La Habana.

2010 Orihuela, J., 2010. Late Holocene fauna from a cave deposit in Western Cuba: post-Columbian occurrence of the vampire bat *Desmodus rotundus* (Phyllostomidae: Desmodontinae). Caribbean Journal of Science, 46 (2–3) 297–312.

2011 Orihuela, J., 2011. Skull variation of the vampire bat *Desmodus rotundus* (Chiroptera: Phyllostomidae): Taxonomic implications for the Cuban fossil vampire bat Desmodus puntajudensis. Chiroptera Neotropical, 17 (1): 863–876.

2012 Orihuela, J., y Tejedor, A., 2012. Peter's ghost-faced bat *Mormoops megalophylla* (Chiroptera: Mormoopidae) from a pre-Columbian archeological deposit in Cuba. Acta Chiropterologica, 14 (1): 63–72.

2012 Rojas-Consuegra, R., Jiménez-Vázquez, O., Condis-Fernández, M.M., y Díaz-Franco, S., 2012. Tafonomía y paleoecología de un yacimiento paleontológico del Cuaternario en la cueva del Indio, La Habana, Cuba. Espelunca digital, 12: 1–12.

2013 Morgan, G.S., y Albury, N.A., 2013. The Cuban crocodile (Crocodylus rhombifer) from late Quaternary fossil deposits in the Bahamas and Cayman Islands. Bulletin of the Florida Museum of Natural History, 52 (3): 161–236.

2013 Orihuela, J., 2013. Fossil Cuban Crow *Corvus* cf. *nasicus* from a Late Quaternary cave deposit in northern Matanzas, Cuba. Journal of Caribbean Ornithology, 26: 12–16.

2014 Arredondo-Antúnez, C., y Rodríguez Suárez, R., 2014. Evidencias directas de herbivorismo en coprolitos de perezosos extintos (Mammalia: Pilosa: Megalonychidae) de Cuba. Revista del Jardín Botánico Nacional, 34–35, 67–73.

2014 Arredondo-Antúnez, C., y Villavicencio-Finalet, R., 2004. Tafonomía del depósito arqueológico solapa del *Megalocnus* en el NE de Villa Clara, Cuba. Revista de Biología, 18 (2): 160–170.

2014 Brochu, C.A., y Jiménez-Vázquez, O., 2014. Enigmatic crocodyliforms from the early Miocene of Cuba. Journal of Vertebrate Paleontology, 4 (5): 1094–1101.

2014 Jiménez Vázquez, O., L. W. Viñola y A. Sueiro Garra. 2014. Una mirada al pasado de los cocodrilos de Cuba. En: Los Crocodylia de Cuba. Publicaciones Universidad de Alicante, 336 pp.

2014 Orihuela, J., 2014. Endocranial morphology of the extinct Antillean shrew *Nesophontes* (Lipotyphla: Nesophontidae) from natural and digital endocasts of Cuban taxa. Palaeontologia Electronica, Vol. 17, Issue 2; 22A; 12 pág.

2015 Soto-Centeno, J.A., y Steadman, D.W., 2015. Fossils reject climate change as the cause of extinction of Caribbean bats. Nature. Scientific Reports, 5: 7971.

2015 Speer, K.A., Soto-Centeno, J.A., Albury, N.A., Quicksall, Z., Marte, M.G., y Reed, D.L., 2015. Bats of the Bahamas: natural history and conservation. Bulletin of the Florida Museum of Natural History, 53 (3): 45–95.

2015 Suárez, W., y Olson, S.L., 2015. Systematics and distribution of the giant fossil barn owls of the West Indies (Aves: Strigiformes: Tytonidae). Zootaxa, 4020 (3): 533–553.

2016 Sato, J.J., Ohdachi, S.D., Echenique-Diaz, L.M., 3, Borroto-Páez, R., Begué-Quiala, G., Delgado-Labañino, J. L., Gámez-Díez J., Alvarez-Lemus, J., Truong Nguyen, S., Yamaguchi, N., y Kita, M., 2016. Molecular phylogenetic analysis of nuclear genes suggests a Cenozoic over-water dispersal origin for the Cuban *Solenodon*. Scientific Reports, 6:31173, https://doi.org/10.1038/srep31173.

2017 Aranda, E., J. G. Martínez López, O. Jiménez Vázquez, C. Alemán Luna, L. W. Viñola López. 2017. Nuevos registros fósiles de vertebrados terrestres para Las Llanadas, Sancti Spíritus, Cuba. Novitates Caribaea, 11: 115–123.

3.4.2.6 Another Works Related

Here are included other works related, directly or collaterally, with the Paleontology of Cuba, which contain contributions on different general aspects, or some details without specifying a certain age.

Álvarez-Conde, J., 1957. Historia de la Geología, Mineralogía y Paleontología en Cuba. Publicaciones Junta Nacional de Arqueología y Etnología. La Habana, 248 pp.

Ceballos-Izquierdo, Y. e Iturralde-Vinent, M. 2018. Biblioteca Digital Cubana de Geociencias. Actualizado hasta 2018, en la Red Cubana de las Ciencias, https://www.redciencia.cu/geobiblio/inicio.html.

De Albear, J.F., e Iturralde-Vinent, M., 1985. Estratigrafía de las provincias de La Habana. En: Contribución a la Geología de las provincias de La Habana y Ciudad de la Habana. Editorial Científico-Técnica, La Habana, pág. 12–54.

De Golyer, E., 1918. The geology of Cuban petroleum deposits. American Association of Petroleum Geologists Bulletin, 2: 133–167.

De Huelbes Alonso, J., (editor) 2013. Léxico Estratigráfico de Cuba. Instituto de Geología y Paleontología de Cuba, La Habana, 573 pág.

Delgado Carballo, I., Pérez Arias, J.R., Díaz Otero, C. 2000. Foraminíferos planctónicos índices del Cretácico de Cuba. En: Gutiérrez Domech, M. R. et al. Atlas de Fósiles índices de Cuba: Microfósiles. La Habana: CNDIG. P. 5–52.

Domínguez Samalea, Y., Iliana L. Delgado Carballo, R. Bolufé Torres, A. Oliva Martín, M. R. Gutiérrez Domech y D. Hidalgo Griff, 2017. Catálogo de equinodermos fósiles de Cuba. XII Congreso de Geología (GEOLOGÍA'2017). IGP, CD-ROM.

Efremov, J. A. 1940. Taphonomy: new branch of paleontology. Pan-American Geologist, 74: 81–93.

Fernández de Castro, M., 1876. Catálogo de los fósiles de la Isla de Cuba. Anales de la Academia de Ciencias de la Habana, 13: 320–326.

Fernández de Castro, M., 1877. Fósiles de la Isla de Cuba, pertenecientes al género Asterostoma. Anales de la Academia de Ciencias de la Habana, 13: 549–553.

Fernández-López, S. 2000. Temas de Tafonomía. Dpto. Paleontología, Universidad Complutense de Madrid. 167 pp.

Fernández-López, S. 1989. La materia fósil. Una concepción dinamicista de los fósiles. En: Nuevas tendencias: Paleontología (Ed. E. Aguirre). Consejo Superior de Investigaciones Científicas, Madrid: 25–45.

Franco-Álvarez, G.L., Acevedo-González, M., Álvarez-Sánchez, H., Artime Peñeñori, C., Barriento-Duarte, A., Blanco-Bustamante, S., Cabrera, M., Cabrera, R., Carassou Agragan, G., Cobiella-Reguera, J.L., Coutin Lambert, R., De Albear, J.F., De Huelbes Alonso, J., De la Torre y Callejas, A., Delgado Damas, R., Díaz de Villalvilla, L., Diaz-Otero, C., Dilla Alfonso, M., Echevarría-Hernández, B., Fernández-Carmona, J., Fernández-Rodríguez, G., Flores-García, R., Florez-Abín, E., Fonseca, E., Furrazola-Bermúdez, G., García-Delgado, D., Gil-González, S., Gonzalez Garcia, R. A., Gutiérrez-Domech, R., Linares-Cala, E., Milián García, E., Millán-Trujillo, G., Moncada Ferrera, M., Montero Zamora, L., Orbera, L., Ortega-Sastriques, F., Peñalver-Hernández, L.L., Perera, C., Pérez Arias, J.R., Pérez Lazo, J., Pérez Rodriguez, E., Pifieiro Pérez, E., Recio Herrera, A. M., Sánchez-Arango, J.R., Saunders Pérez, E., Segura-Soto, R., Triff-Oquendo, J., Zuazo Alonso, A., Pszczółkowski, A., Brezsnyánszky, K., Slavov, I., y Myczyński, R., 1992. Léxico Estratigráfico de Cuba. Centro de Nacional de Información Geológica, La Habana, 658 pág.

Franco-Alvarez, G.L., y Delgado-Damas, R., 1997. Sistema Neógeno. En: Furrazola-Bermúdez, G., y Nuñez-Cambra, K., (Editores). Estudios sobre Geología de Cuba. Instituto de Geología y Paleontología, La Habana.

Furrazola-Bermúdez y Núñez-Cambra (ed.). 1997. Estudios de Geología de Cuba. Instituto de Geología y Paleontología, Centro Nacional de la Información Geológica, La Habana. 527 p.

Furrazola-Bermúdez, G., 1997. El Sistema Jurásico en Cuba. En: Furrazola-Bermúdez, G., y Nuñez-Cambra, K., (Editores) Estudios sobre Geología de Cuba. Instituto de Geología y Paleontología, Centro Nacional de la Información Geológica.

García Gonzales, A., 1989. Antonio Parra en la ciencia hispanoamericana del siglo XVIII. Editorial Academia. La Habana, 172 p.

Goto, K., Tada, R., Tajika, E., Iturralde-Vinent, M., Matsui, T., Yamamoto, S., Nakano, Y., Oji, T., Kiyokawa, S., García-Delgado, D., Díaz-Otero, C., y Rojas-Consuegra, R., 2008. Lateral lithological and compositional variations of the Cretaceous/Tertiary deep-sea tsunami deposits in northwestern Cuba. Cretaceous Research, 29 (2): 217–236.

Gradstein, F. M. et al. 2004. A Geologic Time Scale 2004. Sitio web de la Comisión Internacional de Estratigrafía (ICS). https://www.stratography.org.

Haczewski, G., 1987. Reconocimiento sedimentológico de la Formación San Cayetano: Un margen continental acumulativo en el Jurásico de Cuba occidental. En: Pszczółkowski, A., (Ed.) 1987. Contribución a la Geología de la Provincia Pinar del Río. Editorial Científico-Técnica, La Habana, 228–247.

Hatten, C.W., 1967. Principal features of Cuban geology: discussion. American Association of Petroleum Geologists Bulletin, 51 (5): 780–789.

Iturralde-Vinent, M., (Editor) 2004. Origen y evolución del Caribe y sus biotas marinas y terrestres. Editorial Centro Nacional de Información Geológica, CD-Rom. La Habana.

Iturralde-Vinent, M., 1981. Nuevo modelo interpretativo de la evolución geológica de Cuba. Ciencias de la Tierra y del Espacio, 3: 51–90.

Iturralde-Vinent, M., 2005–2006. La paleogeografía del Caribe y sus implicaciones para la biogeografía histórica. Revista del Jardín Botánico Nacional, 25–26: 49–78.

Iturralde-Vinent, M., 2009. Origen de la biota cubana actual. En: Iturralde-Vinent, M., (Editor) 2009. Geología de Cuba para todos. Editorial Científico-Técnica, Instituto del Libro, La Habana, 150 p.

Iturralde-Vinent, M., 2012. CD-Rom Multimedia: Dinosaurios y otros reptiles del Caribe. Editorial CITMATEL, La Habana.

Iturralde-Vinent, M., y MacPhee, R.D.E., 2004. Los mamíferos terrestres en las Antillas Mayores. Su paleogeografía, biogeografía, irradiaciones y extinciones. Publicaciones de la Academia de Ciencias de la República Dominicana, Editora Buho, Santo Domingo, 30 p.

Judoley, C.M., Furrazola, G., 1965. Estratigrafía del Jurásico Superior de Cuba: Publicación Especial del Instituto Cubano de Recursos Minerales, 126 pp.

Judoley, C.M., y Furrazola-Bermúdez, G., 1968. Estratigrafía del Jurásico Superior de Cuba. La Habana, 300 p.

Kantshev, I., Boyanov, I., Popov, N., Cabrera, R., Goranov, A., Iolkicev, I., Kanszirski, M., y Stancheva, M., 1976, 1978. Geología de la provincia de Las Villas. Resultado de las investigaciones y levantamiento geológico a escala 1: 250 000. Academia de Ciencias de Cuba y Bulgaria, Instituto de Geología y Paleontología. Oficina Nacional de Recursos Minerales, Ministerio de Energía y Minas, La Habana (Inédito).

Kartashov, I.P., Cherniakhovsky, A.G., y Peñalver-Hernández, L.L., 1981. Antropogene of Cuba. Geological Institute Transaction 356, 145 pp., Editorial Nauka, Moscú.

Kartashov, I.P., y Mayo, N.A., 1972. Algunas particularidades de las estructuras de los depósitos del Cuaternario continental de Cuba central y occidental. Instituto de Geología y Paleontología, Academia de Ciencias de Cuba, La Habana. Serie Geológica, 10: 1–9.

Kartashov, I.P., y Mayo, N.A., 1974. Principales rasgos del desarrollo geológico de Cuba Oriental en el Cenozoico tardío. En: Contribución a la Geología de Cuba, Publicación Especial Número 2, Instituto de Geología y Paleontología, Academia de Ciencias de Cuba, La Habana, pág. 165–174.

Kauffman, E. G., 1988. The dynamics of marine stepwise mass extinction. Revista Española de Paleontología (No. extraordinario): 57–71.

Kauffman, E. G. y C. C. Johnson, 1988. The Morphological and ecological evolution of middle and upper cretaceous reef-building rudistids. Palaios 3 (Reefs Issue): 194–216.

Kauffman, E. G. y N. F. Sohl, 1974. Structure and Evolution of Antillean Cretaceous Rudist Frameworks. Verhandl. Naturf. Gas. Basel. 84 (1): 399–467.

Kerr, A.C., Iturralde-Vinent, M., Saunders, A.D., Babbs, T. L., y Tarney, J., 1999. A new plate tectonic model of the Caribbean: Implications from a geochemical reconnaissance of Cuban Mesozoic volcanic rocks. Geological Society of America Bulletin, 111 (11): 1581–1599.

Khudoley, K.M., y Meyerhoff, A.A., 1971. Paleogeography and geological history of Greater Antilles. Geological Society of America Memoirs, 129: 1–199.

Kiyokawa, S., Tada, R., Iturralde-Vinent, M., Matsui, T., Tajika, K., Yamamoto, S., Oji, T., Nakano, T., Goto, K., Takayama, H., García-Delgado, D., Díaz-Otero, C., y Rojas-Consuegra, R., 2002. Cretaceous-Tertiary boundary sequence in the Cacarajícara Formation, western Cuba: An impact-related high-energy, gravity flow deposit. En: Koeberl, C., y MacLeon, K.G., (Editors) Catastrophic events and mass extintions: Impacts and Beyond. Geological Society of America, Special Paper 356: 124–145.

López-Almirall, A. 2005. Nueva perspectiva para la regionalización fitogeográfica de Cuba: Definición de los sectores. En Regionalización biogeográfica en Iberoamérica y trópicos afines. Primeras Jornadas Biogeográficas de la Red Iberoamericana de Biogeografía y Entomología Sistemática.(Eds. J. Llorente Bousquets & J. J. Morrone). Universidad Nacional Autónoma de México, México D.F., pp. 417–428.

López-Almirall, A. 2015. Particularidades de la flora neotropical relacionadas con su origen. Biogeografía, 8: 45–51.

López-Matínez, N. y Truyols-Santonja, J. 1994. Paleontología. Ciencias de la viva, 19. Editorial Síntesis. 234 p.

MacPhee, R.D.E., e Iturralde-Vinent, M., 2000. A short history of Greater Antillean land mammals: biogeography, paleogeography, radiations, and extinctions. Tropics, 10 (1): 145–154.

MacPhee, R.D.E., e Iturralde-Vinent, M., 2005c. The interpretation of Caribbean paleogeography: reply to Hedges. En: Alcover, J.A., y Bover, P., (Editores). Proceedings of the International Symposium: Insular Vertebrate Evolution: The Paleontological Approach Monografies de la Societat d'Historia Natural de les Balears, 12, 175–184.

MNHNC, 2015. Colección Paleontológica y Grupo de Paleogeografía y Paleobiología del Museo Nacional de Historia Natural de Cuba. Agencia de Medio Ambiente. CITMA.

Michelotti, 1855. Sur des Fossiles par lui dans l'ile de Cuba, pres de la Habana. Bulletin of the Geological Society of France, (2) 12: 676–678.

Morgan, G. S. and Albury, N. A. 2013. The Cuban crocodile (*Crocodylus rhombifer*) from late Quaternary fossil deposits in the Bahamas and Cayman islands. Bulletin Florida Museum Natural History vol. 52(3): 161–236.

Nagy, E., Brito, A., Jakus, P., Gyarmati, P., Brezsnyánszky, K., Franco-Alvarez, G.L., Radocz, G.Y., Pérez Nevot, N., Formell-Cortina, F., De Albear, J.F., y De la Torre y Callejas, A., 1983. Contribución a la geología de Cuba Oriental. Editorial Científico-Técnica, La Habana, 273 p.

Nakano, Y., Tada, R., Kamata, T., Tajika, E., Oji, T., Kiyokawa, S., Takayama, H., Goto, K., Yamamoto, S., Toyoda, K., García-Delgado, D., Rojas-Consuegra, R., Iturralde-Vinent, M., y Matsui, T., 2002. Origin of Cretaceous-Tertiary boundary in Moncada, western Cuba and its relation to K/T event. En: Koeberl, C., y MacLeon, K.G., (Editores) Catastrophic events and mass extintions: Impacts and Beyond. Geological Society of America Special Paper, (356).

Núñez-Jiménez, A., 1998. Geología. Londres. Editorial Mec Graphic. (Cuba: La Naturaleza y el Hombre), 435 p.

Oliva-Martín, A., 2015. Actualización de la sistemática de los gasterópodos (Mollusca, Gastropoda) reportados en el registro fósil de Cuba. Parte IV: Heterobranchia. En: Memorias VI Convención Cubana de Ciencias de la Tierra (Geociencias'2015), Sociedad Cubana de Geología, La Habana, CD-Rom.

Oliva-Martín, A., 2016. Sistemática de Heterobranchia (Gastropoda) en el registro fósil de Cuba. Ciencias de la Tierra y el Espacio, 17 (1): 112–122.

Oliva-Martín, A., I. Delgado Carballo, Y. Domínguez Samalea, R. Bolufé Torres, M. R. Gutiérrez Domech, D. Hidalgo Griff, 2017. El catálogo de ammonites de Cuba: un producto del instituto de geología y paleontología/servicio geológico de cuba para la divulgación del conocimiento geocientífico. En: Memorias VI Convención Cubana de Ciencias de la Tierra (Geociencias'2015), Sociedad Cubana de Geología, La Habana, CD-Rom.

Orbigny, A. d'., 1839. Foraminifères. En: Ramón de la Sagra, Histoire physique, politique et naturelle de l'ile de Cuba, 224 p.

Orbigny, A. d'., 1840. Foraminiferas. En: Ramón de la Sagra, Historia física, política y natural de Cuba, 180 p.

Ortega-Sastriques, F., 1983. Una hipótesis sobre el clima de Cuba durante la glaciación de Wisconsin. Ciencias de la Tierra y el Espacio, 7: 57–68.

Ortega-Sastriques, F., 1984. Las hipótesis paleoclimáticas y la edad de los suelos de Cuba. Ciencias de la Agricultura, 21: 45–59.

Ortega-Sastriques, F., y Arcia, M.I., 1982. Determinación de las lluvias en Cuba durante la glaciación de Wisconsin, mediante los relictos edáficos. Ciencias de la Tierra y el Espacio, 4: 85–104.

Pajón-Morejón, J.M., 1999. Paleoclima del Cuaternario cubano: Una caracterización cuantitativa. Monografía, La Habana, 362 p.

Palmer, R.H., 1934. The geology of Habana, Cuba, and vicinity. Journal of Geology, 42 (2): 123–145.

Palmer, R.H., 1945. Outline of the Geology of Cuba. Journal of Geology, 53: 1–34, 6 figuras.

Palmer, R.H., 1948. List of Palmer Cuban Fossil localities. Bulletin of American Paleontology, 31 (128): 178 p.

Parra, 1787. Descripción de diferentes piezas de historia natural, las más del ramo marítimo, representadas en sesenta y cinco láminas. Imprenta de la Capitanía General, La Habana.

Peñalver-Hernández, L.L., Lavandero, R., y Barriento-Duarte, A., 1997. El sistema Cuaternario. En: Furrazola-Bermúdez, G., y Nuñez-Cambra, K., (Editores) Estudios sobre geología de Cuba, Instituto de Geología y Paleontología, La Habana, pág. 165–178.

Pérez Gil, W., Cobiella Reguera, J.L., y Díaz Díaz, S.P., 2011. Estudio de un corte del Cretácico Temprano (Berriasiano Superior-Valanginiano Inferior), en el Miembro Tumbitas de la Formación Guasasa, Sierra del Infierno, Sierra de los Órganos, Cuba Occidental. En: Memorias, Trabajos y Resúmenes. IV Convención Cubana de Ciencias de la Tierra (Geociencias'2011). Centro Nacional de Información Geológica, Instituto de Geología y Paleontología de Cuba, La Habana, CD-Rom.

Perkins, S., 2005. Caribbean extincions. Science News, 168: 275–276.

Peros, M.C., Graham, E., y Davis, A.M., 2006. Stratigraphic investigations at Los Buchillones, a Taíno site on the North Coast of Central Cuba: evidence from geochemistry, mineralogy, paleontology, and sedimentology. Geoarchaeology, 21: 403–428.

Peros, M.C., Gregory, B., Matos, F., Reinhardt, E.G., y Desloges, J., 2015. Late Holocene record of lagoon evolution, climate change, and hurricane activity from southeastern Cuba. The Holocene, 1–15, https://doi.org/10.1177/0959683615585844.

Peros, M.C., Reinhardt, E.G., y Davis, A.M., 2007. A 6000 cal yr record of ecological and hydrological changes from Laguna de la Leche, north coastal Cuba. Quaternary Research, 67: 69–82.

Pregill, G., y Olson, S.L., 1981. Zoogeography of West Indian vertebrates in relation to Pleistocene climatic cycles. Ann. Rev. Ecol. Syst., 12: 75–98.

Pszczółkowski, A., 1976. Stratigraphic-facies sequences of the Sierra del Rosario (Cuba). Bulletin of the Polish Academy of Sciences. Earth Science, 24 (3/4): 193–203.

Pszczółkowski, A., 1978. Geosynclinal sequences of the Cordillera de Guaniguanico in western Cuba, their lithostratigraphy, facies development and paleogeography: Acta Geologica Polonica, 28, 1–96.

Pszczółkowski, A., 1986. Megacapas del Maestrichtiano de Cuba occidental y central. Bulletin of the Polish Academy of Sciences. Earth Science, 34 (1): 81–87.

Pszczółkowski, A., 1987a. Paleogeography and tectonic evolution of Cuba and adjoining areas during the Jurassic-Early Cretaceous. Annales Societatis Geologorum Poloniae, 57: 127–142.

Pszczółkowski, A., 1987b. Desarrollo paleotectónico y paleogeográfico de Cuba en el Jurásico. Boletín de Geociencias, 2 (1): 46–55.

Pszczółkowski, A., 1999. The exposed passive margin of North America in western Cuba, in Mann, P. (Editor), Caribbean Basins: Amsterdam, Elsevier Science B.V., Sedimentary Basins of the World, 4: 93–121.

Pszczółkowski, A., Myczyński, R., 2010. Tithonian-Early Valanginian evolution of deposition along the proto-Caribbean margin of North America recorded in Guaniguanico successions (western Cuba): Journal of South American Earth Sciences 29: 225–253.

Pszczółkowski, A., Piotrowska, K., Piotrowski, J., Torre y Callejas, A., Myczyński, R., Haczewski, G., 1987. Contribución a la geología de la provincia Pinar del Río. Editorial Científico-Técnica, La Habana, Cuba, 255 p.

Roca, A.L., Kahila Bar-Gal, G., Eizirik, E., Helgen, K.M., Maria, R., Springer, M.S., O'Brien, S.J., y Murphy, W.J., 2004. Mesozoic origin of the West Indian Insectivores. Nature, 429: 649–651.

Rodríguez Ferrer, M., 1871. Del archipiélago de las Antillas y de si Cuba estuvo unida o no al continente americano. Revista de España, 19, 332 p.

Rodríguez Ferrer, M., 1882. La Isla de Cuba estuvo unida un día al continente americano. Int. Cong. Americanistas, Madrid, Actas 1: 95–113.

Rojas-Consuegra, R. 1999. Productividad de Carbonatos en el Dominio del Arco Volcánico Cretácico. Tesis de Maestría. Facultad de Ciencias Técnicas, Universidad de Pinar del Río. Portal de la Ciencia Cubana, IDICT (Publ. Electr.).

Rojas-Consuegra, R. y K. Núñez Cambra, 2017. Guía para la excursión al límite K-Pg en Cuba occidental. Excursiones post convención No.4. Sociedad Cubana de Geología. La Habana. En: Memorias de VII Convención de Ciencias de La Tierra. CD ROM, 34 pp.

Rojas-Consuegra, R. y N. Alabarreta-Pérez. 2009. Sinopsis del registro fósil de Cuba. En: Sitio web sobre Paleontología de Cuba (2009–2019). [en línea]. https://www.redciencia.cu/webpaleo/.

Rojas-Consuegra, R., 2014. Columna ilustrada del registro macrofósil de Cuba. Anuario de la Sociedad Cubana de Geología, No. 2: 13–18. ISSN 2310-0060.

Rojas-Consuegra, R., y Denis-Valle, R., 2013. Influencia climática en los sistemas turbidíticos del Paleógeno cubano, según el registro estratigráfico cubano. Anuario de la Sociedad Cubana de Geología, No. 1: 83–94.

Rojas-Consuegra, R., y Viñola López, L., 2013. La Región Paleontológica Matanzas: un caso patrón. Savia (III) 18: 2–5.

Rosenberger, A.L., 1984. Fossil New World Monkeys Dispute the Molecular Clock. Journal of Human Evolution, 13: 737–742.

Rosenberger, A.L., Cooke, S.B., Halenar, L.B., Tejedor, M. F., Hartwig, W.C., Novo, N.M., y Muñoz-Saba, Y., 2015. Fossil Alouattines and the Origins of Alouatta: Craniodental Diversity and Interrelationships. En: Kowalewski, M.M., et al. (Editores), Howler Monkeys, Developments in Primatology: Progress and Prospects, https://doi.org/10.1007/978-1-4939-1957-4_2.

Rosenberger, A.L., Klukkert, K.S., Cooke, S.B., y Rímoli, R., 2013. Rethinking Antillothrix: The Mandible and Its Implications. American Journal of Primatology, 75: 825–836.

Rosenberger, A.L., Pickering, R., Green, H., Cooke, S.B., Tallman, M., Morrow, A., y Rímoli, R., 2015. 1.32 ± 0.11 My age for underwater remains constrain antiquity and longevity of the Dominican primate Antillothrix bernensis. Journal of Human Evolution (2015), https://doi.org/10.1016/j.jhevol.2015.05.015.

Sagra y Huerta, Ramón de la, 1855. Paleontología. En: Historia física, política y natural de la Isla de Cuba, Tomo VIII.

Sánchez-Arango, J.R., 1985. Ostrácodos de algunas formaciones en las provincias de la Habana. En: Contribución a la Geología de las provincias de La Habana y Ciudad de La Habana. Editorial Científico-Técnica, La Habana, pág. 116–125.

Sánchez-Roig, M., 1923a. Nuevas especies de equínidos fósiles cubanos. Memorias de la Sociedad Cubana de Historia Natural "Felipe Poey", 7: 83–192.

Sánchez-Roig, M., 1923b. Revisión de los equínidos fósiles cubanos. Memorias de la Sociedad Cubana de Historia Natural "Felipe Poey", 6: 6–42.

Sánchez-Roig, M., 1926a. Contribución a la paleontología cubana. Los equinodermos fósiles de Cuba. Boletín de Minas, 10: 1–179.

Sánchez-Roig, M., 1930. Rectificaciones y adiciones al mapa geológico de Cuba. Instituto Nacional de Investigaciones Científicas, 1: 99–139.

Sánchez-Roig, M., 1949. Los equinodermos fósiles de Cuba. Paleontología Cubana, 1: 1–330.

Sánchez-Roig, M., 1952a. El género Cubanaster (Equinidos fósiles irregulares). Torreia, 16: 3–8.

Sánchez-Roig, M., 1952b. Nuevos géneros y especies de equinodermos fósiles Cubanos. Memorias de la Sociedad Cubana de Historia Natural "Felipe Poey", 21 (1): 4–30.

Sánchez-Roig, M., 1952c. Nuevos géneros y especies de equinoideos fósiles Cubanos. Torreia, 17:1–18.

Sánchez-Roig, M., 1952d. Paleontología cubana: revisión de los equinodermos fósiles del grupo Cassiduloida. Memorias de la Sociedad Cubana de Historia Natural "Felipe Poey", 21 (1): 47–57.

Sánchez-Roig, M., 1953a. Dos nuevos géneros de equinoideos cubanos: Lambertona y Neopatagus. Memorias de la Sociedad Cubana de Historia Natural "Felipe Poey", 21 (3): 257–262.

Sánchez-Roig, M., 1953b. Nuevos equinoideos fósiles de Cuba. Anales de la Academia de Ciencias Médicas, Físicas y Naturales de la Habana, 91 (2).

Schubert, C., 1989. Paleoclima del Pleistoceno tardío en el Caribe y regiones adyacentes: Un intento de compilación. Ciencias de la Tierra y el Espacio, 15–16: 40–58.

Schulte, Alegret, Arenillas, Arz, and 28 more authors, 2010. The Chicxulub asteoid impact and mass extinction at the Cretacous—Paleogene boundary. Science, 327.

Seiglie, G.A., 1961. Contribución al estudio de las microfacies de Pinar del Río. Revista de la Sociedad Cubana de Ingenieros, 61 (3–4): 87–109.

Steadman, D.W., Alburyb, N.A., Kakukc, B., Meadd, J.I., Soto-Centenof, J.A., Singletona, H.M., y Franklin, J., 2015. Vertebrate community on an ice-age Caribbean island. PNAS, www.pnas.org/cgi/doi/10.1073/pnas.1516490112.

Tada, R., Iturralde-Vinent, M., Matsui, T., Tajika, E., Oji, T., Goto, K., Nakano, Y., Takayama, H., Yamamoto, S., Kiyokawa, S., Toyoda, K., Garcia-Delgado, D., Díaz-Otero, C., y Rojas-Consuegra, R., 2003. K/T boundary deposits in the Paleo-western Caribbean basin. En: Bartolini, C.,

Buffler, R.T., y Blickwede, J., (Editores) The Circum-Gulf of Mexico and the Caribbean: Hydrocarbon habitats, basin formation, and plate tectonics: American Association of Petroleum Geologists Memoir, 79: 582–604.

Tada, R., Nakano, Y., Iturralde-Vinent, M., Yamamoto, S., Kamata, T., Tajika, E., Toyoda, K., Kiyokawa, S., Garcia-Delgado, D., Oji, T., Goto, K., Takayama, H., Rojas-Consuegra, R., y Matsui, T., 2002. Complex tsunami waves suggested by the Cretaceous-Tertiary boundary deposit at the Moncada section, western Cuba. En: Koeberl, C., y MacLeod, K.G., (Editores) Catastrophic Events and Mass Extinctions: Impacts and Beyond: Boulder, Colorado, Geological Society of America Special Paper, 356: 109–123.

Takayama, H., Tada, R., Matsui, T., Iturralde-Vinent, M., Oji, T., Tajika, E., Kiyokawa, S., Garcia-Delgado, D., Okada, H., Hasegawa, T., Toyoda, K., 2000. Origin of the Peñalver Formation in northwestern Cuba and its relation to K/T boundary impact event. Sedimentary Geology, 135: 295–320.

Vaughan, T.W., 1922. Stratigraphic significance of the species of the West Indian fossil echini. Carnegie Institution of Washington, 306: 107–122.

Vaughan, T.W., 1933. Report on species of corals and larger foraminifera collected in Cuba by O.E. Meinzer. Journal of the Washington Academy of Sciences, 23: 352–355.

Vaughan, T.W., 1933. Report on species of fossils collected in Cuba, by O.E. Meinzer in november and december, 1915. Journal of the Washington Academy of Sciences, 23: 261–263.

Vélez-Juarbe, J., Martin, T., MacPhee, R.D.E., y Ortega-Ariza, D., 2014. The earliest Caribbean rodents: Oligocene caviomorphs from Puerto Rico. Journal of Vertebrate Paleontology, 34 (1): 157–163.

Veltkamp, C.J, y Donovan, S.K., 2001. Squeezing the fossil record: unravelling the evolution of stalked crinoids in the Caribbean. Geology Today, 18 (3).

Vesa, A. 1909. Relación de ejemplares fósiles que proceden de Viñales. Anales de la Academia de Ciencias de la Habana, 46: 259–261.

Vidal y Careta, F., 1894. Fitofósiles de Cuba. Anales Instituto de Segunda Enseñanza. La Habana.

Villegas Martín, J., y Rojas-Consuegra, R., 2012. Ichnology of Cuba: present state of knowledge. En: Guimarae, R.N., et al. (Editores). Ichnology of Latin America. Selected papers. Monografias da Sociedade Brasileira de Paleontologia, 2: 99–106.

Vishnevaskaya, S., Chejóvich, V.D., y De Albear, J.F., 1982. Edad y condiciones de formación de silicitas de la zona de Camajuaní (Cuba). Ciencias de la Tierra y el Espacio, 5: 113–116. Radiolarios y esponjas!

Weaver, P.F., Cruz, A., Johnson, S., Dupin, J., y Weaver, K. F., 2006. Colonizing the Caribbean: biogeography and evolution of livebearing fishes of the genus *Limia* (Poeciliidae). Journal of Biogeography, https://doi.org/10.1111/jbi.12798.

Weyl, R., 1965. Die palaogeographische Entwicklung des mittelamerikanischwestindischen Raumes. Geol. Rdsch., Bd. 54, 1213–1240.

Whidden, H.P., y Asher, R.J., 2001. The origin of the Greater Antillean insectivorans. En: Woods, C.A., y Sergile, F.E., (Editores) Biogeography of the West Indies: patterns and perspectives, p. 237–252. CRC. Press, Boca Raton.

White, J.L., y MacPhee, R.D.E., 2001. The sloths of the West Indies: a systematic and phylogenetic review. En: Woods, C.A., y Sergile, F.E., (Editores) Biogeography of the West Indies Patterns and Perspectives. Ed. 2. CRC Press: Baton Rouge.

Yamamoto, S., Hasegawa, T., Tada, R., Goto, K., Rojas-Consuegra, R., Díaz-Otero, C., García-Delgado, D., Yamamoto, S., Sakuma, H., y Matsui, T., 2010. Environmental and vegetational changes recorded in sedimentary leaf wax n-alkanes across the Cretaceous-Paleogene boundary at Loma Capiro, Central Cuba. Palaeogeography, Palaeoclimatology, Palaeoecology 295: 31–41.

Zenkovich, V.P., 1965. Costas coralinas de Cuba. Vokrug Sveta, 12: 12–13.

In general, most of the works compiled (more than 625) have been about fossil vertebrates (207), mainly Jurassic marine and terrestrial Quaternary (Fig. 3.39). They are followed by works dedicated to invertebrates (197), the largest numbers on Cretaceous groups (70), followed by Jurassic (57), and then those of the Paleogene (27), the Quaternary (25), and finally, those of the Neogene (18). Thus, there are a few works on fossil plants, pollen, and spores (25); and on ichnofossils (10) whose studies have been increasing in recent years (Fig. 3.40).

3.5 Conclusions

- The most characteristic fossils of the Jurassic period in Cuba are the petrified bones of marine reptiles, the shells and molds of the ammonites, the skeletons and molds of the ganoid fish, bivalve mollusc shells, and the carbonized fronds and stems of plants, mainly, ferns.

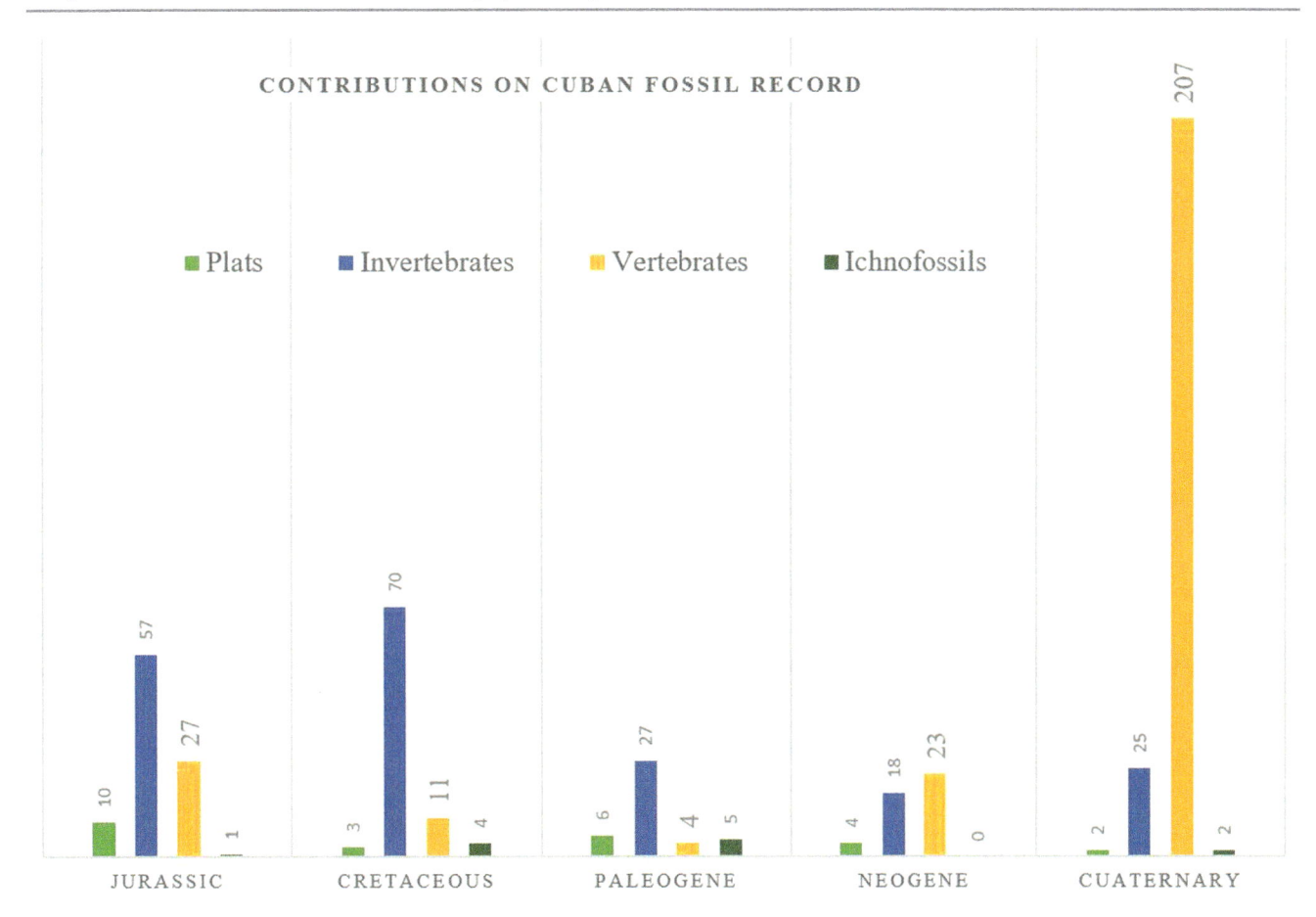

Fig. 3.40 A large scientific work has been carried out on the fossils of Cuba, although there are still several aspects in which to deepen the investigations

- The main fossils that appear in the Cuban Cretaceous are the shells of the rudists, the ammonites and its aptics, the endoskeletons and radioles (spines) of the echinoids, the shells of the acteonelids, nerineids, naticids, turritellids, and ostreids. In addition, corals and algae, fossil traces, very diverse foraminifera, both planktonic and benthic.
- The fossils that characterize the Paleogene in Cuba are remains of echinoderms, shark teeth, fossil traces, shells and molds of turritellids and ostreids, and the abundant and varied planktonic and benthic foraminifera, especially large orbitoids.
- The Cuban Neogene is rich in fossil remains of a large group of animals, both vertebrates and invertebrates, mainly marine, but also has the first terrestrial fossil mammals. The rocks of this period contain abundant shells and molluscs of bivalves and gastropods, and mineralized endoskeletons of echinoderms are common. Usual are also the corals, and frequent, the skeletons and molds of marine crustaceans. Among the vertebrates fish remains are abundant, mainly the teeth of sharks and rays, less frequent are the bony fish and rare whales. Scarce, but very important, is the discovery of remains of terrestrial mammals, such as monkey, rodent, and megaloniquid.
- The greatest wealth, which shows the fossil record of the Cuban Quaternary, is perhaps the peculiar fossil material produced by the diverse megafauna of terrestrial vertebrates, which inhabited our territory in the last hundreds of thousands and thousands of years ago. Likewise, we highlight the bones and teeth of the large sloths, the various rodents, the giant predatory birds, and gunboats, the small and giant insectivores, the numerous bats, reptiles, and amphibians, among other animals that disappeared in the recent past.
- In general, the fossil record of Cuba is approximately the last 200 million years of life on Earth. It is rich in very varied fossils, witnessing a wide diversity of organisms, both animals and plants, which inhabited the Antillean region and Caribbean, and that constitute the inheritance of the biological diversity that the current Cuban Archipelago exhibits.
- In a quantitative approach to the CFR, approximately a total of 1270 taxoregisters of macrofossils are counted: 986 (78%) invertebrates, 178 (14%) vertebrates, 55 (5%)

plants, and ichnofossils 38 (3%). These are figures that are constantly changing as the degree of study of the different fossil groups in the country increases (Rojas Consuegra 2007, 2009, 2013, 2014).

- The fossils are part of the Cuban natural heritage, and as such, they deserve to be studied, conserved, and protected, as a legacy to future generations, to contribute to a better understanding of our origins and to the full enjoyment of our island nature. No less important, today as part of the geological heritage of the country, the preserved elements of paleontological heritage, are revealed as georecursos susceptible to be put into social use, for the purposes of the economic and territorial support of their tenants.
- The current state of knowledge about several groups of fossils in Cuba, advises the realization of scientific research using modern methods, techniques, and technologies, in the fields of taxonomy, biogeography, and paleogeography, with a view to a better understanding of the distant past and recent of our archipelago; solid foundation to contribute to a more effective management in the conservation of Cuban flora, fauna, and geodiversity; and as an ultimate goal, to the welfare of society itself.

Acknowledgements Throughout more than 32 years of work in the activity of Geology, as a geologist and curator of Paleontology, I have worked and learned with numerous people, some considered as teachers, other colleagues, students, and most of them, also friends. To all, my sincere and profound gratitude.Mentioning each name makes the present message too extensive, so from now on, I offer my apologies to those who do not appear on the list. These mentions also do not intend to follow a strict order determined by any criteria, although it has come out with a certain chronological sense, there they go: Kenia E. Núñez Cambra, Manuel A. Iturralde Vinent, Osvaldo Jiménez Vázquez, Peter Larson, Susan Hendrickson, José María Pons, Peter William Skelton, Jorge Isaac Mengana, Gloria Alencáster, Blanca E. Buitrón, María del C. Perrillat, Eulalia Gilli, Dinorah Karell Arrechea, Alfredo Bonzoño Gonzales, Riuji Tada, Kazuiza Goto, Shinji Yamamoto, Dora García Delgado, José A. Arz , Ignacio Arenillas, Alfonso Meléndez, Eustoquio Molina, Ana R. Soria, Laia Alegret, José Manuel Grajales Nishimura, Carmen Rosales, Rafael A. López Martínez, Ricardo Barragán, Osmany Ceballos, Esther Pérez Lorenzo, Gilberto Silva Taboada, Jorge Villegas Martín, Lázaro W. Viñola López, Rodolfo Corona Esquivel, Jesús M. Pajón Morejón, Alberto Arano Ruiz, Carlos Borges Serem, Anabel Olivia Martín, Carlos Díaz Guanche, Manuel E. Pardo Echarte, and many others. I also apologize, if I included the name of someone, who would not have wanted to have been included!I thank the family and friends who have always supported, in the most diverse ways, our scientific and social work! My apologies for not naming them explicitly, but they always go in my heart.RRC.

Reference

1. All the works cited are included in the previous sections on the analyzed bibliography. However, to avoid increasing the present work excessively, on the contrary, not all the communications listed have been cited in the text.

Stratigraphy of Cuba

Evelio Linares Cala and Dora Elisa García Delgado

Abstract

The lithostratigraphic units of the Triassic, Jurassic, Cretaceous, Paleogene, Neogene, and Quaternary periods participate in the geological structure of the Cuban archipelago. Precambrian rocks have been dated to singular outcrops in the southern area of Corralillo and La Teja in the provinces of Villa Clara and Matanzas, respectively. The formations of these periods are represented by sedimentary, volcanic, volcanic-sedimentary, and metamorphic rocks, which are registered in the subsoil or forming the depressions and mountainous areas of the national territory. For a better understanding of this chapter, the stratigraphy has been described through the Paleogeographic Domains: Synrift, North American Continental Margin, Cretaceous and Paleogenic Volcanic Arcs, and Superposed Basins with their corresponding Petrotectonic Set. The metamorphic rocks, whose protolith were sedimentary rocks, and those of the Neogene and Quaternary systems have been briefly treated. Most of the mentioned lithostratigraphic units are registered in the Third Version of the Léxico Estratigráfico de Cuba (2013), but some do not appear. They are described by a large number of samples studied, of use and importance in the gaso-petroleum search works of different Cuban regions, especially in the Northwest Belt of Hydrocarbons of Cuba. This territory includes the northern provinces of Havana, Mayabeque, and Matanzas. When making the stratigraphic description, the maxim of Gignoux (Stratigraphic geology: W.H. Freeman and Co., San Francisco 682 Gignoux (Stratigraphic geology: W.H. Freeman and Co., San Francisco. English edition (Fourth French edition trans: Woodford GG) p 682 [1])) has been followed, which considers that "Stratigraphy and Tectonics are two inseparable branches of the geological sciences." So that, the Global Tectonics or Plates Tectonics has been considered when recognizing the processes of folding and thrust that occurred during the Cuban orogeny. The main overlapping basins of Cuba and their lithostratigraphic units are described.

Keywords

Cuba • Paleogeographic domain • Petrotectonic set • Hydrocarbons • Tectonics • Synrift

---

E. L. Cala (✉)
Centro de Investigación del Petróleo, Churruca No. 481 entre Washington y Vía Blanca, Cerro, 12 000 La Habana, CP, Cuba
e-mail: bello@ceinpet.cupet.cu

D. E. G. Delgado
Sherritt International, Edificio Trade Center Quinta Avenida esquina 78 Miramar, Municipio Playa, La Habana, Cuba
e-mail: dgarcia@sherrittogp.com

© Springer Nature Switzerland AG 2021
M. E. Pardo Echarte (ed.), *Geology of Cuba*, Regional Geology Reviews,
https://doi.org/10.1007/978-3-030-67798-5_4

Abbreviations

NBHC	Northwest Belt of Hydrocarbon of Cuba
FMI	Formation Microresistivity Image
PD(s)	Paleogeographic Domain(s)
IGP/SGC	Instituto de Geología y Paleontología/Servicio Geológico de Cuba
IGP/ACC	Instituto de Geología y Paleontología/Academia de Ciencias de Cuba
TSU(s)	Tectonic Stratigraphic Unit(s)
CAME	Consejo de Ayuda Mutua Económica
My	Million years
U–Pb SHRIMP	Uranium–Lead Sensitive High-Resolution Ion Microprobe
TOC	Total Organic Carbon
MFS	Maximum Flood Surface
CVA	Cretaceous Volcanic Arc
PS(s)	Petrotectonic Set(s)
BS	Boundary Sequences
IUGS	International Union of Geological Sciences

4.1 Introduction

Stratigraphic studies in Cuba have a long history. The reports on Stratigraphy and Paleontology became known at the beginning of the second half of the nineteenth century. In 1869, the first "Geological sketch of Cuba" appeared by Manuel Fernández de Castro and Salterain [2], followed in 1877 by the Catalog of fossils of the Island of Cuba belonging to the genus Asterostroma [3]. In 1895, the Bulletin of the Commission of the Geological Map of Spain was published with a new Geological Map of Cuba prepared by Ramón Adán de Zarza. When the book Geology of Cuba was written [4], the authors referred dozens of geologists and paleontologists from various countries of the world, who had elaborated their stratigraphic schemes before 1959. The splendor of geological cartography and oil industry drillings occurred in 1959–1992 throughout the national territory [5]. It increased stratigraphic studies. In 1985, after multiple stratigraphic generalizations and field studies by provinces, the Geological Map of the Republic of Cuba, scale 1: 500,000 [6] was published. The geologists of the Academies of Sciences of Cuba, Poland, Bulgaria, Hungary, and the former USSR, made works in the Cuban provinces with the objective of editing the Geological Map of Cuba scale 1: 250,000 [7]. Other works were made on a scale of 1: 100,000, including 1: 50,000. Numerous explanatory texts and monographs were published, where the lithostratigraphic units reign. On the other hand, the Cuban oil geologists and the extinct USSR, in 1971–1975, made a stratigraphic generalization of all of Cuba by making numerous stratigraphic columns by regions.

Until 2018, although in a smaller volume, the investigations on stratigraphy of the oil zones have continued and work is underway for a geological map of Cuba scale 1: 50,000 by the Instituto de Geología y Paleontología-Servicio Geológico de Cuba.

The Stratigraphy of Cuba has always been said to be very complicated, but we should not exaggerate. This is explained by the complex structure of the Cuban Folded and Overthrusted Belt. However, the voluminous information in technical files and publications is available to researchers who overcome this obstacle every day. Serve this chapter to get an idea of what was raised.

4.2 Theoretical Framework

Stratigraphy is the branch of Geology that deals mainly with the study and interpretation of stratified sedimentary rocks, although it should be noted that stratigraphic principles and procedures are currently applied to all the bodies of rocks in the earth's crust, stratified or not. Through this discipline, rock aggregates are identified and described, both vertically and horizontally, which allows their geological mapping and regional correlation with other existing lithological units.

It is surprising the data provided by the rocks after being subjected to special studies, allow to make several types of maps of geological content: tectonic, metallogenic, hydrocarbon, hydrogeological, lithological–facial, paleogeographic of any scale; fundamental bases for forecasts of useful mineral deposits, oil, hydrogeological works, and geological engineering. These complex investigations require, above all of Stratigraphy.

4 Stratigraphy of Cuba

The sedimentary, igneous, and metamorphic rocks, are differentiated with a particular practical purpose in lithostratigraphic units, that is, in divisions or parts of the totality of the regional or local stratigraphic set, either for the purpose of finding useful, mineral, hydric, or hydrocarbons resources. The stratigraphic classification consists of the definition and systematic ordering of rock bodies in units of different rank, such as groups, formations, members, layers, among others.

This chapter will summarize the Stratigraphy of Cuba governed by a theoretical framework that prioritizes the tectonic concepts of Paleogeographic Domains and Tectono-Stratigraphic Units (TSU).

A Paleogeographic Domain (PD) "is a region of the earth's surface of considerable dimensions in the present or geological past, individualized by a geodynamic settlement of a plate tectonics" [8]. In Cuba, the Paleogeographic Domains are differentiated: Synrift, North American Continental Margin, Volcanic Arcs (Cretaceous and Paleogenic), Superposed and Frontal Basins. A synthesis of the metamorphic sequences whose protolith was sedimentary rocks will be included in this chapter. In Fig. 4.1a, b (Legend), in summary, the distribution of these domains and tectonostratigraphic units is schematized.

The rocks formed in these domains, by subsequent overflow phenomena, were integrated into what has been called

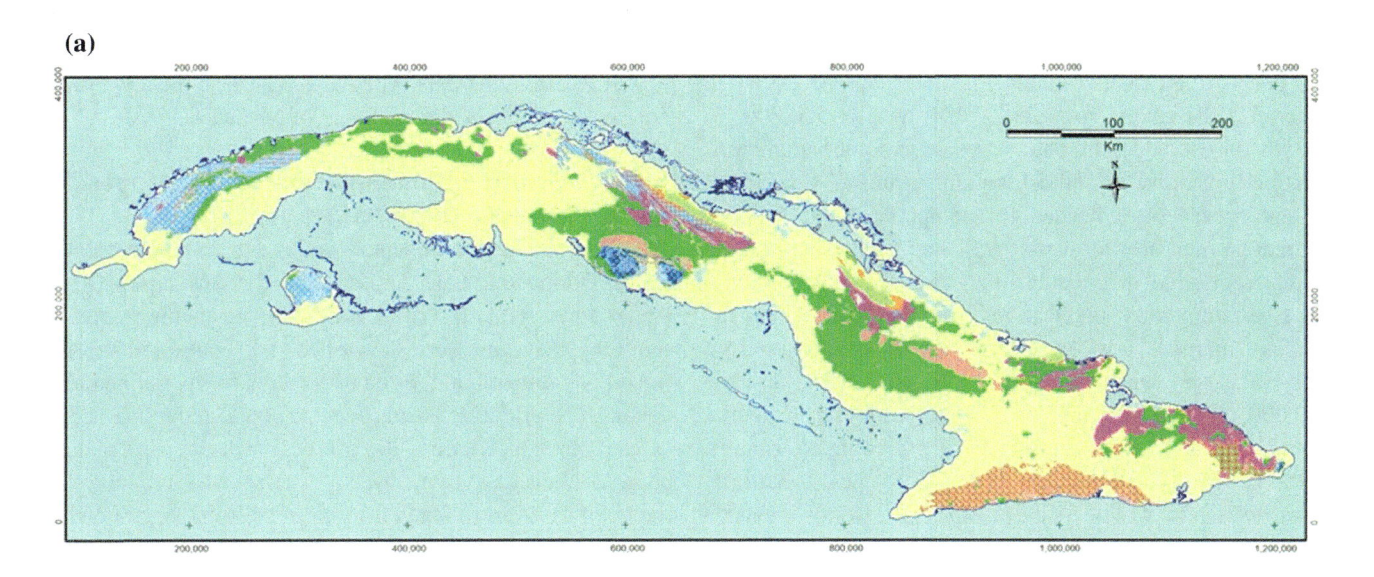

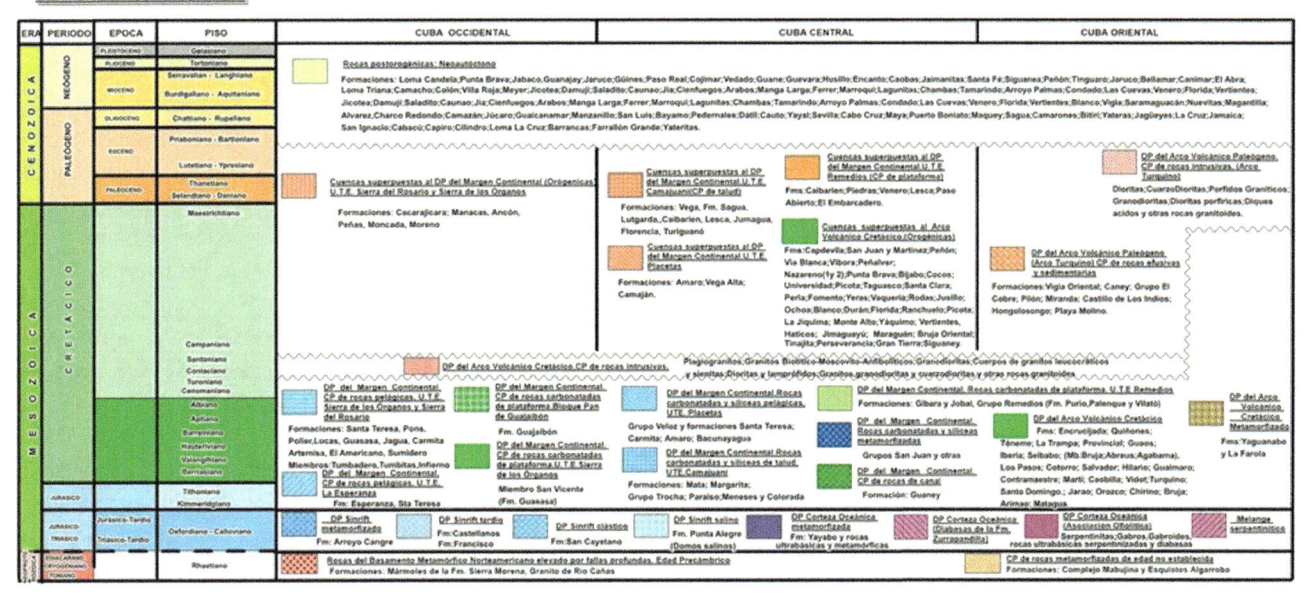

Fig. 4.1 a Scheme of paleogeographic domains and UTEs used for this chapter. b Legend of Fig. 4.1

the Fold and Overthrusted Belt of Cuba and its postorogenic coverage. In this strip, several Tectono-Stratigraphic Units (TSUs) are juxtaposed. A TSU, according to the connotation given by those who defined it, Howell [9, 10]; Hatten et al. [11]: "is a set of rocks with distinctive stratigraphy, limited by faults of regional extension, characterized by a geological history that differs from neighboring units."

4.3 Materials and Methods

After 1959, numerous geological mapping works were carried out in the Republic of Cuba with different objectives and different details. This required the study of dozens of lithostratigraphic units. By 1985, several areas were covered with scales 1: 50,000 and 1: 100,000, and hundreds of wells had been drilled for the search for oil. It was necessary to generalize the studies to scales of less than 1: 500,000 [6] and 1: 250,000 [7]. In both cases, due to the numerous existing inventory of lithostratigraphic units, new stratigraphic works were needed for comparison and support. A great amount of information was accumulated in addition to that registered in the subsoil by the oil wells. Simultaneously, numerous publications, projects, and thematic studies emerged. For the elaboration of the generalized maps, the geological cartography by provinces scale 1: 250,000 of the Academies of Sciences of Cuba, Bulgaria, Hungary, and Poland was used. For the Geological Map of Cuba 1: 500,000 the surveys of the Geological Territorial Enterprises and works of the CAME (Consejo de Ayuda Mutua Económica) were also used. These materials have been considered for the present review.

Its objective is to present, in a summarized way, the Cuban stratigraphy, based on the theoretical framework discussed above and considering the new approaches to the oil search: petroleum systems, which are in recent times the newest and most important amount, since for several years, no new investigations have been undertaken in superficial geological surveys in the national territory.

A valuable material used for the nomenclature of lithostratigraphic units, validated for this chapter, is recognized in the third version of the Léxico Estratigráfico de Cuba [12], but some, because of their connotation and use by oil geologists—being well developed in the subsoil—they have been included even if they have not been registered by it.

As a method when conceiving the stratigraphic description, the authors consider it important to follow the axiom of the French geologist Maurice Gignoux [1], who expressed: "Stratigraphy and Tectonics are two inseparable branches of the geological sciences. A geologist, a structuralist who is not a stratigrapher, is only a geometer, because he reasons on abstract surfaces and volumes, forgetting geological history and, on the other hand, a stratigrapher who has never worked tectonics alone, will only produce a stratigraphy dead."

4.4 Results and Discussion

4.4.1 Rocks of the Metamorphic Basement Raised by Faults

In 1977, the first data of an age as old as Precambrian was published in a sample from the town of Socorro southwest of the town of Sierra Morena, Corralillo, province of Villa Clara. These investigations were later extended by Renne et al. [13]. It was a phlogopite marble that, by means of the potassium–argon method, gave values of 945 ± 2.5 and 910 ± 2.5 million years. They also appear in the town of La Teja in the north of the province of Matanzas.

4.4.1.1 Synrift Sediments (Late Triassic-Upper Jurassic Oxfordian)

They are the oldest sedimentary rocks dated to the present in Cuba. During the Late Triassic and the lower part of the Upper Jurassic, as a result of the nascent opening episode, semigraben basins were developed that controlled the terrigenous sedimentation, in environments from continental to deltaic. The primitive basins became connected to the ocean, as can be deduced from the neritic limestones in the later phase of their development [14]. The set of sandy-clay terrigenous rocks most known in western Cuba is grouped in the San Cayetano Formation, whose sediments are derived from a source of sialic nature [15].

At this time, there was a large deposition area divided into smaller depressions, located on the supercontinent Pangea. In them, terrigenous elements from different parts of the supercontinent were deposited, which would explain some facial variations in the lithic units that are reported associated with this PD in Isla de la Juventud, Guamuhaya Massif, in the Sierra de Guaniguanico in Pinar del Río, and in the Sierra Verde in Guantánamo. Also, we can distinguish a stage that could conventionally be called "Late Synrift," as a transition to the drift event. The transitional units are the Castellanos, Francisco, and Constancia formations (Oxfordian- Kimmeridgian lower part). In addition, evaporitic sedimentations [16] have been reported to the north of the province of Ciego de Ávila and adjacent keys. During the Early Jurassic and the Oxfordian, in a large region, salt deposits formed as those originated in the Michigan Basin, in the United States, and those reported in the basin of southeastern Mexico. In Central Cuba, the Punta Alegre and "Cunagua Salt" formations are known, which emerge or are recorded by oil drilling.

4.4.1.2 San Cayetano Formation [17]

Before beginning to describe the lithostratigraphic units of Cuba, it is not idle to make the following clarification. Professional ethics indicates that each author when referring to a non-original work is forced to cite the source of their data. In the case at hand, we would be forced to refer to field geologists, paleontologists, stratigraphers, and other participants. They are hundreds of works. It will be understood that the chapter would have more pages of bibliographical citations than of text. There is a Léxico Estratigráfico de Cuba [12] and other digital works [18], where the interested party can consult the synonyms and amendments of each lithostratigraphic unit, the list of hundreds of authors and publications for many decades. In any case, in some cases, references will have to be made.

The San Cayetano Formation was released as Cayetano Fm., in 1918. It is distributed in the La Esperanza, Los Órganos, and Sierra del Rosario TSUs. It is demonstrated both by surface data, such as those of oil drilling, that lie beneath the rocks of the North American Continental Margin and their coverage or, frequently, lie tectonically on these rocks.

In Sheet 10 (F 17–5) Pinar del Rio, from the Geological Map of the Republic of Cuba of the Cuban Academy of Sciences IGP, scale 1: 250,000 [7], there are large areas of distribution of what they separated as San Cayetano Formation upper part (which would correspond to the Fm. Castellanos) and lower part or "Unit a," mainly in the Sierra de Los Órganos of the Cordillera de Guaniguanico. In smaller areas, it crops out in the Sierra del Rosario. In the Geological Map of the Republic of Cuba scale 1: 500,000 Sheet No. 1 [6] this unit is widely distributed in the Sierra de Los Órganos following a northeast–southwest direction, only that in this work, of the area previously mapped as San Cayetano Fm., a part that was geologically mapped as Castellanos Fm., corresponding to the youngest deposits of this stratigraphic interval, is separated. They were only mapped in the westernmost area of the province of Pinar del Río, while deposits of this same age to the Sierra del Rosario region, have not yet been differentiated, but probably correspond to this formation. In the La Esperanza TSU, the San Cayetano Fm. is known by several oil wells from the area of Los Arroyos de Mantua to Puerto Esperanza. It was registered under the Neoauthocton by oil wells in the areas of Guane and Isabel Rubio. This unit is much folded, sometimes with folds lying down (Fig. 4.2).

In general, it is only possible to identify different mantles or tectonic scales of the San Cayetano Formation, without being able to distinguish stratigraphically by their age. There are mantles where thin layers of siltstones (60–70%), schistose siltstones, and sandstone silts of dark gray to black (charred and pyritized) predominate. They usually exist, quartz sandstones of fine grains, light gray feldspathic quartz sandstones that disperse greenish, carbonaceous grayish clays, greenish siltstones in very thin layers, reddish-brown siltstones (by weathering).

In other mantles, psamites predominate (60%): white-yellowish quartz sandstones, medium and fine grains with siltstones, stratified, and sandstone silt in alternations. Some horizons of quartz and microquartzite conglomerates and quartz sandstones of coarse grains are recognized [19]. Areas are recorded, where tectonic mantles of great thickness appear, of violet schists, purple and dark gray siltstones, phyllites, sandstones of fine grains, greenish, micaceous–calcareous sandstones, greenish-gray, with remains of charred plants. In contrast to these tectonic mantles, there are others where sandstone prevails (90–95%). These are white, quartz-feldspathic quartzites with quartz, greenish and violet

Fig. 4.2 Folds lying in the San Cayetano formation. Km. 13 of the highway Consolación del Norte (La Palma) to Santa Lucía, province of Pinar del Rio

siltstones, and fine quartz conglomerates interspersed. In the probably upper part of the unit, microorganic and detritic carbonaceous micritic layers, calcareous sandstones, and carbonaceous shales begin to be determined. In calcareous sandstones and carbonate rocks, brachiopods and pelecypods have been determined. The accessory minerals in this formation are pyrite, chalcopyrite, magnetite, hematite, pyrrhotite, galenite, zircon, rhombic pyroxene, biotite, ilmenite, leucoxene, spheena, muassanite, chromite, garnet, spinel, rutile, apatite, and tourmaline. The foregoing testifies in favor of the fact that the eroded rocks were acidic: granitoids, gneiss, magmatic, and metamorphic. By studies of well cores from the north of the province of Pinar del Río, the lithological composition of the San Cayetano Formation was defined as siliclastites which are without exception, quartziferous sandstones of fine and medium grains, rarely coarse-grained, which transitions from sandstones to grauvacas as the matrix/grain ratio grows. Also included are the quartziferous siltstones very frequent in the cuts and pelites: that in the flyschoide set, they contribute significantly to the lamination of the same. It ranges from claystones to argillites and sericitic schists at the thresholds of dynamometamorphism. Often, they contain silty quartz fraction, being able to pass transitionally to silt argillites. They have a higher content of organic matter, macroscopically they are recognized by their charcoal appearance and smudge when touched [15].

In recent years, attention has been paid to the study of fossil flora essentially Jurassic. There are important works by Fernández Carmona y Areces A in [20], Dueñas y Linares in [21], and Aliena Flores in [22]. Especially, it was investigated the upper part of the San Cayetano Formation when there is already evidence of the beginning of the penetration of the sea (Fig. 4.3).

According to Dueñas y Linares [21], the association of palynomorphs recovered from these sediments is very varied, with the predominance of Gymnosperms of small size. The presence of Bisaccates is restricted to two grains of Alisporites. The presence of dinoflagellates is relatively low as well as their diversity. Within the recovered palynomorphs, forms that probably correspond to new species are presented. From the regional point of view, the recovered palynological association is similar to the palynoflora reported from North Africa.

The terrestrial palynomorphs present good preservation while the dinoflagellates present poor conservation. Good recovery of organic matter was obtained from the samples, which, like the recovered palynomorphs, indicate that these sediments have not suffered the effects of elevated diagenesis. Taking into account the association of recovered palynomorphs, these sediments are assigned a Middle Jurassic age. The presence of dinoflagellates is indicative that these sediments were deposited in partially marine environments. The recovered organic matter is predominantly continental which suggests that these rocks were deposited in environments close to the coastline (Internal Neritic). As it has been verified by geochemical studies of Moretti et al. [23] in these samples and other surveys from well samples from the north of Pinar del Río, the San Cayetano Fm. in its upper part has source rocks that, under favorable conditions, can generate liquid and gaseous hydrocarbons.

Fig. 4.3 Rocks from the upper part of the San Cayetano Formation, where studies of palynomorphs were carried out, south of Cinco Pesos, province of Artemisa

The studies of Aliena Flores [22], consider the age of the San Cayetano Formation between the Upper Triassic (Rhaetian)/Lower Jurassic boundary and Callovian. They were determined 39 genera of spores, 11 genera of pollen, with 15 species each, 3 of acritarchs (*Veryachium, Michrystridium, Leiosphaeridia*), and 1 of algae (*Cymatiosphaera*), these last two groups are reported for the first time in the San Cayetano Formation, determining a total of 44 genera, of them, 30 originally determined in Cuba. Of the identified species, 19 are reported for the first time in this formation and in Cuba. As a novelty, redepositions of palynomorphs of Carboniferous–Permian age were reported. For the work that we are dealing with, ten microfloristic associations (A–J) were determined, which allowed us to determine the sedimentation environments corresponding to the coastal plain and the delta-pro delta front, which coincides with the deposition sequences determined in the subsoil in the Los Arroyos No. 1 well. A warm and humid subtropical climate was indicated for the deposition of the San Cayetano Formation, which was subject to repeated cycles of increase and decrease in sea level until a general marine transgression began the deposition of carbonates. The pollen Ephedripites, associated with evaporites, was presented in most of the samples, which argues the existence of hypersaline basins in areas near the deposition site since the Late Triassic (Rhaetian). One aspect that this research meant is the type of organic matter that is fundamentally type III, capable of generating gaseous hydrocarbons. Type II organic matter is present in a lower proportion, which may indicate the presence of liquid hydrocarbons. The seismic sequences determined in the Central Basin of the Exclusive Economic Zone of the Gulf of Mexico, analogous to the San Cayetano Formation, plus the total organic carbon richness in samples corresponding to this formation and similar in offshore, which could be mature, constitute premises for the existence of a new petroleum system, where the most favorable sequences to constitute source rocks are those corresponding to argillites. At the same time, the abundance of argillites and other packages of clay rocks gives the San Cayetano Formation characteristics of sealant rock.

In relation to the age of the Synrift base, there are no conclusive data. Most researchers point to the Lower Jurassic or Late Triassic, without further details due to the lack of paleontological data. Based on radiometric studies U–Pb SHRIMP (Uranium–Lead Sensitive High-Resolution Ion Microprobe), from zircons in the sandstones of the San Cayetano formation, Rojas Agramonte et al. [47], propose several sources of contribution, of rocks with very old ages not younger than the Devonian, probably from the massifs of Colombia and Venezuela and also of Yucatan in Mexico, before the separation of Pangea.

4.4.1.3 *Late Synrift: Middle Jurassic Callovian—Upper Jurassic* Kimmeridgian

Previously, it was meant a stage that will conventionally be called Late Synrift, as a transition toward the drift event. In the north of Pinar del Rio to these floors belong the transitional units of the Castellanos and Francisco formations. In the depth, to the north of the provinces of Havana, Mayabeque, and Matanzas, and in Central Cuba, the Constancia Fm outcrops.

There are geological and geophysical data, which show that between the Callovian and the Oxfordian, in the transition stage between the synrift and the drift stage, the sea was occupying some of the spaces that were being created at the beginning of the drift event in the proto-Caribbean and proto-Gulf of Mexico paleogeographic settlements.

4.4.1.4 Middle Jurassic Callovian—Upper Jurassic Oxfordian, Castellanos Formation

It was proposed in 1980, with Callovian–Oxfordian age. It comprises a wide strip in the northwestern region of the province of Pinar del Río, mapped between 1981 and 1988. The unit bears a strong resemblance to some parts of the San Cayetano Formation but has attributes that distinguish it. They are rocks of finer grain, more clay, and carbonated, high carbon. Among the sandstones, quartz—feldspathic with a high content of clayey and carbonate—clayey elements predominate. Slime sandstone and siltstone with flyschoid stratification predominate. Four sub-formations were distinguished: The first sub-formation is of white quartz sandstones, gravelites, and sedimentary breccias, carbonaceous siltstones, and carbonaceous schists, and finally carbonaceous limestones. The second sub-formation consists of siltstones, sandstone silts, quartz sandstones, and feldspathic quartz of fine grains, low-carbonaceous clay schists, and carbonaceous limestones. The third sub-formation consists predominantly of sandstones, siltstone sandstones, and siltstones in the lower part. In the middle part, carbonaceous siltstones, schists, and limestones alternate. In the highest part, it ends with carbonaceous limestones, schists, and siltstones, predominating limestones. The fourth sub-formation is an alternation of quartz sandstones of fine to medium grains, carbonaceous calcareous schists, argillites, and carbonaceous limestones (up to 20%).

4.4.1.5 Middle Oxfordian Upper Part—Upper Oxfordian Lower Part, Francisco Formation

The Francisco Formation, by the current level of study, is only known in the Sierra del Rosario. Previously, the deposits assigned to it were described as transitional between the San Cayetano and Artemisa formations. Lithologically, it consists of argillites and siltstones, clay schists, micrites, and fine intercalations of calcareous and quartziferous sandstones. Occasionally, there are calcareous concretions in the shales, similar to those found in the Jagua Fm. The deposits contain some ammonites, rare pelecypods, fish remains, and plant remains. Eventually, *Globochaete* spores occur. The ammonites indicate the upper part of the Middle Oxfordian and perhaps also the lower part of the Upper Oxfordian. The thickness of this unit does not exceed 25 m. It has a similarity with the Constancia Formation of Central Cuba and subsoil of the north of the provinces of Havana, Mayabeque, and Matanzas.

4.4.1.6 Oxfordian–Kimmeridgian Lower Part, Constancia Formation

It extends forming narrow strips on the northeast and southwest slopes of Sierra Morena in the area of Corralillo, province of Villa Clara. Outcrops are recognized by the hills of Santa Fe, Camajuaní area, and, by the southern slope of the Loma de Santa Cruz, northwest of the town of Placetas province of Villa Clara. The Constancia Formation is exposed in a wide and long strip along the northern slopes of the hills of Sabana Nueva south of Jarahueca, province of S. Spíritus. Notable outcrops occur in the area from the old Central Constancia to the south of Encrucijada where a Lectostratotype and an auxiliary profile are currently located.

With the discovery of several oil and gas deposits in the Northwest Belt of Hydrocarbons of Cuba (NBHC) since 1969, this lithostratigraphic unit was recognized in numerous wells from the area of Guanabo to Varadero. Their rocks were studied by drill cores and wash samples. In this way, descriptions of hundreds of thin sections were accumulated and there are also materials by geophysical (FMI, diagrams, and others) and geochemical methods.

They are considered as diagnostic facies of the Constancia Formation: quartziferous sandstones with intercalations of claystones. The composition of the fundamental minerals (quartz and feldspar) as well as the accessories (silicite, muscovite, zircon) places the sources of supply similar to that established for the San Cayetano Formation, evidencing in both cases, at least, two cycles of deposition to from the disintegration of sialic massifs. Such as those that had to be eroded in the Pangea supercontinent in a Paleogeographic Domain of conditions close to those of the Continental Margin in the Late Synrift, when the Tethys was communicating with the Pacific Ocean and Western Gondwana was separated of Laurasia. The depositional environment was internal neritic with a depth of 50–100 m with continental influence.

The upper limit of this formation is concordant under "Packages IV and V" of the Cifuentes Formation of Late Kimmeridgian age. Numerous sporomorphs of Oxfordian and lower Kimmeridgian ages are described as fossils.

The paleoenvironment ranges from continental to medium neritic, approximately up to about 100 m in its youngest part. In the deposits of the north of Havana—Matanzas a variation is suggested from internal neritic, very close to the coast, up to medium to external neritic. Among the wells that register these sediments are the Varadero No. 23, 31, 101, 201, Marbella Mar No. 1 and 2, Cupey 1-X, LPC No. 1, LPN No. 1, and Litoral Pedraplén No. 21. The average thickness reaches about 200 m.

4.4.1.7 North American Continental Margin Paleogeographic Domain in the Westernmost Part of Cuba

From the Oxfordian–Kimmeridgianbegins to differentiate the sedimentation in the late synrift, initiating the deposition of rocks mainly carbonated, typical of a Continental Margin, naming different lithostratigraphic units in each of the TSUs of the Cuban territories, so it is considered convenient to describe them briefly within each one of the corresponding ones.

In spite of the similarity of the Petrotectonic Set of the Paleogeographic Domain of the North American Continental Margin in the different regions, the researchers have named different TSUs. In the northwestern strip at least from Mantua in the province of Pinar del Rio, to the north of the province of Artemisa, to the North American Continental Margin, the following TSUs are attributed: La Esperanza, Sierra de Los Órganos, and Sierra del Rosario. From the Martin Mesa, deposit to the north of the province of Artemisa and following the deposits of the Northwest Belt of Hydrocarbons of Cuba (provinces of Havana, Mayabeque, and Matanzas) in the subsoil by oil wells, the formations of the Placetas TSU and in Varadero and also the Camajuaní and Colorados TSUs are described. From oil drilling and outcrops, in Central Cuba (north of the provinces of Villa Clara, S. Spíritus, and Ciego de Ávila), Cayo Coco, Remedios, Colorados, Camajuaní, and Placetas TSUs with their corresponding coverages of Paleogenic rocks are described from north to south. In the Sierra de Cubitas in the province of Camagüey, Remedios TSU emerges, as well as small exhibitions of the Placetas TSU occur in the Sierra de Camaján, northeast of the city of Camagüey. Lastly, in the Sierra de Gibara in Holguín, rocks from the Remedios TSU and its coverage of Paleogenic rocks emerge. From geological and geophysical works, the presence of the Camajuaní and Placetas TSUs is assured under the ophiolitic and the Cretaceous Volcanic Arc rocks [24].

4.4.1.8 La Esperanza TSU [25]

Originally, it was called La Esperanza Structural–Facial Zone. In it, rocks of the PDs of the Synrift and the North American Continental Margin and their orogenic coverage emerge or are recorded in the depth. The rocks of the

Esperanza Group are assigned to the PD of the North American Continental Margin. A good stratigraphic study of this Group would allow the separation of lithostratigraphic units similar to those of the Placetas TSU, such as the Cifuentes, Ronda, and Morena formations. Additionally, they are very similar to the coevals that are recognized in the neighboring Sierra de Los Órganos and Rosario. The Castellanos, Esperanza, and San Ramón formations have been proposed within the set that is being treated, but these last two are synonymous and the Castellanos Formation was considered in the PD of the Late Synrift, reasoning it transitional between the rocks of this Dominion and those of the PD of the North American Continental Margin. So, the Esperanza Fm. would be the first representative of the North American Continental Margin PD in this TSU. Only part of the Esperanza Formation contains Kimmeridgian rocks, reaching other lithological assemblages at the Neocomian age. In addition to this same character, this strip has been recognized as Cretaceous rocks of the Santa Teresa Formation and deep sediments of the Maastrichtian of the Cacarajícara Fm. and also of the Paleogene of the Manacas Fm. (Los Arroyos No. 1, Pinar No. 2 and Rio del Medio No. 1 wells), marking the existence of different tectonic mantles.

Upper Jurassic Oxfordian—Lower Cretaceous Neocomian, Esperanza Group, Esperanza Formation

Its name comes from the Esperanza No. 1 and Esperanza No. 2 oil wells, drilled in the vicinity of the town of Puerto Esperanza, province of Pinar del Río. As a type area, the northwest region of the province of Pinar del Río is recognized, in a narrow strip limited by deep faults that separate it from the Sierra de los Órganos and that extends from the village of Arroyos de Mantua to Puerto Esperanza, forming a large part of the La Esperanza TSU.

Upper Jurassic Kimmeridgian of the Esperanza Formation

In the lower part of the unit, representative fossils of a shallow seas paleoenvironment have been reported (San Ramón well, Core No. 21), evidenced by fragments of algae and representatives of the Miliolidae family, also in Core No. 42 (3154–3157 m) Echinoderma and Textularides thorns in the Core No. 39 (2937–2940 m). By studies of the wells of the north of Pinar del Río, Kimmeridgian was established mainly by the presence of Favreinas specimens. In wells Esperanza No. 2 and 3, it is dated by several cores. Also in the San Ramón No. 1, Dimas No. 1, and Los Arroyos No. 1 wells. The deposits are characterized by quartz sandstone, limestone mudstone with silty grains of quartz, argillite, floadstone, wackestone/wackestone peloidales. The fossil complex is

Globochaete alpina, *Cadosina parvula*, *Favreina joukowskyi*, *Favreina* sp., Crustaceans Coprolites, Ophalmididae, and few small benthic foraminifera. At the Los Arroyos No. 1 well, it was studied for palynology, discovering *Inaperturispolenites* sp., *Cycadoites* sp. In the strata, microstratification and the presence of carbonaceous matter occur. An internal neritic paleoenvironment is then determined, with very shallow waters of a restricted platform. It may be a high to moderate energy level, with a sedimentation rate that ranges from high to moderate. These rocks are similar to those of "Packages IV and V" of the Cifuentes Fm. of the Veloz Group of the Placetas TSU in Central Cuba.

Upper Jurassic Tithonian of the Esperanza Formation

On this floor, sediments are observed where alternating carbonates, siliclastites, and pelites, so that the beginning of a deepening of the basin can be induced. They are located with certainty, packages of this type in several oil wells. The lithology of this floor in almost all the probes becomes more carbonated, which greatly differentiates it from those of the Kimmeridgian. The microfauna is varied, which fixes the Lower Tithonian Age Upper Part. This can be associated with an external neritic paleoenvironment of moderately deep waters (100–200 m), with a low to moderate sediment precipitation rate. During this time, a maximum transgression occurred with the explosion of microorganisms. Similar rocks are described in part of the Cifuentes Formation of Veloz Group and in the members El Americano of the Guasasa Formation and La Zarza of the Artemisa Formation. Lithologically, they are composed of microcrystalline, argillaceous and organogenic limestones, siltstones, argillites, quartz sandstones, and silicites.

Lower Cretaceous Berriasian–Valanginian, Esperanza Formation

In the Berriasian–Valanginian of this belt, which refers to the middle part of the Esperanza Formation; sediments are observed where carbonates, silicates, and pellets alternate. All the wells, except Los Arroyos No. 1, cut sequences from these floors. The lithology is schistose limestone with organic matter, little recrystallized mudstone, relicts of sandy quartz floadstone, bioclastic and peloidal wackestone/packstone, sandstone, sandstone–siltstone, frequent silty-sandy fraction of quartz. The fossils are abundant and fix the Berriasian–Valanginian, similar to the rocks and age of the Ronda Fm. of the Placetas TSU. This microfaunal set is characteristic of deep waters, in a neritic environment external to batial, depth greater than 1500 m, the energy is very low and the sedimentation velocity too. In the Valanginian, the *Calpionellidae* family became extinct. They are also comparable coeval rocks, those of the Sumidero

Formation and the Tumbadero and Tumbitas members of the Guasasa Formation of the Los Órganos TSU.

Lower Cretaceous Hauterivian–Barremian of the Esperanza Formation

In the Hauterivian–Barremian Upper Part, the sedimentation of the Esperanza Formation ended. Pelagic sedimentation of fine-grained, deep-water carbonates were generalized. Abundant biota are reported in the limestones: calpionellids, radiolarians, *Stomiosphaera proxima,* and *Nannoconus* s.l. —printing the limestones appearance of nannoconic limestones—which indicate depths that exceed 200 m above the level of compensation of carbonates, which during the Jurassic and Cretaceous did not exceed 2000 m. This biochemical deposition was interrupted by distal turbiditic currents, typical of the basin, which carried terrigenous material. As in the Tithonian intervals of the Esperanza Formation, these deposits originated in anoxic background conditions, as evidenced by the absence of benthic fossils, very thin laminar stratification, well preserved and, abundance of syngenetic pyrite. In the Placetas TSU in this age, the Morena Formation is described. The upper limit of the Esperanza Fm. is transitional to the Santa Teresa Fm., this contact is observed in several locations.

Currently, several authors taking into account the similarities between the La Esperanza TSU and its coevals in the Cordillera de Guaniguanico, have considered that it is part of the Rosario Zone, more precisely of the Rosario Norte stratigraphic sequence.

Cretaceous Aptian–Albian, Santa Teresa Formation

The Santa Teresa Formation emerge in several areas of the La Esperanza TSU. The main expositions occur to the north and northwest of Consolación del Norte, forming small elongated bodies that stand out in the relief as hills on the Esperanza Formation, with a transitional contact. There are excellent outcrops in old quarries such as El Tres por Ciento, Sitio Morales (Fig. 4.4), and Malas Aguas, in the region of Puerto Esperanza; in the town of Santa Lucía and in the El Cocuyo quarry southwest of Santa Lucía. It is also exposed in several zones in the area between Dimas and Santa Lucia. Other outcrops are exposed in the Sierra del Rosario TSU, mainly in the Soroa area, around Rancho Mundito, Sabanilla, Valdés, Las Terrazas-Soroa, and Mameyal roads, among others. There is a large block included within the chaotic sequence of the Manacas Fm. in the Pons Valley where the unit retains all its characteristics and even the fossiliferous association found in some layers of intercalated limestones confirm the age of the formation. It is the most conspicuous formation of the Placetas TSU so its description will be made later.

4.4.1.9 Tectonostratigraphic Unit Sierra De Los Órganos [25]

Middle Jurassic Late Callovian–Upper Jurassic Early Oxfordian, Pan de Azúcar Formation

The name Azúcar with the rank of formation, was proposed by Charles W Hatten in 1957 and rewritten and published by A Pszczolkowski et al. 1975, with the hierarchy of members of the Jagua Formation. In a different way, Fernández Carmona [26] analyzed the Pan de Azúcar Member, which considered that the rocks of the Pan de Azúcar Member mark the sedimentation of carbonates in the drift regime, which started between the Late Callovian–Early Oxfordian and, approved this unit, once more, with the formation category. According to his opinion, it is dated based on the occurrence of *Coniscospirillina basilliensis*, ammonites, brachiopods, and pelecypods.

Upper Jurassic Oxfordian, Jagua Formation

The Jagua Formation consists mainly of gray and black micrites, marly limestones, sandstones, and argillites, grouped into four different lithological assemblages that allowed them to separate four members: Pan de Azúcar— raised to the category of formation by Fernández Carmona [26], and under this category is considered in this chapter, by what was described independently—Zacarías, Jagua Vieja, and Pimienta. It is then considered the Zacarías Member that groups the argillites with fine intercalations of cocks and siltstones, for their fossils indicating the Middle Oxfordian age. The Jagua Vieja Member would comprise the middle part of the formation, with limestones, argillites, and calcareous concretions that have Ammonites and fishes also of Middle Oxfordian age and, the Pimienta Member in the upper part, where there are limestones interbedded with siltstones. The age by several genera of Ammonites of these members is Middle Oxfordian and possibly reach the lower part of the Upper Oxfordian (Myczynski 1976).

The outcrops of this unit in the region of Viñales and Consolación del Norte, province of Pinar del Rio, are widespread. The rocks of the Jagua Formation were deposited in a shallow neritic environment within the initial platform stage. The thickness can reach up to 160 m.

In the Zonal Legend of the Geological Map of the Republic of Cuba, scale 1: 500,000 [6], the Jagua Formation is located in the Oxfordian and its geological cartography is

Fig. 4.4 Radiolarian silicites, clays, and interbedded tuffites. Santa Teresa formation in the locality of Sitio Morales, province of Pinar del Rio

in Sheet No. 1 of this map. Also with this age is drawn on Sheet 10 (F 17–5) of Geological Map 1: 250,000 of the Academy of Sciences of Cuba [7].

Upper Jurassic Kimmeridgian–Lower Valanginian, Guasasa Formation

The Kimmeridgian in the Sierra de Los Órganos TSU begins with the sedimentation of the Guasasa Formation, which has five members: San Vicente, El Americano, Tumbadero, Tumbitas, and El Infierno. The San Vicente Member is composed of carbonate rocks of marine facies with shallow depths between 50 and 100 m. The rocks have extensive development forming mogotes and other elevations in the Sierras de Viñales, La Guasasa, Lomas de San Vicente, the Sierras Ancon, and San Vicente, among other areas. Until the recent reports of neritic rocks from the internal platform in the lower part of the Cifuentes Formation, it was believed that the San Vicente Member was the only representative of such a paleoenvironment in the Kimmeridgian of Cuba. It is even an argument supported by many researchers, who deny the marked similarity that is demonstrated between the rocks of the Paleogeographic Domain of the Continental Margin of Central and Western Cuba. There is a close similarity between the San Vicente Member and the "packages IV and V" of the Cifuentes Formation, well studied in the oilfields from Varadero to Guanabo [27], and [14]. This member congregates light gray to black limestones, with massive stratification or thick layers, usually karstified. In some parts, they are stratified and are exposed totally or partially dolomitized. They can have nodules and dark chert lenses. Between the Jagua Formation and the San Vicente Member, there is a sedimentary calcareous breccia, which serves as a separation horizon. The following types of microfacies of the member we are treating are distinguished: micrites with coprolites, pelsparites, oomicrites and oosparites, biosparites, biomicrites, oncolytic limestones, intraoosparites, and intrabiosparites.

The studies corroborate the geodynamic and paleogeographic evolution model that is being described in this chapter. At the end of the Oxfordian, carbonated sedimentation began in shallow seas. Its sedimentary structures and the characteristics of microfacies indicate a moderately agitated paleoenvironment. In the upper part of the San Vicente Member and within the transitional strata toward the El Americano Member, there are some micrite and calcilutitic packages in fine strata, which may indicate an interdigitation of basin facies and shallow waters.

Later, it will be observed that the description of the characteristics of the "packages IV and V" of the Cifuentes Formation of the Placetas TSU is a solid foundation for its comparison with the San Vicente Member of the Guasasa Formation of Pinar del Río.

Upper Jurassic Tithonian, El Americano Member

At the beginning of the Tithonian an increase in sea level occurred in Proto-Caribbean, evidenced by the deposition of pelagic carbonates with radiolarians, Chitinoidella spp.,

Saccocoma sp., Embryonic ammonite chambers, Globochaete alpina, which suggests that the deposition depth exceeded 100 m, in the environment from external neritic to batial, in anoxic conditions. These are manifested by the conservation of a fine stratification, almost total absence of benthic fossils, abundance of pyrite and syngenetic organic matter, resulting in source rocks potentially generating oil. This was an episode on a global scale. However, in some areas, shallow deposit conditions were still preserved, which is demonstrated in the singular outcrops of the Sierras de Los Órganos and del Rosario. In the Sierra de Los Órganos TSU the Tithonian rocks are represented by the El Americano Member of the Guasasa Formation. The El Americano Member was recognized in the middle part of the Guasasa Formation on the San Vicente Member deposits. The deposits comprise, dark gray to black, granular, well-stratified limestones, with occasional intercalations of schistose clayey limestones. Dolomitic limestones and dolomites are also reported. The unit in question is of Tithonian age and has been dated both by microfossils and by the abundance of ammonites, in its type locality. It is comparable with the "Packages I, II, III" of the Cifuentes Formation of the Placetas TSU.

Lower Cretaceous Berriasian, Tumbadero Member

It characterizes the limestones that overlap the Tithonian (El Americano Member) and underlie the Tumbitas Member. Diagnostic lithology consists of micrites and biomicrites, often laminated, well stratified (0.1–0.3 m), and calcilutites with intercalations of black chert. The Tumbadero Member, emerges in several sections of the Sierras de Viñales, Guasasa, San Vicente, Infierno, and Guacamaya among others. In Hacienda El Americano, the thickness of the member reaches about 40 m, while in other sections, it ranges between 20 and 50 m. The most notable fossils are calpionellids of the genera *Calpionella* and *Calpionellopsis* that fix the Berriasian age. It is meant that the set of fossils described in several areas of western Cuba, is identical to that found in part of the Ronda Formation of Cuba Central and is present in the subsoil of gaso-petroleum deposits in the provinces of Havana and Matanzas. The petrographic description of its rocks coincides with those of the Ronda Formation, including the presence of micrite and biomicrite banding and the presence of interspersed black cherts. This member lies accordingly on the El Americano Member and is covered accordingly by the Tumbitas Member.

Lower Cretaceous Upper Berriasian–Lower Valanginian, Tumbitas Member

This lithostratigraphic unit has been distinguished in the upper part of the Guasasa Formation, over the Tumbadero Member deposits. Lithologically, it consists of light and dark gray biomicrites, even black, well stratified, where bioturbation phenomena are observed, which causes the rocks to be mottled frequently. Unlike the Tumbadero Member, there are no cherts here. As in the underlying member, in this unit, the ammonites are poorly conserved and hardly observed. However, the *calpionelids* are abundant. In the Pinar No. 1 well between 3440 and 3515 m, cores No. 58 and 59, deposits are described that do not contain calcareous calpionelids showing a lithology similar to that described for the Tumbitas Member but that have a Hauterivian–Barremian age.

Lower Cretaceous Upper Valanginian–Lower Cretaceous Aptian, Pons Formation

The Pons Fm. according to the most recent proposals, it has the Upper Valanginian–Lower Aptian age, based on nannokonids determinations, contained in the rocks of the type locality. In principle, the formation was proposed to distinguish a set of rocks that emerge in the Valle de Pons in the Sierra de Los Órganos. These deposits, with relationships not very clear yet, are overlain by the Peñas Formation that distinguishes the rocks of the Upper Cretaceous. In fact, the limestones and cherts that are grouped in both units, only differ in the content of fossils and do not have the amount of siliciclastic rocks that have been attributed to it. In the Pons Quarry, micrites and biomicrites have been described from light gray to almost black, rich in calcareous nannofossils that have thin layers, lenses, and interbedded chert nodules. They correspond to deposits of a pelagic environment. Its limestones are rich in Total Organic Carbon (TOC).

Upper Cretaceous Campanian–Maastrichtian, Peñas and Moncada Formations

They belong to the orogenic deposits. These are the formations with the lowest surface distribution among all the units that emerge in the area. The Peñas and Moncada formations up to the present have been reported in isolated localities, the first, in the Las Piedras River south of the town of Pons and, the Moncada Fm., on the Viñales-Pons road at the junction with the El Moncada road. Probably both units have been eroded during the catastrophic phenomena that occurred in the Cretaceous-Tertiary limit or during the strong tectonic processes of the Lower Eocene that caused the formation of tectonic mantles and the consequent formation of olystostromal deposits (Vieja Member of the Manacas Fm.), where blocks and fragments of equal composition and age of these units have been found.

The Peñas Fm., consists of dark gray to black limestones, well stratified with abundant intercalations of thin layers and black chert lenses. The reported fossiliferous association clearly defines a Campanian–Maastrichtian Upper Cretaceous age and pelagic sedimentation, batial environment below the compensation level of the aragonite without terrigenous input. Its thickness can reach up to 80 m.

The Moncada Formation was studied in detail during the works of the Cuban–Japanese project called "Paleoenvironmental and Paleoecological Events that occurred in the Cretaceous-Tertiary Limit in Cuba" [28] and [29]. This set of rocks is only 184–191 cm thick. The studies have established a sequence of calcareous sandstones, interspersed with calcareous shales and fine-grained sandstones in the upper part of the cut. The layers have laminar, crossed, and wavy stratification. The highest part of the unit is a calcareous shale of dark color, interspersed with very thin calcareous sandstone of 3–5 cm thick of dark color finely interspersed with fine-grained sandstone. The main mineral components of the formation are calcite, quartz, plagioclase, and clay minerals such as chlorite, illite, and smectite, only the pumpellyite and hematite being observed as traces.

It is determined that the Moncada Fm. in western Cuba is a sandstone complex of the K/T Limit that is characterized by abundant ejection materials such as shock quartz and fragments of molten impact rocks throughout the formation. These scholars found a layer of clay rich in Iridium at the top of the complex, and they consider that the occurrence of wave traces showing inversion of the paleocurrents between the units, as well as the absence of hemipelagic clays between them, makes them similar to other sandstones of the K/T Limit in Mexico and that has been interpreted as being formed by tsunamis caused by the meteorite impact. These characteristics, together with the biostratigraphy that allow dating it with an estimated age between the Upper Maastrichtian and the Paleocene, show that the sandstone complex of the Moncada Formation was formed by several series of tsunami waves caused by the K/T impact.

4.4.1.10 Tectonostratigraphic Unit Sierra Del Rosario [25]

Jurassic Magmatism (Callovian (?)–Oxfordian) Associated with Rocks of the North American Continental Margin, El Sábalo Formation

In western Cuba, several cuts of Jurassic mafites are recorded. The best-known case is that of the El Sábalo Formation. The unit consists of diabases and basalts with intercalations of varying thicknesses of carbonated and siliciclastic sediments. Representative cuts occur on the mountain road northwest of Soroa where its type locality is located; there are basalts in pads (pillow) and diabases predominantly, with intercalations of argillites, siltstones, and limestones. From its fossil content and stratigraphic position under the Artemisa Formation and over the San Cayetano Formation (upper part), the El Sábalo Formation must have a Callovian age?–Oxfordian; more likely Oxfordian (Fig. 4.5). In the northern part of the province of Pinar del Río, in the Esperanza Fm., basic effusive rocks have been described, essentially affiric basalts and hyalobasalts.

In Sierra Camaján, in Camagüey, emerge the Nueva Maria Fm. of similar composition, which is associated with rocks of the North American Continental Margin, but there is attributed to a Tithonian age.

Fig. 4.5 Basalts with pad structure, diabases with thin layers of carbonate rocks. El Sábalo Fm., mountain road, on El Sábalo hill

Upper Jurassic Tithonian, Artemisa Formation

The increase in sea level in Proto-Caribbean in the Tithonian is evident as in the Sierra de Los Órganos TSU for the deposition of pelagic carbonates. Some shallow deposits were also preserved, which is demonstrated in some outcrops and wells (Martín Mesa No. 2, CHD-1X and Cayajabos No. 3), where even the palynomorphs are reported: *Classopollis* sp., *Cyathidites australis*, woody debris, that confirm depositions on an internal platform of shallow water, within a protected environment.

E Linares and collaborators, in 1985, considered that the rocks of this unit comprise all the time of the Tithonian, placing them in the middle part of the Los Órganos Group. Linares since 2003, in his doctoral thesis, equates them with the "packages" I, II, and III of the Cifuentes Formation, recognized in the subsoil of the northern deposits of the provinces of Havana, Mayabeque, and Matanzas. In his last works, only consider the Artemisa Fm., in its equivalence with the La Zarza Member in this unit, since it accepts the proposal of Fernández Carmona [26], to elevate the Sumidero Member to the category of formation.

With the Tithonian age, the Artemisa Formation was mapped on the Geological Map of the Republic of Cuba, scale 1: 500,000 [6], forming a wide strip north of the Pinar Fault, from the north of Los Palacios to the Cayajabos neighborhoods in the provinces of Pinar del Rio and Artemisa. Minor bodies emerge to the west and southeast of La Palma (Fig. 4.6).

In the part corresponding to the Lower Tithonian in the Sierra de Los Órganos TSU, the Artemisa Formation integrates the rocks formed in deep waters, although not as much as those of the Middle and Upper Tithonian, since this section represents the transition between the shallow water facies of the San Vicente Member of the Guasasa Formation and the deep waters of the La Zarza Member and their coevals of the El Americano Member of the Guasasa Formation. The predominant facies correspond to that of calcareous mudstone that gradually or abruptly transitions to bioclastic wackestone, whose third part is composed of radiolarians. The rocks occur finely banded or in medium layers, sometimes they are black limestones very rich in organic matter.

The Middle Tithonian of the Artemisa Formation is lithologically similar to the Lower Tithonian, but with few silicite horizons. As main facies, fossiliferous calcareous mudstone and bioclastic wackestone are distinguished. This set is stratified in flyschoid form due to the intercalations of fine bituminous argillites. Secondarily, dolomites have been described.

In the Upper Tithonian of the unit, we are dealing with the proportion of bioclastic wackestone is almost double that of calcareous mudstone. Subordinately, microfacies of radiolaria packstone and sporadic calcarenites, fine sandstones, and dolomites are described.

For the Lower Tithonian of the Artemisa Formation, the most notable fossils are Saccocoma sp., Aptychus, and Radiolarians.

It is important to make a comparison of the Artemisa Formation with similar rocks from the Northwest Belt of Hydrocarbons of Cuba, where there are important gaso-petroleum deposits.

Fig. 4.6 Carbonate rocks well stratified in thin to medium layers, typical of the Artemisa Fm., Vegas Nuevas quarry, province of Pinar del Rio

The Tithonian in the Martín Mesa block corresponds to the Artemisa Formation (La Zarza Member), which has been correlated in the exploration wells of this region with the "packages I, II and III" of the Cifuentes Formation. These sediments are recorded in several cores of Cayajabos No. 3 and CHD-1X wells and, in wells of the Martín Mesa deposit. The description of "packages I, II and III" of the Cifuentes Formation will be made later.

Lower Cretaceous, Berriasian–Valanginian, Sumidero Formation

Previously, it had the category of member of the Artemisa Fm. Fernández Carmona [26], considering the comparisons made by oil stratigraphers [30] in similar rocks from wells and deposits from Soroa to Varadero and, reasoning that, during the Berriasian–Valanginian interval, abundant and well-characterized associations of fossils—greater diversity of calcareous *calpionelids*—as well as calcareous nannoplankton and radiolarians of external neritic marine environment to basin, proposed to elevate the Sumidero Member to the formation category because these rocks constitute an important source of organisms for the generation of oil, evidenced by its good values of TOC. The intense banding of this stratigraphic set is due precisely to the fact that the content of organic matter marks the stratification. The numerous cores of drilling and the increasingly refined divisions in the petroleum stratigraphy made this division immediately accepted by the community of researchers of this activity in Cuba.

The unit that we are dealing with, emerges well in the Sierra del Rosario both north and south of it. The Sumidero Formation has its facial and temporal equivalent in the Sierra de Los Órganos, in the so-called Tumbadero and Tumbitas members of the Guasasa Formation, with which it has close similarity. The age of the Sumidero Formation marks an event of maximum transgression with the extinction of the *Calpionellidae* family, where calcareous calpionelids, calcareous nannoplankton, and radiolarians also proliferate. These indicate an external neritic marine environment—basin, bathymetry greater than 1500 m, where the main producer of organic matter was phytoplankton, which is an important source for oil generation. The most striking aspect in the Pinar del Río region is the intense banding of the clay rocks with organic matter, interspersed in the limestones, which clearly notice the Neocomian stratification (valid in parts for the Tithonian). The group of Berriasian–Valanginian rocks, well studied in all the oilfields of the northern strip from Guanabo to Varadero, show their similarity to those of Pinar del Río, considering that: the same Calpionellid biozones are defined, which from bottom to top are Calpionella alpina (which marks the Berriasian), Calpionellopsis oblonga, and Calpionellites darderi (note the Valangianian). These rocks reflect the same lithological set, that the one named for the Sumidero Fm. in Pinar del Río and, Ronda Fm. in the subsoil of the Vía Blanca, Boca de Jaruco, Puerto Escondido, and Canasí oil fields, among others. They represent rocks from a low energy anoxic environment.

In the Martín Mesa deposit, the Berriasian–Valanginian rocks have been referred to the Sumidero Formation (Martín Mesa well No. 4 between 2260 and 2490 m and Martín Mesa No. 2 well between 2230 and 3105 m). Additionally, in the Chacón No. 2 (1700–2040 m) and CHD-1X wells. They are rocks similar to those of the Ronda Formation of the Placetas TSU. It is wanted to stand out, that from 1985, these rocks were described in the deposits Varadero and Varadero Sur, with the informal appellatives: producing horizons "Artemisa," "Sumidero," and "Layers of Intercalations." The calcareous and calcareous—siliceous—clay horizons, which had fossils from the areas of *Crassicollaria, Calpionellopsis,* and *Calpionellites darderi*, were distinguished with them, which highlights that since more than 30 years ago, analogies were observed between the outcrops of the North American Continental Margin of Pinar del Río, Artemisa, and the subsoil rocks of Varadero.

Lower Cretaceous, Hauterivian-Barremian, Polier Formation

The Polier Formation consists of micrites stratified in thin layers, of gray to dark gray colors, interspersed by argillites and sandstone of calcareous cement that show organic and inorganic petroglyphs in the bases of the layers (currents molds among others). Also, gradational stratification, horizontal and oblique lamination are presented. Exceptionally, there are layers of sandstones a meter thick. One criterion, which unfortunately was becoming widespread, was the exaggeration of the sandstone content in the composition of the Polier Formation. It is meant that the proportion of sandstones is not predominant, prevailing micrites. The micrites of this lower part have some ammonites and molds of fish have been reported in the outcrops near Luis Carrasco. There are abundant Nannoconus species. These species are also present in the Morena Formation, described in Central Cuba and in the subsoil of the oilfields, from Guanabo to the Varadero area; there they form the nannoconic limestones of the Morena Formation.

The Polier Formation is checked in Martín Mesa No. 1, 2, 3, and 4 wells and, in Chacón No. 2, Caridad No. 4, Cayajabos No. 3, Pinar No. 2, and CDH—1X wells. The previous stratigraphic revisions in the subsoil of the Martín Mesa deposit reached the conclusion that part of the rocks grouped previously with Aptian–Albian, Cenomanian–Turonian, and some Neocomian, could refer to the Polier Formation. For the latest data, the first ones can be attributed to the Carmita and Santa Teresa formations, which even emerge in small scales in the Martín Mesa Block and, the last ones to the Polier Formation. The Polier Formation is exposed in small scales southeast of Mariel and in the vicinity of the La Jutía hill. The predominant cut in this interval is carbonate-terrigenous and terrigenous-carbonated, subordinately clayey.

Lower Cretaceous, Upper Hauterivian–Lower Barremian, Lucas Formation

Lithologically, in the lower part finely stratified micrites are described, of gray to black colors, with abundant aptychus and ammonites poorly conserved. In the upper part, they are also micrites but interspersed by hard calcareous argillites, showing similarly aptychus and scarce impressions of ammonites. There are also calcified radiolarians. Age is Upper Hauterivian–Lower Barremian. The total thickness ranges between 200 and 300 m. This unit gradually transitions upward, passing to the schists and silicites of the Santa Teresa Formation. It is distributed about 20 km in length between Pinalilla and the Loma del Cable, province of Artemisa.

Cretaceous Cenomanian–Turonian, Carmita Formation

Currently, many geologists describe the presence of the Carmita Fm. in the provinces of Pinar del Río and Artemisa, with the same characteristics that is revealed by outcrops in Central Cuba and in the depth of the oilfields of northern Havana, Mayabeque, Matanzas, and Villa Clara. In several reports, the Carmita Fm. Is examined in the depth of the sectors from Soroa to Martín Mesa, from the Guanabo Field to the Yumurí Field, and from this to the Corralillo region. When comparing the revelation of this unit in the subsoil, it is appreciated that it is poorly recognized in the areas of the Puerto Escondido, Seboruco, and Yumurí deposits; however, this is not the case in the areas of Boca de Jaruco, Vía Blanca, Guanabo (Guanabo No. 19, Boca de Jaruco No. 29 and 35, wells), Varadero (several wells), and the areas of Cantel, Camarioca, Martí, Guadal, and Bolaños. Toward the west, in Pinar No. 2 Well, and in isolated outcrops, this lithostratigraphic unit is verified. The lithological characteristics, fossil indicators, and paleoenvironment of the sedimentation of the Carmita Fm., are perfectly correlated between central Cuba and western Cuba, showing that at this time distal turbiditic processes occurred in an anoxic environment from an external to a batial platform. As a result of the petrological, paleontological, and paleoenvironmental descriptions, it is judged that the Carmita Fm. is equated, in all the characteristics outlined in its type area in central Cuba, with those that are revealed in-depth by oil wells between Corralillo and the area of Soroa (Pinar No. 2) and the outcrops of the Valle de Pons in Pinar del Río. Typical outcrops of this unit associated with the Santa Teresa Fm., are exposed in tectonic contacts in the locality of Mameyal, west of Rancho Mundito, and in Candelaria del Aguacate in the province of Artemisa.

Deposits of carbonate bank, Lower Cretaceous Albian–Cretaceous Cenomanian, Guajaibón Formation

A stratigraphic singularity occurs in the highest altitude of western Cuba, the Pan de Guajaibón. They are deposits of carbonate bank of Albian–Cenomanian age known with the name of Guajaibón Fm. It only emerges in this elevation and its surroundings as well as in the heights of the Sierra Azul. Its relations are of tectonic character with the surrounding units and an approximate thickness of 500 m is calculated. Lithologically, it is composed of massive limestones or thick layers of medium to light gray to cream or dark gray to black with sporadic intercalations of shales and dolomites. Toward the base, the limestones tend to be massive or in thick banks with intercalations of dolomite and abundant microfauna. In the unit, bauxite bodies are observed.

Cretaceous Cenomanian–Turonian, Pinalilla Formation

It owes its name to the locality of Pinalilla, to the north of the road that connects the town Asiento de Cacarajícara, in the elevations of Sierra Azul, in the northern part of the Sierra del Rosario, province of Artemisa. It is distributed in the form of a belt with southwest-northeast direction from the east of Pan de Guajaibón to the east of Luis Carrasco (Quiñones) in the province of Artemisa. It is shown in Sheet 10 (F 17–5) Pinar del Rio of the Geological Map of Cuba 1: 250000 of the IGP / ACC [7].

It is composed mainly of strata of gray-greenish limestones, of fine to massive stratification, and biomicrites, which contain planktonic foraminifera. Locally, fine intercalations of siltstones and argillites may appear between the limestones of this unit. In the lower part, it almost always contacts the Santa Teresa Formation and eventually the limestones and cherts of the Carmita Formation. The upper limit is discordant with the siliciclastic and carbonate rocks of the Moreno Formation. This formation is of great interest to petroleum geologists since the manifestations of hydrocarbons are frequent in the fractures of their limestones. This is observed in the quarry located in the vicinity of Luis Carrasco (Quiñones). In samples of this locality and Sierra Azul, calcareous mudstone was described with foraminifera of the Cenomanian–Turonian.

Orogenic Deposits, Upper Cretaceous Campanian–Maastrichtian

The rocks deposited in the 17.6 million years that lasted these two floors are well represented in the Sierra del Rosario TSU by the distribution of the Moreno and Cacarajícara formations. In the subsoil, some of these units are also recognized in wells drilled in the La Esperanza and Sierra del Rosario TSUs and in the Martin Mesa deposit.

Upper Cretaceous Campanian–Maastrichtian, Moreno Formation

It consists of a set of rocks that are exposed limitedly in the Sierra del Rosario TSU, being a temporary equivalent of the Peñas Fm. of the Sierra Los Órganos TSU. It differs considerably from the same by the high content of siliciclastic rocks: limestones, sandstones, shales. It appears in narrow strips of sub-latitudinal direction in the neighborhoods of Rancho Lucas; in the western part of the Luis Carrasco quarry (Quiñones) in Los Cayos and, by Valdés, in the northeastern part of the Sierra del Rosario. Also, it is observed north of Loma del Mulo, Santo Domingo, and north of Mango Bonito, where its holoestratotype is located.

It is considered that similar deposits are registered in several of the wells of the Martín Mesa deposit: Martín Mesa No. 1 (450–760 m), Martín Mesa No. 3 (1125–1302 m), and Martín Mesa No. 21 (595–830 m); lying, generally, below the Paleogene sequences.

The fossil association is extensive. Age is Campanian Upper Cretaceous–Upper Maastrichtian. The sediments were deposited in a pelagic environment at shallow depths with the influence of shallower environments.

The thickness of this formation is variable, being the maximum measured in the surface of 45 m, while in the wells of the Martin Mesa deposit can reach up to 310 m (Martin Mesa well No. 1).

Upper Cretaceous Upper Maastrichtian, Cacarajícara Formation

It is very similar to the Amaro Formation of Central Cuba. It crops extensively in the Sierras Chiquita (type area) and Rosario, to the south of the Pan de Guajaibón, extending sub-latitudinal direction in a belt of about 50 km long and up to one wide, with thicknesses ranging from a few tens to more than 700 m. These deposits are considered megaturbidite calcareous or megalayer clastic—carbonated. A Lower Member (Brecha Los Cayos) and the Upper Member (Calcarenitas) were separated. The Lower Member is constituted by a breccia of about 500 m of thickness and contains blocks and fragments of metric size strata, of the underlying formations, as well as soft gravel and fragments of smaller rocks. There are also fragments of volcanic and metamorphic rocks. The Upper Member is constituted by calcarenites decreasing grain, until becoming a calcilutite at the top of the unit. The thickness of this member is almost 400 m and, in its lower and middle parts, large columnar structures of water escape are observed. It may be sediments of a colossal turbiditic current linked to the events of the K/T limit. The fossiliferous association indicates Upper Cretaceous Upper Maastrichtian age. This formation is detected in several oil wells to the east in the Martin Mesa field.

Paleogene in the Sierra de los Órganos and Sierra del Rosario TSUs

The Upper Paleocene–Lower Eocene Ancón and Manacas (Paleocene–Lower Eocene) formations are distinguished.

Ancón Formation

In the Ancón Formation of the Upper Paleocene–Lower Eocene, two breccia members have been established for this unit, the "basal member La Güira" and the "upper member La Legua." A possible terrigenous package has been mentioned at the base of the unit. In addition, a group of limestones was recognized, which was denominated as "a sequence of greenish, pinkish and violet gray limestones" and renamed as a member of "micrites and marly limestones"; package of limestones, which constitutes its largest volume, and the one that emerges in its type locality (Fig. 4.7).

Manacas Formation

The Manacas Formation is comparable to the Vega Alta Fm. of the Placetas TSU in the Northwest Belt of Hydrocarbons of Cuba and other zones of the national territory. In the Sierra de Los Órganos and Rosario TSUs are constituted by siliciclastic and olystostromal sequences of lower Paleocene–Eocene age. This unit is widely distributed on the surface all along the Cordillera de Guaniguanico at the westernmost end of the Folded and Overthrusted Belt in Cuba, at the La Esperanza, Órganos, and Rosario TSUs (considered within the latter up to the Martín Mesa Block). Additionally, it is recorded in oil drilling and geological mapping of the region.

At the moment, two sequences are differentiated in the Manacas Formation: a siliciclastic, the Pica Pica Member, and the Vieja Member.

The Pica Pica Member occupies the lower part of the unit and is composed of polymictic sandstones of brownish-gray, brown when weathering, argillites, calcarenites with tuffitic material, yellowish-gray siltstone to brownish, micritic limestones, intercalations of finely stratified cherts and clay.

Fig. 4.7 Micritical and marly limestones, violaceous, of the Ancón Fm., north of Valle del Ancón, province of Pinar del Rio. It has manganese minerals between the layers

Fig. 4.8 Policomponent Olistostrome of the Manacas Fm. in the vicinity of Rancho Mundito, province of Pinar del Rio. Observe large olistolites of limestones

Fine intercalations of calcareous breccias with small fragments of limestone and grayish chert may occur.

The "Vieja Member" is a policomponent olistostromic sequence, constituted by olistolites, and fragments of different types of limestones, cherts, serpentinites, volcanic rocks, and in smaller amount metamorphic rocks, in a matrix that varies laterally and vertically in short spaces, and that it is made up of sandstones of different types, including arches, siltstones, and argillites.

Olistolites and blocks come from the oldest formations (San Cayetano, Jagua, Francisco, Guasasa, Pons, El Sábalo, Artemisa, Sumidero, Polier, Santa Teresa, Carmita, Pinalilla, Guajaibón, Moreno, Cacarajícara and Ancón); from the ophiolites and their exotic metamorphic rocks and; from rocks of the Cretaceous Volcanic Arc, among others. Good exposures can be seen on the road that communicates Quiñones (Luis Carrasco) with the Los Cayos locality, in the province of Artemisa (Fig. 4.8).

The oil wells that have registered the Manacas Formation are Dimas No. 1, Pinar No. 1, Pinar No. 2, Puerto Esperanza No. 2, and numerous wells of the Martín Mesa Block (Martín Mesa No. 1, Martín Mesa No. 2, Martín Mesa No. 3, Mariel Norte No. 1-X, Cayajabos No. 3, Martín Mesa No. 20A, Alpha No. 1, Chacón No. 2, Chacón No. 1-X). In these, oil and gas demonstrations have been found in some cases or regional seal properties are manifested in others, such as in the Martin Mesa No. 1, Martín Mesa No. 2, Martín Mesa No. 3, Mariel Norte No 1-X. At the La Esperanza TSU, outcrops of the Manacas Formation could be recognized, one of them at the junction of the Dimas road with the Mantua-Santa Lucia road.

The most representative fossils are *Acarinina spinuloinflata, Acarinina pseudotopilensis, Globigerina* sp, *Morozovella aequa, M. angulata, Chiloguembelina cubensis, Igorina broedermanni, Acarinina pentacamerata, Parasubbotina.* cf. *P. pseudobulloides, Globanomalina compressa, Pseudohastigerina* cf. *P. wilcoxensis, Acarinina soldadoensis, Globigerina* sp., spines of echinoderms. Many samples were sterile and others have abundant redeposition and sometimes total absence of Paleogenic specimens, so they have been dated as Campanian–Maastrichtian or Turonian–Maastrichtian. The age that throws the fossil set in surface is Lower Paleocene–Lower Eocene.

The fossil associations reported in the wells in the Martin Mesa Block area are composed of: *Globigerina* sp, *Morozovella aequa, M. angulata, Chiloguembelina cubensis, Igorina broedermanni, Parasubbotina* cf. *P. pseudobulloides, Globanomalina compressa, Pseudohastigerina* cf. *P. wilcoxensis, Acarinina soldadoensis*, spines of echinoderms, dating unit as well as on surface as Upper Paleocene–Lower Eocene. In the Martín Mesa well No. 33 (799–802 m.), in the upper part of the unit, it was determined by Planorotalites palmera indicative of the Lower Eocene.

The lithoclastic composition described on the surface coincides with that of the subsoil. In many cores of wells studied (Martín Mesa No. 1, 3, 20, 0.20-A, 21, 24, and 33 wells) were reported the following redeposited fossils: *Stomiosphaera* sp., *Pithonella* cf., *P. sphaerica, Pithonella ovalis, Bonetocardiella cardiiformis, Peudotextularia* sp., *Heterohelix striata, Rugoglobigerina* sp., *Globotruncana* sp, *Globotruncanita conica, Rugoglobigerina* sp., *Sulcoperculina* cf. *S. globosa, Sulcoperculina vermunti, Sulcoperculina* cf. *S. dickersoni, Pseudorbitoides* sp., *Vaughanina cubensis, Orbitoides* sp., *Siderolites* sp., Miliolidae, *Nummoloculina heimi*.

The presence of *Morozovella* as a keel morphotype and of the genera *Globigerina* and *Globoconusa* as globular morphotypes evidences the deposition of sediments in an external neritic environment lower to the Oceanic Basin. The diversity in fossils found in clasts coming from the continental margin sequences prior to the unit's origin, characterize it as a policomponent deposit accumulated according to its indigenous fauna in a pelagic environment of batial depths.

The total thickness of the formation in the outcrops has been estimated at around 200 m. In the deep wells drilled in the provinces of Pinar del Río and Artemisa, their thickness oscillates considerably, in some even different thicknesses were cut in independent scales, the greater thickness reported occurs in Dimas No. 1 well with a package in the final part of the well of 1680 m. Also notable are the thicknesses reported in the wells of the Martin Mesa area, in the Martín Mesa No. 3 of up to 1125 m, and in the Mariel Norte 1-X of 960 m.

4.4.1.11 North American Continental Margin North of the Island of Cuba and Its Marine Zone from the North of Artemisa to the North of Holguín

To understand the stratigraphy of the northern half of the island of Cuba and its northern seas, in addition to the descriptions made on the northern portion of the provinces of Pinar del Rio and Artemisa, it is necessary to consider the outcrops and hundreds of oil drilling from the so-called Martin Mesa block in the province of Artemisa, up to the Sierra de Gibara in the province of Holguín. In this territory, the Martin Mesa oil and gas deposits are recognized, those of the Northwest Belt of Hydrocarbons of Cuba (NBHC): Guanabo, Vía Blanca, Boca de Jaruco, Yumurí, Puerto Escondido, Seboruco, Varadero and its peripherals; Majaguillar. Additionally, the oil wells isolated in the provinces of Villa Clara, S. Spíritus, Ciego de Ávila, Camagüey, Las Tunas, and Holguín. In this territory are included in the Paleogeographic Domain of the North American Continental Margin from north to south, five Tectonostratigraphic Units with their corresponding Petrotectonic Sets: Cayo Coco, Remedios, Colorados, Camajuaní, and Placetas. Due to the similarities between the TSUs described in Pinar del Rio (La Esperanza, Sierra de Los Órganos, and Sierra del Rosario) with the Placetas TSU, the characterization will begin with it. In Fig. 4.9 the position of the different stratigraphic units is shown, before being subjected to the compressional phenomena.

4.4.1.12 Placetas TSU [31]

It consists of carbonate deposits and, in a large proportion, silicites from very deep waters of the Santa Teresa Formation (Fig. 4.10). There are manifestations of magmatism of the Continental Margin and the Cretaceous Volcanic Arc. The sediments of the Paleogenic cover are mainly olistostromic and terrigenous.

The Cifuentes, Ronda, Morena (Grupo Veloz), Santa Teresa, Carmita, Bacunayagua, and Amaro formations are distinguished. Its covers are the Vega Alta and Peñón formations. The limits of this TSU are complicated by different overthrust faults, which put it in contact with mantles of different nature from Zaza, Camajuaní, and the Ophiolitic Association TSUs. It is therefore one of the most deformed tectonically in the region. Due to its nature of having rocks formed essentially in deep and very deep waters, it has excellent source rocks with high contents of TOC.

Detailed studies from oil drilling have found it very practical to propose the Veloz Group with the Cifuentes,

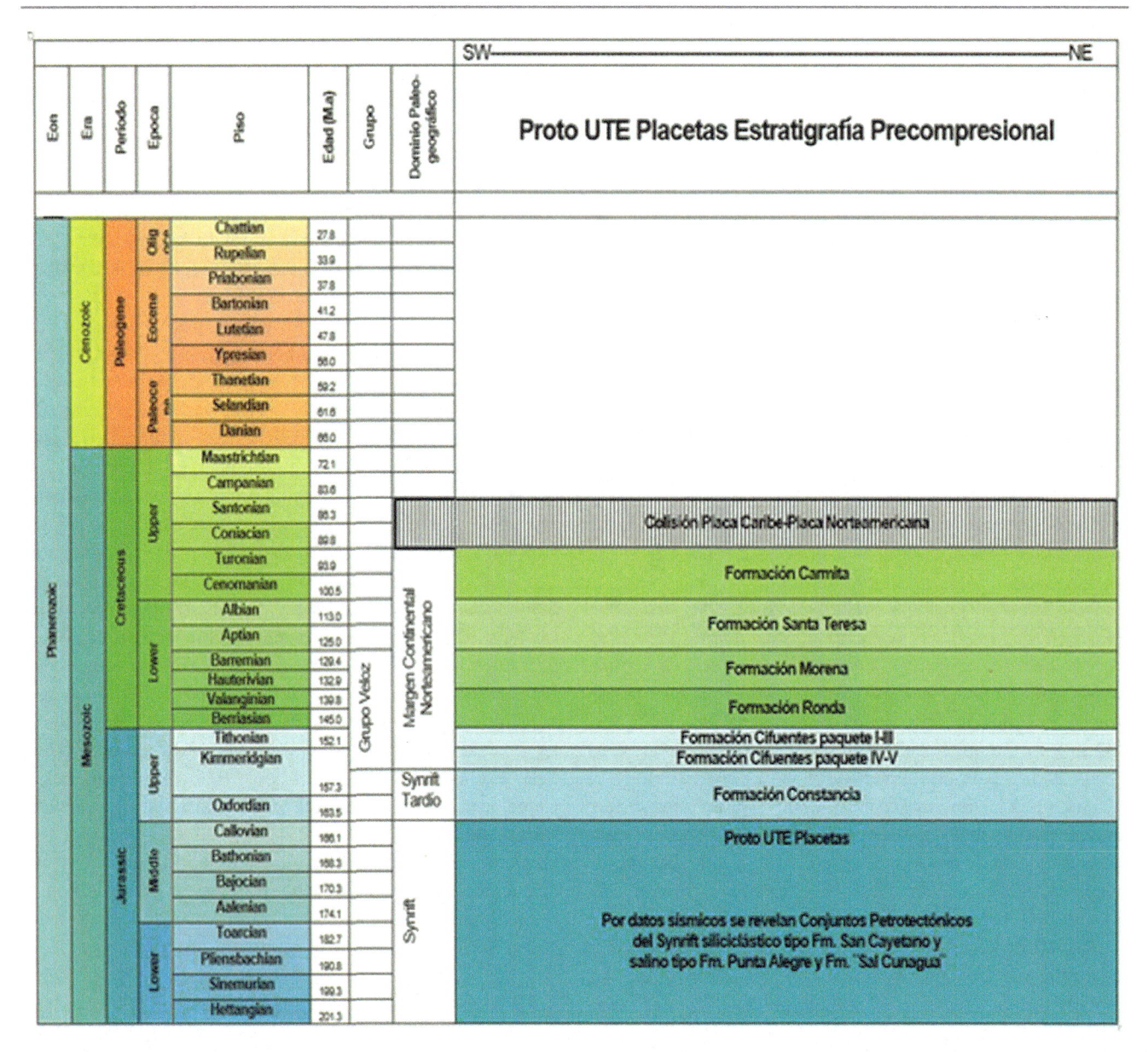

Fig. 4.9 Original position of lithostratigraphic units of the North American continental margin and its substrate of synrift in placetas proto TSU (Precompresional)

Ronda, and Morena formations. This proposal is due to Sánchez Arango and Attewel (1993). See Fig. 4.11.

Upper Jurassic Late Kimmeridgian—Tithonian, Cifuentes Formation of the Veloz Group

The Cifuentes Formation is the oldest formation of the Veloz Group. It has been informally divided by oil geologists Sánchez Arango and Attewell (1993); Fernández Carmona [26] and, Linares [27], in five packages that have been numbered from top to bottom with Roman signs (I–V). They are well represented in the subsoil of the northern provinces of Havana, Mayabeque, and Matanzas and in some outcrops in Central Cuba.

Packages V and IV (Late Kimmeridgian)

To the Late Kimmeridgian correspond the "packages V and IV" of the Cifuentes Formation, attributed to the Placetas TSU and are registered in several wells between Guanabo and Varadero. The "Package V" of the Cifuentes Formation was deposited in shallow waters. The main microfacies are micrites (47–60%) and in a lesser proportion wackestone of bioclasts and peloids and fortuitously of intraclasts or peloids. Occasionally, there are grainstone microfacies of intraclasts and peloids, sometimes silty. The secondary microfacies are dolomite, gypsum, and anhydrite, which increase toward the region of the Yumurí deposit.

Fig. 4.10 Radiolarian silicites, clays, and tuffites of the Santa Teresa Formation in the locality of La Sierra, south of Sierra Morena, province of Villa Clara

This package is very well represented in some wells of the Boca de Jaruco Oilfield. The biofacies are represented by the association of *Globochaete alpina, Didemnoides moreti,* and algae. The "Package IV" was also deposited in shallow waters. The microfacies are very similar to those of "Package V," predominating calcareous mudstone (micrite 45.6–48.5%) that may have a clay content and variable sandy–silt fraction. There are microfacies subordinated to wackestone of bioclasts or peloids and pellets. The age of this package is assigned to the Late Kimmeridgian, on the basis of *Favreina salevensis* and *Globochaete alpina*. This lithological set is detected in the oilfields of Vía Blanca, Boca de Jaruco, Yumurí, and Varadero. So, the sedimentation environment for the two packages is the same: restricted platform neritic, shallow waters with a depth of 0–50 m, with moderate circulation indicated by scattered or agglutinated pellets and peloids and the few strata of fine siliciclastic. The presence of *Favreina*, is conditioned to more protected areas, with low energy, to be related to microfacies of calcareous mudstone. Note the similarity between these deposits and those of the San Vicente Member of the Guasasa Formation of the province of Pinar del Rio.

Package III (Lower Tithonian)

These are rocks formed in deep waters, although not as much as those in "packages I and II," since they represent the transition between shallow-water facies ("IV and V" packages) and deep water ("packages I and II"). The predominant facies corresponds to that of calcareous mudstone that gradually or abruptly transitions to bioclastic wackestone, whose third part is composed of radiolarians. The rocks occur finely banded.

The most conspicuous fossils are *Saccocoma* sp., *Aptychus* sp., and Radiolarians. It is well characterized in the Guanabo well No. 19, in several wells of the Boca de Jaruco deposit, and in wells of the Yumurí and Puerto Escondido fields.

Package II (Middle Tithonian)

Lithologically, it is similar to the previous one, but with few silicites. As main facies, fossiliferous calcareous mudstone and bioclastic wackestone are distinguished. This set is stratified in flyschoid form due to the intercalations of fine bituminous argillites. Dolomites have been described secondarily.

The fossils that allow to date the packet as Middle Tithonian are *Chitinoidella boneti, Dobeniella bermudezi,* and *Saccocoma* sp. It is well characterized in wells of the Boca de Jaruco and Yumurí fields.

Package I (Upper Tithonian)

It is a sequence that in its highest part, transitions to the Ronda Fm. By the degree of study, or because it can constitute condensed levels with the Ronda Fm., its occurrence is erratic, that is, it does not always appear in the wells. It is

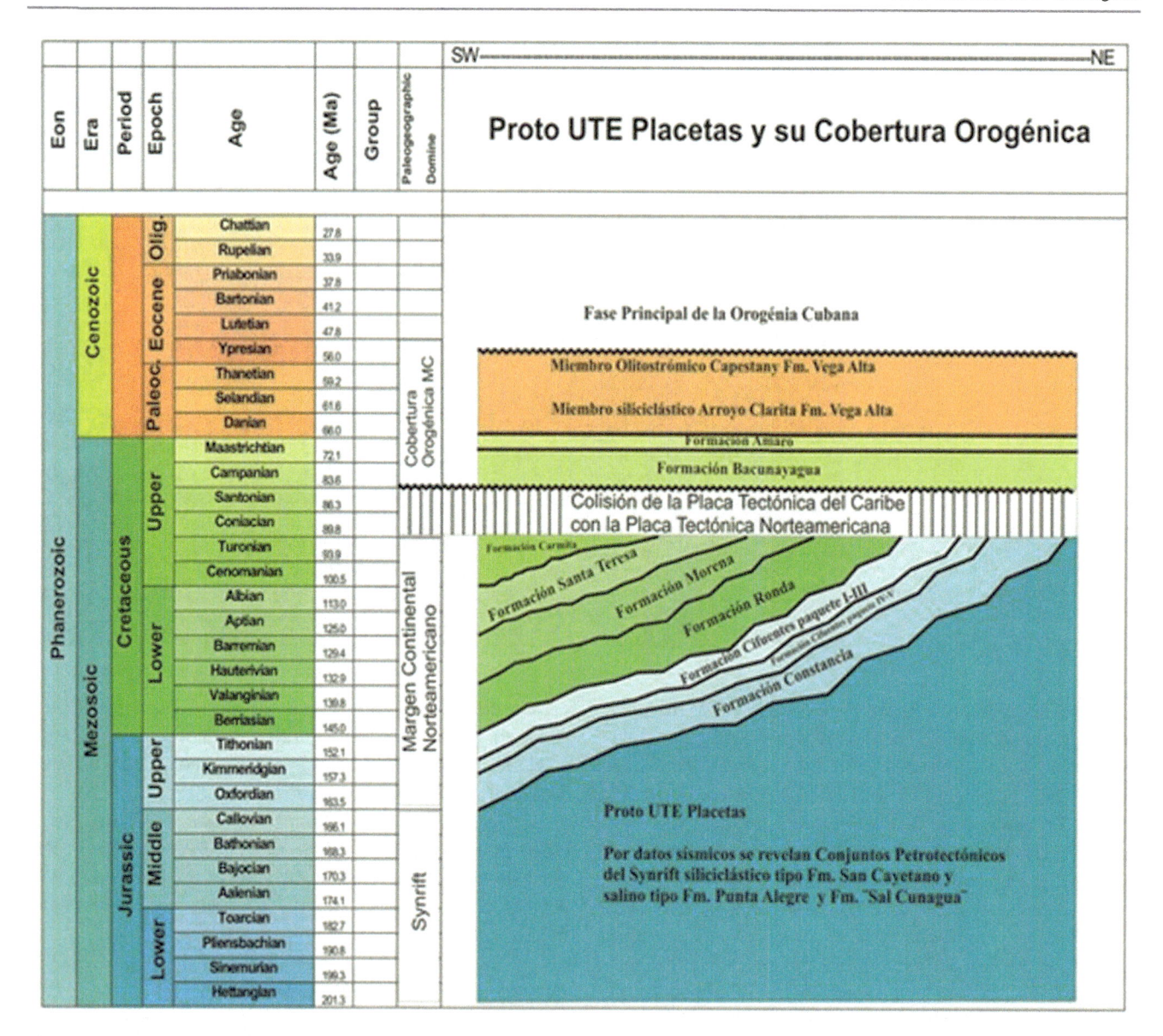

Fig. 4.11 Scheme of Placetas proto TSU and its orogenic coverage after the collision of the Caribbean plate with the North American one. Stratigraphic position of the Bacunayagua formation

also true that there are few index fossils that define age. The proportion of bioclastic wackestone is almost double that of calcareous mudstone. Subordinately, microfacies of radiolaria packstone happen and sporadically one of the dolomites. The biostratigraphic events that fix the Upper Tithonian age are defined by several species of the genus *Crassicollaria* and the large forms of *Calpionella alpina*. It has been possible to distinguish at Puerto Escondido wells No. 2 and some wells of the Yumurí deposit. The rocks of the Upper Tithonian indicate an upper batial paleoenvironment, with depths of the sea below 200 m, in pelagic and hemipelagic conditions. The upper part of the Cifuentes Fm. represents the radiolaric facies that accuses the maximum level of marine transgression of the coastlines, corresponding to BS 138 of the global eustatic charter.

Lower Cretaceous Neocomian

To Neocomian refers the Ronda and Morena formations of Veloz Group.

Berriasian–Valanginian, Ronda Formation

It consists of alternations that include microfacies of calcareous mudstone, bioclastic wackestone, cherts, and argillite. Occasionally, these rocks are dolomitized and recrystallized. Although the paleoenvironment is of deep waters, three microfacies have been recognized for their study, attending to the evident differences of lithofacies with characteristics of oil storage rocks. According to the presence of fossils, three intervals are defined based on the Calpionellid biozones: the lowermost *Calpionella alpina*, indicates the Berriasian, above that of *Calpionellopsis oblonga* and on the roof of the unit the *Calpionellites darderi* that marks the upper part of the Valanginian. The Ronda Fm. has been crossed by several deep wells from the Vía Blanca, Boca de Jaruco, Puerto Escondido, and Pozo Canasí 1-X fields. In the subsoil, between the area of Yumurí and Corralillo, the similarity between the Sumidero (Pinar del Rio) and Ronda formations is very marked, including the same types of black radiolaric cherts and micrites and biomicrites banded with siltstones and bituminous siltstones, similar to those that emerge in the Caimito River through the area of Mil Cumbres in Pinar del Río.

Hauterivian–Barremian, Morena Formation

Until February 2000, this unit had not been reported in the wells of the north of the Havana–Matanzas region. In the Seboruco No. 2 well in two intervals: 2550–2800 m and 2998–3038 m (it is an inclined well) the Morena Fm. was announced in that year. In the first interval, there are abundant white and gray, clayey limestones, cherts of black, light gray, greenish colors; they are frequently radiolarians. Also described are mudstones, pyritized light greenish claystone; black lamellar argillites. Sliding phenomena (Slickensides) are observed. In the second interval, calcareous mudstone, frequently laminated argillites and cherts occur. Between 2600 and 2800 m, abundant Nannoconus species were reported, where *Nannoconus steinmanni, N. globulus, N. colomi, N. bermudezi, N. wassalli*, radiolarians, and *Colomisphaera* cf. *C. heliosphaera* recognized, confirming the Hauterivian–Barremian age for this formation. It is well developed in the Guanabo, Vía Blanca, Boca de Jaruco, Yumurí, Puerto Escondido, and Varadero oilfields in the northern provinces of Havana, Mayabeque, and Matanzas. The notable decrease of the turbulence, as well as the tight lamination of the rocks, indicate conditions of greater tranquility in the deposition, that is to say, greater distance of the hydrodynamic influence of the carbonated platform, in relation to the coeval sequences of the Camajuaní and Colorados TSUs.

Aptian–Albian, Santa Teresa Formation

The Santa Teresa Formation in the subsoil of the north of the provinces of Havana, Mayabeque, and Matanzas and in its outcrops in the north of the province of Villa Clara (Fig. 4.10), has the same characteristics as in the outcrops of the provinces of Pinar del Río and Artemisa. Exotic outcrops exist in the Yumurí Valley in the province of Matanzas. In the subsoil of the oilfields of Varadero and the northern provinces of Mayabeque and Matanzas, in addition to the silicites, clays, and siltstones, in the upper part of the unit, limestone occurs with the presence of fossils: *Nannoconus* sp. and planktonic foraminifera such as *Globigerina infracretácea, Ticinella* sp., and Hedbergellidae of age Aptian–Albian. As a general rule, in the areas studied by wells, the scales of the Placetas TSU, show a level of the takeoff of thrusts by the base of the Santa Teresa Fm. Its existence is also known in areas such as Cantel, Varadero Sur, and Guásimas, arranged in scales that are expressed by the repetition of the rocks of the Santa Teresa and Carmita formations, well separated or in imbricate forms. The Santa Teresa Fm. is evident among others, in the following wells: Varadero No. 201 of 1853–1500 m, together with the Carmita Fm. in the Varadero No. 41 between 1525 and 700 m; Varadero No. 31 between 1284 and 910 m; Martí No. 5 interval 3092–2616 m and; Bolaños No. 1 between 2647 and 2562 m. In this region, it consists mainly of siliciclastic turbidites, which are arranged as rhythmic alternations of radiolaric silicites, clays, and limonites; subordinately marl and mudstone. Occasionally, fine-grained quartz sandstones and siliceous cement occur. A characteristic of this unit is the pronounced folding of its layers, forming flattened folds, as well as small faults that cause displacement and fracturing of the rocks.

This is observed both in the province of Pinar del Rio and in Central Cuba (Figs. 4.4 and 4.10). The dating of this unit by radiolarians is difficult. Diagnostic fossils such as *Nannoconus* s.l., *Ticinella* sp., *Muricohedbergella* sp., *Schackoina* sp., *Macroglobigerinelloides* sp., come from interspersed mudstones; such planktonic foraminifera allow assigning Aptian–Albian age for Santa Teresa Fm., where these rocks are best recognized is in the depth of the Yumurí deposit. There, in fact, are abundant radiolarian silicites and siliceous–clayey rocks. The most representative wells are the Yumurí No. 1, 7A, 8, 12, 18, and 20. It has also been possible to differentiate in the Boca de Jaruco No. 3 well (core No. 5).

Upper Cretaceous Cenomanian–Turonian, Carmita Formation

The origin of the name comes from the old sugar mill Carmita, currently, it is the deactivated Luis Arcos Bergnes from the province of Villa Clara. But in reality, the outcrops there are poor and this unit is better studied in other locations and by oil wells. It is preferred to observe its hypoestratotype that is a profile to the southwest of Loma Berraco, to about 3 km to the south of the town of Sitiecito, to the southwest of the city of Sagua la Grande, province of Villa Clara. In this cut, it lies on the Santa Teresa Fm. It is established by L Dodekova and V Zlatarski, in Kantchev et al. 1978. Before indicating this unit in the Northwest Belt of Hydrocarbons of Cuba, it is good to highlight the opinions of Linares [27], already indicated on its presence in western Cuba. This author said that the Carmita Formation has been recognized and mapped recently in the province of Pinar del Río. In Sierra del Rosario, the deposits corresponding to the Carmita Formation were initially assigned to the informal unit "Member of limestones and silicites" of the Buenavista Formation. In the detailed geological mapping work, at 1:50,000 scale, by Burov, Martínez, and collaborators, the Santa Teresa, Carmita, Moreno, and San Miguel formations were included in the Buenavista Group. They are well reflected in the San Cristóbal and Mariel topographic sheets and in the descriptions of the report. Previously, Cobiella Reguera, in a geological study of the Sierra del Rosario, in the area between Soroa and Cayajabos, highlighted the presence of the Santa Teresa and Carmita formations in the referred area. Pszczolkowski, published the presence of the Santa Teresa and Carmita formations in the north and south sequences of the Sierra del Rosario, including them in the Buenavista Group. In the aforementioned work, a column of Cretaceous rocks exposed in San Diego de Tapia between Quiñones and La Palma is shown, describing a section of the Carmita Formation, composed of the alternation of micrites, violet clayey limestones, interspersed by cherts and calcareous schists greenish.

Other reference sections are located in the vicinity of Polier hills, on the road between Soroa and San Diego de Núñez; to the south of Quiñones (or Luis Carrasco) near the San Cristóbal–Bahía Honda Highway and, in the San Diego de Tapia section south of Rancho Lucas. In general, the description of the Carmita Formation in the Sierra del Rosario does not differ from that recognized in Central Cuba and in some areas of the Sierra de Camaján in Camagüey. The Carmita Formation was recorded in Martín Mesa No. 2 and 7 and in Chacón No. 2 wells, also observed as olistolites in the Manacas Formation. Regarding its presence in the subsoil at the NBHC, the oil wells register the Carmita Formation with the same characteristics that have been reported on the surface. It is an eminently carbonated formation. It consists of several types of well-stratified limestones in thick layers predominantly gray, violet, and greenish colors. Siltstones and calcareous sandstones with fine grains are intercalated to the limestones. Sometimes intraformational breccias occur, although not all breccias are of these characteristics; frequently the breccia is due to fracture due to tectonic causes. A distinctive character of the Carmita Formation is the presence of some thin layers of amber or black cherts that sometimes form nodules. Between the limestones, the bioclastic, radiolaric and nannoconic wackestones prevail. The index microfauna is composed of *Archeoglobigerina* cf. *A. cretacea, Globotruncana* ex. gr. *linneiana, Favusella washitensis, Rotalipora evoluta, R. appenninica, Rugoglobigerina* sp., *Schackoina cenomana, S. multispinata, Praeglobotruncana* spp., *Rotalipora* spp., *Muricohedbergella* spp., *Nannoconus* spp., Pithonellidae, Stomiosphaeridae, and abundant radiolarians for a Cretaceous age Lower Albian–Upper Turonian Cretaceous; more likely Upper Cretaceous Cenomanian–Turonian (VB 5 N6-8: 1425–1580; VB 20 N2-5: 1420–1500). In the area we are describing, the Carmita Formation has been traversed by both the deep wells of the Vía Blanca field (VB 8: 1715–1825, VB 10: 1872–2186, VB15: 1572–1702, VB 31: 1438–1575; VB 35: 1248–1445; VB 100: 1430–1660) as well as those of the Guanabo field. It is poorly developed towards the Puerto Escondido, Seboruco, and Yumurí oilfields. As more representative wells among others, we can list Guanabo 19 (2765–2920 m) and Boca de Jaruco 29 and 35.

It has discordant stratigraphic relationships with overlying formations of the Upper Cretaceous (VB 5: 1425, VB 8: 1715, VB 10: 1872, VB 15: 1572, VB 20: 1420, VB 28: 1790, VB 31: 1483; VB 35: 1445; VB 100: 1430) and with the Vega Alta Formation of the Lower Paleocene-Eocene with which it can eventually have tectonic contacts (VB 5: 1580, VB 8: 1785, VB 10: 2186; VB 28: 1790; VB 31: 1575; VB 35: 1445; VB 100: 1660).

Campanian–Maastrichtian at the NBHC, Bacunayagua, and Amaro Formations

Bacunayagua Formation

The interaction of the Caribbean Tectonic Plate with the North American Tectonic Plate occurred between the Coniacian and the Campanian. The process lasted approximately 6.2 million years (International Chronostratigraphic Chart, International Commission Stratigraphy IUGS, 2016). It has been shown that between the Campanian and the Maastrichtian the Bacunayagua Fm. was deposited in the southern part of the North American Continental Margin [32].

The lithological composition given by the studied samples of both surface and subsoil relates: conglomerates and arcosic breccia-conglomerates, subarcosic, arcose, lithic arcose, and lithic sandstones (Fig. 4.11). Also noteworthy is the predominance of lithoclastic carbonate rocks, where all the textural types of limestone including crystalline carbonate are represented, as well as the abundance of the accessory minerals of quartz, potassium feldspar, and plagioclase. So the so-called arcosic rocks are not exclusively the predominant lithology since there are many carbonate and other types. This suggests that this set was deposited on the rocks of the North American Continental Margin and are the result of a source of contribution evidently constituted by some of the units of that Margin and its crystalline basement, probably the carbonated rocks that today are grouped in the Placetas TSU and the so-called "Granitos Rio Cañas" and the "Mármoles Flogopíticos de Sierra Morena."

The fossiliferous association shows fossils of various ages. The age of the carbonate clasts can be Upper Jurassic–Upper Cretaceous, Tithonian–Berriasian, Tithonian, Kimmeridgian–Berriasian, Lower Cretaceous Valanginian–Berriasian, Lower Cretaceous Aptian–Upper Cretaceous, Lower Cretaceous Albian–Upper Cretaceous, and Lower Cretaceous Albian–Upper Cretaceous Cenomanian. The age according to the fossils of the matrix is Upper Cretaceous (Campanian–Maastrichtian). The rocks of the Bacunayagua Formation were deposited in the southern edge of the Paleogeographic Domain of the North American Continental Margin, being part of its orogenic coverage.

Upper Maastrichtian, Amaro Formation

In the oil wells of the NBHC is where this lithostratigraphic unit has been studied. It is a unit partially similar to the Cacarajícara Fm. of western Cuba. They are rocks of fragmentary carbonate nature, essentially breccia-conglomerate and fine conglomerates that include poorly selected fragments of different limestones, usually banded and bituminous and, cherts. The matrix is clayey–calcareous and sometimes with a slight secondary process of dolomitization. There are also calcarenites levels. The limestones that make up the fragments are described as floadstone (intramicrudite) and mudstone of extraclasts and lithoclasts. To the Boca de Jaruco field, the extraclasts are mostly from the Cifuentes Fm. (55%), of the Ronda Fm. (40%) and the rest can be attributed to the Carmita and Santa Teresa formations. In some areas, almost all the fragments come from the Ronda Fm. So, the age of the fragments of limestones, oscillates between Kimmeridgian to the Cenomanian, but in the matrix, they are foraminiferous such as *Abathomphalus mayaroensis, Omphalocyclus macroporus, Globotruncanita stuarti, Contusotruncana contusa, C. fornicata, Vaughanina cubensis, Lepidorbitoides* sp., and *Sulcoperculina* sp., indicating upper Maastrichtian age.

Paleogenic Deposits as Sealant Rocks, Vega Alta Formation

The Vega Alta Formation is a policomponent olistostrome. It was formed in a foreland basin and on the substrate of the formations that form the Placetas TSU, constituting an important element for the oil systems of the Northwest Belt of Hydrocarbons of Cuba, since it is considered the regional seal that covers with a great discordance erosive, the producing horizons of the Cretaceous and Upper Jurassic of the North American Continental Margin, from the north of the provinces of Havana, Mayabeque and up to Matanzas. On the surface, it has been observed in small areas in the Cantel, Martí, and La Teja structures. It forms broad strips from the south of Corralillo, in the eastern part of the province of Matanzas, to the north of the provinces of Villa Clara and Sancti Spíritus. It is estimated that the area occupied by the Vega Alta formation on the surface must be much larger than the one shown in the current geological maps and that it extends even along the length of the Cañas river valley, where it is partly covered, by Quaternary alluvial deposits. Two members have been proposed: The Arroyo Clarita Member, for the lower part of the unit and, the Capestany Member, for the chaotic part. The Arroyo Clarita Member is a finely stratified carbonated sandy clay loam where greenish-gray and brown shades, gray loams, thin layers, and laminar shales are interspersed, all carbonated. To the top, polymictic sandstones appear in layers of 10–12 cm between the shales, at the end of the visible section the cream-colored, calcareous shales predominate, where the silicified limestone blocks of greenish color and sandstone begin to appear, all surrounded by loamy clayey material. The fossils of this member indicate Lower Paleocene Danian age–Upper Paleocene.

The Capestany Member is in general a chaotic sequence, with a policomponent olistostrome property, comprising blocks and fragments of limestones of different types and ages, silicites and serpentinites, volcanic and vulcanomictic rocks, breccias, and metamorphic rocks. Its matrix is predominantly clayey greenish gray, with intercalations of fine sandstones, siltstones, and polymictic conglomerates.

In the subsoil in the oilfields discovered on the north coast of the provinces of Havana, Mayabeque, and Matanzas, this formation has been found in all the wells and, is composed, as well as on the surface, by chaotic sequences with clasts and calcareous blocks of Cretaceous and Jurassic formations, of polymictic sandstones, and siliceous-clayish greenish rocks, clays, siltstones, and calcarenites.

The age of this sequence is Lower Paleocene-Eocene with abundant redeposition of the Upper Jurassic and Cretaceous. The thickness of this unit, both on the surface and in the wells, can be greater than 200 m. It is comparable with the Manacas Formation of western Cuba.

4.4.1.13 Paleocanal Cayo Coco TSU and Remedios TSU [33]

Cayo Coco TSU

This unit testifies to the development of a carbonate scenario, whose sedimentary thickness from the Upper Jurassic to the Cretaceous, reached about 3600 m. The Cayo Coco TSU models a paleocanal of carbonate and silicite rocks from the Aptian Cretaceous to the Campanian–Maastrichtian. There are no traces of volcanism and yes strong dolomitization processes. The Cayo Coco and Perros formations, common to the Cayo Coco and Remedios TSUs, are recognized; the Guaney Fm. at the Coco TSU and Palenque Vilató and Purio formations at the Remedios TSU. Crowning both TSUs, the Sagua, Jumagua, Vega, Caibarién, and the Turiguanó formations. In very local areas, there is a carbonate-dolomite slope breccia called the Florencia Formation.

Kimmeridgian-Tithonian at Cayo Coco and Remedios TSUs

Cayo Coco Formation [33]

It is common to the Remedios and Cayo Coco TSUs. Age it is Kimmeridgian to Tithonian. It is a thick sequence of micrites and pelmicrites, crystalline brownish dolomites, with frequent intercalations of white and gray anhydrites, towards the base of the formation. The dolomites have excellent porosity, with values between 15 and 24%. The holoestratotype of this lithostratigraphic unit is at Cayo Coco No. 2 well from 2153 m to the final depth of 3202 m. The real thickness drilled was 870 m, making corrections based on 30° inclination of the layers. In the Gloria No. 1 well, north of Esmeralda, province of Camagüey, the actual thickness reached 1,400 m, without crossing the entire unit. The microfauna determined in drilling controls is composed of *Favreina salevensis*, *Favreina* sp., *Saccocoma angulata*, and *Nauticolina* spp., the oryctocenosis represents a neritic paleoenvironment, from the internal type (intertidal or post-reef), according to the presence of Favreina sp. to external (Saccocoma). The oolitic zones also suggest somewhat agitated waters. The presence of stratified anhydrite indicates evaporitic conditions, where the deposits were formed by precipitation of the sulfates from concentrated solutions in a basin, whose waters were of restricted circulation. It is considered that the rocks that underlie the Cayo Coco Fm. are those of the Punta Alegre Fm. that includes the "Cunagua Salt" of the Domain of the Late Synrift. The roof of the formation has been defined by the appearance of the stratified anhydrites in the highest part of the section. The wells that penetrated the Cayo Coco Fm. are Cayo Fragoso No. 1, Cayo Coco No. 2, Collazo No. 1, Gloria No. 1, and Cayo Felipe 1-X.

Neocomian at Cayo Coco and Remedios TSUs

Perros Formation [33]

They are light gray to grayish dolomites—grayish, dolomitized limestones, stratified in 2–6 cm thick layers, and very fractured. Relictic oolitic and peletal structures are observed. The lower part that appears contains coprolites of Favreina sp. that indicates that this organism manages to extend to the Neocomian when the conditions of deposition are propitious. In the upper part of the unit, only *Nannoconus steinmannii* of Upper Tithonian age–Aptian have been described. The section is laminated, the cause being the changes of colors and the differentiation in the size of the grains. Some areas of dolomitic and micritic breccias are described. The thickness measured on the surface is about 500 m, but in the wells, it is more than 1000 m. For example, in the Mayajigua well in the province of S. Spíritus, it is 1500 m, even considering the inclination of its layers with an average of 35°. The lower limit of the Perros Fm. is concordant with the Cayo Coco Fm. and the upper one is transitional to the sediments of the Middle Cretaceous. There are good outcrops in the Sierra de Jatibonico through the Lomas de Mabuya, provinces of Ciego de Ávila, and S. Spíritus.

In the Gibara No. 1 well, in the area of Velasco, in Holguín, in the interval 2150–4527 m a dolomitic sequence comprising the cores of No. 17–42 is described. Lithologically they are represented by rhombohedral grain dolomites from fine to coarse, with fine fractures and stylolites sealed by calcite and old bitumen. In some cases (cores of No. 21–39), they reach the brecciation, even in cores such as No. 22, 23, and 29, where pores and caverns are observed. The fossils are scarce and very recrystallized, where the presence of milliolids, *Opthalmidium* spp., Ostracod molds, and Textulariidae are determined. The interval in question is related to the Perros Formation, which is the oldest unit known in the Remedios TSU. In the hills of Mabuya, in the province of Ciego de Ávila, were reported in several samples of this formation the age Upper Jurassic, Upper Tithonian–Lower Cretaceous Neocomian, Valanginian, based on the fosilifera association: *Favreina salevensis*, *Globochaete alpina*, *Tintinnopsella carpathica*, and *Calpionelids* s.s. In calcareous fragments of the gypsum breccia of the Punta Alegre Formation, these fossils are also described in rocks of the Perros Formation.

Aptian–Maastrichtian, Rocks of Paleocanal Cayo Coco TSU

Guaney Formation (Iturralde Vinent y Roque, in [34])

The type area is located around the Loma Guaney, Loma de Miranda, and Loma Paso Abierto elevations, northeast of the town of Esmeralda, province of Camagüey. In the Gloria No. 1 well, the rocks of this lithostratigraphic unit are well described. The Guaney Formation is reported in Collazo No. 1, Manuy No. 1, Cayo Coco No. 1 and No. 2, Felipe 1-X wells, and in Gloria No. 1 well, located in the northern part of the provinces of Camagüey and Ciego de Ávila. It was geologically mapped in Sheets B2 of Geological Map of Camagüey scale 1: 100,000 and, Geological Map of the Republic of Cuba, Sheets 22 (F 18–9) CAMAGÜEY, and 21 (F 17–12) CIEGO OF ÁVILA, scale 1: 250,000, IGP/ACC (1988).

This formation is conventionally divided into a lower section (Aptian–Cenomanian) and an upper one (Turonian–Maastrichtian). The lower sequence is cut by Collazo No. 1, Manuy No. 1, Cayo Coco No. 1 and No. 2 wells, and Gloria No. 1 well. The upper sequence is crossed by the Cayo Coco No. 1 and No. 2 wells, although only thin sections of the top cores could be studied. In spite of this, the information provided by the Cayo Coco No. 1 and No. 2 wells and the Gloria No. 1 well, allow us to know quite accurately the evolution of a paleocanal. In the Felipe 1-X well, several microfacies were established, which delimited nine packages with their microfaunal and lithological characteristics. In "package 1," the limestones are very clayey, marls, and limestones with abundant chert intercalations are common. The clay content is somewhat lower in "packages 2 and 3," but already in "packages 5 and 6," there are microfacies with intraclasts and peloids, safe indicators of higher energy environment. In the limestones of "packages 2 to 9," there are few stylolytic fractures, commonly stained by dead bitumen and fine lamination is frequent due to bituminous impregnation. In "Package 5," fine bands of syngenetic bitumen were reported. The Felipe 1-X well confirmed that the Aptian was the period of the first great retrogradation of the platform, that is, it represented an important regional transgression, with a maximum flood surface (MFS). In the seismic studies, it is observed as the main event of high and continuous amplitude, from the Great Isaac well through the Doubloon Saxon sounding to the Block L. There is no evidence to establish discordances within the pellets of the Guaney Formation. The uniform sedimentological rhythms in it, rather indicate continuity in its sedimentogenesis. The lack of biomarkers in the Coniacian and Santonian is explained as condensed intervals, although in the Middle Cretaceous there were probably slumping currents evidenced by a certain amalgamation of the fossil associations, like several Albian species of Nannoconus in the Cenomanian–Turonian.

The study of the succession of the fossil microfauna allowed to differentiate ten biozones of planktonic foraminifera from the Aptian to the Cenomanian, both in the Gloria No. 1 well and in the Cayo Coco No. 1 and No. 2 wells. According to the fossils, the age of the layers is therefore established as Lower Cretaceous (Aptian)–Upper Cretic (Maastrichtian), deposited in a sedimentation environment that gradually transitions from external neritic to batial. The thickness exceeds 670 m.

The Guaney Formation lies stratigraphically concordant on the Perros Fm. It is covered discordantly by the Sagua and Venero formations.

The Sagua, Jumagua, and Vega formations are common to the Coco, Remedios, Camajuaní, and Colorados TSUs and were deposited after a great unconformity. Therefore, they will be described below as coverage for all.

Sagua Formation (Pardo, in [35])

This formation is one of the most distinctive regionally and its outcrops extend from the provinces of Matanzas to S. Spíritus to be exhibited again in the Sierra de Cubitas in Camagüey and in the area of Gibara in the province of Holguín. (In Gibara, as in Camagüey, similar rocks are attributed to the El Embarcadero Formation). The Sagua Formation is a slope deposit, adjacent to the area where the carbonate bank of the Cretaceous and partially the rocks of the deep-water basin were eroded. The composition of this lithostratigraphic unit is similar to that described in other regions where it is also covered by the Remedios, Colorados, Camajuaní, and Cayo Coco TSUs. These are calcareous breccias and calcareous conglomerates with some calcarenite horizons. The breccias are fragments very variable in size, while the conglomerates reach the fine fraction. They include in their fragments, micrites, biomicrites, silicified micrites, oolitic limestones, and cherts. In the fragments, the redeposited fossils of the Lower and Upper Cretaceous of the shelf and the Lower Cretaceous of pelagic facies predominate. Calcarenites are fine to coarse grains, with siliceous-carbonated cement, sometimes somewhat dolomitized. The matrix of the breccias is essentially marly and that of the conglomerates micritic-clay. In addition to numerous redeposited fossils of bank and pelagic facies of various ages, including the Paleocene, in the micritic-marly matrix are related: *Morozovella Formosa, M. aragonensis, M. wilcoxensis, Acarinina* spp., *Morozovella* spp., *Discocyclina* sp., *Amphistegina lopeztrigoi, Boreloides cubensis,* and *Tribrachiatus orthostylus*. This confirms the age of the Sagua Formation which is Lower Paleocene–Eocene. This sequence is identified in the Colorados No. 1 well between 860 and 960 m. Also with a considerable thickness is detected in the Colorados No. 2 well between 1030 and 2093 m, covering the cores from number 5 to number 28.

The lower contact is discordant with a "hardground" surface of the roof of the Colorados Formation and the Lutgarda Fm. The upper limit is discordant with the Oligocene sediments of the Neoauthocton, described in the Colorados No. 1 well or with the Miocene in the Colorados No. 2 well.

Jumagua Formation

It is recognized in the Cayo Coco wells and some of Punta Alegre (Collazo No. 1 and Tina No. 2), also by the Sierra de Jatibonico in S. Spíritus-Ciego de Ávila and in the Sierra de Cubitas in Camagüey. It includes carbonated breccias, microcrystalline gray-bluish micrites, with intercalations of argillites and shales in thin-layered sheets and lenses of diagenetic cherts 1–10 cm thick; calcarenites with abundant organic detritus, especially of pelecypods. The shales have carbonaceous organic matter. Towards the upper part of the unit, somewhat arcosic sandstones cemented by calcium carbonate with greenish-gray minerals occur. In the argillites, *Discoaster lodoensis* and *Tribrachiatus orthostylus* of the Lower Ypresian NP-12 biozone have been described. In the limestones *Acarinina topilensis, Acarinina pseudotopilensis, Pseudohastigerina wilcoxensis, P. micra, Morozovelloides crassatus,* and *Morozovella aragonensis* are reported. Thus, the age of the formation is Lower Eocene to Early Middle Eocene, from the biozone of *Morozovella aragonensis* to *Acarinina bullbrooki*. The paleoenvironment is deep neritic, representing a transgressive event. It lies discordantly on the Sagua Formation. The lower contact is discordant or tectonic with older formations. The thickness is about 120 m. It has been recognized in Marbella Mar No. 2 well, forming scales imbricated with the Sagua and Vega formations. Very similar rocks appear in the Sierra de Jatibonico at the Campismo de Boquerón and the Valle del Alunado.

Vega Formation (Pardo, in Bronnimann y Pardo 1954)

With the name Vega Formation, an alternation of sandstones and clay loams with intercalations of conglomerates was determined. Because this unit has horizons with characteristics of regional seal, some localities will be proposed in this work, for a future proposal of hypoestratotype, following the norms of the International Stratigraphic Nomenclature Code, additionally, its description will be expanded with recent determinations. In the place Los Barriles, Florencia municipality, province of Ciego de Ávila, there is an excellent outcrop of the Vega Formation.

It is an outcrop of a flysch composed of siltstone and dark gray argillites and brown–brownish sandstones with abundant clay matrix containing abundant radiolarians and foraminifera keel from the Lower Eocene; among them, several species of the genera *Morozovella* and *Acarinina* are observed.

Additionally, the Vega Formation has excellent exhibitions in the Valle del Alunado in the province of S, Spíritus, in the area northeast of Vega Alta and, south of the Loma de Sinaloa in the province of Villa Clara, where the most argillaceous rocks emerge.

In the Valle del Alunado, the sandstones have abundant components of the destruction of the Cretaceous Volcanic Arc, lying stratified with greenish-gray limolites, which are polymictic, with grains of plagioclase, amphibole, pyroxene, quartz, and, in a lesser proportion, epidote, potassium feldspar, biotite, and chlorite, with a poor, carbonated-hydromicaceous cement of porous type. The grains are angular, semi-rounded, and rarely rounded. The argillaceous rocks are those that more fossil fauna has that allows to date it like Paleocene–Middle Eocene.

Florencia Formation (Proposed by Hatten et al. in 1958)

It is a formation of little development, it constitutes a polygenic breccia with clasts of several formations among them Alunado, Sagua, Jumagua, Perros, Guaní, and Mabuya. It presents slight porosity by dissolution of the microgranular type and good porosity by organic dissolution of coral detritus. The clasts of these breccias are 1. Intrabiosparite with Operculinidae and Miliolidae 2. Oobiosparita with agglutinating forms and Gastropods and 3. Biosparite with *Discocyclina* sp., *Distichoplax,* and Nummulitidae. The cementizer contains *Acarinina bullbrooki* from the Early Middle Eocene. In the breccia, there are large fragments of dolomite and dolomitized limestones. There is an excellent exhibition of this formation in the quarry for construction materials located in Los Barriles, near Florencia. There, it lies discordantly on stratified carbonate rocks of the Remedios TSU. In the contact part, a synsedimentary breccia with a biomicritic matrix is observed, showing bitumens. In the clasts, *Heterodictyoconus americanus, Pseudophramina havanensis, Ethelia alba,* and Amphisteginidae were determined. The cement contains *Globigerina* sp. and *Globorotalia* sp. definitely, the age of the Florencia Formation is later than that of the Jumagua Formation, that is, the Late Middle Eocene. The thickness can reach 200 m. The paleoenvironment corresponds to an upper slope, with a water depth of 80–120 m.

Turiguanó Formation (Proposed by Truitt in 1955)

Presents, the same as the Florencia Fm., little distribution. This unit occupies the lower part of the Middle Eocene. There is a cut in a small hill called El Cerrillo located north of Las Lomas de Turiguanó, province of Ciego de Ávila, which is the only place where it is known on the surface. It consists in the base of calcirudites of large fragments composed of dolomites, dolomitized limestones, calcarenites, organogenic limestones, and other rocks. The matrix is scarce fragmentary-carbonated. Above these thick calcirudites, calcirudites of small fragments are presented, interspersed with calcarenites of pink color and, above, limestones of different grayish-brown and whitish coloration.

Caibarién Formation (Ortega and Ros 1931. Redescribed by Popov, in Kantchev et al. 1978)

It is developed to the north of the province of Villa Clara in long stretches between the towns of Remedios and Caibarién and, to the northeast of the latter, province of Villa Clara. This lithostratigraphic unit of a pelagic environment is transgressive both on the platformic rocks of the Remedios TSU and on those of the Cayo Coco TSU, where its lower limit is older. About the Remedios TSU, it is observed crowning the heights of the Sinaloa hill and represents a discordant event.

These are white, lazy biomicrites with fine intercalations of calcareous argillites. The marls and limestones have rounded fragments of silicified limestones of various shapes and silicified lenses. The fauna is of planktonic foraminifera composed of *Acarinina topilensis*, *Globorotalia* spp. and Radiolarians, of Middle-Late Eocene age. The paleoenvironment corresponds to lower batial, with a depth of water between 800 and 1000 m. It can reach a few tens of meters thick. The upper limit is discordant with the Miocene deposits and the lower limit, with the same character, with the Vega Formation and with the units of the Remedios Group. It was reported in the Caibarién No. 2 well, in the interval 30–123 m.

4.4.1.14 Remedios TSU [33]

The formations of the Remedios TSU are very well observed in the Central Cuba region. They are rocks formed on a carbonate platform. The deformations are not as marked as those of the TSUs south of it. The layers tilt smoothly, however, retro-thrusts are recognized and areas with notable dips near the faults. There are no manifestations of magmatism. The deposits are mainly carbonated, practically no silicites exist. Sedimentation has occurred almost continuously from the Upper Jurassic, probably before, to the Upper Cretaceous. The Cenomanian rocks are recognized. The Cayo Coco, Perros, Remedios Group (Palenque, Vilató, and Purio formations) formations are described. In the geological cartography, the Remedios TSU is delimited from the Camajuaní TSU due to overthrust faults and the southernmost outcrops of the Purio, Palenque formations, and the Paleogenic ones Sagua, Caibarién, and Vega [6, 7].

Aptian–Maastrichtian, Remedios Group [36]

It is currently composed of the Palenque, Vilató, and Purio formations. As a group, it is geologically separated in the Geological Map of Cuba 1: 500,000 [6]. In the Gibara Sheet of the Geological Map 1: 250,000 [7], the aggregate of these rocks is mapped as Gibara and Jobal formations, but without doubt, the Palenque, Vilató, and Purio formations are contended in them.

This petrographic set of platform rocks are developed in the form of discontinuous and elongated bands to the north of Cuba, in the provinces of Villa Clara, Sancti Spíritus, Ciego de Ávila, Camagüey, and Holguín.

They are distinguished as bioesparites, micrites, and dolomites. To a lesser degree, calcareous turbidites and very locally laminar limestones. At different levels of the cut of its formations, thicknesses up to 30 m of calcareous breccias with the calcareous matrix are observed.

The fossils of the different formations that compose it allow to date the unit as of the Lower Cretaceous Aptian–Upper Cretaceous (Maastrichtian).

In the deposits that make up the Remedios Group, three fundamental environments can be differentiated: lagoons and retro-reef lows, biostromal banks, and the open sea environment (the least widespread). Adding the thickness of all the units of the group, this ranges between 800 and 2000 m but in some areas exceeds 2600 m.

Early Cretaceous, Aptian–Cretaceous Cenomanian, Palenque Formation

It is developed in Lomas El Palenque, Sierra de Jatibonico, Sierra de Cubitas, and Sierra de Gibara in the provinces of Villa Clara, Sancti Spíritus, Camagüey, and Holguín. The holoestratotype is a profile in the San Pedro ravine, approximately 1 km southeast of the El Palenque quarry, north of the Remedios-Camajuaní highway, and about 6 km west of the town of Remedios, province of Villa Clara. The unit is composed of a diverse group of limestones represented by micrites, biomicrites, intrabiosparites, biopelmicrites, and dolosparites. There are calcarenite levels and breccias with limestone and dolomite clasts, with micrite–bioclastic matrix (Fig. 4.12). The intrabiosparites show packstone textures with a matrix of pellets. The biopelmicrites and microsparites exhibit textures of calcareous peletal wackestone, with numerous fissures and fractures and detritus of rudists. The latter constitutes a stratigraphic horizon Cenomanian age marker in the outcrop of the El Yigre hill and can be correlated with the upper part of the

Fig. 4.12 Carbonate rocks of the palenque formation in the Viet Nam Heroico quarry, sierra de Cubitas, Camagüey province

Palenque Formation in Chambas 1 and Cayo Lucas 1 wells. There are also biomicrites with "bird's eye" structures and calcareous peletal wackestone texture, with numerous stylolites with bituminous argillite. Calcarenites have also been described as bioclastic grainstone, coquines, and stromatolitic contexts, as well as karstic levels and dog tooth topography. The fauna (macro and micro) is abundant, being able to mention the following index forms that argue the age: *Nummoloculina heimi; Dicyclina schlumbergeri, Orbitolina concava texana, Dictyoconus walnutensis, Thaumatoporella parvovesiculifera*, Rudists.

The paleoenvironment, in general, corresponds to intertidal deposits, finding reef patches of rudists that served as a barrier for the lagoons. However, the waves and floods reached the lagoons, because they are interspersed with pelagic sediments rich in intraclasts, as seen in the Blanquizal, Cayo Lucas, and Chambas wells. This event is related to the Cenomanian, as it is verified in the Cayo Lucas well for the presence of *Rotalipora* sp., *Muricohedbergella* sp.

The greater and non-traversed thickness is found in the Cayo Romano No. 1 (1800 m) well, although on the surface, in the Puntilla anticline, 2660 m were measured. The lower and upper contacts are concordant with the Perros and Purio formations, respectively.

Cretaceous Cenomanian–Turonian, Vilató Formation (Iturralde Vinent y Díaz [37])

At the current level of study, it is only recognized in the Sierra de Cubitas in its type area, around the Paso Paredones, Hoyo de Bonet, and the area of Jiquí, in Esmeralda, province of Camagüey, but in future studies can be found in other regions where the Remedios TSU emerges.

In the Vilató Formation, micrites and lamellar biomicrites predominate, alternating with calcirudites lying concordant on the Palenque Fm. Its fossil content is represented by *Dicyclina schlumbergeri, Muricohedbergella* sp., *Praeglobotruncana* sp., *Pithonella sphaerica*, fossil set that argues Cenomanian–Turonian age.

The sediments are of pelagic facies within the Remedios TSU and are only known in the Sierra de Cubitas and in the subsoil it seems to be present probably in the Camagüey 1 well (2250–2264 m).

Cretaceous Turonian–Upper Cretaceous Maastrichtian, Purio Formation [33]

It develops to the north of Central Cuba, Sierra de Cubitas, and Sierra de Gibara. It was registered in several oil wells in the region. It lies concordantly on the Vilató Formation and discordantly on the Palenque Fm. It is discordantly covered by the El Embarcadero formation equivalent to the Sagua Formation.

Lithologically, intercalations of intramicrites and biomicrites with wackestone/peletal and peloidal mudstone texture, biomicrites with calcareous packstone texture with Rudists fragments in a micritical matrix, cavernous limestones, thickly stratified. Biosparites and biointrasparites are also described with fractures and caverns containing bitumen. These rocks at the base of the formation are pale pink (El Purio hill); El Yigre quarry is dominated by orange, pink gray, and light gray varieties.

In the new quarry of El Purio, province of Villa Clara, brittle white marls are abundant, at the base of the formation. Dolomites abound in the wells, as in the cases of the Blanquizal drilling and the Cayo Fragoso well, apparently corresponding to the pre-Maastrichtian interval.

There is a wide variety of benthic foraminifera described in the carbonate sediments in the outcrops. In the wells, the index forms are scarce, with the exception of Chubbina cardenasensis, which indicates the Maastrichtian. *Nummoloculina* sp., *Kathina* sp., *Pseudochrysalidina* sp., *Rhapydionina* sp., and others are also reported in samples from the Sinaloa quarry, San Antonio de las Vueltas, southwest of the city of Remedios.

Paleogenic rocks covering the Remedios TSU in the Sierras de Cubitas and Gibara in the provinces of Camagüey and Holguín

Sagua Formation

This set was recognized and mapped in different parts of the Sierras de Cubitas and Gibara under the name of Embarcadero Formation, which later became known as El Embarcadero due to the works of the Léxico Estratigráfico de Cuba [12]. They are equal to the Sagua Fm. already described for the Cayo Coco, Remedios, Colorados, and Camajuaní TSUs and this is how it is accepted for this chapter.

Lesca Formation (Iturralde Vinent 1981, in [34])

This formation is very similar to the Caibarién Formation of Central Cuba and the El Recreo Formation of the province of Holguín. In the future, with a detailed study, only one lithostratigraphic unit can be approved, complying with the rules of the Stratigraphic Nomenclature Code. Its rocks were deposited in the North Foreland basin of Cuba. Lithologically, it includes gradational biodetritic limestones, from microgranular to micritic and biomicritic calcirudites, clayey limestones, and intercalations of siltstones that transitions to sandstones and silicites. Its outcrops take place on the southern flank of Sierra de Cubitas, Camagüey province, and is recorded in the Pontezuela No. 1 well of 2330–2410 m. It lies concordantly on the El Embarcadero Formation. It is covered discordantly by the Nuevitas Fm. and the Senado Formation. For a set of numerous fossils, the unit is dated as of the Middle Eocene with redepositions of the Lower Eocene. It was deposited in environments that oscillate between external neritic depths to batial. Thickness: 150 m.

Senado Formation [38]

At the beginning, it was called "Olistostroma Senado" so it is an informal lithostratigraphic unit; however, it has a certain degree of development in outcrops and is registered by some oil wells south of the Sierra de Cubitas and, several researchers treat it as a formation. It develops east of the Sierra de Cubitas and is detected in the Pontezuela No. 1 well near the Banao farmhouse, Camagüey province.

The diagnostic lithology comprises a policomponent olistostromic sequence, constituted by serpentinite olistolites and limestones in a breccia-conglomeratic matrix with clasts of volcanites, gabbroids, limestones, sandstones with layers of intercalations and sandstone, grauvacas, sandstone, and serpentinitic siltstones packets. It lies on the El Embarcadero, Lesca, and the Remedios Group formations. It is covered discordantly by the Nuevitas Formation.

Only fossil fauna redeposited from the Cretaceous has been reported. Due to its well defined stratigraphic position, it has been assigned a Middle Eocene age upper part.

4.4.1.15 Camajuaní TSU [31]

They are mainly carbonated sediments where there are important horizons of silicites. The ages of its rocks are from the Upper Jurassic to the Upper Cretaceous Maastrichtian, with terrigenous sediments of the Paleogene that cover it. The Sagua, Jumagua, and Vega formations are common with the Remedios and Cayo Coco TSUs (Fig. 4.13). There are also no manifestations of magmatism and signs of deformation are more marked than those of the Remedios TSU. The southern limit is a large overflow fault that puts it in contact with the Placetas TSU and those of the Zaza TSU, in addition to mantles of the Ophiolitic Association.

From bottom to top, the Trocha Group is described (Jagüita, Colorada, and Meneses formations); Margarita Fm.; Alunado Fm., Mata Fm., and Lutgarda Fm. (Fig. 4.13).

Kimmeridgian-to the Tithonian, Trocha Group TSU

The Trocha Group name has been used by the geologists of the Centro de Investigaciones del Petróleo [39, 14,27], for some years, but not with the sense used by the authors of the unit [40], since the oil stratigraphers include the Jagüita, Colorada, and Meneses formations of ages from the Kimmeridgian to the Tithonian (Fig. 4.13).

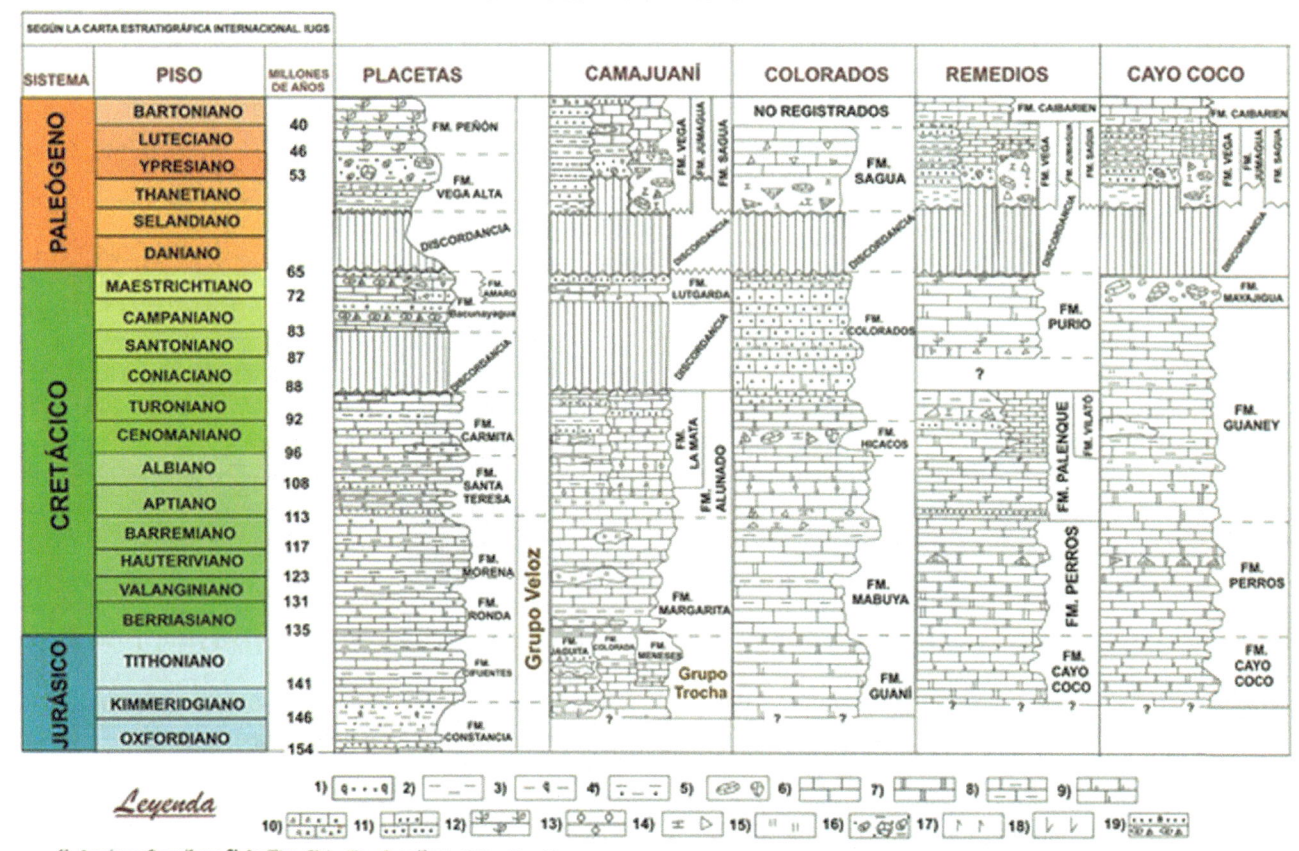

Fig. 4.13 Generalized stratigraphic columns of the PD North American continental margin and its coverage central Cuba-Havana. Current position of the TSUs

Kimmeridgian–to the Tithonian, Jagüita Formation

It mainly appears between the towns of Rancho Veloz and Sagua La Grande. To a lesser extent, it extends between the localities of Sitio and Encrucijada, in the province of Villa Clara. The Jagüita Fm. is geologically mapped on the Cuban Geological Map of the Cuban Gulf Oil Co., approximate scale 1: 40,000, Sheet No. 11 (C). These are micrites and fragmentary biomicrites, of dark gray colors with shades from gray to brown, finely stratified, contain bitumen in the fractures. Rocks sometimes have a primary breccia character, presenting some levels of calcareous breccias. In the lower part known from the cut, there are oolitic and pseudo-oolitic limestones. In the area of the town of Mata, in the province of Villa Clara, the limestones are more fragmentary and ferruginous, showing thin layers of cherts. The association of fossils is of Kimmeridgian age to the Tithonian. The paleoenvironment corresponds to medium batial, with a water depth of 500 m, with an anoxic confined basin character. They are excellent source rocks. The lower contact is still unknown and the upper one is transitional with the Neocomian limestones of the Margarita Fm. The Jagüita Formation is well identified in the wells of the Varadero region to Corralillo such as Chapelín No. 1, Marbella Mar No. 2, Majaguillar No. 1, La Manuy No. 1, Corralillo No. 1, and Guayabo No. 1. Thickness can reach 400 m according to data from oil wells.

Upper Tithonian, Colorada Formation

The Colorada Fm. was proposed by Shopov in 1982. It consists of biomicrites and intramicrosparites with pyrite, phosphates, and spicules of calcareous sponges, tetractines, and hexactines from gray to yellowish cream, which in depth vary from dark gray to black, strongly bituminous. Fine intercalations of cherts and argillites mineralized with pyrite occur. In several samples in Loma Las Azores, the paleontologist Silvia Blanco, has determined Miliolidae, small benthic foraminifera, abundant molds of radiolarians, *Saccocoma* sp., *Crustocadosina semiradiata olzae*, *Commistophaera pulla*, *Didemnoides moretti*, and fragments of pelecypods, embryonic chambers of ammonites, dating the Upper Tithonian.

Upper Jurassic Tithonian, Meneses Formation

The Meneses Formation was also proposed by Shopov in 1982. It adopts the name of the Sierra de Meneses in the province of S, Spíritus. It corresponds to the most northeastern outcropping area of the Camajuaní TSU, in what is called the Jatibonico subzone, where the Camajuaní TSU presents considerable facial changes. It develops to the east of Meneses and to the south and southeast of Yaguajay in the Sierras de Meneses and Jatibonico. It has not yet been recognized in oil drilling. It is constituted by a disorderly alternation of gray limestones, microgranular, massive, compact and stained by oxidized oil and gray-brownish limestones, banded, of thin layers. In addition, calcareous or dolomitic breccias participate. Dolomitization is a much-observed process, especially in the upper parts. The fossil fauna is scarce and recrystallized, being able to recognize *Favreina* sp. Age is Upper Jurassic Tithonian. The paleoenvironment corresponds to the middle neritic. Its thickness ranges between 110 and 250 m. Transition gradually to the Margarita Formation.

Berriasian–Barremian, Margarita Formation

The Margarita Formation consists of gray micrites and biomicrites, light gray and dark gray, with bands of diagenetic brownish-yellow chert, sometimes black. Between the limestones there are fine intercalations of laminar siltstones in thin layers of about 5 mm thick. In some parts of the formation, the micrites are very compact, gray-bluish to black, with bands of very consolidated bituminous clay limestones. There are calcareous breccias and oolitic detritus. Biomicrites are sometimes lumpy, with pyrite concretions. The age is Berriasian–Barremian. The paleoenvironment is batial, of deep slope with a water level of 1500 m. There is a coastal influence, represented by blue algae, milliolids, and brachiopods. In some levels of the formation, as it is observed in Majaguillar No. 1 well, corresponds to Middle-Upper Slope, with bathymetry of 600 m. The lower contact is gradual with the formations of the Trocha Group, and the superior one is concordant with the Aptian part of the Alunado Fm. The Margarita Fm. outcrops very well in the Margarita hill west of Sagua la Grande, in the quarry of Aguada La Piedra, and between the towns of Cifuentes and Mata. In the subsoil, it is well represented in the Chapelín No. 1, Marbella Mar No. 1 and 2, La Manuy No. 1, Majaguillar No. 1, Corralillo No. 1, and Guayabo No. 1 wells.

Aptian–Albian in the Camajuaní, Colorados, Remedios, and Cayo Coco TSUs (Central Cuba)

The Aptian–Albian is where the division of the TSUs that we are dealing with really happened. In the Aptian–Albian in the Camajuaní TSU the Alunado Fm. began to be deposited. Together in the Albian began the deposition of the Mata Fm. In the Colorados, TSU continued the development of the Mabuya Fm., while at the Remedios TSU the Palenque Fm. began to be deposited. In the paleocanal of the Cayo Coco TSU, the lower packages of the Guaney Fm. were sedimented. After the deposition of the Perros Fm. in the carbonate platform, two parallel events originated: (1) A deep water channel was differentiated to the north that allowed the pelagic sedimentation: calcareous mudstone with typical planktonic fossils and even with intercalations of silicites and pelites; we are dealing with the Guaney Fm. (2) On the adjacent platform, the temperature, oxygenation, and salinity factors proper for the development of bioconstructions of rudists as carbonated banks concur, thus differentiating the Palenque Fm. At this time, evaporites disappeared completely, although dolomitization persists, but less intense than before. At the Colorados TSU, calcareous mudstones, wackestones, and a few grainstones that constitute the upper part of the Mabuya Fm. were deposited, characteristic of proximal slope. The Camajuaní TSU had a change in sea level (or in the subsidence rate) since the carbonates are interspersed with thick layers of silicites (radiolaric cherts). Carbonates are radiolaric calcareous mudstones. They correspond to the Alunado Fm. whose environment is a distal slope.

Aptian–Turonian, Alunado Formation

The Alunado Fm. has two facies. To the south, there are micrites of fine stratification with chert in thin layers, containing *Nannoconus* spp., *Colomiella recta*, *C. mexicana*, *Ticinella roberti*, *Schackoina* sp., *Rotalipora* sp., *Favusella washitensis*, and *Globotruncana* spp. The northern facies consist of biointramicrites and oolitic limestones, calcareous carbonate schists and calcirudites, reporting *Nummoloculina* sp., *Orbitolina* sp., *Choffatella* sp., *Cuneolina* sp., *Pseudocyclammina hedbergi*, *Textularia* sp., Miliolids, Algae, Briozoos, and remains of rudists. The age of the formation is Aptian–Turonian. In the Calienes Quarry in the Mayajigua area, fossils from the upper Albian have been reported: *Muricohedbergella delrioensis*, *Planomalina* sp.,

Herbergella gorbachikae, Macroglobigerinelloides cf. *M. bentonensis, Muricohedbergella simplex,* recrystallized molds of radiolarians, *Ticinella* sp., *Muricoherbergella* spp., *Macroglobigerinelloides* sp., *Rotalipora* sp. (Sample by E Linares EL-213-98). The paleoenvironment corresponds to a deep neritic basin that towards the north interdigitates with deposits of the platform edge whose representatives were dragged by storm currents. The thickness can reach 300 m.

Albian–Turonian, Formation Mata

The Mata Fm. comprises intercalations of marly micrites with limonite concretions, biomicrites laminated with phosphates, calcarenites, and clays. The color of these rocks ranges from creamy or yellowish tones to brownish-brown and gray. A distinctive feature of this lithostratigraphic unit is the intercalations of bands and lenses of black, wine, and gray cherts. Carbonate rocks are laminated micritic mudstone, with vertical and subvertical fractures filled with bitumens or sealed by calcite. The cherts are radiolaric wackestones with some spicules of sponges. In the rolling cherts have been identified an abundant fauna of the Cenomanian and Turonian of more than 50 species of radiolarians identified in the type locality of the hill El Guayabo, between the villages of Mata and Calabazar de Sagua in Central Cuba. The age of the Mata Formation is Albian–Turonian but probably the lower layers may be of the Aptian. The sedimentation environment shows variations. The most carbonated part of the section attributable with doubts to the Aptian–Albian, oscillates between slope relatively close to the continental break, with weak sedimentation rate until lower batial, with a water depth of 1500 m, in a confined environment and sedimentation rate higher. The sedimentation of silicites is of particular interest. The purity of its lithological composition, its fine microstratification, the mixture of non-generalized spicules, and, above all, the conservation of fine acicular forms of radiolarians, was possible by a slow and quiet sedimentation, at a depth of 3000–4500 m outside the action of the stationary currents, that is, abyssal basin.

Upper Maastrichtian, Lutgarda Formation

The Lutgarda Fm., by the lithological studies, it is constituted by calcirudites, calcarenites, and micrites, with bands of red chert and to a lesser extent some intercalations of calcareous clays. The breccias contain exogenous, angular lithoclastic, and biomicritic matrix of

(a) Medium to internal platform bar, with large fragments of Rudists and micrite remains with Bonetocardiella maestrichtiense (Maastrichtian).
(b) Shallow medium platform, biointrasparite with micritical bioclasts, containing *Dictyoconus walnutensis, Salpingoporella* sp., *Cuneolina* sp., Miliolids, and Ophalmidiidae (Albian).
(c) Intrabiomicrite with Miliolidae from lagoon to internal platform.
(d) Micrites with planktonic fossils from external platform to slope.
(e) Dolosparites with platform Gastropods.
(f) Radiolarites from slope to basin.

In the biomicritic matrix of the breccias, fossils have been described; they bear witness to the Upper Maastrichtian age for the Lutgarda Formation.

The paleoenvironment of this formation corresponds to the middle neritic (deeper carbonate—platform fore—slope), with water depth of 20–40 m. The thickness is approximately 120 m. Its lower contact is discordant erosive with the Mata Fm. and the upper is discordant with the Sagua Fm.

Possible Rocks of the Placetas and Camajuaní TSUs in the Area of Gibara in the Province of Holguín

In the north of the province of Camagüey the rocks of the Placetas and Remedios TSUs evidently emerge and, by geophysical data, it is supposed that those of the Camajuaní TSU lie in the depth. They must also lie under the young rocks; the ophiolites and the Cretaceous Volcanic Arc (CVA) to the north of Holguín; where the Remedios TSU is emerging throughout the Sierra de Gibara, as well as; its Paleogenic coverage. Figure 4.14 shows these considerations by multidisciplinary studies [40].

4.4.2 Paleogeographic Domain of the Cretaceous Volcanic Arc, Zaza TSU [39]

This Tectono-Stratigraphic Unit comprises the Arc of Volcanic Islands of the Cretaceous and the belts of intrusive granitoid rocks of Central Cuba and the provinces of Camagüey and Las Tunas, mainly. The volcano-plutonic rocks of arc of islands can be characterized as an ancient archipelago of volcanic islands, whose stratigraphic and tectonic limits are quite well defined. These rocks emerge in the area of Bahía Honda (north of the province of Artemisa), in the area of Sabana Grande (Isla de la Juventud), in the provinces of Havana, Mayabeque, Matanzas, Cienfuegos, Villa Clara, Sancti Spíritus, Ciego de Ávila, Camagüey, Las Tunas, Holguín, Granma, and Guantánamo. Additionally, some oil wells register them.

In the rocks of the Cretaceous Volcanic Arc, it is possible to identify the Petrotectonic Sets (PSs), volcanic,

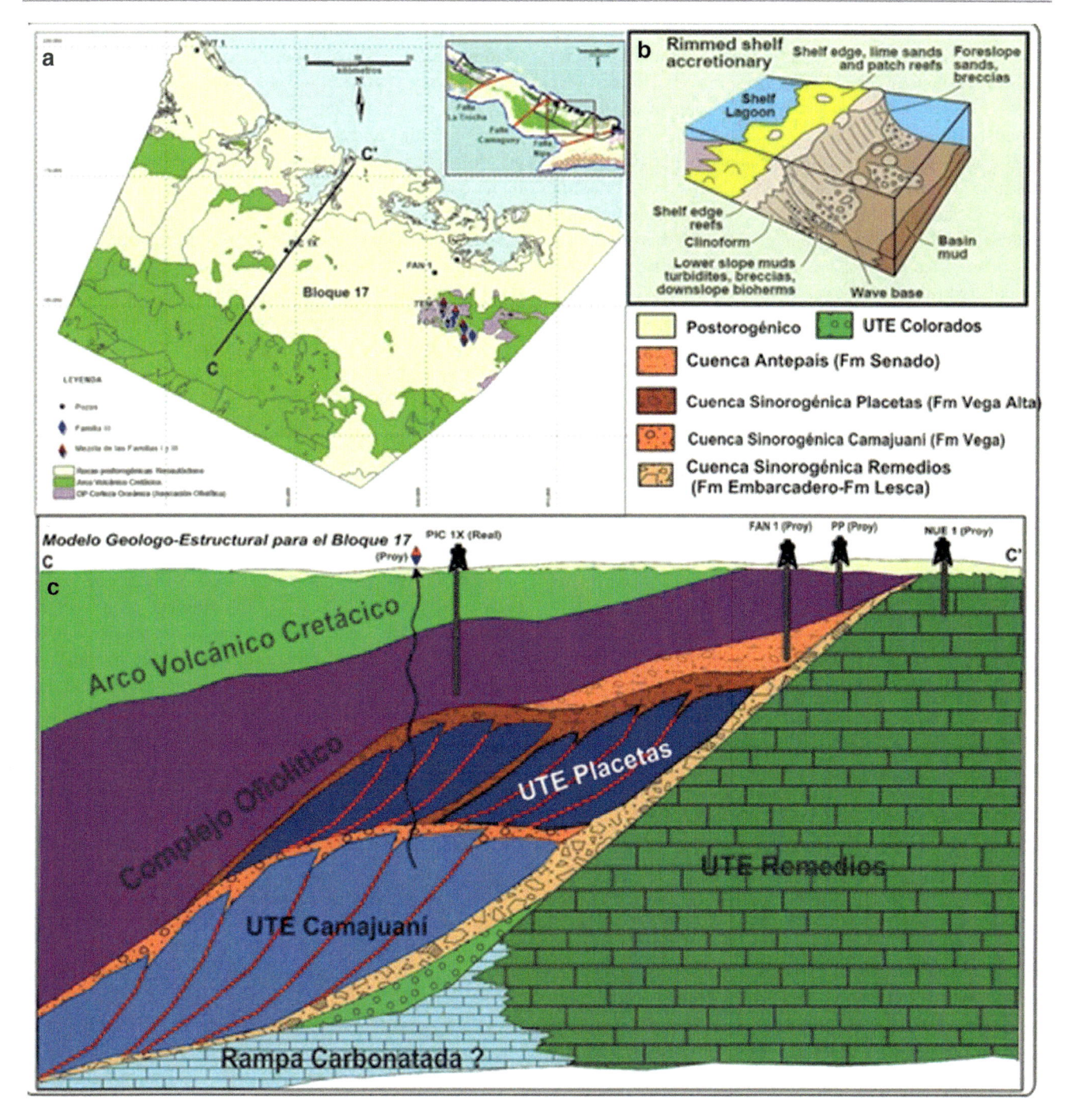

Fig. 4.14 Geological–structural model for block 17 (Puerto Padre-Gibara Region), provinces of Las Tunas and Holguín [40]

sedimentary volcanic, the plutonic, and the metamorphosed volcanic rock complex. These sets are present in different localities of the national territory, having lateral and vertical variations in their chemical and lithological composition. The volcanic, sedimentary volcanic and plutonic volcanic PSs correspond to the effusive, intrusive, pyroclastic, and sedimentary rocks that form in the volcanic islands and in the seas that surrounded them. These PSs, due they include fossiliferous strata, can be dated by paleontological means. By means of rudists and benthic foraminifera, the presence of limestones of Upper Albian, Santonian, and Campanian has been determined. By means of other methods and criteria, the antiquity of the volcanites of the arc has been extended to the Lower Cretaceous Neocomian, bearing in mind that beneath the fossiliferous horizon of the Albian there are hundreds of meters of volcanic rocks. Such is the case of the Los Pasos Formation, which, like the "Olds volcanics", could belong to an aborted primitive arc. By radioactive dating, the Los Pasos Fm. is Hauterivian, (pre ~ 133 My), also considering it as part of a primitive bimodal arc.

The age of the youngest volcanites of the Cretaceous Volcanic Arc has been considered by some authors as the Upper Cretaceous Maastrichtian, but the most accepted tendency is to date the upper part of the CVA as the upper part of the Late Campanian starting from the fact that they are covered discordantly throughout the entire Cuban territory by different units belonging to deposits of overlapping basins of the Upper Campanian–Maastrichtian and Upper Maastrichtian (Linares Cala et al. 2015). The volcanic-sedimentary sequence is cut by numerous and sometimes very large intrusive bodies of different composition and age. The intrusive bodies or massifs can have a length greater than 100–120 km and a width of about 20 km, like the massif of granitoids of Camagüey-Las Tunas. The ages of the intrusives are determined by numerous absolute age analyzes.

Depending on the geological cartography in each province, many names are found for the lithostratigraphic units of the Cretaceous Volcanic Arc. In Artemisa, the Orozco, Quiñones, Encrucijada formations. In the Havana-Matanzas region, La Trampa and Chirino formations are developed. In Central Cuba, the Los Pasos, Porvenir, Mataguá, Cabaiguán, Jarao, Dagamal, La Rana, Pelao, Arimao with its Moscas Member, Hilario, Las Calderas, Guaos, Brujas, Seibabo, Provincial, among others formations. In the provinces of Ciego de Ávila and Camagüey, the Guáimaro, Caobilla, Crucero de Contramaestre, Vidot, and Marti formations were named. Finally, in the provinces of Las Tunas and Holguín, the Iberia, Buenaventura, and Loma Blanca formations are developed. All these units can be consulted in the third version of the Léxico Estratigráfico de Cuba [12].

Regarding the age and thickness of the rocks of the Cretaceous Volcanic Arc, the data are not homogeneous throughout the national territory. From several investigations, the rocks of a contrast volcanism of the Los Pasos Formation refer to the Neocomian by absolute age dating. Already in the Aptian and the Albian, there are some fossils that allow dating. There is a short recess of volcanism in the Cenomanian and, the Late Campanian is fixed as its end in the Upper Cretaceous. By their nature, these rocks have few fossils; to this it, is added that they are imbricated with mantles of different composition in the Folded and Overthrusted Belt of Cuba. With such unfavorable factors, it is not easy to measure their thickness in geological survey work. In the oil wells of western Cuba, several hundred meters are recorded, for example, in the deep wells of Vega No. 1, Ariguanabo No. 2, Mercedes No. 1 and No. 2, Cochinos No. 1; the wells of Guanabo, Vía Blanca, Boca de Jaruco and Yumurí oilfields; and several oil drillings of the Madruga area. In conclusion, the apparent thickness in this region is not less than 1,500 m. In Central Cuba in the provinces of Villa Clara, S. Spíritus, Ciego de Ávila, and Camagüey, the study of the formations that make up this group has been very remarkable. This is explained because numerous wells have been drilled in and out of the Cristales, Jatibonico, Pina, and Jarahueca oilfields. Taking into account the mantle structure, Hatten and collaborators calculate between 6000 and 6700 m, not including the lavas flows of the basalts and more acidic rocks of the Los Pasos Formation. This thickness is exaggerated. In many wells of the Central Basin, thicknesses of the order of the first thousands of meters are evident (Vega Grande No. 2, greater than 1000 m; Jatibonico Sur No. 1, Reforma No. 1 and Guayos No. 1, higher than 1500 m; Jatibonico No. 78, greater than 2000 m).

The rocks of the Cretaceous Volcanic Arc, contact tectonically with the ophiolites of the northern belt in different locations along the Island between Havana and Holguín. The tectonic relations have been represented by numerous geologists in the geological mapping works and in the reports of the oil drilling. Often the contacts coincide with very fissured and foliated areas, or with chaotic masses that contain a mixture of blocks of ophiolites, volcanites, and arc plutonites, that represent in most cases overflow faults and their corresponding melánges.

4.4.3 Late Aptian–Campanian, Intrusive Magmatic Rocks of the Cretaceous Volcanic Arc

In the so-called Belt of Manicaragua and in the provinces of Ciego de Ávila, Camagüey and Las Tunas, numerous magmatic bodies appear, above all, granitoid intrusive rocks. Minor bodies are detected in the south of Camagüey, altering in their contact the formations of the basin of the Vertientes-Amancio Rodríguez basin, such as the Guáimaro and Vidot formations. The thirteen intrusive bodies of Las Tunas-Camagüey are the following: Sibanicú-Las Tunas with 800 km^2, Siboney with 48 km^2, Ignacio 9 km^2, Camagüey 132 km^2, La Larga 15 km^2, Algarrobo 44 km^2, Las Parras 40 km^2, Santa Rosa 26 km^2, Florida–Cespedes–San Antonio of more than 220 km^2. There are other smaller bodies, such as Guáimaro about 2.5 km^2, Piedrecitas 6 km^2, La Presa 11 km^2, and Norte of 7 km^2, these last three are located northwest of the intrusive Céspedes–Florida–San Antonio, in the province of Camagüey. Four magmatic "formations" have been established that correspond to the different stages of development of the arc: gabbro-plagiogranitic, of the Aptian-Albian, sienitic, of the Albian–Cenomanian, granodioritic-granitic, of the

Turonian–Campanian and, gabbro-monzonitic, probably of the Late Campanian. In the subsoil, notable thicknesses of granitoids have been recorded in the oil wells of Tinajón de Oro 1-X, province of Camagüey and, Picanes 1-X, in the northern edge of the basin in the province of Las Tunas.

4.4.4 Cuban Metamorphic Rocks

There are important areas where metamorphic rocks are developed in Cuba, mainly in Arroyo Cangre Strip in Pinar del Rio, Isla de la Juventud Massif, and Guamuhaya Massif in the provinces of Cienfuegos and S. Spíritus, La Corea in Holguin province and, in the Maisí area, in the province of Guantánamo. But minor outcrops are frequent in several Cuban areas: Rancho Veloz, Arroyo Blanco, Camagüey, Las Tunas, Velasco, and others. With the exception of some isolated outcrops such as those in the south of Sierra Morena and La Teja in the provinces of Villa Clara and Matanzas, these metamorphites have rocks of Jurassic and Cretaceous age as a protolith, correlated with different lithostratigraphic units that emerge in our country, whose ages and geological position are well established. The metamorphism of all these Mesozoic formations occurred essentially during the Cretaceous, under various geological and tectonic conditions, with different types of metamorphism by its temperature–pressure relationship.

The main studies of Cuban metamorphites are due to Guillermo Millán Trujillo who has made them known in different works [41]. They have been the basis for this chapter.

Metamorphites of a Premesozoic Sialic Basement

It treats, as it was said previously, of a marble silicic phlogopitic that by means of the argon potassic method gave values of 945 ± 2.5 and 910 ± 2.5 million years. They appear in La Teja and Socorro in the provinces of Matanzas and Villa Clara, respectively.

Metamorphites Whose Protolith are the Rocks of the Synrift of the San Cayetano Fm

They emerge in Pinar del Rio, Arroyo Cangre Fm.; on the Isla de la Juventud, Cañada, and Agua Santa formations; in the Guamuhaya Massif, La Llamagua, Herradura, La Chispa y Loma La Gloria formations. They are quartziferous metasandstone phyllites, quartziferous metaterrígenos schists, and micaceous quartz, metapelitic schists rich in muscovite and graphite. In the Escambray (Guamuhaya Massif), there is a lower degree of metamorphism; they are quartziferous metasandstone, with glossy phyllites conserving their primary traits, quartz, and quartz-moscovitic metaterrigenic schists, moscovitic shales rich in graphite.

Polymeric crystalline schists, sometimes calcareous, some marbles, calcareous schists, and metasilicitic quartzites are described.

In the mountains of Sierra del Purial forming a belt that underlies the Chafarina Fm., occurs a succession of glossy phyllites and fine metapsamites, rich in graphitic matter with intercalations of gray limestones and radiolaric metasilicites. They are very similar to the San Cayetano Fm. from Pinar del Rio.

Metamorphites Whose Protoliths are the Carbonate Rocks of the PD North American Continental Margin

On the Isla de la Juventud the rocks of the group Gerona emerge. They are marbles of four well-defined formations: Playa Bibijagua (conchiferous marbles of black color); Colombo (gray stink marbles, with dolomitic marbles); Sierra Chiquita (very fine dolomite marbles, clear, banded, metachert thin layers; Sierra de Caballos, is the highest part of the Gerona Group (they are gray, medium grain, fetid marbles, metacherts, and dolomite marbles). The "Sierra de Casas marbles" which are massive, coarse-grained marbles, are described. These are an independent tectonic scale.

In the Guamuhaya Massif, the San Juan Group is described, which is a succession of marbles of dark blue-gray color, marbled limestones, generally stratified. The basis of the group is the Narciso Fm. where ammonites of the Middle Oxfordian upper part were found. The Sauco Fm. are dark gray marbles, stinky, grossly stratified. Mayarí Fm. (Fig. 4.15), it is the one that best emerges, which are dark gray marbles with thin layers of metachert quartzites. It has Tithonian ammonites. The top of the group is the Collantes and Vega del Café formations. Graphitic black marbles and gray marbles stratified with clay shale zones are described. In a sample taken at Topes de Collantes in one of these black clay schists the content of 11% Total Organic Carbon (TOC) was determined. In addition to the San Juan Group, the formations of Cobrito, of Jurassic-Cretaceous schistose marbles and, Boquerones, with schistose marbles and calcitic schists, crystalline limestones with radiolarians, have been studied. The Loma Quivicán, Los Cedros, La Sabina and Yaguanabo formations are also described. The first are marbles of light shades of various colors, metapsammitic calcareous schists and intraformational metabreccias. The Sabina Fm. is a succession of metasilicical quartzites and metasilicitic schists and well-stratified moscovitic quartz. The Los Cedros Fm. are light gray to dark marbles, laminated, interspersed quartzite metasilicitic, sometimes manganiferous. The Sabina Fm. comprises a succession of metasilicitic quartzites and metasilicitic quartz- moscovitic schists, well stratified and banded, often granatiferous.

The Yaguanabo Fm. includes basic metavolcanic green schists with isolated interspersed marbles and metasilicitic

Fig. 4.15 Marmorized black limestones of the Mayarí Formation, homonymous locality, province of S. Spíritus

quartzites. The El Tambor Fm. groups green schists, sometimes calcareous, whose protolith was a flysch rhythmically stratified. These rocks crown the Escambray stratigraphic column.

In the region of Maisí, the Chafarina Fm. it is recognized. It consists of dark schistose gray marbles, well stratified, creamy and pink marbles. They are bituminous. There are sometimes slightly altered limestones with foraminifera possibly from the Upper Jurassic.

Metamorphic Formations Whose Protoliths were the Rocks of the PD Cretaceous Volcanic Arc

In several places in Cuba, metamorphosed rocks emerge from the sequences of the Cretaceous Volcanic Arc. The main areas are located north and south partially of the Escambray, in the provinces of Cienfuegos and S. Spíritus; in the Sierra de Rompe in the region of Las Tunas and in the province of Guantánamo in the Sierra del Purial.

It has been informally called "Complejo Mabujina" amphibolites of low to medium pressure widely exposed in the south of Central Cuba that contact tectonically with El Escambray. Its protolith according to Millán [41], are the inferior parts of the CVA and by different components of the Ophiolitic Association, which underlies the CVA, metamorphosed and folded together. By radiometric data amphibolites are attributed to the Lower Cretaceous.

The "Metamorphites of Sierra de Rompe" are a belt of Cretaceous volcanites with typical low pressure amphibolites. They are to the south of the "Granitoides de Camagüey-Las Tunas." Amphibolites are common from basalts. Occasionally it is possible to see the original stratification of the layers.

Millán [41], mentions metamorphic formations of different origins such as the Güira de Jauco Fm., in the Sierra del Purial; " Perea Metamorphic," in the province of S. Spíritus; the so-called "Esquistos La Suncia," east of the city of Camagüey; " La Corea Metamorphites," in Sierra Cristal; "Mateo Metamorphites," in the area of Velasco in the province of Holguín; " Macambo Metamorphites," in the Sierra del Convento, province of Guantánamo; "Daguilla Amphibolites," on the Isla de la Juventud and, finally; "Sierra de Agabama Metamorphites," southeast of Santa Clara, province of Villa Clara.

4.4.5 Paleogene System in Cuba

In the geological constitution of our country, the Paleogene rocks have a great development by areas, with considerable thicknesses. The substrate on which they formed is varied: different TSUs of the Paleogeographic Domain of the North American Continental Margin; PD Cretaceous Volcanic Arc and; rocks of the PD of the Oceanic Crust. They form large parts of the overlying and frontal basins. In the southeastern region, the Turquino Volcanic Arc (Paleogenic Volcanic Arc) is described. In the Sierra Maestra, thick volumes of volcanic, volcanic-sedimentary and pyroclastic rocks emerge, cut by different intrusive rocks. The imprint of this volcanism has been observed in places as far away as the Cuenca de Vertientes-Amancio Rodriguez in the province of

Camagüey and the Zaza Formation (also named the Bijabo Formation) of Central Cuba.

Paleogene Developed on the PD of the North American Continental Margin

The Manacas and Ancón formations have already been described in the Esperanza, Sierra de Los Órganos and Sierra del Rosario TSUs (4.4.4.3). About the Placetas TSU, it is detailed the Olistostrome of the Vega Alta Fm. (4.4.5.1). Sagua, Jumagua and Vega formations they are common to the Cayo Coco, Remedios, Camajuaní and Colorados TSUs (4.4.5.2). In some regions, the Florencia, Turiguanó, Caibarién, Lesca (also El Recreo Formation) and Senado formations, all of the Paleogene, were deposited. It shows a great lithological variability over time. At the base, siliciclastic sediments predominate over calcareous ones, a proportion that varies vertically in favor of the carbonate rocks that predominate at the end with the deposits of the Upper Eocene and Oligocene.

Paleogene Developed in Overlapping Basins on the Cretaceous Volcanic Arc

Los Palacios Basin

It is located to the south of the Cordillera de Guaniguanico with great thicknesses of rocks of the Upper Cretaceous, Paleogene and Neogene. The Mariel Group, composed of the Madruga and Capdevila formations of mainly siliciclastic rocks, belong to the Paleogene. At its base in the region of the provinces of Havana and Mayabeque, the Mercedes or Apolo formations can be found indistinctly. Another unit is the Universidad Group with its Toledo and Principe formations. On the Universidad Group and the Capdevila Fm. was transgressively deposited the Loma Candela Formation.

In the upper part of the Upper Eocene the Consuelo Fm. it is deposited and, in the Upper Oligocene, the Guanajay Fm., of essentially carbonated nature.

Some of these lithostratigraphic units were overthrusted over the North American Continental Margin during the Cuban orogeny and form with others, the Folded and Overthrusted Belt of the northern provinces of Havana, Mayabeque and Matanzas.

Vegas Basin

Arranged to the east of the Los Palacios Basin. Here it is described the Bejucal-Madruga-Limonar structure, located in the central part of the provinces of Mayabeque and Matanzas. It includes Capdevila, Perla, and Tinguaro formations and the Nazareno Group. Carbonate rocks predominate over terrigenous rocks.

Santo Domingo Basin

The cut of this basin is one of the most complete and continuous from the Danian Paleocene (Cocos and Santa Clara formations) to the Oligocene (Tinguaro and Jía formations). The Yeras, Ochoa, Damují and Jicotea formations are also recognized. There are calcareous rocks, marly, sandstones, clays (Cocos Fm.); fragmentary limestones, marls and clayey limestones (Santa Clara Fm.); marl, clay clayey limestones and siltstones (Tinguaro Fm.); limestones and organogenic carbonated breccias (Jía Fm.); limestones (Yeras Fm.); clays and mainly marls and sandstones with some siltstones (Ochoa Fm.); limestones (Damují Fm.) and marls, siltstones, polymictic sandstones, organogenic limestones and conglomerates (Jicotea Fm.).

Cienfuegos Basin

It borders the Cienfuegos Bay. It presents a continuous cut from the Upper Cretaceous to the base of the Oligocene. The Upper Cretaceous-Paleogenic Vaquería and Caunao formations are described. The first is of flyschoid stratification and consists of limestones of various types and marls with fossils that allow dating from the Upper Maastrichtian to the Middle Eocene. The Caunao Formation are sandstones and polymictic conglomerates with limestones of various types, very fossiliferous. Age is Upper Eocene to Lower Oligocene.

Trinidad Basin

It has a unique location between the domes of S. Spíritus and Trinidad. Their sources of contribution were very different from the basins mentioned above. The Paleogene begins in the lower part of the Middle Eocene and lasted until the Oligocene, recognizing the Meyer, Condado and Las Cuevas formations. The rocks are mainly terrestrial with sources of contribution of the metamorphic massif of the Escambray and the CVA.

Central Basin of Cuba (Cabaiguán Basin, Partially)

The Central Basin of Cuba (Fig. 4.16), with its salient (apophysis, Roche Basin or Gálata, also called the Cabaiguán Basin), is known in large areas and in the subsoil by hundreds of oil drilling wells. The Paleogenic formations are listed: Fomento, Jucillo, Taguasco, Loma Iguará, Siguaney, Zaza (also Bijabo), Arroyo Blanco, Marroquí, Chambas, Tamarindo and Jatibonico. Some of them are olystostromal in nature (Taguasco Fm.), others terrigenous (Zaza Fm.) and,

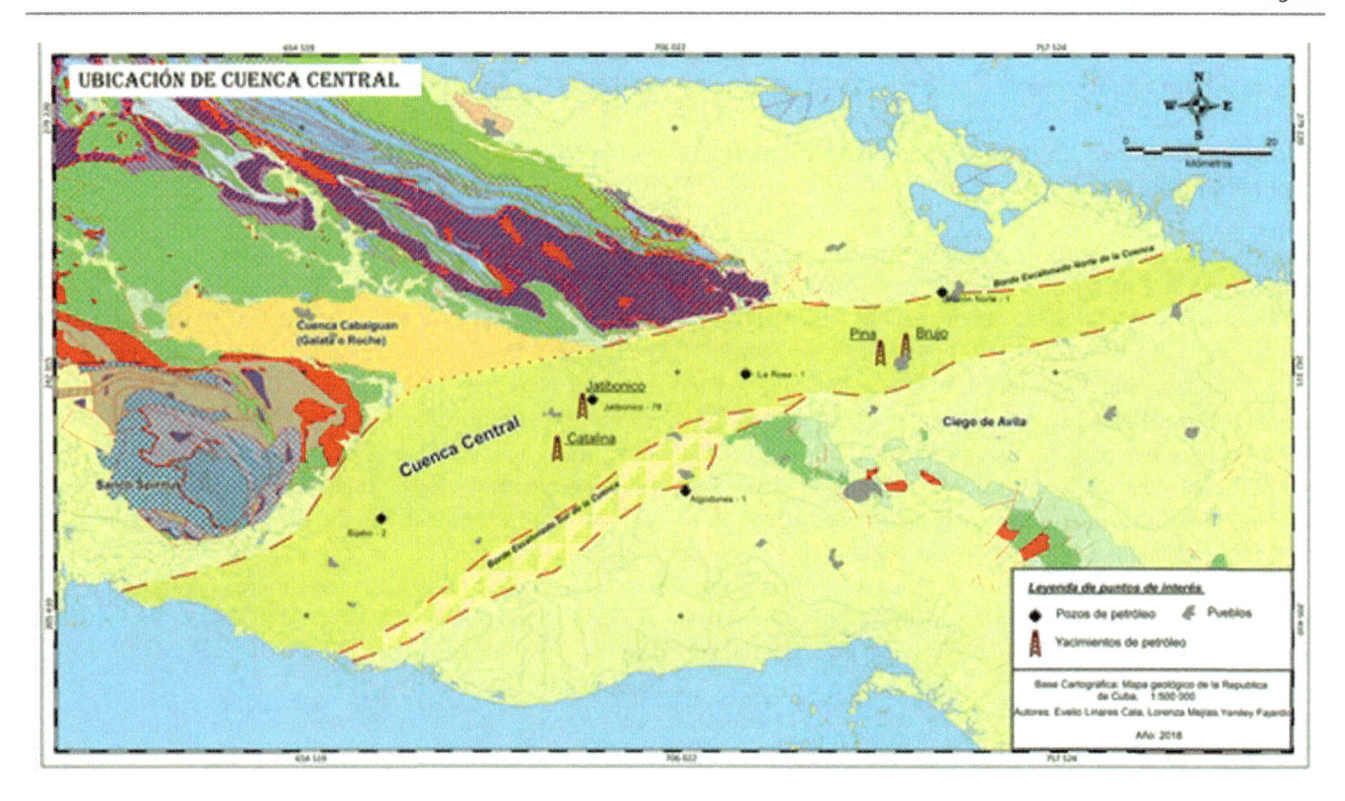

Fig. 4.16 Scheme with the location of the Central Basin of Cuba, provinces of S. Spíritus and C. de Ávila. Notice some of the oil and gas fields

to a lesser extent, carbonated (Chambas and Jatibonico formations).

Vertientes-Amancio Rodriguez Basin

Paleocene rocks began their development with the Vertientes Fm. of the Lower Eocene (sandy-argillaceous flyschoide). After, many formations have been described: Florida (Middle Eocene lower part) of rocks mainly of carbonated breccias, limestones, marls and siltstones; Maraguán (Middle Eocene) that are polymictic sandstones, marls, siltstones, clays and conglomerates; Saramaguacán (Middle Eocene) consisting of limestones and marls with sandstones and siltstones; Nuevitas (Upper Eocene) formed mainly by marls of various types and limestones and; Pastelillo (Upper Oligocene) where limestones and marls predominate. Ultimately, in the basin, rocks of a siliciclastic nature and also carbonated predominate.

Gulf of Guacanayabo Region-Cauto-Nipe Basin

The numerous oil companies that investigated in this region, named countless lithostratigraphic units, which caused excessive synonyms. Structurally, Alsina de la Nuez et al. [42], on a map of relative gravity, highlighted the maximum of the Sierra Maestra-Lewiston and the Cauto-Guacanayabo Minimum; on a seismic structural map, the tentative horizon of the Eocene, in the Manzanillo-Bayamo and Cauto El Paso-Palmira areas. Additionally, they showed in the area of Guacanayabo the horizon named 2. Finally, they annexed an Aeromagnetic Map of the region. According to its data, the Cauto Basin is a depression within the Cauto Tectonic Unit, bounded to the north by the so-called Tectonic Unit of the East and to the south by the Bartlett Trench. Within this Cauto Tectonic Unit there are three well-defined tertiary depressions: Ana María, Cauto-Nipe and San Luis-Guantanamo, with the Cauto-Nipe being the largest area. This unit is characterized by being the only one in Cuba with a marked volcanic activity during the Paleogene. At the same time, with the orogenic movements, faults that formed graben structures arose, with intense sedimentation coming from the areas surrounding the graben. The aforementioned researchers mentioned three periods of great folding: that of the Cuban Orogeny, in the Middle Eocene, to which they attribute the emersion of the Sierra Maestra; that of the Upper Eocene, with strong folds, judging by the lying of rocks of that age on other older rocks of different ages. They also suppose tangential movements in the Lower Oligocene with the direction of the Cuban NO-SE direction, with intense folds evidenced in the Santa Regina No. 1 and Manzanillo No. 1 wells. During the Miocene the tectonics was moderate, with soft folds.

Shein and collaborators in their Tectonic Map of Cuba (1985), consider that this basin forms part of the Eastern

Cuban Basin Region, which they divide into three: The Cauto Basin, the Nipe Basin and the San Luis Basin. The Cauto Basin is limited to the south by the Oriente deep transcortical fault; by the north, the Tunas Fault. The Cauto and Nipe depressions are separated by an uplift of the oceanic crust, covered by the Neoautochtonous sedimentary cover, which in turn, is separated from the San Luis basin by the Charco Redondo uplift.

4.4.5.1 PD of the Paleogene Volcanic Arc (Arco Turquino)

Sierra Maestra Area

It extends completely occupying the Sierra Maestra, its foothills and, in the subsoil, has been described in several oil wells under the Neogene-Quaternary of the Gulf region of Guacanayabo, the Cauto-Nipe Basin and the San Luis-Guantánamo Basin. The Paleogenic stratigraphic column has volcanic-sedimentary Cretaceous rocks in the regions of El Uvero, La Plata, Gran Piedra and others.

The main lithostratigraphic unit is the El Cobre Group (Paleocene-Middle Eocene, lower part), with its Pilón (Paleocene-Lower Eocene) and El Caney (Middle Eocene, lower part) formations.

El Cobre Group is made up of different types of volcanic and volcanic-sedimentary rocks in different correlations and alternating combinations, very variable, both vertically and laterally. The transitions between them are sometimes abrupt and other gradual and, in many cases, it is practically impossible to establish boundaries between them. The most abundant rocks are: tuffs, aglomeric tuffs, lavas and aglomeric lavas, of andesitic, andesite-dacitic and dacitic composition, rarely rhyolitic, rhyodatic and basaltic (Fig. 4.17).

These rocks are interspersed with tuffites and limestones, and hipabisal bodies and dikes of different composition are associated with this volcanic-sedimentary complex. Also participating in its structure are cineritic tuffs, tuffites, calcareous tuffs, tobaceous limestones, polymictic and volcanic sandstones and grauvacas.

The described index fossils are foraminifera that allow to date the Group with Paleocene-Middle Eocene lower part age. The deposition environment fluctuates between open platform to slope deposits with normal salinity.

The Pilón Fm. occupies a large part of the Group and comes out mainly in the western part of the Sierra Maestra. Its thickness can reach 2000 m. The Caney Fm. has a fossiliferous association that is restricted to the Middle Eocene lower part.

Also in the Middle Eocene and, on the El Cobre Group, the Charco Redondo (Middle Eocene), Puerto Boniato (Middle Eocene upper part), Farallón Grande (Middle Eocene upper part) and, San Luis (Middle Eocene upper part-Upper Eocene) formations develop. In the Upper Eocene the Camarones Fm. was deposited. In these formations, carbonate and siliciclastic sediments predominate, with the imprint of the rocks of the Paleogenic Volcanic Arc.

San Luís-Guantánamo Basin

It is a basin which extends for more than 100 km in length from the west of the town of San Luís in the province of Santiago de Cuba to the east of Guantánamo, with a width of 50 km. It is located geographically in the southern part of the provinces of Guantánamo and Santiago de Cuba.

The rocks that are appropriately assigned to the San Luís-Guantánamo overlapping basin, began to be deposited in the Late Upper Cretaceous, having as substratum the

Fig. 4.17 Basalts and andesite-basaltic rocks of El Cobre Group. Observe the columnar arrangement. Location Puerto de Moya, Carretera Central, near El Cobre, province of Santiago de Cuba

oldest rocks of the Cretaceous Volcanic Arc, attributed in the region to the Santo Domingo Formation-also called Bucuey previously- (See Third Version of the Léxico Estratigráfico de Cuba 2013) and its equivalent, the Sierra del Purial Formation-previously La Farola Formation, which shows the imprint of an incipient metamorphism. Participate as a substrate also, the rocks of the Ophiolitic Association.

In the area considered as the San Luís-Guantánamo basin, two different processes were developed that contributed to the formation of this thick sedimentary sequence: first, the deposits of a superposed basin in its first cycle developed on the Cretaceous Volcanic Arc, with formations of the Upper Campanian to the Danian Paleocene (Mícara, La Picota, Gran Tierra formations) and, then, a retroarc basin that overlapped due to the emergence of a Paleogenic volcanic arc in the region of the Sierra Maestra between the Paleocene and the Middle Eocene, very well represented in the Sabaneta Fm. After the extinction of this arc and on this, the normal deposition of the rest of the sedimentary units that occupy this great basin continued. Paleogene lithostratigraphic units that make up the San Luís-Guantánamo basin are: the Mícara, Gran Tierra, El Cobre Group (Pilón, and El Caney formations); Sabaneta, Charco Redondo (Fig. 4.18), Puerto Boniato, San Luís, Camarones, San Ignacio, Maquey, Yateras and Cilindro formations.

The Mícara Fm. is composed of a terrigenous sequence of polymictic and grauvaric sandstones and polymictic conglomerates, siltstones, shales, limestones and eventually marls. Its age is Paleocene Danian (basal). The El Cobre Group has the same characteristics as in the Sierra Maestra. The Gran Tierra Fm. lithologically consists of brecciated limestones, vulcanomictic conglomerates, breccias, marls, tuffs, organic-detrital limestones, vulcanomictic sandstones of calcareous cement, siltstones and tuffites. It lies concordantly on the Mícara Fm. and discordantly on the Santo Domingo and Sierra Verde formations. The age is Lower Paleocene (Danian). It was deposited in a marine environment of medium depths. The approximate thickness is 200 m. The Sabaneta Fm. is a sequence of volcanic-sedimentary and sedimentary rocks from the Paleocene age up to the Middle Eocene. The unit can reach up to 1100 m thick. The Puerto Boniato Fm. lithologically is represented by massive limestones, finely stratified or lenticular, biodetritic, algaceous limestone, marly, micrites and marls with fine intercalations or nodules of black silicites, brown or creams, in smaller proportion siltstones and shales. The studied fossils allow to date it as Middle Eocene upper part. It was deposited in an environment of moderately deep sea waters and its thickness does not exceed 50 m.

In the stratigraphic interval of the Middle Eocene in this basin is the Charco Redondo Fm. of mainly carbonate rocks. The San Luís Formation, eminently siliciclastic, of Middle Eocene upper part -Upper Eocene age, is also developed.

The San Ignacio Fm. is a polymictic breccia with fragments of metamorphic rocks predominantly green schists, phyllites and serpentinites, in a clay matrix. Its index fossils are foraminifera: they date the unit as Middle Eocene.

Fig. 4.18 Limestones of the Charco Redondo Formation, middle Eocene, on the Guantánamo-Santiago de Cuba highway, west of Guantánamo

The Camarones Formation is made up of polymictic conglomerates of subrounded and rounded edges and coarse-grained polymictic sandstones. The matrix of the conglomerate is sandy, and also of polymictic composition. Its development is limited to the province of Santiago de Cuba. Age is Upper Eocene. It was deposited in an open neritic environment. It can reach up to 500 m thick.

During the Oligocene and part of the Miocene the Maquey and Yateras formations were deposited. Maquey Fm. is an alternation of sandstones, siltstones and calcareous clays, white to cream marls that contain intercalations of varying thickness of limestones. The stratification is fine to medium, less frequently thick or massive. Some horizons, particularly of siltstones and biodetritic limestones, are fossiliferous, with large "lepidocyclines" being abundant. Other horizons contain plaster, lignite and lignitized plant remains. The thickness can exceed 700 m. Age is Upper Oligocene-Lower Miocene lower part.

The Yateras Formation lithologically are alternations of detrital, biodetritic and biogenic limestones of fine to coarse grain, thin to coarse or massive, hard stratification, of varying porosity, sometimes aporcelanated, often containing large "lepidocyclines." The coloration usually is white, cream or pinky, less frequently brownish. Age is Lower Oligocene-Lower Miocene lower part. They are reef deposits, covering different varieties of the reef complex. The thickness ranges between 160 and 500 m.

4.4.5.2 Neogene and Quaternary Systems

If you look at the Geological Maps of the Republic of Cuba 1: 500,000 [6] and 1: 250,000 [7] the wide spatial distribution of the formations of these systems will be observed. Its outcrops are abundant. It is registered in dozens of oil wells in its first meters. Carbonate and eventually terrigenous and evaporite rocks predominate. The fossiliferous content is very rich and varied. Not only stand out the foraminifera, but also the ostracods, mollusks, corals and echinoderms. It is divided into the Miocene and Pliocene series. The lithostratigraphic units are grouped into Oligo-Miocene, Late Lower Miocene-Early Upper Miocene, and Late Upper Miocene-Lower Pliocene. The Pliocene extends from the Upper Pliocene to the Lower Pleistocene of the Quaternary system.

Upper Oligocene—Early Lower Miocene

The tectonic movements were characterized mainly by subsidence, the uplifts were not remarkable. The most important occurred in the Los Palacios Basin: the Candelaria No. 1 well registered 2050 m thick of Miocene rocks, while the Mangas well crossed 1555 m. In the third version of the Léxico Estratigráfico de Cuba (2013) from east to west, the following formations are described: Paso Real; Jaruco; Colón; Banao; Lagunitas; Báguanos; Camazán; Bitirí; Sevilla Arriba; Yateras; Maquey; Cilindro and Cabacú. Carbonate rocks predominate and, in a lesser proportion, the terrigenous ones.

Late Lower Miocene—Early Upper Miocene

It points towards a greater development of coral reefs as well as the predominance of benthic fossil fauna and greater contribution of biodetritic materials and clay materials. The cut began to become more carbonate, which differentiates it from the previous ones that are more terrigenous and carbonate terrigenous. From the west to the eastern region the following formations are described: Güines, Cojímar; Santa María del Rosario; Arabos; Loma Triana; Baitiquirí; Rio Jagüeyes and Vázquez. The Güines Formation is the most distributed in the country in the Miocene.

Late Upper Miocene-Lower Pliocene

At the beginning of the Upper Miocene there was an almost total uprising of the eastern area, much of the province of Camagüey, as well as part of the provinces of Pinar del Rio, Artemisa, Mayabeque and Matanzas. In most of the national territory, the sedimentations occur in shallow waters, except the Nipe region.

The formations recognized from east to west are: Bellamar, La Cruz; Manzanillo Cabo Cruz; Júcaro; Punta Imías; Baracoa.

Quaternary System

The stratigraphic investigations of the Quaternary in the Cuban archipelago have reached a broad development in recent years. They can be separated into studies on Upper Pliocene-Early Pleistocene; Lower Pleistocene, Middle Pleistocene and Upper Pleistocene.

Glacio-eustatic transgressions and regressions of sea level played an important role in the history of Cuba's quaternary sedimentation. During the marine transgressions, huge areas of the current island of Cuba were covered by water and, sediments were deposited, which later in each regression, were subjected to tropical subaerial weathering agents.

The following formations are mapped; Vedado; Rio Maya; Punta del Este; Guane, Dátil and Bayamo. The Vedado Fm. consists of bio-thermal, organogenic, consolidated limestones. The Fm. Rio Maya are also biohermic, coral, micritic limestones, polymictic conglomerates. The Punta del Este, Guane, Dátil and Bayamo formations have mainly friable rocks and some are somewhat consolidated. There are conglomerates, gravels, sands and weakly cemented clays. The Dátil Fm. has boulders with blocks and sandy pebbles. In the Bayamo Fm. describes sand, sandstones and conglomerates, poorly consolidated clays.

Fig. 4.19 Organogenic limestones of the Cabo Cruz, Maya and Jaimanitas formations, in the terraces of Maisí to the south in the coastal zone, province of Guantánamo. Upper Pliocene—Early Pleistocene

Lower Pleistocene

The Guevara and Cauto formations are described. They are formations of sediments mainly friable.

The Guevara Formation emerges in the provinces of Pinar del Rio, Artemisa, Mayabeque, Matanzas, Cienfuegos, Ciego de Ávila and Camagüey. It consists of plastic clays, sands, gravels and boulders of oligomictic composition.

The Cauto Formation takes place in the Cauto river valley, in the provinces of Granma, Holguín and Santiago de Cuba. They are described, clays, silts, sands, polymictic gravels, with horizontal and cross stratification.

Middle Pleistocene

The following formations are recognized: Villarroja; Cayo Piedras; Guanabo and Versalles.

The Villarroja Fm. is distributed in the provinces of Havana, Matanzas, Cienfuegos, S. Spíritus and Camagüey. These are mainly sandy clays, sandy loams, quartz sands and gravels. The Fm. Cayo Piedras are biocalcarenites, pseudoolitic calcarenites, argillaceous and sandy limestones. In the province of Havana, the Guanabo Formation emerges, consisting of cross-stratified biocalcarenites. Finally, the Versalles Fm. is distributed in the province of Matanzas and are biodetritic limestones and calcarenites.

Upper Pleistocene

It includes the following formations: Camacho; Siguanea; Jamaica; Jaimanitas and Playa Santa Fe. These are unconsolidated or poorly consolidated sediment formations. Sometimes they are recognized as eolianites with crossed stratification.

The Camacho formation consists essentially of sandy silts, clays and clayey silts. The Siguanea Formation contains quartz sand, clay gravels and peat. The Jamaica Formation contains polymictic conglomerates, sandy rocks and silts. The Jaimanitas Fm. (Fig. 4.19) is a unit that is widely distributed along the coasts of Cuba. They are very organogenic limestones, organic-detrital rich in corals and fossils that mark the Pleistocene. Finally, the Playa de Santa Fe Formation is well recognized in the northwestern part of Cuba and is used in quarries where blocks for building materials are extracted. They are lamellar calcarenites, biocalcarenites that show crossed stratification.

There are Quaternary deposits not subdivided as they are: elluvial karst alluvial; elluvial-colluvial-prolluvial; marines; swamps, biogenic and caverns.

4.5 Conclusions

- Stratigraphic data on the Cuban archipelago, recognize as older stratified rocks, those of the Synrift Paleogeographic Domain, deposited between the Late Triassic–Upper Jurassic Oxfordian.
- From Late Upper Jurassic Kimmeridgian to the Coniacian, mainly carbonate rocks from the Paleogeographic Domain of the North American Continental Margin were deposited, and their corresponding orogenic coverage was deposited on them.
- Two volcanic arcs are separated, one from the Cretaceous, possibly Neocomian to the Campanian

(CVA) and, another, from the Paleogene of the Daniano to the lower part of the Middle Eocene (Turquino Arc).

- The youngest rocks are grouped in the Neogene and Quaternary Systems that have ages from the Upper Oligocene to the Recent. They are rocks mainly carbonated and in smaller quantities terrigenous.
- When considering the tectonic data, it is shown that by interacting the rocks of the Cretaceous Volcanic Arc—located at the front edge of the Caribbean Plate—with the southern edge of the Passive Margin of the North American Continental Crust, was formed, in the Early Paleogene, the Cuban Folded and Overthrusted Belt.
- After the Middle Eocene, the post-orogenic rocks mainly carbonated, that have not undergone the phenomena of folding and overpressure, were deposited; these are observed little inclined and without manifestations of volcanism.

Acknowledgments The authors wish to express our gratitude to Dr Sc. Manuel Pardo Echarte for entrusting us with the writing of this chapter and his constant occupation and invaluable encouragement in the revision of it. To the geologists Osmany Pérez Machado Milán, Yeniley Fajardo Fernández, and Lorenza Mejías Rodríguez for the making of the figures.

References

1. Gignoux M (1950) Stratigraphic geology: W.H. Freeman and Co., San Francisco. English edition (Fourth French edition trans: Woodford GG) p 682
2. Fernández de Castro M, Salterain P (1869–1883) Croquis geológico de la Isla de Cuba. Bol Mapa Geol España, Madrid, 8
3. Fernández de Castro M (1877) Fósiles índices de la Isla de Cuba pertenecientes al género Asterostroma. Anales Acad Cienc Med Fis Nat Habana 13:549–553
4. Furrazola Bermúdez G, Judoley CM, Mijailovskaya MS, Miroliubov YS, Novojatsky IP, Nuñez Jiménez A, Solsona JB (1964) Geología de Cuba. Inst Nac Rec Miner, Minist Indust, Editorial Nacional de Cuba. Havana 1–239
5. Linares Cala E, Rodríguez R (1997) Grado de estudio geológico y geofísico de Cuba. Libro Estudios sobre Geología de Cuba. pp 479–490. CNDIG-IGP. ISBN. 959-243-002-0
6. Linares E, Osadchiy PG, Dovbnia VA, Gil S, García DE, García LM, Zuazo A, González R, Bello V, Brito A, Bush WA, Cabrera M, Capote C, Cobiella JL, Díaz de Villalvilla L, Eguipko OI, Evdokimov JV, Fonseca E, Furrazola G, Hernández J, Judoley CM, Kondakov LA, Markovskiy BA, Pérez M, Peñalver L, Tijomirov YN, Vtulochkin AN, Vergara F,. Zagoskin AM, Zelepuguin VN (1985) Mapa geológico de la República de Cuba a escala 1:500000. Minist Ind Bas Fábrica Cartográfica, Instituto de Investigaciones Geológicas A. P. Karpinski, Leningrado
7. Albear JF, Boyanov I, Breznyanszky K, Cabrera R, Chejovich V, Echevarría B, Flores R, Formell F, Franco G, Haydutov I, Iturralde Vinent M, Kantchev I, Kartashov I, Kostadinov V, Millán G, Myczynski RE, Nagy V, Oro J, Peñalver L, Piotrowska K, Pszczolkowski A, Radocz J, Rudnicki J, Somin M (1988) Comisión de unificación del mapa geológico de la República de Cuba escala 1: 250000, 40 Hojas. Academia de Ciencias de Cuba. Instituto de Geología y Paleontología, Edición Instituto de Geología de la URSS
8. Longoria JF (Julio–Diciembre, 1993) La terrenoestratigrafía: un ensayo de metodología para el análisis de los terrenos con un ejemplo de México: boletín de la asociación Mexicana de geólogos petroleros vol XLVIII(2): 30–48
9. Howell DG (1989) Tectonic of suspect terranes. Chapman and Hall, London, New York, Mountain building and continental growth
10. Howell DG et al (1985) Tectonostratigraphic terranes of the circum pacific region. Council for Energy and Mineral Resources, Houston, pp 3–30
11. Hatten CW, Somin M, Millán G, Renne PR, Kistler RW, Mattinson JM (1989) Tectonostratigraphic units of central Cuba. Trans Eleventh Caribe Geol Conf Barbados 35:1–14
12. Franco GL, Colectivo de redactores (2013) Léxico estratigráfico de la República de Cuba. Instituto de geología y Paleontología servicio geológico de Cuba. Ministerio de energía y minas. Oficializado en 1986, editado en 2002 y 2013 en su tercera versión. ISBN: 978-959-7117-58-2. Havana
13. Renne PR, Mattinson JM, Hatten CW, Somin M, Onstott TC, Millan G, Linares E (1989) 40 Ar 39 Ar and U-Pb evidence for late proterozoic (Grenville age) continental crust in north-central Cuba and tectonic implications. Precambric Res 42: 325–341
14. Linares E, García DE Delgado López O, López JG, Strazhevich V (2011) Yacimientos y manifestaciones de hidrocarburos de la República de Cuba. Centro nacional de información geológica. IGP–Ceinpet. p 480. ISBN 978-959-7117-33-9. Imprenta PALCOGRAF, Havana
15. Pszczolkowski A (1986) Composición del material clástico de las arenitas de la formación San Cayetano, en la Sierra de Los Órganos, provincia de Pinar del Rio. Ciencias de La tierra y del espacio 11/86
16. Iturralde Vinent M, Roque F (1982) Nuevos datos sobre las estructuras diapíricas de punta alegre y turiguanó, Ciego de Ávila. Rev Ciencias de la Tierra y del Espacio 4:47–55
17. De Golyer E (1918) The geology of Cuban petroleum deposits: American assocation petroleum geology bulletin, vol 2. Tulsa. Okl. USA, pp 133–167
18. Iturralde VM (ed) (2012) Compendio geología de Cuba y del caribe. Segunda Edición CITMATEL DVD-ROM, Havana, Cuba
19. Haczewski G (1976) Sedimentological reconnaissance of the San Cayetano Formation. An accumulative continental margin in the jurassic. Acta Geologica Polonica 26(2):331–353
20. Fernández Carmona J, Areces A (1987) Estratigrafía del área Los Arroyos. Archivo del CEINPET, Havana, Provincia de Pinar del Río
21. Dueñas H, Linares E (2001) Asociaciones palinológicas de muestras de la formación San Cayetano: O-1674, Archivo del CEINPET, MINEM, Havana (Inédito)
22. Flores Nieves Aliena (2011) Estudio palinológico de la formación San Cayetano y su vinculación con la exploración de hidrocarburos. Archivo Universidad de Pinar del Rio y CEINPET, MINEM
23. Moretti I, Tenreyro R, Linares E, López JG, Letousey C, Magnier CF, Gaumet CF, Lecomte JC, López JO, Zimine S (2002). Petroleum system of the Cuban North-West offshore: Gulf of México. AAPG Memoir
24. Valdivia Tabares C, Veigas C, Martinez E, Delgado O, Dominguez Z, Pardo M, Jiménez L, Cruz R, Gómez J, Rosell Y, Rodríguez Morán O (2015) Informe de los resultados de la evaluación del potencial de hidrocarburos del Bloque 17. Archivo Técnico del Ceinpet, Havana
25. Pszczolkowski A et al (1987) Contribución a la geología de la provincia de Pinar del Rio. Ed. científico-técnica. ACC. Havana

26. Fernández Carmona J (1998) Bioestratigrafía del jurásico superior —cretácico inferior neocomiano de Cuba occidental y su aplicación en la exploración petrolera: tesis doctoral. Archivo del CEINPET, Havana
27. Linares E (2003) Comparación entre las secuencias mesozoicas de aguas profundas y someras de Cuba Central y Occidental. Significado para la exploración petrolera. Tesis de Doctor en Ciencias Geológicas, Archivos CUJAE y Ceinpet, Havana
28. Kiyokawa S, Tada R, Iturralde Vinent MA, Matsui T, Tajika E, Garcia DE, Yamamoto S, Oji T, Nakano Y, Goto K, Takayama H, Rojas Consuegra R (2000) Cretaceous-tertiary boundary sequence in the Cacarajicara formation, Western Cuba: an impact-related, high-energy, gravity-flow deposit. Geol Soc Am Spec Pap 356:125–144
29. Tada R, Nakano Y, Iturralde Vinent MA, Yamamoto ST, Kamata E, Tajika K, Toyoda Kiyokawa S, García Delgado DE Goto K, Takayama H, Rojas Consuegra R, Matsui T (2002) Complex tsunami waves suggested by the Cretaceous-Tertiary boundary deposit at the Moncada section, western Cuba geological society of America special Paper 356
30. Blanco Bustamante S, Fernández G, Fernández J, Flores E y Sánchez J (1985) Zonaciones cubanas de los principales grupos fósiles de importancia estratigráfica. Grupo I: escala paleontológica única. Proy. 165. P. I. C. G. 2a Reunión Internacional, Havana (inédito)
31. Ducloz C, Vuagnat M (1962) A propos de l'age des serpentinites de Cuba. Arch Sci, Soc et d'Hist Nat 15(2):3O9–332
32. Linares Cala E, García DE, Blanco Bustamante S, Fajardo Fernández Y, Perez Estrada L (2015) Precisión de la edad de la Formación Lindero y su correspondencia con el final del Arco de Islas Cretácicas. Anuario SCG Número 3
33. Hatten CW, Schooler OE, Giedt N, Meyerhoff AA (1958) Geology of central Cuba, eastern Las Villas and western Camaguey Provinces, Cuba. Unpublished report, Standard Cuban Oil Co., p 174
34. Belmustakov E, Dimitrova E, Ganev M, Haydutov Y, Kostadinov Y, Ianev S, Ianeva J, Kojumdjieba E, Koshujarova E, Popov N, Shopov V, Tcholakov P, Tchounev D, Tzankov T, Cabrera R, Diaz C, Iturralde Vinent M y Roque F (1981) "Geología del territorio Ciego de Ávila-Camagüey-Las Tunas. Resultados de las investigaciones y levantamiento geológico a escala 1: 250,000". Academias de ciencias de Cuba y Bulgaria. 940 páginas y mapas (Inédito). ONRM, MINBAS, Havana
35. Bronnimann P, Pardo G (1954) Annotations to the correlation chart and catalogue of formations (Las Villas province). Centro Nac Fondo Geol, Minist Indust Bas, Havana (inédito)
36. Bermúdez PJ (1950) Contribución al estudio del cenozoico Cubano. Mem De La Soc Cubana De Historia Nat 19(3):205–375
37. Iturralde Vinent M, Díaz C (1986) Nueva unidad litoestratigráfica del cretácico de Camagüey. Encuentro de geólogos en la escuela de Cuadros del MINBAS, Havana (res. pub.)
38. Flores G, Auer WF (1949) Geology of the northwestern Camaguey province, Cuba. Bi-weekly report # 17. Centro nacional del fondo Geológico, MINBAS, Havana (inédito)
39. Rutten MG (1936) Geology of the Northern part of the province Santa Clara. Cuba Geogr Geol Mededdel, Utrecht Geol Phys Reeks 11:1–60
40. Truitt P, Pardo G (1956) Pre tertiary stratigraphy of northern Las Villas province and northwestern Camaguey province, Cuba. Geologic memorandum PT-47. Unpublished report, Cuban Gulf Oil Co., p 76
41. Millán Trujillo G (1997) Posición estratigráfica de las metamorfitas cubanas. Estudios sobre geología de Cuba. Compilación de: Gustavo Furrazola Bermúdez y Kenya E. Núñez Cambra. Centro nacional de información geológica. IGP. Havana
42. Alsina de la Nuez P, Álvarez Castro J y Ramírez G (1968) Consideraciones geológicas acerca de las posibilidades de producción comercial de hidrocarburos en el área del Cauto. Rev. Tecnol., Havana, 6(1–2):33–57
43. International Commission Stratigraphy IUGS (2016) International chronostratigraphic chart
44. Kantchev Il, Boyanov Y, Popov N, Goranov Al, Iolkichev N, Cabrera R, Kanazirski M, y Stancheva M (1978) Geología de la provincia de Las Villas. Resultados de las inves-tigaciones geológicas y levantamiento geológico a escala 1:250 000, realizado durante el período 1969–1975. Brigada cubano-búlgara. Inst Geol Paleont, Acad Cienc Cuba, Havana (inédito)
45. Myczynsky, R. 1976. A new ammonite fauna from the Oxfordian of the Pinar del Rio province, western Cuba. Acta Geológica Polaca. 26 (2): 261–299, Warszawa, 1976.
46. Ortega y Ros P (1931) Informe geológico presentado al gobierno provincial de Santa Clara sobre el registro petrolero "Carco", denunciado por la compañía petrolera CARCO en la provincia de Santa Clara. Oficina nacional de recursos Minerales, MINBAS, Havana (inédito)
47. Rojas Agramonte Y, Kroner A, Pindell J, Garcia Delgado DE, Dunyi L, Yusheng W (2008) Detrital zircon geochronology of jurassic sandstone of Western Cuba. (San Cayetano Formation): implications for the jurassic paleogeography of the NW proto-caribbean. Am J Sci 308: 639-656
48. Sánchez Arango J, Attewell R (1993) Stratigraphy. In: The geology and hidrocarbon potential of the Republic of Cuba. Proprietary report, simon petroleum technology an CUBAPETRÓLEO eds. Llandudno, U.K., Chapter 3 and Box. No. 3
49. Shein VS, Konstantín A, Klischov Jain VE, Dikenshtein GE, Yparraguirre JL, García E, Rodríguez R, López JG, Socorro R y López JO (1985) Mapa Tectónico de Cuba escala 1: 500 000. Centro de investigaciones geológicas del ministerio de la industria básica. Edición ICGC, 4 Hojas
50. Shopov V (1982) Estratigrafia y subdivisión de las zonas placetas y Camajuaní en la antigua provincia de Las Villas. (Cuba Central). Ciencias de La Tierra y del espacio. No. 4 Academia de Ciencias de Cuba. pp 39–46
51. Truitt P (1955) Memo PT-34. Geology of the Punta alegre-cayo coco-Turiguanó area (1955) ONRM. MINEM, Havana (inédito)

An Overview to the Tectonics of Cuba

Jorge Luis Cobiella Reguera

Abstract

Two main structural levels or stages can be distinguished in the geological architecture of Cuba. The lower stage is the socle, a great rock complex, mainly formed by Jurassic–Eocene rocks, unconformably resting below the cover. The socle is divided into three major complexes, according to their litho-structural features and rock age: (a) the Proterozoic–Paleozoic basement, (b) the Mesozoic basement, and (c) the Paleogene folded and thrust belt. A small part of the socle is represented by Precambrian (Grenvillian) rocks, only known by tiny outcrops in the northcentral Cuban mainland, whereas the Paleozoic eratem is only known in the sea floor of the Cuban Exclusive Economic Zone of the SE Gulf of Mexico. The Mesozoic basement consists of four rock complexes of very different nature (a) The Mesozoic passive paleomargin of the SE North American plate (NAP). Whereas the NAP contains autochthonous or parauthoctonous massifs, the remaining Mesozoic units are tectono-stratigraphic terranes, separated by tectonic contacts between them and with the NAP. These terranes are (b) the northern ophiolite belt (NOB), (c) the Cretaceous volcanic arcs (KVA), (d) the southern metamorphic terranes (SMT). The Mesozoic basement attains a wide distribution both in outcrops and subsurface. The Paleogene folded and thrust belt contains four regional structures: the foreland basin, the piggyback basins, the Sierra Maestra–Cayman Ridge volcanic arc, and the Middle and the Late Eocene Eastern Intramontane Basin. However, complicated, their mutual relationships are quite much clearer than that prevailing in the Mesozoic basement. The Eocene–Quaternary cover contains little disturbed beds, without magmatic or metamorphic rocks. The nearness of SE Cuba to the Caribbean/North American plate boundary prints its cover with some special features.

Keywords

Tectonics • Gulf of Mexico • Caribbean • North American Paleomargin • Cuba • Ophiolite

J. L. Cobiella Reguera (✉)
Instituto de Geología y Paleontología-Servicio Geológico de Cuba, Vía Blanca No. 1002 y Carretera Central, San Miguel del Padrón, 11000 La Habana, Cuba
e-mail: jcobiellar@upr.edu.cu; jorgecobiella45@gmail.com

J. L. Cobiella-Reguera
Departamento de Geología, Universidad de Pinar del Rio, calle Martí final No. 270, 20100 Pinar del Rio, Cuba

Abbreviations

NAP	North America Paleomargin
NOB	Northern Ophiolitic Belt
KVA	Cretaceous Volcanic Arcs
KVAT	Cretaceous Volcanic Arc Terrane
SMT	Southern Metamorphic Terranes
Fm	Formation
TSU	Tectono-Stratigraphic Unit
IGP	Instituto de Geología y Paleontología
CEEZ	Cuban Exclusive Economic Zone
APS	Alturas de Pizarras del Sur tectono-stratigraphic unit
SO	Sierra de los Órganos tectono-stratigraphic unit
SR-APN-E	Sierra del Rosario-Alturas de Pizarras del Norte tectono-stratigraphic unit
PgB	Paleogene Basin
N-Q	Neogene–Quaternary
PG	Pan de Guajaibón
R	Remedios tectono-stratigraphic unit
CC	Cayo Coco tectono-stratigraphic unit
P	Placetas tectono-stratigraphic unit
Cj	Camajuaní tectono-stratigraphic unit
DSDP	Deep Sea Drilling Project
K	Potassium
Ar	Argon
GNL	Guacanayabo–Nipe Bay lineament
MC	Mabujina Complex
HT/MP	High-Temperature/Mid-Pressure metamorphism
MTU	Main Tectonic Unit
PFB	Paleogene Foreland Basin
FZ	Fault Zone
T-Cr	Turquino–Cayman Ridge volcanic arc
COEB	Central Oriente Eocene Basin
PBB	Piggyback Basins
OFZ	Oriente Fault Zone
PFZ	Pinar Fault Zone
TFZ	La Trocha Fault Zone
SDB	Santiago Deformed Belt
OFZ	Oriente Fault Zone
SWB	Southwestern Cuban Basin
My	Millions years
U-Pb	Uranium–lead

5.1 Introduction

During the last 25–30 years the publications on the geology of Cuba have been very limited, particularly, those related to the regional tectonics research suffered a remarkable decrease during the 90s and first years of the twenty-first century. However, in the 80s of the twentieth century, several important achievements in the field of the regional geology, especially in its cartography were attained: two geological maps, embracing all the Cuban territory at scale

1:500,000 [112] and 1:250,000 [110], and the tectonic map 1:500,000 [111] stand as major achievements. Several books devoted to some results on the knowledge of the tectonics and stratigraphy of almost 50% of its territory appears between 1983 and 1987. In spite of these results in the advance of the geological knowledge, from 1970 to 1990, a significant part of this information attained limited divulgation in journals, even in Cuba.

This dramatic contrast in only a few years is tightly related to the difficulties and severe stress in the national economy after the international events developed between 1987 and 1992.

In the next pages, the author presents and discusses some general ideas on several items of the Cuban regional tectonics developed during almost 50 years.

5.2 Materials and Methods

From 2013, the current author works on a project of the Instituto de Geología y Paleontología (IGP), devoted to assemble a tectonic map to support the information for the Metallogenic Map of Cuba at scale 1:250,000 [24].

In the preparation of this chapter, we devoted a special attention to the reworking of the information obtained by field Cuban and foreign geologists in many localities of the national territory, in the search of a plate tectonics interpretation from the old data. There is much excellent information disseminated, including reports more than 80 years old, that still can be used, i.e., the fine reports of the geologists from Utrecht University (Holland) in the early decades of the twentieth century, the reports of the U.S. Geological Survey during the 40s and 50s, several isolated papers on thematic researches developed by the IGP during the last 60 years, and many others. A special place corresponds to the reports and maps related to the program for the geological cartography of the Cuban territory developed by the IGP from 1971 to 1989, especially the Geological (1988) and Tectonic (1989) maps of Cuba. The author has worked more than 50 years in different places of the country, a fact very important to assume this task but, obviously, he does not know every place and also the quality of the geological information from different territories is not homogenous, as it occurs in all countries.

In our approach, a particular attention is devoted to integrate the relationships between the Cuban territory and its surroundings, a subject that will be visited several times along the paper. Another subjects with many questions are the age of several main tectonic events, because of their poor datation and, therefore, the uncertain correlation between distinct territories. In some cases, the information on the nature and composition of the sedimentary record as well as the sedimentary structures present, fundamental for paleogeographic interpretations is scarce. Some cases are discussed.

There are many interesting items on the tectonics of Cuba but, within the limits of the current publication, only a small part of the problems could be visited. We will present an overview of the subject and try to introduce a new approach to some of them. Obviously, our experience is not uniform in all the fields, and several items will not receive the attention that each specialist demands.

5.3 Results and Discussion

Cuba and its surroundings are a geological mosaic located in the southern border of the North American plate. Rocks from many different origins, with Proterozoic to Quaternary ages, belong to this puzzle belt, extended along the southern border of the plate, from northern Central America to the Virgin Islands. From the Middle Eocene, this belt has been dissected by several great faults, related to the development of some oceanic depressions (Cayman Trough and Yucatan Basin) in the Mesoamerican area.

Two main structural levels or stages can be distinguished in the geological architecture of Cuba. The lower stage is the socle, a great rock complex, mainly formed by Jurassic–Eocene rocks, unconformably resting below the cover.

The socle is divided into three major complexes, according to their litho-structural features and rock age: (a) the Proterozoic–Paleozoic basement, (b) the Mesozoic basement, and (c) the Paleogene folded and thrust belt.

A small part of the socle is represented by Precambrian (Grenvillian) rocks, only known by tiny outcrops in the northcentral Cuban mainland, whereas the Paleozoic eratem is only known in the sea floor of the Cuban Exclusive Economic Zone of the SE Gulf of Mexico. The Mesozoic basement consists of four rock complexes of very different nature: (a) the Mesozoic passive paleomargin of the SE North American plate (NAP). Whereas the NAP contains autochthonous or parauthoctonous massifs, the remaining Mesozoic units are tectono-stratigraphic terranes, separated by tectonic contacts between them and with the NAP. These terranes are (b) the northern ophiolite belt (NOB), (c) the Cretaceous volcanic arcs (KVA), (d) the southern metamorphic terranes (SMT). The Mesozoic basement attains a wide distribution both in outcrops and subsurface.

The Paleogene folded and thrust belt contains four regional structures: the foreland basin, the piggyback basins, the Sierra Maestra–Cayman Ridge volcanic arc, and the Middle and the Late Eocene Eastern Intramontane Basin. However complicated, their mutual relationships are quite much clearer than those prevailing in the Mesozoic basement.

The Eocene–Quaternary cover contains little disturbed beds, without magmatic or metamorphic rocks.

5.3.1 The Premesozoic Basement. Precambrian and Paleozoic Rocks

A main achievement in the regional geology of the Greater Antilles, attained during the last 25 years of the twentieth century, is the discovery of Precambrian and Paleozoic rocks in Cuba and its surroundings.

Some of the main papers, dealing with the Cuban Precambrian rocks, are as follows: [81, 109, 113, 119, 121, 122]. As it is evident, however, the theoretical interest, because these rocks are the unique reference of very old events in the northern Caribbean, the problem of the Precambrian rocks in Cuba received very little attention during the last 30 years. The Precambrian basement is exposed in several small outcrops in northern central Cuba [101, 109, 122]. The Precambrian rocks are represented by phlogopitic marbles. Somin and Millán [121] obtained K–Ar radiometric ages of 910 and 945 My in Sierra Morena. Later Renne et al. [113] obtained $^{40}Ar/^{39}Ar$ ages of 903 My from phlogopite in marbles of the Socorro complex. The marbles are possibly intruded by Jurassic (172 My) granitoid plutons [113]. An arcosic regolith, containing granite and marble clasts, rests at the floor of the overlying Upper Jurassic–Lower Cretaceous sequence. Later, additional data on this subject will be displayed.

Meanwhile, there are some Precambrian outcrops in Cuba, we do not have evidences on Paleozoic rocks inland. The direct information on a Paleozoic basement comes from the DSDP wells 537 and 538A, located in the Cuban Exclusive Economic Zone (CEEZ, Fig. 5.1) of the SE Gulf of Mexico Basin, drilled in January 1981 [119]. According to these authors, in the first well, the basement rocks were phyllites with Ar/Ar ages of circa 449 and 456 My. In the hole 538A, the basement is represented by biotitic gneisses yielding radiometric ages of 496 and 501 My. These Lower Paleozoic rocks are intruded by diabase dykes of, at least, two generations of radiometric ages. The younger intrusive event (165–163 My) correlates with the mafites of El Sábalo Formation in western Cuba. The oldest one intrusions are 190 My [119].

Pszczolkowski [103] found another evidence on the existence of Paleozoic rocks very near to Cuba in the clasts of silicified limestones with Upper Carboniferous and Permian *fusulinids* (benthic foraminifera) from the Upper Jurassic beds of the San Cayetano Formation, near Cinco Pesos, in Sierra del Rosario Mountains. We will see that this unit is part of a Jurassic Paleo Delta, accumulated at the southwestern margin of Laurasia supercontinent (the North American plate).

According to Hutson et al. [51], who studied the radiometric age of 67 muscovite detrital grains in four samples from the Jurassic San Cayetano Formation, the abundance of grains with Paleozoic ages (79%) suggests a provenance from a Taconic orogenic source, probably located in

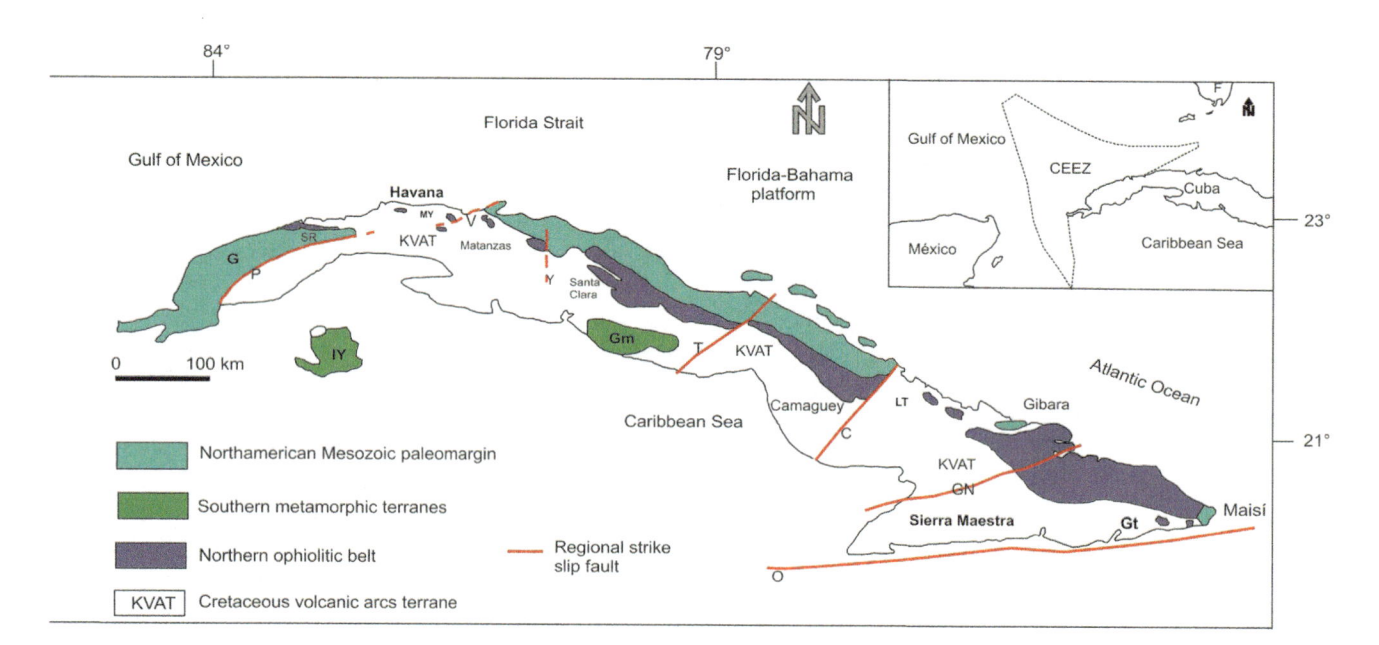

Fig. 5.1 Major strike-slip faults of Cuba (red lines) and Precenozoic regional structures G: Guaniguanico Mountains, SR: Sierra del Rosario Mountains, Gm: Guamuhaya Massif, IY: Isla de la Juventud, V: Varadero lineament, P: Pinar fault zone, Y: Yabre lineament, T: La Trocha lineament, C: Camagüey lineament, O: Oriente fault zone, CEEZ: Cuban Exclusive Economic Zone in the Gulf of Mexico, MY: Mayabeque Province, LT: Las Tunas Province, GT: Guantánamo, F: Florida Peninsula

Yucatan, because rocks of such ages are unknown in Florida and northernmost South America, the other probable sources, according to different precedent paleotectonic models.

Finally, Rojas-Agramonte et al. [116] studied the radiometric ages from the zircon grains in two sandstone samples of the same lithostratigraphic unit in Alturas de Pizarras del Norte. These authors also consider that part of the zircon grains could be derived from Paleozoic source rocks in Yucatán.

5.3.2 The Mesozoic Basement

Structurally, the units of the Mesozoic basement extend parallel to the axis of the main island of the Cuban Archipelago (Fig. 5.1). It represents the most complicated socle element. The northernmost unit is the North American passive continental paleomargin (NAP) now outcropping as part of a fold-thrust belt, along the northern edge of the Cuban mainland. The northern ophiolite belt (NOB) and the Cretaceous volcanic arcs terrane (KVAT) lie with tectonic contacts upon the paleomargin rocks, whereas farther southward outcrop the southern metamorphic terranes.

5.3.2.1 The Passive Mesozoic Continental Margin of North America (NAP)

The NAP includes a variegated group of deposits and some mafic tholeitic igneous rocks, accumulated upon an extensional continental margin, from the Jurassic to the Late Cretaceous (locally Paleocene). Three areas with a particular development of the NAP stratigraphy can be distinguished: the mountains of Guaniguanico Cordillera, in westernmost Cuba; northcentral Cuba, from La Habana to NW Holguin Province and Maisí, in the eastern tip of the island (Fig. 5.1).

The Guaniguanico Cordillera. In the low mountains of westernmost Cuba, a great nappe pile, several kilometers thick, outcrops. Its tectonic style was discovered during the field researches at Sierra de los Órganos in the twentieth century (50s) [46, 114]. Along the southern fringe of Guaniguanico Cordillera the thrust sheets are cut by the Pinar fault zone, separating the Jurassic–Cretaceous rocks from the Eocene–Quaternary cover (Los Palacios Basin, Fig. 5.2). Five tectono-stratigraphic units (TSU), each with its own stratigraphy and structural style, are represented in the Cordillera (Figs. 5.2)There are three lower units: Sierra de los Órganos (SO), Alturas de Pizarras del Sur (APS), and Sierra del Rosario-Alturas de Pizarras del Norte-Esperanza (SR-APN-E). Here essentially the structure of the socle is a great antiform made by nappes, with its western two-thirds forming a NW convex arc [17, 19, 30, 61, 92, 97, 99, 107, 104, 108]. In the eastern third (Sierra del Rosario mountains), the SR-APN-E TSU presents a general picture of thin north dipping tectonic sheets, except locally, where south-dipping structures outcrop [29].

The oldest Mesozoic deposits belong to the San Cayetano Formation [39], a thick siliciclastic complex, accumulated in a large Jurassic delta. Oxfordian diabase sills and basaltic

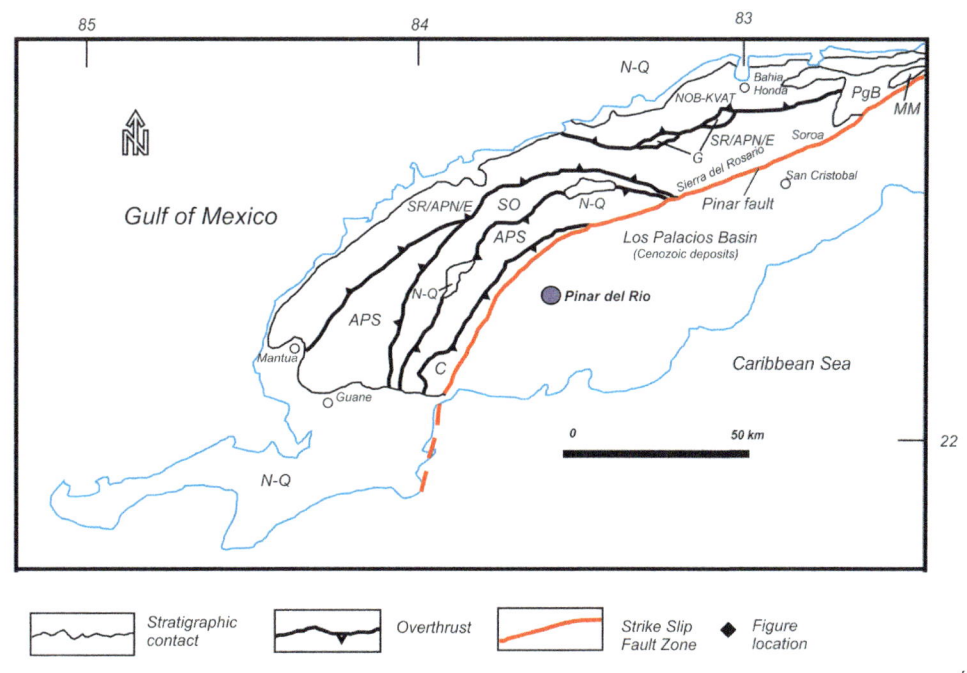

Fig. 5.2 The regional structures below the cover in western Cuba. Modified from Cobiella Reguera [19]. SO: Sierra de los Órganos TSU,APS: Alturas de Pizarras del Sur TSU; SR-APN-E: Sierra del Rosario-Alturas de Pizarras del Norte-Esperanza TSU; G: Pan de Guajaibón TSU; C: Cangre Belt TSU; NOB-KVAT: Northern Ophiolite Belt and Cretaceous volcanic arc terranes; PgB: Paleogene Basin (Cover); MM: Martin Mesa uplift (outlier of SR-APN-E); Cover: N-Q: Neogene–Quaternary deposits

pillow lavas (Fig. 5.4, El Sábalo Formation) appear frequently near their transitional contact with the Oxfordian carbonate beds of the overlying sequence [14, 14, 107, 108]. Evidences about the basement lying below the San Cayetano Formation still remain unknown and, probably, a huge decollement surface marks this lower contact. Haczewski [45], in a detailed sedimentological research, distinguishes nine facies in San Cayetano Basin and considers that their deposition occurs in the frame of a widespread deltaic environment (see also [4]), where the SO and APS contain alluvial and litoral marine deposits and SR-APN-E beds contain the deeper marine turbidites. More recent studies, supported on (a) the decrease in the quartz grain percent and the increase in feldspar + rock grain content in the San Cayetano sandstones, (b) the areal distribution of the Haczewski's facies model in the Cordillera nappe pile, and (c) the systematic structural evidence in their beds of a northward vergence [17, 28, 100], conclude that the lower units (SO and APS) represent the original northern part of the Jurassic delta and SR-APN-E belong to the primary southern locations. Therefore, the sediment source was located toward the N-NW of the delta.

Above the siliciclastic beds rests an Oxfordian–Cenomanian sequence (Fig. 5.3), whose lower beds conform a transitional terrigenous-carbonate middle–upper Oxfordian packet. In Sierra de los Órganos TSU, they belong to a 160-m-thick carbonate ramp, the Jagua Formation, whereas in Sierra del Rosario-Alturas de Pizarras del Norte-Esperanza unit, the same interval corresponds to deeper water deposits (Francisco Formation), represented by argillites and well-bedded micritic limestones, only a few meters thick, accumulated at neritic depth under restricted environments [30, 108]. An unconformity appears at the Oxfordian/Kimmeridgian contact. In Sierra de los Órganos TSU, a 650-m-thick Kimmeridgian-lower Tithonian carbonate bank (the Guasasa Formation) began a great sedimentary sequence [98, 77]. The overlying lower Tithonian–Cenomanian strata are deep water limestones with some cherty interbeds showing a clear trend to deeper water sediments with time. In SR-APN-E (Fig. 5.5), a similar phenomenon appears, but terrigenous and carbonate turbiditic beds are frequent up to the Aptian or Albian. In this TSU, thin and very scarce tuff beds appear from the Aptian [17]. This suggests a possible geographic connection with the Greater Antilles Cretaceous volcanic arc, active at that time. Thin Turonian strata are reported in some isolated localities [99, 108]. A conspicuous and well-developed Coniacian–Maastrichtian hiatus (the Upper Cretaceous Unconformity) marks the stratigraphic column of the Mesozoic paleomargin in the Guaniguanico Cordillera, the SE Gulf of Mexico and the NAP in northcentral Cuba [20]. Only in a few locations in the eastern SR-APN-E unit, Campanian beds have been recorded (Moreno Formation, [42, 104, 108]). In this lithostratigraphic unit, together with carbonate sediments, there are some tuffs and volcanoclastic beds, again suggesting the proximity of a volcanic source.

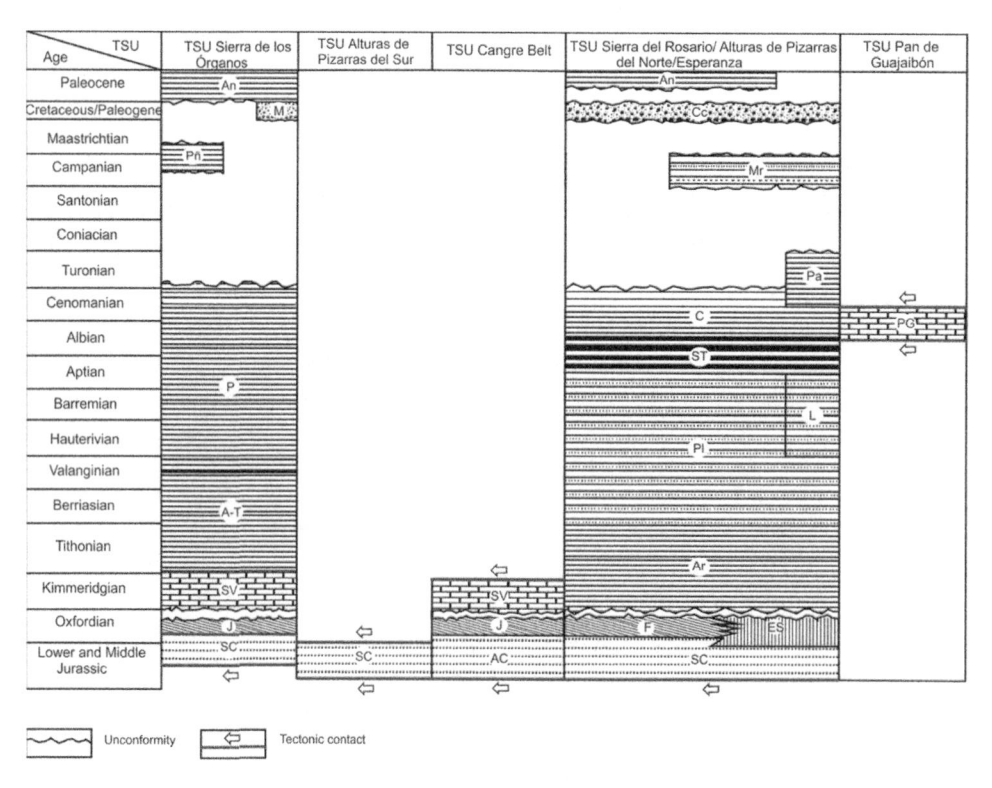

Fig. 5.3 Lithostratigraphic columns of the Guaniguanico Cordillera TSU. An: Ancon Fm.; M: Moncada Fm.; Peña Fm.; Pons Fm.; A-T (El Americano, Tumbadero and Tumbitas members of Guasasa Fm.); SV: San Vicente Fm.; J: Jagua Fm.; SC: San Cayetano Fm.; Arroyo Cangre Fm.; Mr: Morena Fm.; Pa: Pinalilla Fm.; C: Carmita Fm.; ST: Santa Teresa Fm.; L: Lucas Fm.; Pl: Polier Fm.; Ar: Artemisa Fm.; ES: El Sábalo Fm. Modified from Cobiella Reguera [19]. See also Pszczolkowski [108]

Fig. 5.4 Outcrop of El Sábalo Fm., with a mix of mafic rocks (basalt and diabase-m) and xenoliths-x, of thin-bedded Upper Jurassic limestones and shales, containing complex synsedimentary folds

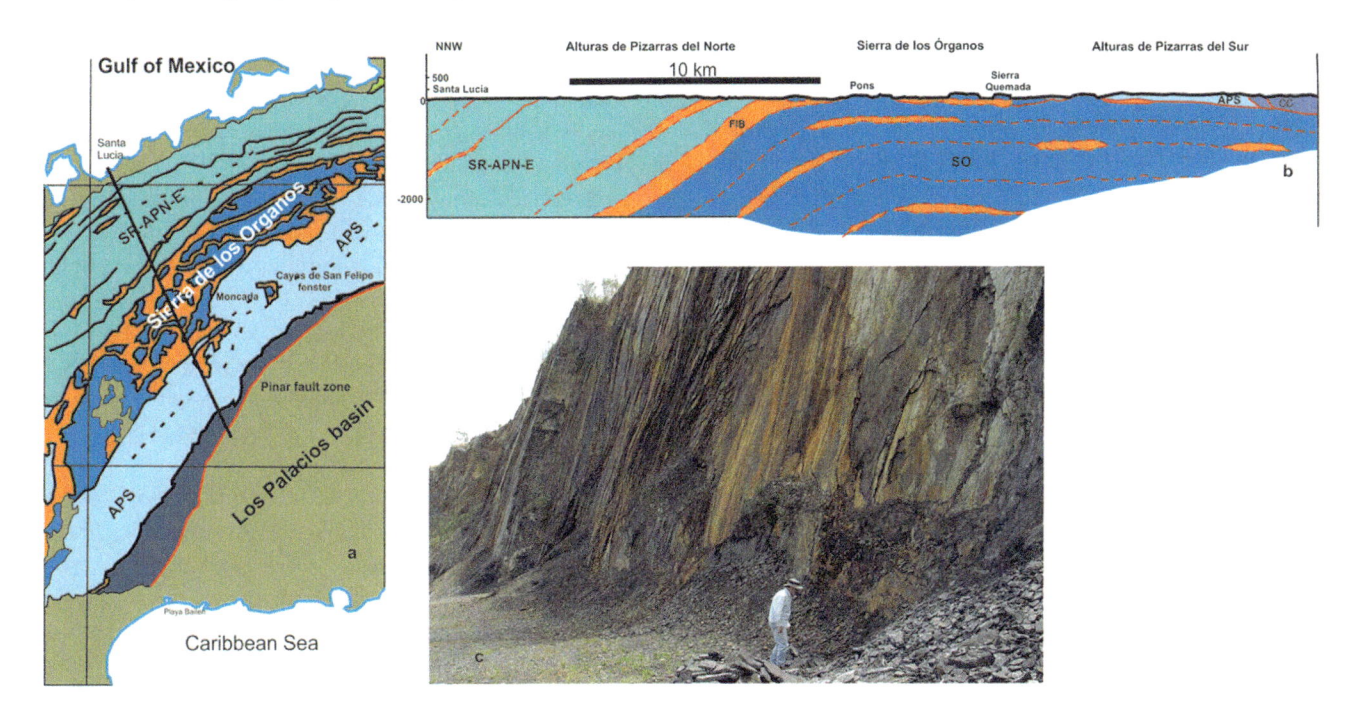

Fig. 5.5 Schematic map (**a**) and tectonic profile (**b**) along the central part of Guaniguanico Mountains, from Pinar fault zone to the northern coast, near Santa Lucia. Compare the flat lying nappes in APS and SO TSU with the north dipping linear structures in the SR-APN-E. **c** North dipping lowermost Cretaceous beds at Vegas Nuevas quarry, near La Palma, SR-APN-E

The western sector of the NAP is only with Paleocene deposits. In SR-APN-E, the lower part is a thick clastic K/Pg boundary deposit (Cacarajícara Formation, Fig. 5.3) with variable thickness (from meters to several hundred meters). The lower part of the unit has very coarse-grained deposits from debris flows, avalanches, and turbiditic currents. Meanwhile, the upper part is a fining upward massive to poor stratified calcarenites to calcilutitic beds, the homogenite of Tada et al. [126], see also Cobiella Reguera et al. [20]. In the SO unit, Cretaceous/Paleogene boundary beds are known only in one locality [5, 126] represented by tsunamic deposits, less than 2 m thick, very rich in ejecta clasts derived from the asteroid impact at Chicxulub, Yucatan, Mexico. The differences in thickness and composition in the K/Pg boundary deposits between both localities are related to the coeval regional geomorphology in the SE Gulf of Mexico–NW Caribbean area [20]. In both areas, the Paleocene to Lower Eocene beds belong to the thin and well-stratified carbonate Ancón Formation. We will return later to the extraordinary K/Pg boundary deposits from Cuba.

The nappes (tectonic sheets) represent the most remarkable structural element in the Sierra del Rosario mountains geology [104]. These structures were discovered during the geological cartography in the 70s and 80s. According to well data, the SR-APN-E thrust pile attains vertical thickness of circa 5 km in the NW Pinar del Rio Province [19]. Because there is not a thick regional rigid bed (as it occurs with the San Vicente Formation in SO, [92, 98], and thin-bedded terrigenous strata are abundant [29, 107], the SR-APN-E represents an excellent example of thin skin tectonics. The deformational features inside each nappe depend on the bed lithologies and thickness, its position with relation to the over thrust sheet floor and the depth of the deformation. In the well-bedded formations (Polier, Artemisa, Santa Teresa and others, Fig. 5.3), tight isoclinal folding frequently develops. In these successions, each strata tends to slide along its bedding planes, particularly where argillites are involved. Therefore, many faults, breccia zones, and tectonic lenses follow the stratification [29] and, frequently, hydrocarbon prints accompany them. In the eastern Sierra del Rosario, the total vertical thickness of the nappe pile is no less than 2 km. In the Alturas de Pizarras del Norte and the northern coastal lowlands, well data reveal a thin skin tectonics pattern from the Earth's surface up to no less than 5 km deep [23, 112].

The Pan de Guajaibón (PG) is the higher and northernmost tectonic unit of the Guaniguanico Mountains tectonic pile. Geographically, it belongs to Sierra del Rosario highlands; however, it is "mogote-type" geomorphology. The unit contains the homonymous lithostratigraphic unit, with circa 500 m of shallow water massive too thick bedded, northward dipping carbonate deposits of Albian–Cenomanian age [35]. The rocks are karstified, with some small bauxite deposits, probably derived from the weathering and erosion of igneous regolith [70].

An important role in the tectonic style of Guaniguanico Mountains plays the lenses of syn tectonic lower Paleogene beds and serpentinitic mélanges imbricated within the Mesozoic paleomargin deposits schematically illustrated in the tectonic map and profile of the central part of Guaniguanico Cordillera (Fig. 5.5). Whereas the serpentinitic mélanges are abundant in the SR-APN-E unit, in the APS unit they are much less frequent and the discrimination of the individual nappes is less effective. This item will be discussed additionally in the epigraph devoted to the lower Paleogene Foreland Basin and the Cuban orogeny.

The North American paleomargin in central Cuba. From Havana to NW Holguin Province, the NAP displays a

Fig. 5.6 Breccia in the San Adrian diapir, Mayabeque Province. According to Meyerhoff and Hatten [78], the main lithologies in the clasts are different types of limestones (those fine grained could contain radiolarian and Nannocunus), metamorphic rocks (marbles and quartz–mica schists), quartzose sandstones, and isolated metric lenticular blocks of serpentinite. Additionally, in the formation appear some grains of pyrite, tourmaline, and apatite

remarkable zonation, discovered by the oil geologists working in Cuba during the 50s of the twentieth century that developed several fine schemes (Fig. 5.1). We will use the most recognized terms in the last decades, proposed in the zonation model by Ducloz and Vaugnat [33]. It includes, from north to south, the following four tectono-stratigraphic units: Cayo Coco (CC), Remedios (R), Camajuaní (Cj), and Placetas (P). Only the last one shows features relating it to the Upper Jurassic–Cretaceous sections in the SR-APN-E TSU of the Guaniguanico Mountains. The essential features of each TSU are discussed in several papers (see [17, 61, 67] and others). A brief review follows.

Cayo Coco and Remedios TSU represent the southern boundary of the Florida–Bahamas platform, and the first is mainly known from subsurface geology. They have very similar pre-Aptian sections, with Middle Jurassic. Evaporites (San Adrian and Punta Alegre formations, Fig. 5.6), probably coeval with the Gulf of Mexico salts [1, 61, 67]. In northern Villa Clara and Sancti Spiritus, about 1800 m of evaporites and carbonates (Cayo Coco Formation) conform the Upper Jurassic–Aptian record in the CC TSU, whereas the evaporites are absent southward, in the neritic carbonates of the R TSU. A dramatic facies change occurs in the upper Aptian–lower Turonian beds of the CC TSU, represented by hemi-pelagic carbonates. A similar deepening is recorded by the Albian–Turonian carbonate turbidites (Vilató Formation), accumulated in narrow basins (grabens) at the southern border of the Remedios TSU, in Camaguey [61]. These events probably correlate with the fracturing of the southern border of the Florida–Bahamas Jurassic–Middle Cretaceous mega platform. At the same time, more than 2000 m of shallow carbonates settled upon most of the Remedios bank (Palenque and Purio formations). Coniacian to Campanian deposits are unknown in the Remedios bank, resting the Maastrichtian carbonate beds on similar middle Cretaceous lithologies.

The southern half of the NAP in central Cuba is occupied by the Camajuaní and Placetas TSU. The first one outcrops are only known in central Cuba, but it is also recorded in deep wells from Havana to Matanzas Province. The Camajuaní TSU outcrops as an Upper Jurassic–Cretaceous belt, about 1200 m thick, of well-stratified carbonates (including turbidites), partially derived from the erosion of the coeval Remedios bank and, in part, fed by hemi-pelagic deposits, including some chert. As it occurs with the Remedios bank, the Turonian–Maastrichtian hiatus is also present. An inlier of shallow water Cretaceous carbonates (Gibara Formation) is present near the homonym city.

The Placetas TSU is known from outcrops and wells from eastern La Habana to Camagüey Province, in eastcentral Cuba. The Tithonian–Cretaceous stratigraphy of the 1300–1700-m-thick Placetas belt is very similar to those in the SR-APN-E unit of the Guaniguanico Mountains [109, 30]. However, in the oldest strata, instead of the thick siliciclastic deltaic succession, a thin shallow upper Oxfordian marine arkosic interval rests below the Tithonian–Cretaceous beds. The arkosic strata (Constancia and Quemadito formations) lie above a paleoregolith containing marble and granitic clasts. Pszczolkowski and Myczynski [109] suggest that the granites (with Jurassic radiometric ages) were torn off from the Proterozoic basement (marbles). The structure of Placetas TSU is very similar to those in SR-APN-E unit. Together, the arkosic strata and the phlogopite marbles are thrust slices caught between Placetas TSU rocks [101]. In the northern Camaguey Province, the lowermost beds of the TSU are a Tithonian sequence with basalts and limestones [55].

It is a remarkable fact that the outstanding similarities between the geological settings of the above studied NAP in northcentral Cuba and the coastal Belize Cretaceous section report by Schafhauser et al. [118] and their relationships with the corresponding neighboring ophiolites and Cretaceous volcanic arc sections. This is a key item for understanding the regional Mesoamerican geology that must be studied in the near future.

Several peculiar deposits related to the asteroid impact at the Cretaceous/Paleogene boundary (the Cacarajícara, Moncada, and Amaro formations) occupy the uppermost NAP. We will return later on this subject.

The North American Paleomargin in easternmost Cuba (Maisí). Dark-colored marbles and metaterrigenous rocks in the eastern tip of Cuba are exposed in Maisí (Fig. 5.1, Asunción complex of [59], see also [122, 26, 27, 81]. The lowermost beds are phyllites and slates, with some marbles and metamaphytes (Sierra Verde Formation), very similar to the metaterrigenous Jurassic beds in western and central Cuba. Above, resting with tectonic contact [27] lies Upper Jurassic marbles and calcareous schists (the Chafarina or Asunción Formation) probably of Upper Jurassic age. KVAT and NOB rocks apparently rest on this small North American paleomargin outlier [81].

5.3.2.2 The Northern Ophiolitic Belt (NOB)

The ophiolitic association is a record of the oceanic lithosphere rocks, emplaced by tectonic processes upon the continental or island arc margins. The Cuban ophiolites belong to the Mesozoic basement. However, some of them were remobilized during the early Paleogene Cuban orogeny and even later been emplaced together with lower Cenozoic rocks. Three different ophiolites can be recognized in Cuba [18]:

– The Northern ophiolitic belt.
– The metamorphic basement of the KVAT [82].
– Tectonic slices in the central Cuba Guamuhaya (Escambray) metamorphic terrane.

The last two units will be visited later, related to the Cretaceous volcanic arcs and the Southern Metamorphic Terranes. More than 90% of the oceanic lithosphere remains in Cuba are included in an almost continuous belt of strongly deformed rocks, northward transported over the NAP during several tectonic events. The NOB (Fig. 5.1) encroaches about 5–6% of Cuba's surface. As in other ophiolite belts, several members of the oceanic lithosphere are present: (a) the ultramafic tectonites, (b) the stratified mafic and ultramafic, (c) the massive gabbros, (d) the diabase complex, and (e) the vulcanogene-sedimentary member (Fig. 5.7). The NOB is essentially a huge mélange (Fig. 5.8), stretching circa 1000 km along the northern half of Cuba, with blocks mainly formed by lithologies of the ophiolitic association (mafic and ultramafic rocks) embedded in a strongly deformed serpentinitic matrix that flows during deformation. This superposed structures strongly mask the original contacts, mixing, crushing, and deforming, not only the ophiolite members, but also their country rocks [18].

Different tectonic settings occur along the strike of the NOB. From west to east, the following main outcrops occur Fig. 5.1: (1) Cajalbana–Bahia Honda,(2) Havana–Matanzas; (3) Villa Clara; (4) Camagüey; (5) Maniabón Highlands; and (6) Mayarí–Baracoa Highlands [18, 38]. Additionally, a peculiar isolated outcrop of brecciated serpentinites near the Caribbean coast in Guantánamo Province reveals the youngest ophiolite emplacement identified in Cuba [18, 23].

Fig. 5.7 Thin bedded chert and shales (dark rocks in the photo) in the vulcanogene-sedimentary member of the western Cuba ophiolites (Encrucijada Formation), south of Bahia Honda, Artemisa. Below, and probably in tectonic contact, rest deeply weathered basalts

Fig. 5.8 Vulcano-serpentinitic mélange at Loma Esmeralda, Matanzas Province. Serpentinite (S), strongly brecciated and with slickensides, contact with tuffs (T)

Few robust age data are known from the NOB rocks: Tithonian, Hauterivian–Barremian, and Aptian–Albian fossils have been reported in sedimentary interbeds [3, 76], whereas K–Ar radiometric ages range from 126 to 52 My [60]. The Early Cretaceous radiometric ages are in a general good agreement with the stratigraphic data, whereas Late Cretaceous–Paleogene radiometric ages record tectonic or thermal events in the most "fortunate" examples. The sum of the geological, geochemical, and petrological data suggests that the NOB rocks probably originated in two distinct tectonic environments [18]:

- as part of the Tethys oceanic basin, from Late Jurassic to Early Cretaceous, during Pangea breakup and later separation of the North and South American plates.
- in an Aptian–Albian Backarc Basin, related to the lower Cretaceous KVAT.

In the next paragraphs, we will briefly discuss how and when the different NOB sectors arrived to their present locations. The oldest ages reported from the NOB are Upper Jurassic. This was the time of the breakup of westernmost Pangea, when the North American plate began its splitting from neighboring continents. Oxfordian, Tithonian, and Berriasian fissure tholeitic magmatism in western and central Cuba [14, 15, 18, 55, 56, 107, 108] support this interpretation. Therefore, it seems likely that a first oceanic lithosphere was created from Late Jurassic to Early Cretaceous (Neocomian) as part of the Tethys Ocean south of the Mesozoic North American margin, as a consequence of the North American/Gondwana separation (see Fig. 14 in [18]).

As we will see, the youngest NOB rocks coeval with the oldest KVA representants, which grew from an oceanic basement, during Aptian–Albian time. According to plate tectonic models, Backarc Basins often develop behind volcanic arcs, and geochemical data suggest that part of the NOB lithologies show such suprasubduction signatures [66]. As convincing evidences of Upper Cretaceous ophiolites in Cuba have not been found, we can suppose that the generation of oceanic lithosphere in the Backarc Basin ended in the Albian (see Fig. 16 in [18]). Some remarkable changes in the rock composition of the KVAT occur in the transit from the Lower to Upper Cretaceous. The most significant is the change to a dominant calcoalkaline signature. This feature is accompanied with frequent volcanoclastic sedimentary interbeds. Some geologists consider that these changes could be related to a change in subduction polarity [32]. After that event, three main tectonic emplacement events can be discriminated in the Cuban ophiolite massifs:

– The late Campanian event.
– The Maastrichtian event.
– The early Paleogene Cuban orogeny.

Additionally, a brief Middle or Late Eocene local emplacement episode, related to the opening of the Cayman Trough, has been identified [18].

The Campanian emplacement event. The abundant Upper Cretaceous volcanic and intrusive rocks, mainly with calcoalkaline signature, but some with a significant alcalic trend, together with some tholeites, and frequent volcanoclastic sedimentary interbeds, points to the development of a second Cenomian to Campanian volcanic arc, more or less parallel to the North American Mesozoic [18]. The simultaneous end of volcanism along all the Late Cretaceous arc during the Campanian suggests that subduction stopped. Several evidences mark a coeval episode of ophiolite emplacement, possibly related to the closing of a small oceanic remnant basin, located between the arc and the North American Mesozoic margin (see Fig. 15 in [18]). Preceding this event, for the first time in its Cretaceous history, to the southeastern North American margin arrived volcanoclastic sediments. This fact could be possible only if the remnant basin was closed at the Campanian decline. In western and central Cuba (from Artemisa to Las Tunas Province), Campanian and/or Maastrichtian terrigenous beds, frequently with isolated serpentinitic clasts, lie on the serpentinites or gabbroids. This unconformity testifies about a major Campanian orogenic event [17, 105, 106]. The isolated clasts of ophiolitic origin in several localities can be explained by minor obduction events from near surface bodies. A coherent explanation to the preceding facts is that the remaining oceanic lithosphere between the Late Cretaceous volcanic arc and the Mesozoic North American Mesozoic margin was finally subducted during the late Campanian and, these formerly isolated tectonic units, juxtaposed (see Fig. 14b in [18]). This episode is concealed by the thrusting related to the early Paleogene Cuban orogeny. However, it was an arc/continent collision, accreting the Paleo-Caribbean oceanic Mesozoic lithosphere plus the Cretaceous volcanic arc terrane, to the North American plate [18]. Such outstanding collision is not included in the most popular regional tectonics versions ([91, 61, 66], and others).

The Maastrichtian emplacement events. These are two remarkable events in the NOB history. Whereas the late Campanian event probably embraces all Cuba, and the Paleocene–Middle Eocene event most of its territory, the Maastrichtian one is limited to two regions in eastern Cuba, where the ophiolite emplacement followed different paths.

The first area contains the NE Cuba Highlands, from Sierra de Nipe to Moa-Baracoa Mountains (Fig. 5.9) where, in several places below the huge horizontal lying ophiolite massifs, rest lenticular Maastrichtian olistostromic deposits (La Picota Formation), attaining tens to hundreds meters thick (Fig. 13 in [23]). The massive chaotic nature, monotonous composition (almost all the clasts belong to the ophiolitic suite) and poor rounding tell us that the ophiolite obduction was an intense and violent process The huge ophiolitic bodies arrived to the Earth's surface in a marine environment and spread, following the regional slope of the sea floor (possibly northward) sliding upon the olistostromic carpet, crushing and pushing it at velocities of ten millimeters per year. The displacement attained, at least, 30 km (probably no least than 60 km in Moa–Baracoa highlands) [12, 21, 18].

The second Maastrichtian episode is recorded at the Maniabón Highlands, in the NW part of the Holguin

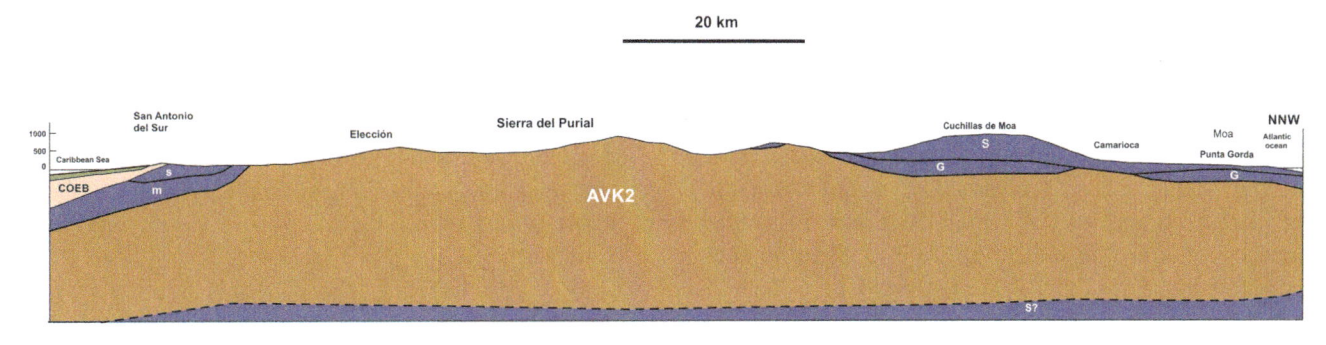

Fig. 5.9 Profile located near the eastern end of Cuba, from San Antonio del Sur, in the Caribbean coast, to Moa, in the Atlantic shore. NOB: G-gabbroids, S: serpentinites, and serpentinized ultramafic rocks, m-melánges. The KVAT is represented by Upper Cretaceous metavolcanics and some metasediments (the Sierra del Purial complex). It probably rests horizontally sandwiched between the ophiolites

Province. The Maniabón Massif consists of anastomosing bands of strongly deformed ophiolites gently convex to the ESE, separated by south-dipping thrust faults of the intervening strongly deformed terrigenous and volcanic rocks "intercalations" (the Iberia Formation, [69, 71]). According to Andó et al. [3] much of "Iberia Formation" are severely deformed greywackes (mélange) probably settled in a forearc. The ophiolites are mainly foliated and brecciated serpentinites. The other members of the ophiolite suite appear as disseminated blocks included in the serpentinites. The relationships among the Maastrichtian events are unclear. Broad Cenozoic outcrops and the wide Nipe Bay separate both regions [110].

Ophiolite emplacement during the Cuban orogeny. This is the last major event affecting the NOB [18, 23, 46, 47, 68, 106]. The deformation developed by steps, in blocks bounded by NE-SW sinestral strike-slip faults. Again, coeval with the ophiolite emplacement olistostromes accumulated. However, in this case, besides serpentinite and other members torn from the ophiolite suite, abundant debris flows with blocks derived from the Mesozoic NAP and the KVAT accumulated on the fast subsiding foreland basin located in front of the northward moving thrust sheets. The structural and stratigraphic evidences indicate that, in the Early Paleogene, those part of the NOB located westward of the Cauto-Nipe lineament traveled northward, together with the Cretaceous volcanic arc, riding the NAP and its overlying foreland basin. Huge wedges, but also thin scales of the well-stratified Jurassic and Cretaceous Placetas TSU were detached, scoured, imbricated, and deformed together with the ophiolitic rocks and the Paleogene olistostrome. In many places, this process conduit to melánge formation ([16], Fig. 5.8).

Meanwhile, the Cuban orogeny, as a whole, extends from middle Paleocene to Middle or Late Eocene, and it doesn't operate simultaneously all along the NOB. During the orogenic event, thrusting migrated progressively eastward along fault-bounded blocks or sectors. In each sector, the orogenic event was a relatively short-lived process. In the western block, flanked by the Yabre lineament, deformations extend from Middle Paleocene to Early Eocene. In the central block, located between La Trocha fault zone and Yabre lineament, folding and thrusting attain the Middle Eocene, and eastward of the last dislocation the orogenic event concluded during the late Middle Eocene or early Late Eocene (see [18], Fig. 16). It is a fundamental item in the interpretation of the tectonics of Cuba to define if there was an ophiolite obduction related to the Cuban orogeny. This will be discussed at item Sect. 5.3.4.

Field and well data indicate horizontal displacements of the NOB attaining several ten kilometers in the most conservative estimations [18, 19, 23, 20, 108, 117].

5.3.2.3 The Cretaceous Volcanic Arc Terrane

In the pre-Plate Tectonics geological literature on Cuba, those geological sections containing Cretaceous volcanic rocks, serpentinites, and igneous Mesozoic rocks, together with the Paleogene foreland and piggyback deposits and, in some schemes, even the Paleogene volcanic arc were referred as the eugeosyncline and included in the classical regional schemes of the 1950 and 1960s. Several oil geologists proposed different local names for the "Cuban eugeosyncline" ([33, 46], and others). The most popular was the "Zaza zone," even used today by some geologists (the Zaza TSU, [75]). However, if we intent to explain the geology of Cuba in plate tectonics terms, the "Zaza zone" is an obsolete concept, because it mingles rocks of very different origins: the ophiolite suite, the KVAT (Fig. 5.1), including its Campanian–Maastrichtian cover, plus some volcanomictic Paleogene successions.

By its volume, the KVAT is the main component of the Mesozoic basement. It's composed of volcanic arc rocks, with more or less outstanding sedimentary interbeds [91]. Intrusive bodies frequently cut the supracrustal rocks. An upper Campanian–Maastrichtian cover unconformably rests on the older rocks. Except in the eastern Cuba outcrops, it rests mainly with tectonic contacts upon the NOB. However, in the great festers of the Isla de la Juventud and Guamuhaya (Escambray) Mountains, the Southern Metamorphic Terranes outcrop below it [110]. The tectonic nature of its basal contact and the lack of stratigraphic contacts with the pre-Maastrichtian rocks of the other Mesozoic basement units allow to consider the KVAT as a tectono-stratigraphic terrane.

Most of the Cretaceous volcano-sedimentary sections lie below hundred to thousand meters of sedimentary Cenozoic rocks in the southern half of Cuba, except in SE Cuba, where the Paleogene volcanic arc developed above the KVAT [110]. Along its strike, the KVAT structure, rock composition, and age modified. The maximum complexity is attained in central Cuba, between the Yabre lineament and La Trocha fault zone. East and westward, the terrane becomes thinner and less varied. Around the Guamuhaya tectonic window (where the SMT outcrops) the following KVAT tectonic assemblage is present (from lower to upper members):

– Mabujina complex: a metamorphic complex mainly built up from mafic protoliths transformed into amphibolites, with a high temperature/pressure ratio. Most of the protoliths belong to a supposed Mesozoic ophiolite basement; however, the complex also includes calcoalcalic rocks derived from the lowest levels of the KVAT [82, 86]. Several deformational events are recorded in the Mabujina complex rocks and three metamorphic zones (all included in the amphibolite facies). The radiometric age data, mainly from K–Ar datation, yield Upper Cretaceous ages, [59]. A U–Pb age in zircons from gneissiod granitoids yields 132 My.

In tectonic contact with the Mabujina complex appears a volcanic non-metamorphosed sequence; 1000–3000-m-thick pile of felsic and mafic rocks (Los Pasos Formation), with some andesite and sandstone intercalations. Abundant subvolcanic bodies and some ignimbritoid rocks point to some subaerial volcanism. However, most of the beds accumulated on submarine settings as a consequence of central and fissure volcanic activity. The petrographic composition is very similar to the Primitive Island Arc suite but geochemical data are insufficient for a robust definition. This bimodal suite is very similar to the Lower Cretaceous Los Ranchos Formation in Dominican Republic [37]. Probably above Los Pasos Formation lie basaltic and andesitic lava flows and tuff beds with marine sedimentary interbeds, conforming a 4000-m-thick packet of calcoalkaline and tholeitic rocks (Mataguá Formation—[34]), of Aptian–Albian age. Vertical and laterally the Mataguá Formation transits to Albian andesitic and felsic tuffs and tuffites with sandstones and andesitic and dacitic lavas (Cabaiguán Formation), 1000–3500 m thick. Thinner Lower Cretaceous andesitic and basaltic tuffs, with some mafic sills and dykes (Chirino Formation), are also recorded in La Habana, Mayabeque, and Matanzas Provinces [110].

A concordant, mainly sedimentary Albian–lower Cenomanian section, with terrigenous and carbonate turbidites (Provincial Formation), covers the Lower Cretaceous beds in central Cuba. Some tuffaceous strata testified about isolated explosive eruptions, whereas shallow carbonates in Cienfuegos Province record coeval banks and a widespread volcanic recess. However, the absence of a widespread unconformity, some local deformations are recorded in central Cuba [17] whereas the petrological features and geological setting of the Upper Cretaceous beds suffer drastic changes. This points to a major tectonic event yet not enough recognized [17].

The overlying volcano-sedimentary Upper Cretaceous section possesses several remarkable features, clearly separating it from the underlying Lower Cretaceous arc:

- In contrast with the tholeitic or tholeitic-calcoalkaline underlying rocks, the Upper Cretaceous magmatic rocks mainly belong to the calcoalkaline suite, sometime with a marked alkaline trend. Only in the NE Cuba Highlands, tholeitic rocks are present [58].
- Volcanomictic interbeds are abundant (Fig. 5.10), pointing clearly to important subaerial volcanic buildings (islands) and erosion [34].
- Whereas the Lower Cretaceous rocks are limited to a discontinuous belt from La Habana to central Camaguey, the Upper Cretaceous volcanic rocks are recognized in outcrops and subsurface all along Cuba.
- The thickness of the volcano-sedimentary Upper Cretaceous section in western Cuba is only of several hundred meters, in central Cuba perhaps attains a maximum thickness of 2000–3000 m, in Camaguey–Las Tunas territory is poorly defined—probably several thousand meters [62, 96].
- A variable volcanic activity is represented in the Upper Cretaceous rocks. Fissure magmatism was active in some localities of western and central Cuba, whereas structures of central volcanism have been identified in eastern Camagüey and Las Tunas Provinces [36]. Coeval intrusive magmatism in the KVAT eastern half is indicated by abundant stratigraphic and radiometric data [61].

In the Maniabón Highlands, an extremely complicated area contains a mélange where the NOB and the KVAT are intimately mingled [3]. The KVAT form narrow, arcuate strips, gently convex toward the SE, alternating with lenses of pervasively crushed and foliated serpentinites. Two kinds of serpentinitic-volcanic mélange coexist. The most widespread is the Iberia mélange, with crushed and foliated basaltic and andesitic rocks and their tuffs and sedimentary packets. The second mélange is the "Loma Blanca Formation," rich in felsic tuffs, often zeolitized. Serpentinitic clasts are abundant in both units.

Contrary to western and central Cuba in the highlands of eastern Cuba (Sierra de Nipe, Sierra Cristal and Moa-Baracoa massif), the Upper Cretaceous volcanic arc rocks lie below the NOB [69], tectonically emplaced during the Maastrichtian [12, 23, 60, 61, 57]. Four lithostratigraphic

Fig. 5.10 Upper Cretaceous conglomeratic and sandstone beds of La Trampa Formation, eastward Havana city. These sediments show that, contrary to the events in the Early Cretaceous, in the Late Cretaceous volcanic arc, some volcanic cones suffer the subaerial erosion

units (the Santo Domingo, Téneme, Morell, and Quibiján Formations) have been distinguished [58]. Except the youngest one (Santo Domingo), the other formations contain Turonian to Coniacian basaltic rocks. The Santo Domingo Formation (Santonian to Campanian) contains andesitic tuffs, with some sills. All these units rest below the large ophiolitic massif of eastern Cuba. Excellent examples of this relationship are the great fenster of the Téneme and Cabonico River Valleys in northern Sierra Cristal [69, 110] and the Mayarí River Valley at Sabanilla, Mayarí Arriba [12]. The third area of Upper Cretaceous volcanic arc rocks is the Sierra del Purial Mountains, located near the eastern tip of the island. In this region, metavolcanic rocks (green and blue schists) of calcoalkaline to tholeitic signature [26, 81, 84]) with some intercalated marbles (Sierra del Purial Formation) record the only metamorphic rocks in the Upper Cretaceous arc. These rocks were sandwiched between the NOB (above) and the NAP Mesozoic metamorphic rocks during the Maastrichtian ophiolite emplacement [122, 26, 58].

The higher structural stage of the KVAT, the Campanian–Maastrichtian cover, is separated from the underlying rocks by a major unconformity. The lower beds are mainly siliciclastics volcanomictic turbidites, sometime with disseminated serpentinite grains. These beds can attain thickness up to several hundred meters. Carbonate lithostratigraphic units (Cantabria, Jimaguayú, Tinajita Formations, and others) tend to occupy the highest stratigraphic positions with thickness from ten to some hundred meters. From Pinar del Rio to Matanzas Provinces, the younger beds of the cover belong to Peñalver Formation, an extraordinary clastic deposit, with abundant spherules, grains of shocked quartz, and other particles, typically related to asteroid impacts [126]. In central Cuba, coeval deposits are known in the Santa Clara Basin (see Sect. 5.3.2.4).

5.3.2.4 The Cretaceous/Paleogene Boundary Event Deposits and Paleogeography in Western and Central Cuba

Several types of extraordinary Cretaceous/Paleogene boundary sediments, with sharp differences in composition and thickness, are widely distributed in western and central Cuba, 500–1000 km eastward from Chicxulub crater in Yucatan: Cacarajícara, Amaro, and Moncada Formations and the DSDP 536 and 540 sites rest on the North American Mesozoic paleomargin, whereas Peñalver Formation and a section of K/Pg deposits, included in the central Cuba Santa Clara Formation, lie on piggyback basins, developed upon the extinct Cuban Cretaceous volcanic arc Fig. 5.11, [2, 20, 44, 102, 126].

According to Cobiella Reguera et al. [20], three types of deposits are present. Type 1 (Cacarajícara, Amaro, and Peñalver Formations) presents in its lower part thick gravity flow accumulations, followed by graded massive finer calcareous debris (homogenite). The depositional area of Cacarajícara plus Amaro formations was circa 25,000 km^2, with estimated original sediment volume of 2500 km^3. For Peñalver Formation, with about 17,500 km^2 of depositional area, gross volume estimates give 1750 km^3 of original sediments. Type 2 (K/Pg deposits in Santa Clara Formation and DSDP sites also contain gravity flow deposits in their lower part, but instead the homogenite, ejecta-rich deposits are present. Type 3 mainly contains reworked ejecta clasts (Moncada Formation). Some features in the deposits can be

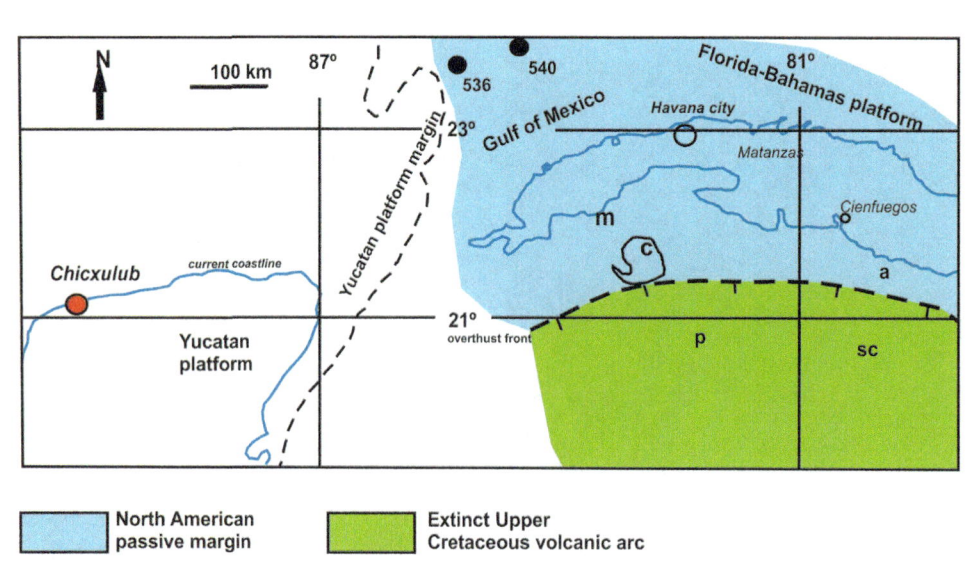

Fig. 5.11 Palinspastic reconstruction of the NW Caribbean–SW Gulf of Mexico region surrounding western and central Cuba at the Mesozoic/Cenozoic boundary (for details consult [20]). m: original area of Moncada-type deposits, c: original area of Cacarajícara-type deposits, a: original area of Amaro-type deposits, p: original area of Peñalver-type deposits, sc: original area of Santa Clara-type deposits

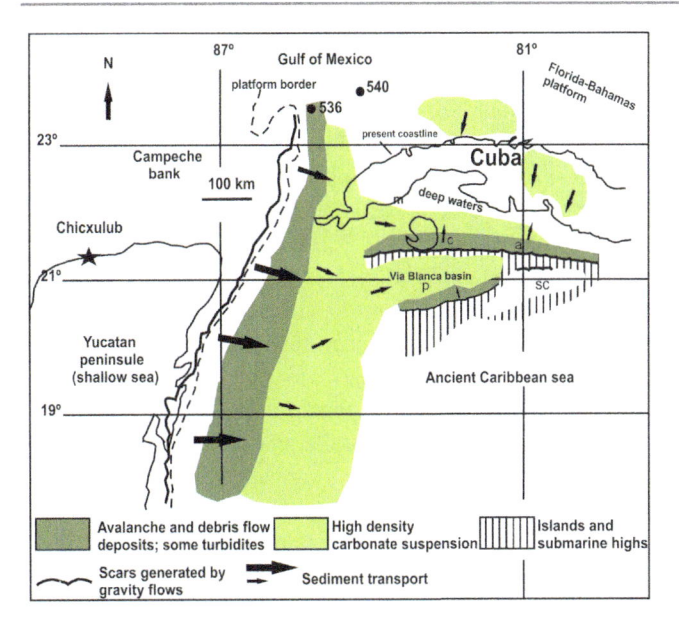

Fig. 5.12 Paleogeographic scheme at the Cretaceous/Paleogene boundary. Observe the complex regional relief in the area of the future western and central Cuba at the moment of the Chicxulub impact [20] modified from . m: original area of Moncada-type deposits, c: original area of Cacarajícara-type deposits, a: original area of Amaro-type deposits, p: original area of Peñalver-type deposits, sc: original area of Santa Clara-type deposits

explained by the travel of several mega tsunami waves during their deposition. A clear zonation of K/Pg boundary beds is present in the Gulf of Mexico. Accumulations similar to Type 1 beds (but without or with poorly developed homogenite equivalents) settled in the northern and southwestern fringes of Yucatan Peninsula, whereas thin-bedded siliciclastic ejecta-rich strata extend northward in the circum-Gulf areas.

The Cretaceous/Paleogene boundary sediments in western and central Cuba record the event chronology related to the asteroid impact in an original area of approximately 90,000 km^2, 500–1000 km east to southeastward of the Chicxulub impact crater. Compared with coeval sediments in other areas of the world, the Cuban K/Pg beds show extreme changes in thickness and composition in short distances, pointing to a complex geography in the northwestern Caribbean, 65,5 My ago (Fig. 5.12). Additionally, some erosion of the underlying beds by currents during the event is recorded in many places, complicating the interpretation of the latest Maastrichtian paleogeography. Finally, the original sediment distribution was severely distorted by the early Paleogene orogenic events in Cuba.

According to different sources, the northward thrusting for distinct tectonic units in western Cuba ranges between 12 and 200 km [19, 108, 117]. Therefore, in order to restore the original K/Pg boundary sediment distribution and the paleogeography of the northwestern Caribbean corner at the time of the asteroid impact, a combination of Palinspastic reconstruction and sedimentological research of such deposits and the underlying upper Maastrichtian and overlying lower Danian beds was necessary [19, 20].

5.3.2.5 The Southern Metamorphic Terranes (SMT)

Iturralde Vinent, in 1996, proposed to distinguish the metamorphic massif of southern Cuba (Isla de la Juventud and Guamuhaya or Escambray) and the Mesozoic sections of western Cuba (Guaniguanico Mountains) as tectonic terranes and named them "The Southwestern terranes." According to Keppie's definition (in [17]), a terrane is "an area characterized by an internal continuity in geology that is bounded by faults, mélanges or a cryptic suture across which neighboring terranes may have a distinct geological not explicable by facies changes or a similar geological record that is bounded by faults, mélanges or a cryptic suture... that may only be distinguished by the presence of the terrane boundary representing telescoped oceanic lithosphere." Initially, several authors followed Iturralde Vinent's proposal ([61, 91, 108], and others). However, Cobiella Reguera [22, 17, 18, 22], and Pszczolkowski and Myczynski [109] questioned the idea. In the current paper, the author only considers as terranes of continental origin in Cuba, the Pinos (Isla de la Juventud) and Guamuhaya terranes. In both areas, the protoliths are of Mesozoic age, and with remarkable similarities, in stratigraphic architecture and age, to the Guaniguanico sequences of western Cuba. Consequently, they are here classified as "proximal terranes."

Most of the Isla de la Juventud (IY in Fig. 5.1) consists of metasedimentary rocks. The lower part of the section contains quartzose graphitic metapelites and metapsamites (Cañada Formation). These lithologies continue in the middle part but with increasing quantities of marbles and calc-silicate beds (Agua Santa Formation). The upper strata consist almost entirely of gray and black dolomitic marbles (the Gerona Group). A Jurassic age for the protolith of the last beds is suggested by the discovery of some cephalopods and foraminifera [84]. Some amphibolites (possibly from a magmatic protolith) are present in Gerona beds. The metamorphism is HT/MP type. Six metamorphic zones were recognized by Millán Trujillo [85], which are (from lower to from lower to higher degree): (1) greenschist, (2) estaurolite/kyanite/garnet (low degree amphibolite), (3) estaurolite/Kyanite (occasional andalusite), (4) garnet/kyanite/biotite; (5) sillimanite/garnet/K feldspar (sillimanitic gneiss); and (6) migmatites.

An elliptical dome, developed in an extensional environment, embraces all the Isle. Extended fracturing of the metamorphic rocks was accompanied by felsic intrusive magmatism and wolfram mineralization. The scarce radiometric data (K–Ar) yield 78–72 My ages for the metamorphites (late Campanian). In a subvolcanic non-metamorphosed body intruding the metamorphic massif, a 78–72 My radiometric age (Campanian) was recorded.

Meanwhile, some nappes are recognized in the Isla de la Juventud Massif, compared with the tectonic setting of Guamuhaya Massif, they are few. This situation probably is related to the absence of serpentinites and mélanges, abundant in the central Cuba Millán Trujillo [86] distinguishes four major regional structures, supported on the lithology and trends of the main foliation: Nueva Gerona and San Juan sinforms and Rio Los Indios and Guayabo antiforms (Fig. 1 in [85]). All the contacts between them are tectonic.

In the Guamuhaya tectonic window, the older rocks are also metaterrigenous beds, but here accompanied by concordant metamafic bodies (La Llamagua, La Chispa, Loma La Gloria, and Herradura Formations). Above rests a metaterrigenous-carbonate interval, followed by marbles with a few Upper Jurassic fossils (San Juan Group, Millán and Myczynski 1978, 1979). Therefore, the general stratigraphic picture is quite similar to those in the Isla de la Juventud and Guaniguanico Mountains, especially to the Sierra del Rosario-Alturas de Pizarras del Norte-Esperanza unit. However, whereas Cretaceous rocks are unknown in the Isla de la Juventud (Isle of Youth) Massif, in the Guamuhaya Mountains the metavulcanogen-sedimentary section structurally above the Jurassic marbles is probably Cretaceous (Los Cedros, La Sabina, Loma Quibicán, Yaguanabo, and El Tambor Formations [86]). Radiometric (K–Ar) ages in the southcentral Cuba metamorphic massif yield mainly Upper Cretaceous ages [86]. In zircons from eclogitic rocks using the U–Pb method, ages from 106 to 102 My have been obtained.

The Guamuhaya Massif (Fig. 5.1) forms two large dome-shaped structures, the Trinidad (west) and Sancti Spiritus cupolas well marked in relief and separated by the Agabama Valley (or the Trinidad Basin, in a tectonic approachment). The dome is also reflected in the KVAT rocks, as is visible in regional geologic or tectonic maps. The age of the Cretaceous arc emplacement is Late Cretaceous, since a basal slip plane (*decollement*) affects the Upper Cretaceous arc rocks but not its Campanian–Maastrichtian cover. Additionally in the Guamuhaya (Escambray terrane), the tectonic contact with the overlying KVAT is crossed by 88–80 My old pegmatites [124] Then, the precedent facts show that the terrane welding in central Cuba seems a Pre-Coniacian event. Millán Trujillo [86] considers that this event was simultaneous with the greenschist facies metamorphism recorded throughout the massif. This same author distinguishes four major tectonic packets ("Main Tectonic Units," MTU). The lowest one is the first MTU and the fourth is the highest. Each MTU is divided into smaller nappes. The first and second MTU are the lower units and the third and fourth the higher ones. The Guamuhaya Massif has a high-pressure metamorphism, with an inverted zonality from the first to the third nappe. The metamorphic peak is reached in the third unit, which records pressure/temperature 15–23 kbar/470–630 °C [124]. The fourth MTU contains rocks metamorphosed under lower temperatures than the third. A subsequent event of greenschist metamorphism affects the whole massif. According to Millán Trujillo [86], the first metamorphic event occurs circa 106–100 My ago (Albian). Contrary to the Pinos terrane, in Guamuhaya Mountains, abundant serpentinite and metamafic bodies are tectonically included or were originally emplaced as magmatic bodies, later metamorphosed together with the sedimentary country rocks [86]. The most remarkable of them are the high-pressure amphibolites of the Yayabo Formation (12–14 kbar/550–580 °C). Probably originally they were part of the Lower Cretaceous volcanic arc, tectonically mixed with the Guamuhaya metamorphites [124]. In our opinion, other metamaphytes, not spatially linked with serpentines (e.g., the Felicidades Schists), could be manifestations of continental margin magmatism, similar to that recorded in the Jurassic beds at Guaniguanico Mountains [14, 107].

A recent study by Despaigne Díaz et al. [31] concludes that the rocks in the Trinidad dome suffered a Late Cretaceous (circa 75 My)–Paleogene (circa 50 My) subduction-exhumation event, with five main episodes (D1–D5). This contradicts all previous results that consider the metamorphic events in Guamuhaya as Cretaceous. In our opinion, the interpretation of the $^{40}Ar/^{39}Ar$ data by the authors is confusing, because of the abundant overlap between D1 and D3 values (all with very similar Paleocene radiometric ages that suggest an overprint derived from the Cuban orogeny events).

The first clasts of metamorphic rocks in the sedimentary basins surrounding Guamuhaya (Escambray) Mountains appear in the Middle Eocene beds [86]. However, Stanek et al. [124] consider that exhumation of the Mesozoic metamorphic complex began about 70 My ago, during the Maastrichtian. The Cenozoic stratigraphic record and the current relief indicate that uplift still continues today (see later).

5.3.3 The Paleogene Folded and Thrusted Belt

The KVAT-UK cover unconformity, together with the tectonic burial of the SMT below the Cretaceous volcanic arc, reveals the arc collision with the North American plate border. This event forced the end of the arc and the reconstruction of the plate boundaries in the early Paleocene. In this complex setting, the following new Early–Middle Paleogene structures are recognized:

- The Early–Middle Paleogene Foreland Basin (PFB).
- The piggyback basins (PBB).
- The Turquino–Cayman Ridge volcanic arc (T-CR).
- The central Oriente intramountainous Eocene Basin (COEB).

5.3.3.1 The Early–Middle Paleogene Foreland Basin

Along northern Cuba, from Pinar del Rio to NW Holguin Province, the Mesozoic NAP is covered by the siliciclastic deposits of a foreland basin. These successions accumulated in front of the thick nappe pile were generated during the early Paleogene Cuban orogeny, as a result of the erosion at the frontal thrust region and the intensive basin subsidence due to the weight of the nappe building. This process creates the space for the accumulation of large volumes of deposits (Fig. 5.13). Therefore, sedimentation is coincident with the orogenic deformation [63]. In the basin, a tight imbrication exists between (a) the debris flows (olistostromes) and turbidites derived from the erosion of the tectonic pile and (b) the southern thin tectonic wedges derived from the Mesozoic basement (ophiolites, KVAT, and the NAP). Northward of this deformed strip, the crust was progressively depressed as the basin developed and the nappes of the Mesozoic basement overrode, crushed and partly mingled with the chaotic breccias, creating mélanges (Fig. 5.14). With time, the Cuban orogeny events migrated eastward. In western Cuba, it developed from late Paleocene to Early Eocene. In this sector, the sediments are mostly chaotic, with clasts from the different units of the Mesozoic basement (except the Southern Metamorphic Terranes). West of La Habana they are known as "Manacas Formation" [16, 97], and lie sandwiched between the Mesozoic units of the Cordillera de Guaniguanico. Eastward from La Habana, the northern sub-basin fill conforms tectonic lenses in between the well-stratified strata of the Placetas TSU, known here as the "Vega Alta Formation."

Between the Yabre lineament and the La Trocha fault zone the tectonic event occurs from the Paleocene to Early Eocene. In this sector, the foreland basin shows two different outcrops (or sub-basins). The southern one contains crushed chaotic olistostromes of the Vega Alta Formation type (Paleocene to Middle Eocene age), with clasts derived from the Placetas TSU and the ophiolite association. In the northern sub-basin, the chaotic deposits (Vega Formation) only contain clasts from the Camajuaní and Remedios TSU. Between La Trocha FZ and Camagüey lineament, the southern foreland sub-basin basement belongs to the Remedios TSU, whereas the depression fills with chaotic sediments, eroded from the ophiolitic association and, in lesser degree, from the Placetas TSU and the KVAT, known as the Senado Formation. The formation is circa 1000 m thick. The external (northern) sub-basin filled with clastic carbonate sediments, torn from the Lower–Middle Eocene Florida–Bahamas platform border deposits. The presence of fine tuff interbeds in the Middle Eocene Lesca Formation is clearly related to the distal volcanic pyroclastic eruptions (probably from the Turquino–Cayman Ridge Paleogene arc. The northern sub-basin fill rests on the Cayo Coco TSU rocks.

The fourth outcrop area of the foreland basin is located around Gibara City in NW Holguin Province (Fig. 5.1). Again, two sub-basins can be distinguished with different

Fig. 5.13 Early Paleogene tectonic profile across central Cuba, including the Turquino–Cayman Ridge arc, in its southern edge to the foreland basin, in the north

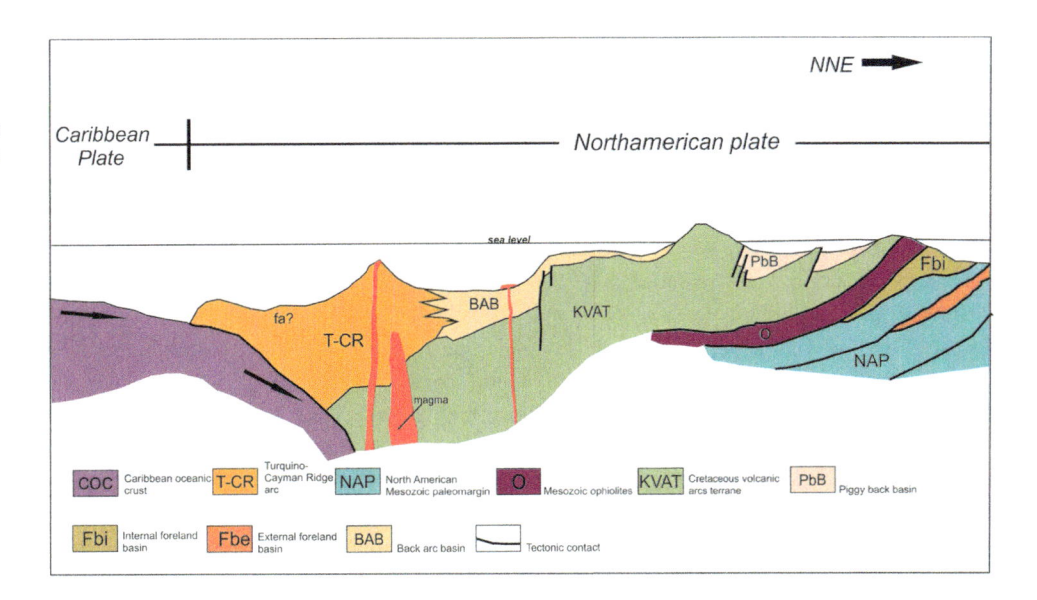

Fig. 5.14 Profile from Los Palacios Basin to the DSDP sites 97 and 540 in the SE Gulf of Mexico (modified from [19]). The profile integrates seismic and DSDP well data in the SE Gulf with data from inland wells and geological cartography in Pinar del Rio Province, Cuba

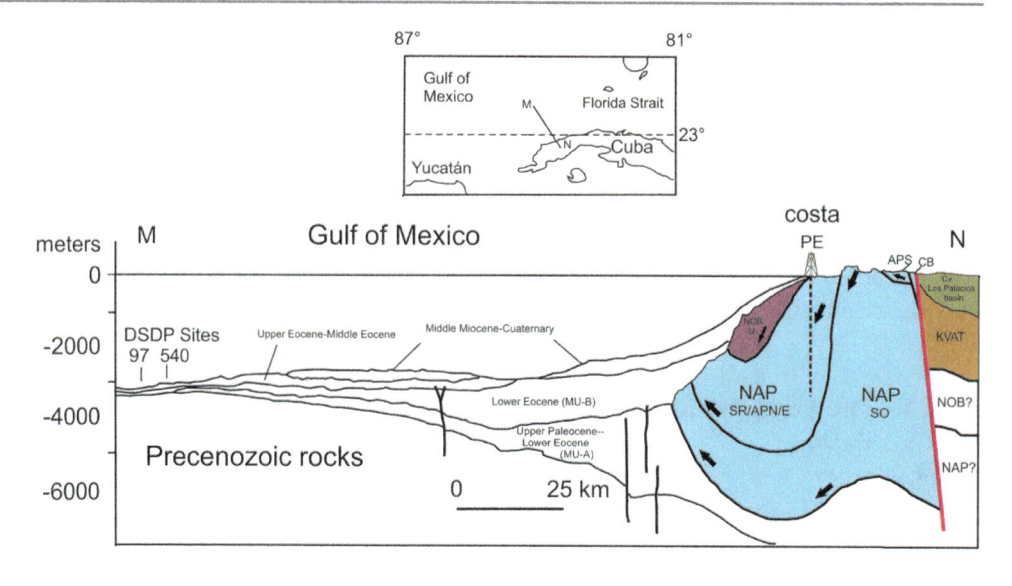

clastic compositions. The southern depression contains the Rancho Bravo Formation, very similar to the Senado Formation from Camagüey. The Rancho Bravo Formation [62] is in tectonic contact with the NOB and the Iberia mélange [110]. Clastic carbonates of Embarcadero and El Recreo Formations fill the northern part of the depression [41].

The northern foreland basin corresponds to the "*backarc collisional foreland basin*" in the model of Miall [79, 89], as will be briefly discussed in Sect 5.3.4.3.

5.3.3.2 The Piggyback Basins (PBB)

Small depressions developed upon the back of great thrust sheets during their advancement were first recognized about 40 years ago (Ingersoll and Busby 1995) and called piggyback basins. Several of these structures, related to the Cuban orogeny, are known. Their main development is attained in central Cuba, between the Yabre and Camaguey lineaments, but several depressions of this kind are recognized in other territories. They are filled with volcanomictic deposits of limited thickness, generally resting upon the KVAT rocks (mainly on its uppermost Cretaceous cover). Less frequently these deposits rest on the NOB and then they are severely deformed, i.e., northwestward of Camagüey City. Tuffaceous interbeds are present in those basins from central and eastcentral Cuba.

In western Cuba, the Mercedes, Madruga, Apolo, and Alcazar Formations of Danian and/or Lower Eocene age belong to this kind of depression. In central Cuba, the piggyback basins attain its maximum development. Three depressions are distinguished in this region, Cienfuegos and Santa Clara, in the west and the Cabaiguán Basin, in the east. In these depocenters, the frontier between the PBB beds and the Cretaceous cover is very diffused, because sedimentation was continuous and without major lithological changes from Late Maastrichtian to Early Danian (Santa Clara Formation, [20]). Therefore, we consider these depressions as inherited PBB. It is very interesting the finding of several tuffs of Maastrichtian–Paleocene ages, evidence of a weak explosive volcanic activity in the Cabaiguán Basin. A younger, Lower–Middle Eocene volcanic explosive event is also recorded in the SE part of the same basin [64]. Other local compositional changes are present eastward. In central Camagüey, the Taguasco Formation rests upon an ophiolitic floor and it is strongly deformed. In southern Camagüey, a few kilometers southward, an increase in pyroclastic beds in the Vertientes Formation (Lower–Middle Eocene) marks the transition to the backarc basin of the Turquino–Cayman Ridge arc. Whereas, in the same province, eastward of Camagüey City, the conglomerates and sandstones, circa 250 m thick of Maraguán Formation occurs, without the report of tuffaceous interbeds. Probably they are part of the fill of a Middle Eocene piggyback basin (see Pushcharovsky et al. [110]). Finally, in the southern fringe of the Maniabón Highlands, the Haticos Formation (Upper Paleocene–Lower Eocene) contains features similar to some parts of the Taguasco Formation but, whereas in the later the clasts were derived from the denudation of a KVAT source, in the Haticos Formation ([62], Garcia Delgado and Torres Silva 1997), the clasts came mainly from outcrops of the ophiolitic suite and beds of altered tuffs are frequently interbedded. Evidently, all these particular features are related to the local composition of the clastic sources, the tectonic stresses, and the arrival (or not) of pyroclastic particles from the Turquino–Cayman Ridge arc eruptions.

5.3.3.3 The Turquino–Cayman Ridge Volcanic Arc

In the Early Paleocene (Danian), a new, almost east–west trending submarine volcanic arc was born, the Turquino–Cayman Ridge arc (T-CR), whose rocks chiefly outcrop in SE Cuba, resting on top of the KVAT (especially on its

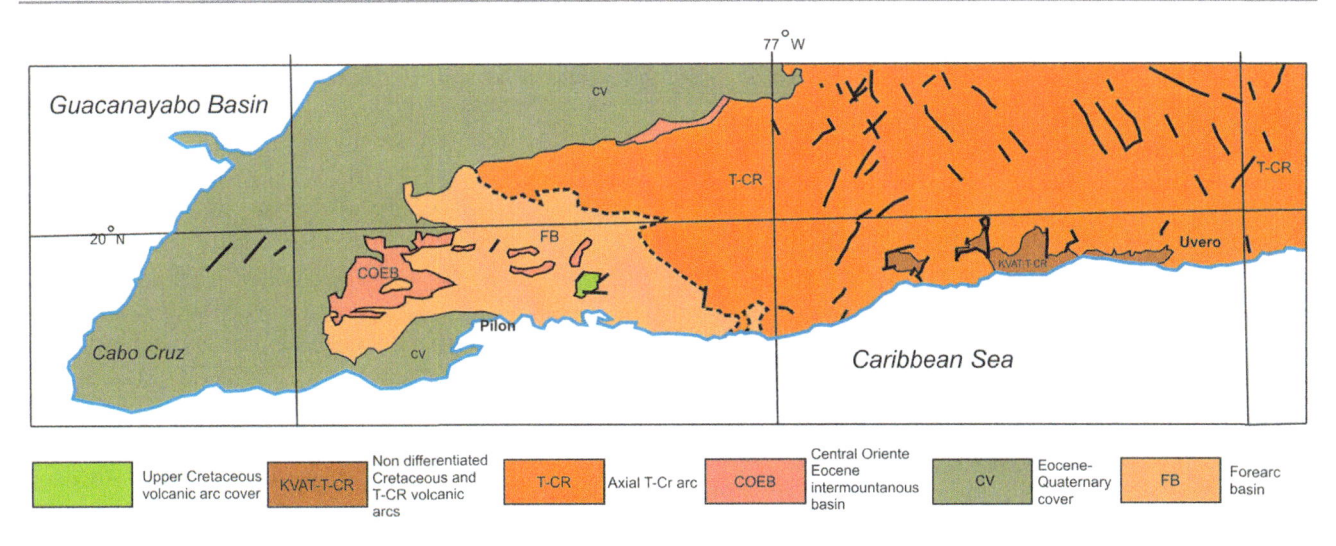

Fig. 5.15 Simplified tectonic map of the western end of the Sierra Maestra Mountains, supported on the interpretation of the Pushcharovsky [110] map and the unpublished Geological Map of Cuba from the Instituto de Geología y Paleontología [53]

Upper Cretaceous cover) and, in lesser degree, on the NOB (Fig. 5.13). Up to several thousand meters of effusive, pyroclastic, and intrusive rocks, ranging from felsic representants to basalts with sedimentary interbeds, crop out mainly in the Sierra Maestra Mountains (the old volcanic arc axis [88],Méndez Calderon et al. 1994, [60]). Its corresponding backarc basin, filled with pyroclastites, volcanogenic turbidites, and some carbonate deposits, lays to the north [13]. Effusive and intrusive bodies are scarce in the backarc. A forearc basin has not been clearly identified in the T-CR arc,however, an abrupt facies change occurs in the lower Paleogene sections, from sandstones with some volcanic interbeds (the Pilón Formation) to thick volcanogenic sections (the El Cobre Formation) at the western end of the Sierra Maestra Mountains, along a ENE tectonic contact (an overthrust or strike-slip fault) eastward from Pilón, Granma (Fig. 5.15).

Intruding along the volcanic axis rocks are a large number of igneous bodies Figs. 5.13, 5.16, ranging in composition from granite to gabbro and in structural setting from dykes and sills to stocks. According to Rojas Agramonte et al. [115], these are calcoalcalic rocks of intraoceanic origin. Pb-U radiometric ages in zircons yield ages from 60 to 48 My (Paleocene–Middle Eocene), very similar to the time span for the volcanic rocks of the T-CR. Westward, coeval volcanic beds were recorded in the Cayman Ridge and Nicaragua Plateau [50, 90]. The record of an early Paleogene volcanic arc, with a northern backarc basin, is present in the Cayman Ridge and Yucatan Basin [120]. With this architecture, a north dipping subduction zone was postulated by these authors. Coeval volcanic rocks occur in the Montaignes Noires and the Northwestern Peninsula in Haiti [8] and Sierra de Neiba in the Dominican Republic [74, 37]. Therefore, we can suspect that all these localities were originally part of an expanded T-CR arc. Probably this almost latitudinal structure was more than 1500 km long (more than the current Lesser Antilles arc) and the subducted Caribbean crust dived with a strong northward component. This was explicitly declared by Siggurdsson et al. [120], but has been largely overlooked. In many recent models on Caribbean region evolution, during the Danian to Middle Eocene interval, the oceanic crust at the leading edge of the Caribbean plate was moving E-NE [91]. In our interpretation, the lithosphere to the north of the T-CR subduction zone was already accreted to the North American plate in the Paleocene and the oceanic Caribbean crust dives below this new margin (Fig. 5.13) and the T-CR arc is essentially an in situ structure. Therefore, the great early Paleogene thrusting event along northern Cuba, from Pinar del Rio to NW Holguin Province (the Cuban orogeny), is not directly related to a supposed obduction of the Caribbean plate upon the North American paleomargin. Compare (Fig. 5.13) with the "backarc collisional foreland basin" from the model in Miall [79] and the "backarc foreland basin" in Nichols [89] (Fig. 5.16).

At the beginning of the Middle Eocene, volcanism almost ceased [13]. Perhaps it was related to the arrival at the subduction suture of a thick and less dense oceanic plateau, docking it, and immediately followed by a change in the regional stress field, with the birth of the great Oriente Fault Zone (OFZ), related to the weakened hot lithosphere of the recently extinct T-CR arc [23, 123].

The axial region of the T-CR arc presents moderate deformations. In a crude approximation, it is a huge monocline with north dipping beds, except in its eastern end, at La Gran Piedra, where the regional strike rotates toward NW–SE. Locally the monocline is complicated by folds, often related to E–W reverse folds that can be followed by several

Fig. 5.16 Basaltic columns at Puerto Moya, Central Road, Santiago de Cuba Province. The rocks belong to the middle–upper part of the axial strata of the Turquino–Cayman Ridge Paleogene arc

kilometers. In general, the structural complication decreases northward. In the Sierra Maestra, these deformations affect the Paleocene to middle Eocene beds, with a trend becoming less complicated northward. However, in the northern foothills of the Sierra Maestra, the volcanoclastic Middle–Upper Eocene beds (San Luis Formation) outcrops, conformably resting upon the underlying T-CR. Supported on this last fact, Refs. [61, 115] believe that the folding and faulting event is Oligocene in age. Dealing with the Central Oriente Eocene Basin (COEB), we will see that the San Luis Formation is related to a regional uplift event, located southward of its depocenter. The first movements along the Oriente fault zone belong to this event, as we will see in the following pages.

5.3.3.4 The Central Oriente Intramountainous Eocene Basin (COEB)

The COEB is an E–W trending synclinorium, with volcanoclastic Middle–Upper Eocene beds that extends from the eastern Guantanamo Valley up to the Sierra Maestra northern foothills, near Bayamo City, in Granma Province. The San Luis (Middle and Upper Eocene) and Camarones (Upper Eocene) Formations represent the bulk of the basin fill, meanwhile Farallón Grande and Mucaral [62, 62, 41] formations are local facies. Except for the Mucaral Formation, almost all the clastic beds were derived from the denudation of highlands located at the southern margin of the basin [27, 65, 73], built by rocks of the then recently extinct lower Paleocene–early Middle Eocene Turquino–Cayman Ridge arc. As we will see, the Central Oriente Intramountainous Basin is a key piece for the solution of some main questions on the geology of the Caribbean/North America plate boundary.

The COEB deposits (Figs. 5.17, 5.18) make a depositional wedge more than 200 km long, by circa 80 km wide (from the Caribbean coast to Sagua de Tánamo Valley), with a maximum thickness circa 800 m in the northern Sierra Maestra foothills, at Mesa de Santa Maria del Loreto Plateau. Northward of its impressive vertical cliffs, the marine coarse-grained deposits of the Camarones Formation transit toward the sandstone–siltstone turbidites of the San Luis Formation. A thickness decrease to some hundreds or tens of meters accompanies this transit at the southern slopes of Sierra de Nipe and Sierra Cristal Highlands. According to the present author interpretation of the Lewis and Straczek's [73] descriptions, probably the Middle–Upper Eocene sedimentary wedge is the sum of many submarine fans where the clastic sediments, sourced by a southern mountain range, were caught (see Fig. 18 in [23]). Keijzer [65] called this highland the "Bartlett Land." As will be seen later, this was an important discovery, not only to explain the Cuban regional geology, but also for the North American–Caribbean plate relationship. However, after 1955, it stood almost forgotten for many years [27, 25]. The Bartlett Land denudation was very active, as the granitoid clasts recorded in Camarones and San Luis Formations show. Evidently they were derived from isolated outcrops of the granitoid plutons intruding the Turquino–Cayman Ridge arc [73, 115].

Another main event recorded in the COEB is the accompanying magmatism. However, it is volumetrically insignificant, it is the youngest igneous event recognized in the Cuban territory. The intrusive magmatism is represented by andesite–basaltic dykes, emplaced at shallow depths, cutting the San Luis and Farallón Grande Formations. The volcanism is represented by felsic tuff beds in the marine Middle–Upper Eocene Barrancas Formation [62].

The youngest pre-cover deformations are recorded in the Turquino–Cayman Ridge and the Central Oriente Basin. The most remarkable by its dimensions is the great Sierra Maestra monocline, followed northward by the Central Oriente Synclinorium [115].

The composition of the San Luis Formation sandstones and conglomerates, outcropping northward of the Sierra Maestra Mountains, is typically volcanomictic, with some carbonate and granitoid clasts and without ophiolitic or metavolcanic clastics. Whereas, in the Cajobabo, Imías and San Antonio del Sur Valleys at the Caribbean Coast, in Guantanamo Province, the clastic rocks of the San Luis Formation, together with their typical volcanomictic debris, show relatively abundant clasts, clearly derived from nearby sources with ophiolitic and metavolcanic rocks (Figs. 5.17 and 5.18). In the easternmost basin locations, at the Cajobabo Valley, upon the well-stratified turbidites of San Luis Formation, lies a small klippen with Lower Eocene rocks

Fig. 5.18 of the Turquino–Cayman ridge volcanic arc [23, 25]. Near the mouth of Cajobabo River, a small serpentinite body occupies a frontal position in the clipper of T-CR rocks, emplaced upon Middle Eocene olistostrome and turbidites of the basal San Luis Formation. All these rocks are unconformably covered by Neogene–Quaternary beds (see Fig. 14 in [23]). The lower Paleogene volcanites are not present in the local autochthon pre-Eocene record and the thrusting probably was a late Middle Eocene event. In the author's criteria, these facts could be explained by the first horizontal movements along the Oriente fault zone of the Caribbean/North America plate boundary (see [72]). Particularly, the geological relationships at the coastal area between San Antonio del Sur and Cajobabo suggest a left lateral (sinestral) displacement of circa 25–30 km during a late Middle Eocene travel along the fault. As a consequence, a tectonic flake (perhaps a southern prolongation of the current Sierra del Convento ophiolitic massif), was caught by the Bartlett Land, displaced several kilometers eastward and finally located in front of the eastern Oriente basin margin, southward Cajobabo. These facts could explain the peculiar clastic composition of San Luis Formation in this area and its relationship with the tectonic emplacement of the Turquino–Cayman Ridge arc rocks.

5.3.4 The Eocene–Quaternary Cover

The cover (Neoauthocton, sensu [61]) embraces the large upper structural stage of the Cuban orogen. From Pinar del Rio to NW Holguín, it includes little deformed strata, accumulated after the Cuban orogeny. In eastern Cuba, south and east of the Guacanayabo–Nipe Bay lineament, the cover

Fig. 5.17 Lower Eocene tuffs outcroping near the Caribbean coast at La Farola road, Cajobabo, Guantanamo.Th strata belong to a tectonic scale with T-CR rocks (El Cobre Formation) and some slivers of serpentinite breccias (see Cobiella et al. [25], Iturralde- Vinent [54], and Cobiella-Reguera [23]). The scale (circa 5 sq. kilometers and perhaps 500 meters of maximum thickness) rests on the basal beds of the Central Oriente Intermontanious Basin

embraces the beds accumulated after the end of the Late Eocene magmatic activity (Figs. 6.19 and 6.20).

As we saw, the Cuban orogeny not concluded simultaneously throughout Cuba. There is a strong evidence of the decisive role played by several narrow strips arranged transversely to the general structural trends generated by the Cuban orogeny. These structures have a linear character (lineaments) and their development is closely related with the nature and age of the tectonic events. Four main lineaments (and several minor) are identified: (a) Yabre lineament; (b) La Trocha lineament (or fault zone); (c) Camagüey lineament; and (d) Guacanayabo–Nipe Bay lineament

Fig. 5.18 Olistostrome of the basal beds of San Luis Formation at Cajobabo. The deposit is a pile of unsorted blocks, the biggest attaining several meters.The most abundant clast lithologiesare lapillitic tuffs and agglomerates with some porfiritic andesites.Obviously they were torn from a near T-CR source rocks. Additionally, some greenschists clasts (from the Upper Cretaceous volcanic arc) and gabbro-pegmatites (from the NOB) are present.

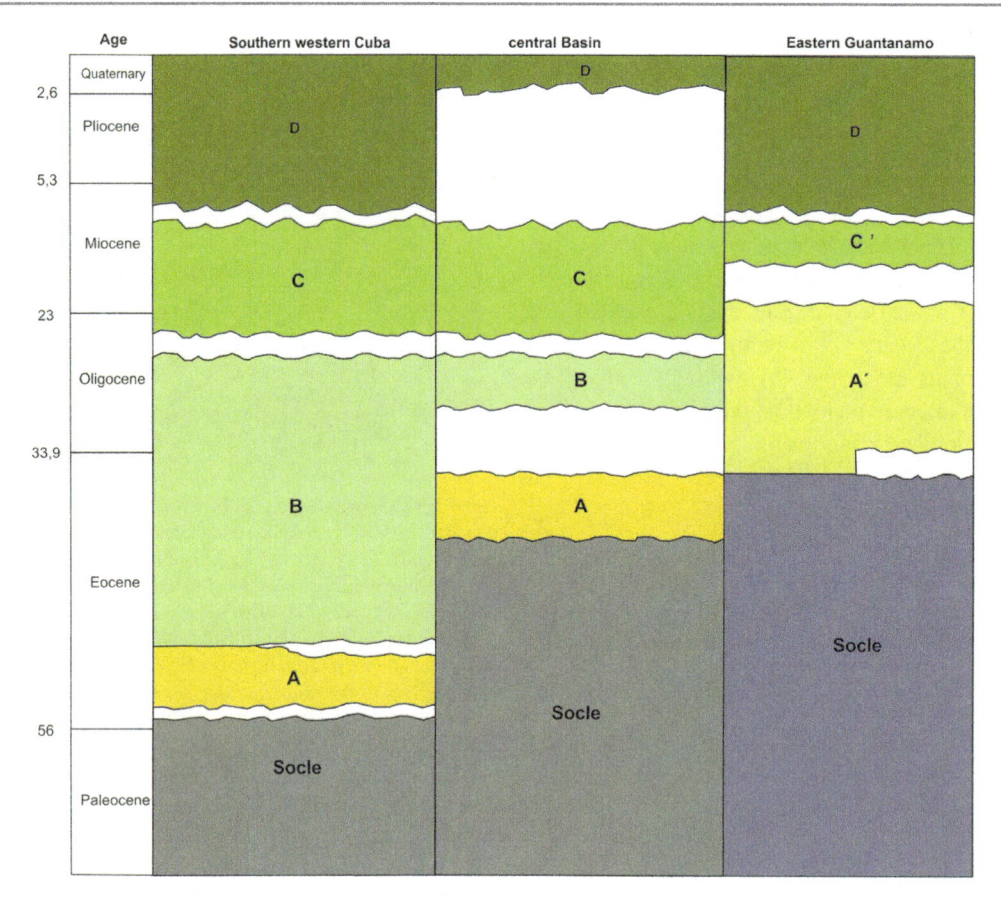

Fig. 5.19 Stratigraphic distribution of the cover sub-stages in different localities. Whereas in western Cuba the oldest cover beds are almost of basal Lower Eocene age, in central Cuba the cover begins with Middle Eocene strata and in easternmost Cuba the lowermost cover rocks settled in the latest Late Eocene. In the figure, the different nature of the socle between eastern and central-western Cuba is emphasized

(Fig. 5.1). As previously noted, the orogenic deformations are genetically linked with the arrival to the foreland basin of remarkable volumes of chaotic terrigenous deposits. Westward of the Yabre lineament these arrivals occur from Lower Paleocene to Lower Eocene. Between Yabre and La Trocha lineaments, the great terrigenous arrival developed between the Paleocene and the beginning of the Middle Eocene. Between La Trocha and Guacanayabo–Nipe Bay lineament, the orogenic events span from Middle Eocene to late Middle Eocene or early Late Eocene. Therefore, the cover bottom is markedly diachronic (from Lower to basal Upper Eocene) and the cover presents beds with ages ranging from Lower Eocene to Quaternary. Eastward of the Guacanayabo–Nipe bay lineament, instead a socle with an alpine tectonic style in its lower Paleogene rocks, a relatively little deformed and basically in situ volcanic arc (the Turquino–Cayman Ridge arc) and the Central Oriente Eocene intramontane basin beds are present. Here the cover embraces rocks from the Uppermost Eocene to the Quaternary (Figs. 5.19, 5.20).

Several major blocks with contrasting movements can be recognized related to the cover tectonics [54]. According to Cobiella (27b, modified), the following main subsiding regional blocks are present in the Eocene–Quaternary history

Fig. 5.20 Unconformity between cover sub-stages C and D at Vía Blanca highway, La Habana. D: sub-stage D, represented by the cross-bedded Pleistocene Guanabo Formation, C: sub-stage C, represented by the shallow water, strongly weathered Lower–Middle Miocene Güines Formation

of the cover: (1) Nipe–Baracoa Basin, (2) San Luis–Guantánamo Basin, (3) Cauto–Guacanayabo Basin; (4) Central Basin; (5) Southwestern Basin; and (6) Northcentral Basin.

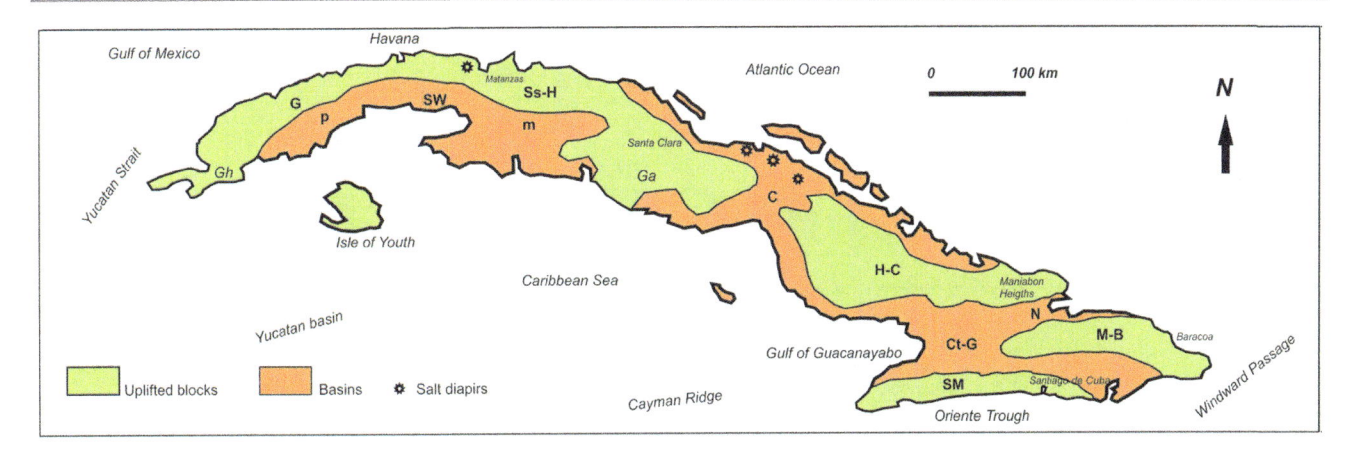

Fig. 5.21 Distribution of uplifted areas and basins in the cover from Eocene to Quaternary. Included are the localities with salt diapirs in Cuba

The coeval regional structures with a dominant rising trend are from east to west: (7) Babiney–Maisí block; (8) the Sierra Maestra Mountains; (9) Holguín–Ciego de Ávila Block; (10) Sancti Spíritus–La Habana Block; (11) Guaniguanico–Guanahacabibes Block; and (12) Isla de la Juventud Block (Fig. 5.21).

The cover outcrops in more than 50% of the Cuban territory. Using different sources, we studied the cover sediment thickness in 40 wells. Its average thickness was about 760 m, but records of 2000–3740 m are known [112]. In western and central-eastern Cuba, we can distinguish four sub-stages, arranged in the following order from bottom to top: A, B, C, and D. Their ages range from Lower Eocene to Quaternary (somewhat similar ideas on the Paleogene tectonic development, but with a different approach, are contained in Refs. [93, 106]. Meanwhile, the proximity of SE Cuba (eastward from Guacanayabo–Nipe Bay lineament) to the Caribbean/North America plate boundary makes its geological development tightly related to the history of the Cayman trough and, therefore, with particular features. In SE Cuba, three sub-stages are distinguished in our research: A', C', and D' (Fig. 5.19).

5.4 Some Comments on Certain Regional Structures

5.4.1 The Oriente Fault Zone

The deep basin that forms the easternmost Cayman Trough is called the Oriente Trough. The first ideas on the geology of the Cayman Trough developed in the 30s of the twentieth century and belongs to Stephen Taber [125] supported on general observations about the geomorphology and geology of the Sierra Maestra Mountains and preliminary information on the bottom topography. In the middle 60s, Bowin [6] presents a first geophysical model of the Cayman Trough showing the oceanic crust nature of its bottom. The picture was later enriched by Perfit and Heezen [90] with drag samples from its walls. At that time, the information on the great linear depression begin to be integrated in the first plate tectonics models related to the Middle American area, when it becomes evident the location of great strike-slip faults along its margins ([11], Cobiella-Reguera et al. [27]). A crucial finding was the location of a little N–S trending, well-defined oceanic crust formation center at the deepest part of the trench, southward of the Cayman Islands. In Cuba, some interesting results attained a group of Cuban geomorphologists working in the Sierra Maestra [48].c

An outstanding contribution to the knowledge of the Cayman Trough geology is the paper by Calais and de Lepinay [10], a seismic research on the structures of the Oriente Fault Zone (OFZ) along the northern Columbus Strait (Fig. 5.22). This study revealed several unknown features on the internal structure of the Oriente Trough. The data collected show that the OFZ is not a single dislocation, but embraces several major fractures, with a general E-NE trend, connected through several minor dislocations. The OFZ is a major sinestral fault zone. As it occurs in all the great strike-slip faults, in some sectors, compressive plicative folds and nappes develop, whereas in others normal faults and grabens are present. The most remarkable structure found is the Santiago Deformed Belt (SDB), a narrow and long compressional belt, extending from 75° to 76° 30′ W, almost in front of the Cordillera de la Gran Piedra Highlands. Some small sedimentary basins are located in the Caribbean Sea bottom, west (Chivirico Basin), and east (Imías Basin) of the SDB. Onshore, the Santiago de Cuba, the Baconao Lake Basin, and the coastal strip between Guantánamo Bay mouth and Cajobabo are recently uplifted marine basins. In these locations, sets of prograding Neogene and Quaternary beds (Fig. 5.23), several ten or a

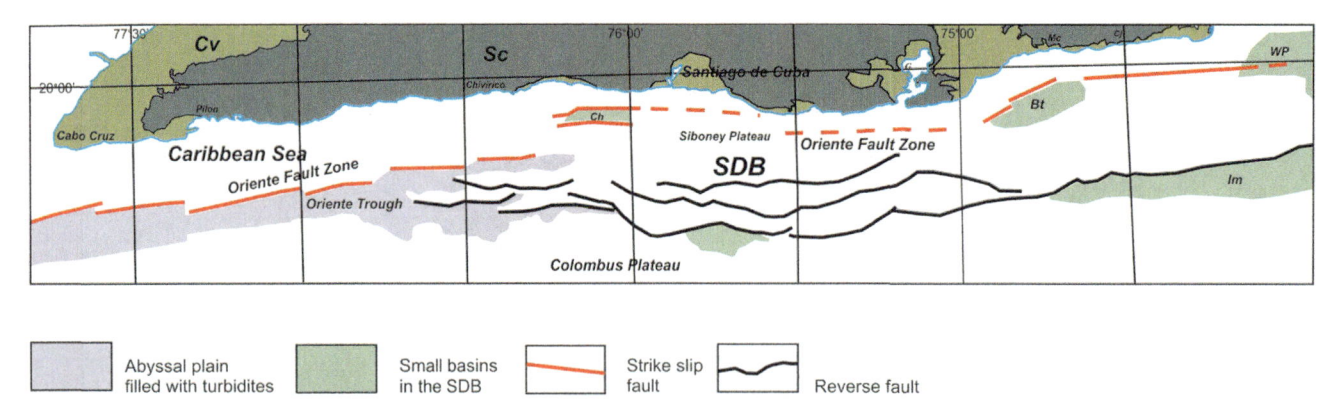

Fig. 5.22 Main structures in the Columbus strait (located between SE Cuba and Jamaica). The geologic setting is explained in the text

Fig. 5.23 Southward prograding Neogene–Quaternary beds (toward the Caribbean Coast) at Macambo, Imías, Guantánamo Province. The study of the bedding reveals at least two sets: magenta and blue lines in the photo. A third (green) set is poorly defined

hundred meters thick, southward tilted toward the shore of the Columbus Strait show that these great geomorphic structures developed at least from the Neogene. Along long tracks of the eastern Cuba's southern coast, extraordinary marine terraces, with cliffs ten meters high, cut the progradational beds.

A narrow abyssal plain, probably filled with turbidites, develops at the deepest portion on the Oriente Trough, almost at 7 km deep [49], and only several kilometers southward of Pico Turquino, the highest summit in Sierra Maestra (circa 2 km high), immediately westward of the SDB. Interesting data and ideas on the tectonic development of the eastern Oriente Trough report [10]. Supported on their interpretation of seismic profiles from the Imías and other small basins located near the Windward Passage, a five-stage tectono-sedimentary succession (correlated with nearby territories, including the Cuban SE tip) was proposed [10]. However, the correlation with neighboring inland Cuba is weak and demands further researches.

5.4.2 A Joint Study of the Regional Neotectonics and Well Data in Western Cuba

The PFZ is a polemic structure, with a general WSW-ENE trend and excellent geologic and geomorphologic contrasts with Los Palacios Basin from El Sábalo, at SW Pinar del Río Coast, to Cayajabos, in Artemisa Province (Fig. 5.2). Eastward of the last locality, the geomorphic evidence is less clear. However, NAP outcrops, with typical lithologies from the SR-APN-E units, appear in the Martin Mesa outlier, located E-NE of Cayajabos Town, and only a few kilometers from the western surroundings of La Habana (Fig. 5.2). In the eastward continuation of the fault trace, wells with NAP are present to the north, in the Mayabeque and Matanzas oil fields, whereas ophiolites and the KVAT outcrop to the south (Pérez Othón and Yarmoliuk 1984) in antiform structures from the Vegas Basin.

A remarkable topographic jump exists along the western PFZ. In the northern block, the Guaniguanico Highlands summit near the FZ attain circa 600 m in the Sierra del Rosario and more than 400 m in Alturas de Pizarras del Sur, whereas the rolling lowland surface at the Los Palacios Basin descends from heights circa 100 m at Guaniguanico foothills, to sea level at the Caribbean shore. Saura et al. [117] studied the northeastern Los Palacios Basin near Candelaria, using the results of recent seismic profiles and information from two old wells. They consider the FZ as two south-dipping fractures (approximately 50°), welded in one at approximately 1.5 km depth. These authors collected kinematic indicators in different places where fault planes outcrop. The oldest evidences record strike-slip motion along the fault. A second deformation stage consists mainly of SE-plunging slickensides and striate that suggest a dip-slip movement. Structurally, the western half of Los Palacios Basin is a great southeast-gently dipping monocline, with some minor unconformities.

In the most complete study on the PFZ, Gordon et al. [43] discriminate five phases in its history. In their opinion, the first is related to the early Paleogene thrusting event during the Cuban Orogeny. The second (compressive) sinestral phase occurs shortly after the settling of Capdevila Formation (Early Eocene), and prior to the Middle Eocene. The third phase is also a compressive ENE–WSW deformation, overprinting older faults with strike-slip features. Probably, after a 30 my quiescence, the fourth phase developed. It is a N–S extensional post early Miocene event. The fifth and last phases, slightly younger than the normal faults, are represented by a wrench faulting episode, related to ESE–WNW compression.

The Los Palacios Basin is the westernmost depression of the Southwestern Cuban Basin (SWB), a group of related Cenozoic Basins located from the southern half of Pinar del Rio and Artemisa Provinces (Los Palacios Basin) that eastward continues in Mayabeque Province (the Vegas and Broa depressions) and continues in Matanzas (Mercedes Basin) ending in the Cienfuegos Basin Fig. 5.21. The Los Palacios Basin contains a relatively thick siliciclastic-carbonate fill. A northern source, located in the Guaniguanico cordillera, is clearly evident in the clastic composition of its younger sediments. However, an older southern source perhaps was present in its earliest development, during the Early Eocene, because siliciclastic grains derived from the erosion of metamorphic and volcanic rocks (the dominant pre-Cenozoic lithologies in the Isla de la Juventud) are abundant in the Lower Eocene Capdevila Formation, but clasts from the NAP are almost absent in the strata older than the Middle Eocene (older than Loma Candela Formation). The thickness of the Eocene–Quaternary fill fluctuates between several meters and more than 2400 m [112].

An integrated study of the main western Cuba neotectonic features Fig. (5.21) and the well data in Perez Othón and Yarmoliuk [112] reveals some basic facts and trends to understand the regional structures. In those wells located in the interior of the southwestern basin, the Mesozoic basement rests below a thick Eocene–Quaternary cover. In Vegas 1, Mercedes 1, and Ariguanabo 2 wells, ophiolitic rocks rest on their bottoms (Fig. 4 in [18]). Above, with tectonic contact, lie KVAT beds, including its Campanian–Maastrichtian cover, and some lower Paleogene piggyback basins. Guanal 1 hole, near La Coloma, Pinar del Rio Province, is a shallow well where the KVAT rocks unconformably rest below the Cenozoic cover.

Northward of the SW basin extends the Sancti Spiritus-La Habana Block (SS-H, Fig. 5.21), an uplifted structure with a thin cover and wide outcrops of different Mesozoic units. In northern Matanzas Province, several antiforms (Fig. 5.24, [95]) show cores that resemble tectonic mosaics, with outcrops from the NAP (Placetas TSU), the NOB, and the KVAT. Westward, at northern Mayabeque and Matanzas, is located the petroleum geologist's "Belt of Heavy Oils." In many wells of the belt, below the cover, nappes of Placetas and Camajuaní TSUs, with tectonic wedges of the foreland basin, are present.

In the Guaniguanico cordillera Block, denudation attains its maximum depth in western Cuba and the deeply covered structures in the Southwestern basin and Sancti Spiritus -Habana Block widely outcrops. Therefore, in the Cordillera, the structural relationships between the distinct socle units are much more visible. In the eastern Sierra del Rosario Highlands, the tectonic units of the Mesozoic basement and the Paleogene fold and thrust belt are essentially located in the following structural positions (from uppermost to lowermost): (a) Paleogene piggyback Basins, (b) Cretaceous volcanic arc terrane (including Campanian–Maastrichtian sedimentary cover), (c) Northern Ophiolite Belt, (d) North American paleomargin, (Pan de Guajaibón TSU, Placetas TSU), and (e) Paleogene Foreland Basin.

It is a crucial fact that the Upper Jurassic–Lower Eocene section in the Belt of Heavy Oils is similar to the coeval beds in SR-APN-E and its overlying foreland basin.

Finally, in central Cordillera de Guaniguanico Mountains, the Sierra de los Órganos karstic landscapes are developed on the mainly carbonate Upper Jurassic–Cretaceous sections where they outcrop the core of the nappe structures of western Cuba (Fig. 5.4) with some foreland wedges (Fig. 5.5).

From the preceding review, we can suggest that the same regional tectonic units could be present along all western Cuba, from Pinar del Río to Matanzas Provinces, a very attractive fact for onshore hydrocarbon search.

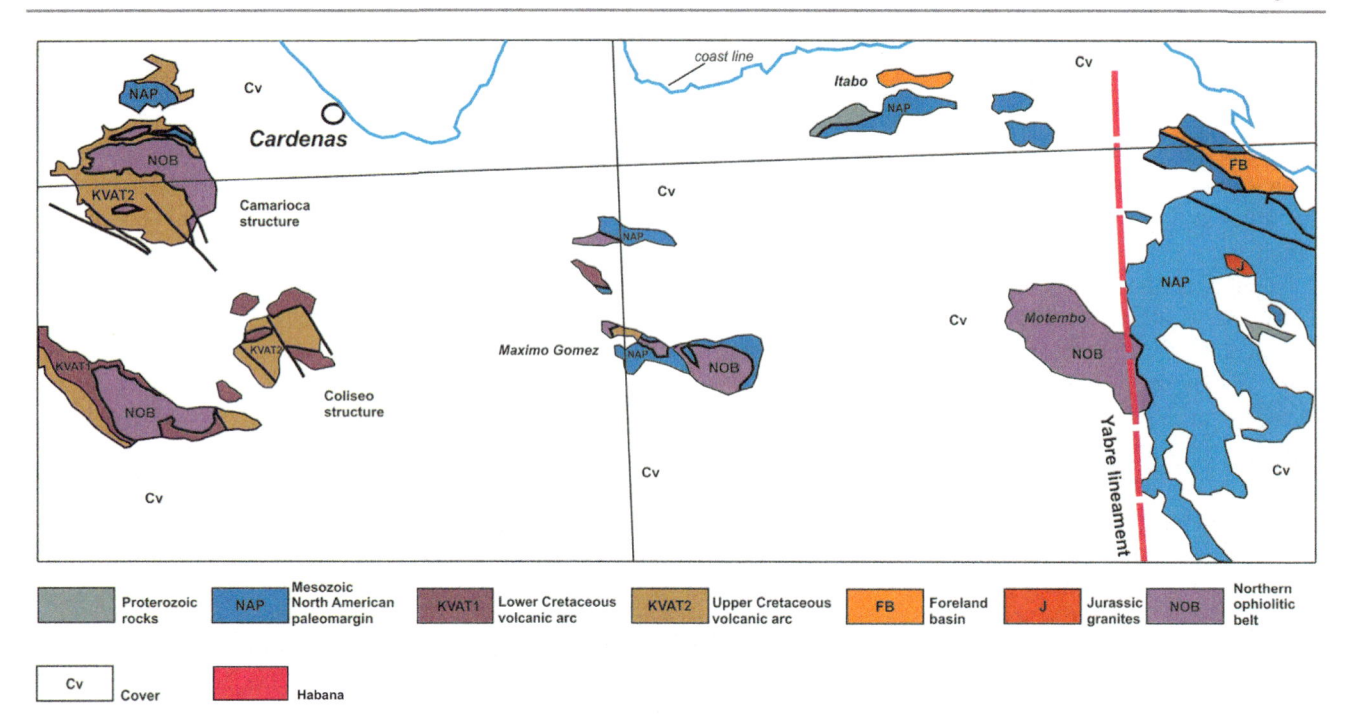

Fig. 5.24 Simplified tectonic map of NE Matanzas and NW Villa Clara Province. The structural setting is explained in the text. Very important are the two small outcrops of Precambrian metamorphic rocks near Itabo (Matanzas) and Motembo (Villa Clara)

5.5 Conclusions

In this chapter, the author presents his approach to several items on the regional tectonics of Cuba and its surroundings. Nobody can deny that the Cuban territory presents a complex geology, additionally increased by its tropical weathering and extensive territories with a discrete relief and relative few outcrops. The proposed tectonic scheme departs essentially from the last geologic (Pushcharovsky ed. [110]) and tectonic (Pushcharovsky ed. [110]) maps of Cuba, plus the main achievements attained in the regional geology studies in the last 30 years. However, we must also return to the study of valuable old data that we consider still not enough exploited, as several cases in the preceding pages show. It is evident that in the next years, the resources for the geological research in Cuba still will be limited. An important alternative way to improve and increase our knowledge on the tectonics of Cuba with minimum resources is extracting the maximum from the abundant old information and confront it with the new one, with a dialectic approach, under the light of plate tectonics, sequential stratigraphy, sedimentary basin analysis, and other paradigms developed in the last decades.

Acknowledgements The author is deeply indebted to Manuel Pardo Echarte for his invitation to work in a project devoted to present the ideas and contributions of Cuban specialists in different fields of the regional geology. Also, I wish to acknowledge the stimulus and support from my colleagues of the Department of Geology at Pinar del Río University and from the Instituto de Geología y Paleontología, Cuba.

References

1. Albear Fránquiz J, Piotrowski J (1984) El enclave yesífero de San Adrián, Cuba. Observaciones sobre su ubicación geólogo-tectónica. Ciencias De La Tierra Y Del Espacio 9:17–30
2. Allegret L, Arenillas I, Arz JA, Díaz C, Grajales Nishimura JM, Meléndez A, Molina E, Rojas R, Soria AR (2005) Cretaceous-Paleogene boundary deposits at Loma Capiro, central Cuba: evidence for the Chicxulub impact. Geol Soc Am 33:721–724
3. Andó J, Harangi S, Szakmany L, Dosztaly L (1996) Petrología de la asociación ofiolítica de Holguín. In: Iturralde-Vinent M (ed) Ofiolitas y arcos volcánicos de Cuba, International Geological Correlation Program, Project 364. Geological Correlation of Ophiolites and volcanic arcs in the Circumcaribbean Realm, Miami, Florida, pp 154–176
4. Areces Mallea A (1991) Consideraciones paleo biogeográficas sobre la presencia de *Piazopteris branneri* (*Pterophyta*) en el Jurásico de Cuba. Rev Española De Paleontol 6(2):126–134
5. Arenillas I, Arz J, Grajales Nishimura J, Melendez A, Rojas Consuegra R (2016) The Chicxulub impact is synchronous with the planktonic foraminifera mass extinction at the Cretaceous/Paleogene boundary: new evidence from the Moncada section Cuba. Geol Acta 14(1):35–51
6. Bowin C (1968) Geophysical study of Cayman Trough. J Geophys Res 73:5159–5173
7. Bralower T, Iturralde Vinent M (1997) Micropaleontological dating of the collision between the North America and Caribbean plates in western Cuba. Palaios 12:133–150

8. Butterlin J (1960) Geologie generale et regionale de la Republique d'Haiti. University of Paris, Travaux et Memories de l'Institute des Hautes Etudes de l'Amerique Latine v. 6
9. Calais E, Mercier de Lepinay B (1991) From transtension to transpression along the northern Caribbean plate boundary off Cuba: implications for recent motion of the Caribbean plate. Tectonophysics 186:329–350
10. Calais E, Mercier de Lepinay B (1995) Strike-slip tectonic process in the Northern Caribbean between Cuba and Hispaniola. Mar Geophys Res 17:63–95
11. Case J, Holcombe T (1980) Geologic-tectonic map of the Caribbean Region. Scale 1:2,500,000. United States Geological Survey. Miscellaneous Investigations, Map I-1100
12. Cobiella Reguera J (1974) Los macizos serpentiníticos de Sabanilla Mayarí Arriba, Oriente. Rev Tecnológica 4:41–50
13. Cobiella Reguera J (1988) El vulcanismo paleogénico cubano. Apuntes para un nuevo enfoque. Rev Tecnológica XVIII 4:25–32
14. Cobiella Reguera J (1996a) El magmatismo jurásico (Calloviano? Oxfordiano) de Cuba occidental: ambiente de formación e implicaciones regionales. Rev De La Asociación Geológica Argent 51(1):15–28
15. Cobiella Reguera J (1996b) Estratigrafía y eventos jurásicos en la cordillera de Guaniguanico Cuba Occidental. Minería y Geología 13(3):11–25
16. Cobiella Reguera J (1998a) Las melánges de Sierra del Rosario, Cuba occidental Tipos e Importancia Regional. Minería y Geología XV 2:3–9
17. Cobiella Reguera J (2000) Jurassic and cretaceous geological history of Cuba. Int Geol Rev 42(7):594–616
18. Cobiella Reguera J (2005) Emplacement of Cuban Ophiolites. Geol Acta 3:273–294
19. Cobiella Reguera J (2008) Reconstrucción palinspástica del paleomargen mesozoico de América del Norte en Cuba occidental y el sudeste del Golfo de México. Implicaciones para la evolución del SE del Golfo de México. Rev Mex De Cienc Geol 25:382–401
20. Cobiella Reguera J, Cruz Gámez E, Blanco Bustamante S, Pérez Estrada L, Gil González S, Peraza Rozón Y (2015) Cretaceous Paleogene boundary deposits and paleogeography in western and central Cuba. Rev Mex De Cienc Geol 32(1):156–176
21. Cobiella Reguera J (1984) Curso de Geologia de Cuba. Editorial Pueblo y Educación, 114p
22. Cobiella Reguera J (1998b) The Cretaceous System in Cuba—an overview. Zentralblatt für Geologie und Paläontologie, Teil I, H.3-6, 431–440.
23. Cobiella Reguera J (2009) Emplacement of the northern ophiolite belt of Cuba. Implications for the Campanian-Eocene geological history of the northwestern Caribbean-SE Gulf of Mexico region [In: James K, Lorente M, Pindell J (eds) The origin and evolution of the Caribbean plate. Geological Society of London Special Publication 328, 313–325]
24. Cobiella Reguera J (2017) Base Estructural-Tectónica del Mapa Metalogénico de la República de Cuba a escala 1:250,000. In: Colectivo de autores, Mapa Metalogénico de la República de Cuba a escala 1:250,000. IGP, Instituto de Geología y Paleontología, Servicio Geológico de Cuba, Centro Nacional de Información Geológica, pp 27–55
25. Cobiella J, Boiteau A, Campos M, Quintas F (1977) Geología del flanco sur de la Sierra del Purial. La Minería en Cuba 3 (1), 54–62
26. Cobiella Reguera JL, Quintas F, Campos M, Hernández M (1984a) Geología de la región central y suroriental de la provincia de Guantánamo. Santiago de Cuba, Editorial Oriente, 125p
27. Cobiella Reguera JL, Rodríguez Pérez J, Campos Dueñas M (1984b) Posición de Cuba oriental en la geología del Caribe. Min Geol 2(2):65–92
28. Cobiella Reguera J, Hernández Escobar A, Díaz Díaz N, Obregón Pérez P (1997) Estudio de algunas areniscas de las formaciones San Cayetano y Polier, Sierra del Rosario, Cuba occidental. Min Geol XIV(3):59–68
29. Cobiella Reguera J, Hernández Escobar A, Díaz Díaz N, Gil González S (2000) Estratigrafía y tectónica de la Sierra del Rosario, Cordillera de Guaniguanico, Cuba occidental. Min Geol XVII(1):5–15
30. Cobiella Reguera J, Olóriz F (2009) Oxfordian—Berriasian stratigraphy of the North American paleomargin in western Cuba: constraints for the geological history of the Proto-Caribbean and the early Gulf of Mexico. In: Memory 90 AAPG. Petrol Geol South Gulf Mexico 421–451
31. Despaigne Díaz A, García Casco A, Cáceres Govea D, Jordan F, Wilde S, Millán Trujillo G (2016) Twenty-five millions years of subduction-accretion-exhumation during the Late Cretaceous-Tertiary in the Northwestern Caribbean: The Trinidad Dome, Escambray Complex, central Cuba. Am J Sci 316:203–240
32. Draper G, Gutiérrez R, Lewis J (1996) Thrust emplacement of the Hispaniola peridotite belt: orogenic expression of the mid-Cretaceous Caribbean arc polarity reversal? Geology 24:1143–1146
33. Ducloz C, Vaugnat M (1962) A propos de l'age de serpentinites de Cuba. Arc SC Geneve 15(2):309–332
34. Díaz de Villalvilla L (1997) Caracterización geológica de las formaciones volcánicas y vulcano-sedimentarias en Cuba central, provincias Cienfuegos, Villa Clara y Sancti Spiritus. En: Furrazola-Bermúdez G y Núñez Cambra (eds) Estudios sobre Geologia de Cuba. La Habana, Cuba. Centro Nacional de Información Geológica, pp 259–270
35. Díaz Otero C, Furrazola Bermúdez e Iturralde Vinent (1997) Estratigrafía de la zona Remedios. En: Furrazola Bermúdez G, Núñez Cambra K. Estudios sobre Geología de Cuba. Centro Nacional de Información Geológica, Instituto de Geologia y Paleontología, pp 221–240
36. Echevarría B, Talavera Coronel F, Tchounev D, Ianev S, Tzankov T (1986) Petrografía y geoquímica de las vulcanitas de la región Guaimaro-Las Tunas. Revista Ciencias De La Tierra Y Del Espacio 11:26–35
37. Escuder Viruete J, Contreras F, Joubert M, Urien P, Stein G, Lopera E, Weiss D, Pérez Estaun A (2007) La secuencia magmática del Jurásico Superior-Cretácico Superior de la Cordillera Central, Republica Dominicana (2007). Boletín Geol Minero 118(2):243–268
38. Fonseca E, Zelepuguin V, Heredia M (1984) Particularidades de la estructura de la asociación ofiolítica de Cuba. Cienc De La Tierra y Del Espacio 9:31–46
39. Furrazola Bermúdez G, Judoley C, Mijailovskaya M, Miroliubov Y, Novajatsky Y, Núñez Jiménez A, Solsona J (1964) Geología de Cuba. Editorial Universitaria, La Habana
40. García R, Fernandez G (1987) Estratigrafía del subsuelo de la cuenca Vegas. Revista Tecnológica XIX 1:3–8
41. García Delgado D, Torres Silva A (1997) Sistema Paleógeno. In: Furrazola Bermúdez G, Nuñez Cambra K (eds) Estudios sobre Geologia de Cuba. Centro Nacional de Información Geológica. Instituto de Geologia y Paleontología, La Habana, pp 115–140
42. Gil González S, Díaz Otero C, García Delgado D (2007) Consideraciones bioestratigráficas de los sedimentos siliciclásticos en Cuba, en cuencas de piggyback del Campaniano-Maastrichtiano. VII Congreso Cubano de Geología, La Habana ISBN: 978-959-7117-16-2 GEO2 P18, pp 1–9
43. Gordon M, Mann P, Cáceres D, Flores R (1997) Cenozoic tectonic history of the North America-Caribbean plate boundary in western Cuba. J Geophys Res 102:10055–10082

44. Goto K, Tada R, Tajika E, Iturralde Vinent M, Matsui T, Takayama S, Nakano Y, Oji T, Kiyokawa H, García Delgado D, Díaz Otero C, Rojas Consuegra R (2008) Lateral and compositional variations of the Cretaceous/Tertiary deep-sea deposits in northwestern Cuba. Cretac Res 29:217–236
45. Haczewski G (1976) Sedimentological reconnaissance of the San Cayetano Formation: an accumulative continental margin in the Jurassic of western Cuba. Acta Geol Pol 26(2):331–353
46. Hatten C (1967) Principal features of Cuban geology: Discussion. Am Asso Petrol Geol Bull 51:780–789
47. Hatten C, Somin M, Millán G, Renne P, Kistler R, Mattinson J (1988) Tectonostratigraphic units of central Cuba. In: Transactions 11th Caribbean Geological Conference, Barbados, pp 1–14
48. Hernández Santana J, Díaz Díaz J, Magaz García A, Lilienberg D (1991) Evidencias morfoestructuro-geodinámicas del desplazamiento lateral siniestro de la zona de sutura interplacas de Bartlett. En: Colectivo de Autores. Instituto de Geografía, Academia de Ciencias de Cuba Morfotectónica de Cuba oriental, pp 5–9
49. Hersey J, Rutstein M (1958) Reconnaissance survey of Oriente deep (Caribbean Sea) with a precision echo sounder. Geol Soc Am Bull 69:1297–1304
50. Holcombe T, Vogt P, Matthews J, Murchison R (1973) Evidence of sea floor spreading in the Cayman Trough. Earths Planet Sci Lett 20:357–371
51. Hutson F, Mann P, Renne P (1998) ^{40}Ar/^{39}Ar dating of single muscovite grains in Jurassic siliciclastic rocks (San Cayetano Formation): constrains in the paleo position of western Cuba. Geology 26(1):83–86
52. Ingersoll R, Busby C (1997) Tectonics of sedimentary basins. In: Ingersoll R, Busby C (eds) Tectonics of sedimentary basins. Blackwell Science, Cambridge, Massachusetts, USA, pp 1–51
53. Instituto de Geología y Paleontología (2011) Mapa Geológico de la República de Cuba, escala 1:100,000
54. Iturralde Vinent M (1978) Los movimientos tectónicos de la etapa de desarrollo plataformico de Cuba. Geol En Mijnbow 57 (2):205–212
55. Iturralde Vinent M (1988a) Consideraciones generales sobre el magmatismo de margen continental de Cuba. Rev Tecnol XVIII 4:17–24
56. Iturralde Vinent M (1988b) Composición y edad de los depósitos del fondo oceánico (asociación ofiolítica) del Mesozoico de Cuba, en el ejemplo de Camagüey. Rev Tecnols XVIII 3:13–25
57. Iturralde Vinent M (2006) Mesocenozoic Caribbean paleogeography: implications for the historical biogeography of the region. Int Geol Rev 48:791–827
58. Iturralde Vinent M, Díaz Otero C, Rodríguez Vega A, Díaz Martínez R (2006) Tectonic implications of paleontologic dating of Cretaceous-Danian sections of Eastern Cuba. Geologica Acta 4 (1–2):89–102
59. Iturralde Vinent M (1996a) Introduction to Cuban geology and geophysics. In: Iturralde-Vinent M (ed) Ofiolitas y arcos volcánicos de Cuba IUGS/UNESCO. Miami, International Geological Correlation Program, Project 364, Contribution No. 1, 3–35
60. Iturralde Vinent M (1996b) Cuba: el archipiélago volcánico Paleoceno-Eoceno Medio. In: Iturralde-Vinent M (ed) Ofiolitas y arcos volcánicos de Cuba IUGS/UNESCO, Miami, International Geological Correlation Program, Project 364 Contribution No. 1, 231–246
61. Iturralde Vinent M (1997) Introducción a la geología de Cuba. In: Furrazola-Bermúdez G, Nuñez Cambra K (comp) Estudios sobre la geología de Cuba. Centro Nacional de Información Geológica, La Habana, pp 35–68
62. Jakus P (1983) Formaciones vulcanógeno - sedimentarias y sedimentarias de Cuba oriental. In: Instituto de Geología y Paleontología (ed.) Contribución a la Geología de Cuba Oriental. La Habana. Editorial Científico – Técnica, pp 17–89
63. Jordan T (1995) Retroarc Foreland and Related Basins. In: Busby C, Ingersoll R (eds) Tectonics of Sedimentary Basins. Blackwell Science, Massachusetts, USA
64. Kantchev I, Boyanov I, Popov N, Cabrera R, Goranov AL, Lolkicev N, Kanazirski M, Stancheva M (1978) Geología de la provincia de Las Villas. Resultado de las investigaciones geológicas y levantamiento geológico a escala 1: 250 000 realizado durante el periodo 1969–1975. Unpublished Report. Oficina Nacional de Recursos Minerales
65. Keijzer F (1945) Outline of the geology of eastern part of the province of Oriente, Cuba (E of 76° W.L.). Utrecht Geogr En Geol Mededeel.Physiogr.-Geol Recks 2(6):1–239
66. Kerr A, Iturralde Vinent M, Saunders A, Babbs T, Tarney J (1999) A new plate tectonics model of the Caribbean: implications from a geochemical reconnaissance of Cuban Mesozoic rocks. GSA Bulletin 111:1581–1599
67. Khudoley K, Meyerhoff A (1971) Paleogeography and geological history of Greater Antilles. Boulder, Colorado, Geological Society of America Memoir 129, 199 pp
68. Knipper A, Puig M (1967) Estructura tectónica de las montañas de Sierra de los Órganos en la zona del pueblo de Vinales y situación en ella de los cuerpos de serpentinitas. Revista de Geología. Acad De Cienc De Cuba, I 1:122–137
69. Knipper A, Cabrera R (1974) Tectónica y geología histórica de la zona de articulación entre el mio- y eugeosinclinal de Cuba y del cinturón hiperbasítico de Cuba. In: Instituto de Geología y Paleontología (ed) Contribución a la Geología de Cuba, pp 15–77
70. Koniev P, Teleguin B, Torshin B, Furrazola G (1979) Criterios litoestratigráficas para la búsqueda de bauxitas en la provincia de Pinar del Rio. La Minería En Cuba 5(4):12–16
71. Kozary M (1968) Ultramafic rocks in thrust zones of northwestern Oriente province Cuba. AAPG Bulletin 52:2298–2317
72. Leroy S, Mauffret A, Patriat P, Mercier de Lepinay B (2000) An alternative interpretation of Cayman trough evolution from a reidentification of magnetic anomalies. Geophys J Int 141:539–557
73. Lewis G, Straczek J (1955) Geology of South-Central Oriente province. U.S. Geol Survey Bull 975D:171–236
74. Lewis J, Draper G (1990) Geology and tectonic evolution of the northern Caribbean margin. In: Dengo G, Case J (eds) The Caribbean Region, Boulder, Colorado, Geological Society of America, The Geology of North America, v. H, pp 77–140
75. Linares Cala E, García Delgado D, Delgado López O, López Rivera J, Strazhevich V (2011) Yacimientos y manifestaciones de hidrocarburos de la República de Cuba. Centro de Investigaciones del Petróleo, La Habana, p 480
76. Llanes Castro A, García Delgado D, Meyerhoff D (1998) Hallazgo de fauna jurásica (Tithoniano) en ofiolitas de Cuba central. Geología y Minería 98. La Habana, Centro Nacional de Información Geológica, Memorias II, pp 241–244
77. López Martínez R, Barragan R, Rehakova D, Cobiella Reguera J (2013) *Calpionellid* distribution and microfacies across the Jurassic/Cretaceous boundary in Western Cuba (Sierra de los Órganos). Geol Carpath 64(3):195–208
78. Meyerhoff A, Hatten C (1968) Diapiric structures in central Cuba. In: Diapirism and Diapirs. American Association of Petroleum Geologists Memoir 8, pp 315–357
79. Miall A (1995) Collision-related foreland basins. In: Busby C, Ingersoll C (eds) Tectonics of sedimentary basins. Blackwell Science, pp 393–424

80. Millán G, Myczynski R (1978) Fauna jurásica y consideraciones sobre la edad de las secuencias metamórficas del Escambray. Acad De Cienc De Cuba, La Habana, Informe Científico-Técnico 80:3–14
81. Millán G, Somin M, Díaz C (1985) Nuevos datos sobre la geología del macizo montañoso de Sierra del Purial, Cuba oriental. In: Millán G, Somin M (eds) Contribución al conocimiento geológico de las metamorfitas del Escambray y Purial. Reporte de Investigación del Instituto de Geologia y Paleontología, Academia de Ciencias de Cuba, La Habana, pp 52–69
82. Millán Trujillo G (1996) Metamorfitas de la asociación ofiolítica de Cuba. In: Iturralde-Vinent M (ed) Ofiolitas y arcos volcánicos de Cuba, International Geological Correlation Program, Project 364. Geological Correlation of Ophiolites and volcanic arcs in the Circumcaribbean Realm, Miami, Florida, pp 131–146
83. Millán G, Myczynski R (1979) Jurassic ammonite fauna and age of metamorphic sequences of escambray. Bulletin de l'Académie Polonaise des Sciences, Série des Sciences de la Terre XXVII (1/2):37–47
84. Millán Trujillo G (1997a) Posición estratigráfica de las metamorfitas cubanas. En: Furrazola-Bermúdez G, Núñez Cambra K (comp) Estudios sobre Geología de Cuba, Centro Nacional de Información Geológica, La Habana, pp 251–258
85. Millán Trujillo G (1997b) Geología del macizo metamórfico Isla de la Juventud. En: Furrazola-Bermúdez G, Núñez Cambra K (comp) Estudios sobre Geología de Cuba, Centro Nacional de Información Geológica, La Habana, pp 259–270
86. Millán Trujillo G (1997c) Geología del macizo metamórfico Escambray. En: Furrazola-Bermúdez G, Núñez Cambra K (comp) Estudios sobre Geología de Cuba. Centro Nacional de Información Geológica, La Habana, pp 271–289
87. Méndez Calderón I, Rodríguez Crombet J, Rodriguez Llanos E, Rodríguez Mejías M, Ruiz Sánchez M, Hernández Ramsay A (1994) Empresa Geominera Oriente, 125 pp
88. Nagy E (1984) Ensayo de las zonas estructuro-faciales de Cuba oriental. Contribución a la Geología de Cuba oriental. Editorial Científico-Técnica, La Habana, pp 9–16
89. Nichols G (2009) Sedimentology and stratigraphy. Willey and Blackwell, Oxford, United Kingdom, 418 p +12
90. Perfit M, Heezen B (1978) The geology and evolution of Cayman Trench. Geol Soc Am Bull 89:1155–1174
91. Pindell J, Kennan L, Stanek K, Maresch W, Draper G (2006) Foundations of Gulf of Mexico and Caribbean evolution: eight controversies resolved. Geol Acta 4:303–341
92. Piotrowska K (1978) Nappe structure of Sierra de los Órganos, western Cuba. Acta Geol Polonica 28:97–170
93. Piotrowska K (1986a) Tectónica de la parte central de la provincia de Matanzas. Bull Polish Acad Sci. Earth Sci 34(1):3–16
94. Piotrowska K (1986b) Etapas de las deformaciones en la provincia de Matanzas en comparación con la provincia de Pinar del Rio. Bull Polish Acad Sci. Earth Sci 34(1):17–27
95. Piotrowski J (1986) Las unidades de nappes en los valles de Yumurí y de Caunavaco. Bull Polish Acad Sci. Earth Sci 34 (1):29–36
96. Piñero Pérez E, Quintana M, Mari Morales T (1997) Caracterización geológica de los depósitos vulcanógeno-sedimentarios de la región de Ciego de Ávila-Camagüey-Las Tunas. En: Furrazola-Bermúdez G, Núñez Cambra K (comp) Estudios sobre Geología de Cuba. Centro Nacional de Información Geológica, La Habana, pp 345–356
97. Pszczolkowski A (1978) Geosynclinal sequences of the Cordillera de Guaniguanico in western Cuba: their lithostratigraphy, facies development and paleogeography. Acta Geol Pol 28(1):1–96
98. Pszczolkowski A (1981) El banco carbonatado jurásico de la sierra de los Órganos, Provincia de Pinar del Rio; su desarrollo y situación paleo tectónica. Cienc De La Tierra Y Del Espacio 3:37–50
99. Pszczolkowski A (1982) Cretaceous sediments and paleogeography in the western part of the Cuban miogeosyncline. Acta Geol Pol 32:135–161
100. Pszczolkowski A (1986) Composición del material clástico de las arenitas de la Formación San Cayetano en la Sierra de los Órganos (Provincia de Pinar del Río, Cuba). Cienc De La Tierra Y Del Espacio 14:71–79
101. Pszczolkowski A (1986) Secuencia estratigráfica de Placetas en el área limítrofe de las provincias de Matanzas y Villa Clara (Cuba). Bull Polish Acad Sci 34(1):67–79
102. Pszczolkowski A (1986) Megacapas del Maastrichtiano en Cuba occidental. Bull Polish Acad Sci 34(1):81–94
103. Pszczolkowski A (1989) Late Paleozoic fossils from pebbles in the San Cayetano Formation, Sierra del Rosario Cuba. Ann Soc Geol Pol 59:27–40
104. Pszczolkowski A (1994b) Geological cross-sections through the Sierra del Rosario thrust belt, western Cuba. Stud Geol Pol 105:67–90
105. Pszczolkowski A, Albear J (1982) Subzona estructuro-facial Bahía Honda, Pinar del Río; su tectónica y datos sobre la sedimentación y paleogeografía del Cretácico Superior y Paleógeno. Cienc De La Tierra Y Del Espacio 5:3–24
106. Pszczolkowski A, Flores R (1986) Fases tectónicas del Paleógeno y Cretácico de Cuba occidental y central. Bull Polish Acad Sci 134:95–111
107. Pszczolkowski A (1994a) Lithostratigraphy of Mesozoic and Paleogene rocks of Sierra del Rosario, western Cuba. Studia Geol Pol 105:39–66
108. Pszczolkowski A (1999) The exposed passive margin of North America in Western Cuba. In: Mann P (ed) Caribbean basins, sedimentary basins of the World, 4 (Series Editor: KJ Hsu), Elsevier, Amsterdam, pp 93–121
109. Pszczolkowski A, Myczynski R (2003) Stratigraphic constraints on the Late-Jurassic-Cretaceous paleotectonic interpretations of the Placetas Belt in Cuba. In: Bartolini C, Buffler R, Blickwede J (eds) The Circum-Gulf of Mexico and the Caribbean: Hydrocarbon habitats, basin formation and plate tectonics. American Association of Petroleum Geologists Memoir 79, pp 545–581
110. Pushcharovsky Y (ed) (1988) Mapa Geológico de la República de Cuba escala 1:250,000: Moscú, Academia de Ciencias de Cuba y Academia de Ciencias de la Unión Soviética
111. Pushcharovsky Y (ed) (1989) Mapa Tectónico de la República de Cuba escala 1:500,000: Moscú, Academia de Ciencias de Cuba y Academia de Ciencias de la Unión Soviética
112. Pérez O, Yarmoliuk V (redactors) (1985) Mapa Geológico de la República de Cuba. Ministerio de la Industria Básica, Centro de Investigaciones Geológicas
113. Renne P, Mattinson J, Hatten C, Somin M, Millán Trujillo G, Linares Cala E (1989) Confirmation of Late Proterozoic age for the Socorro complex of North-Central Cuba from $^{40}Ar/^{39}Ar$ and U-Pb dating. In: Resúmenes y Programa del Primer Congreso Cubano de Geología, pp 118
114. Rigassi Studer D (1963) Sur la Geologie de La Sierra de los Órganos Cuba. Extrait Des Archives De Sciences, Geneve 16 (2):339–350
115. Rojas Agramonte Y, Neubauer F, Bojar A, Hejl E, Handler R, García Delgado D (2006) Geology, age and tectonic evolution of the Sierra Maestra, southeastern Cuba. Geol Acta 4:123–150

116. Rojas-Agramonte Y, Kroner A, Pindell J, Garcia-Casco A, Garcia-.Delgado D, Liu Y (2008) Detrital zircon geochronology of Jurassic sandstones of western Cuba (San Cayetano Fm): Implications for the Jurassic paleogeography of the NW Proto-Caribbean. Am J Sci 308:639–656. https://doi.org/10.2475/04.2008.09
117. Saura E, Vergés J, Brown D, Lukito P, Soriano S, Torrescusa S, García R, Sánchez J, Sosa C, Tenreyro R (2008) Structural and tectonic evolution of western Cuba fold and thrust belt. Tectonics 27, TC 4002. https://doi.org/10.1029/2007TC002
118. Schafhauser A, Stinnesbeck W, Holland B, Adatte T, Remane J (2003) Lower cretaceous Pelagic Limestones in Southern Belize: ProtoCaribbean Deposits in the Southeastern Maya Block. In: Bartolini C, Buffler R, Blickwelde J (eds) The Circum-Gulf of Mexico and the Caribbean: hydrocarbon habitats, basin formation, and plate tectonics. American Association of Petroleum Geologists Memoir 79, pp 624–637
119. Schlager W, Buffler R (1984) Deep sea drilling project leg 77, southeastern Gulf of Mexico. Geol Soc Am Bull 95:226–236
120. Siggurdsson H, Leckie R, Acton G (1997) Proceedings of the ocean drilling program, initial reports 165. College Station (Ocean Drilling Program), pp 49–130
121. Somin M, Millán G (1977) Sobre la edad de las rocas metamórficas cubanas. Informe Cient-Tecn. N.2, La Habana
122. Somin M, Millán G (1981) Geology of the metamorphic complexes of Cuba, Moscow, Nauka, 219 pp [in Russian]
123. Sommer M, Hüneke H, Meschede M, Cobiella Reguera J (2011) Geodynamic model of northwestern Caribbean: scaled reconstruction of Late Cretaceous to Late Eocene plate boundary relocation in Cuba. N Jb Geol Palaont Abh, 299–311
124. Stanek K, Maresch W, Grafe F, Grevel Ch, Baumann A (2006) Structure, tectonics and metamorphic development of the Sancti Spiritus Dome (eastern Escambray massif, Central Cuba). Geol Acta 4:151–170
125. Taber S (1934) Sierra Maestra of Cuba, part of the northern rim of the Bartlett trough. Geol Soc Am Bull 45:567–619
126. Tada R, Iturralde Vinent M, Matsui T, Tajika E, Oji T, Goto K, Nakano Y, Takayama H, Yamamoto S, Kiyokawa K, Toyoda K, Garcia Delgado D, Díaz Otero C, Rojas Consuegra R (2003) K/T boundary deposits in the Paleo-western Caribbean Basin. In: Bartolini C, Buffler R, Blickwede J (eds) The circum-Gulf of Mexico and the Caribbean: hydrocarbon habitats, basin formation and plate tectonics. AAPG Memoir 79, pp 582–604

Geochemical Fingerprinting of Ancient Oceanic Basalts: Classification of the Cuban Ophiolites

Angelica Isabel Llanes Castro and Harald Furnes

Abstract

Mesozoic ophiolites are an important feature of Cuban geology. Although Cuban ophiolites have been studied over the past 40 years, there is still need for systematic studies regarding their internal structure, geochemical characteristics, and emplacement mechanisms. The ophiolites are distributed along the so-called "Northern" and "Eastern" Cuban ophiolite belts, and are strongly dismembered and intermingled mainly with Cretaceous volcanic arc rocks. Ophiolite-associated basalts along the northern Cuban orogenic belt record magmatic history of the ophiolite formation from the Protocaribbean seafloor spreading to subduction initiation stage. We have compiled geochemical data of 15 oceanic basalt samples from previous works, together with data of an analyzed sample during this study. We discuss geochemical criteria based on immobile element proxies for fractionation indices, alkalinity, mantle flow and subduction addition, and field relationships, providing a comprehensive ophiolite classification according to the tectonic setting at which these ophiolites formed. The lavas exhibit three magmatic types. One type has subduction-related fingerprint with dominance of boninite and IAT affinities, likely related to a forearc setting. The second type has a MOR-type (N-MORB and E-MORB) signature, most of them carrying a subtle influence of subduction component, overlapping a border of the backarc field. Subordinately, a transitional type occurs with MOR-type-OIB fingerprint that is considered most likely plume-type ophiolite. The results show then that the studied lavas correspond mostly to subduction-related ophiolite and some of them have incorporated subduction component probably during the time when the Protocaribbean oceanic lithosphere downgoing beneath the Caribbean plate. Some rare remnants of plume-MORB-type ophiolite have also occurred.

Keywords

Cuban ophiolites • Caribbean plate • Oceanic basalts • Protocaribbean oceanic lithosphere • Geochemical fingerprinting

A. I. Llanes Castro (✉)
Instituto de Geología Y Paleontología-Servicio Geológico de Cuba, Vía Blanca no. I002 Y Carretera Central. San Miguel Del Padrón CP, 11 000 La Habana, Cuba
e-mail: isa19111961@gmail.com

H. Furnes
Department of Earth Science, University of Bergen, 5007 Bergen, Norway
e-mail: harald.furnes@uib.no

List of Abbreviations

BA	Backarc
CA	Calc-Alkaline
C-MORB	Contaminated Middle Ocean Ridge Basalt
ECOB	Eastern Cuban Ophiolite Belt
E-MORB	Enriched Middle Ocean Ridge Basalt
FA	Forearc
FAB	Forearc Basalt
Fm	Formation
IAT	Island Arc Tholeites
ICP-EOS	Inductively Coupled Plasma Optical Emission Spectrometer
ICP-MS	Inductively Coupled Plasma Emission Mass Spectrometer
INAA	Instrumental Neutron Activation Analysis
MOR	Mid-Ocean Ridge
MORB	Mid-Ocean Ridge Basalt
My	Millions of years
N-MORB	Normal Middle Ocean Ridge Basalt
NCOB	Northern Cuban Ophiolite Belt
OIB	Ocean Island Basalt
P-type	Plume type
REE	Rare Earth Elements
TAS	Total Alkali Silica
VA	Volcanic Arc
WSU	Washington State University
XRF	X-ray Fluorescence Spectrometry

6.1 Introduction

Ophiolites are interpreted to be remnants of ancient oceanic crust and upper mantle, composed of ultramafic, mafic, and felsic rocks. The oceanic crust preserved in ophiolites may form in any tectonic setting during the evolution of ocean basins, where their geochemical characteristics, internal structure, and thickness vary with spreading rate and location of spreading center relative to a subduction zone [7]. Most ophiolites appear to represent forearc basement, originated at the time of subduction initiation [39]. An ophiolite is emplaced either from downgoing oceanic lithosphere or from the upper plate incorporated into subduction accretion complexes and exhumed in suture zones.

The major ultramafic and mafic complex rocks in the Caribbean are found in Cuba [21]. The ophiolites represent one of the most important features of the Cuban orogenic belt, extending for more than 1000 km along the island (Fig. 6.1). Thus, the knowledge regarding their composition, geochemistry, tectonic setting, and provenience constitute an important issue for reconstruction of the geological evolution of the Caribbean region.

The tectonic position of the Cuban ophiolite allows to separate the Northern Cuban Ophiolite belt (NCOB) and the Eastern Cuban Ophiolite Belt (ECOB). The former over thrusted Bahamas paleomargin continental rocks and comprise ophiolite massif from Cajálbana–Mariel, Habana–Matanzas, Las Villas, Camagüey, and Holguín. The eastern ophiolites lie sub-horizontally over Cretaceous arc rocks and cover the Mayarí-Cristal and Moa-Baracoa ophiolite massifs. The best preserved ophiolites are those from Camagüey and Moa-Baracoa, but as well as the rest, they can be considered dismembered.

Ophiolite-related basalts have been described in both the NCOB and ECOB. These volcanic rocks appear mainly as tectonic blocks, intermingled with other ultramafic and mafic components of the ophiolite, with Cretaceous volcanic arc rocks and with rocks from the subduction mélange.

For geochemical classification of the ophiolitic basalts, we used a diagram combination following ophiolite classification approach of Pearce [29] to discriminate between subduction-related and subduction-unrelated basalts, and to evaluate the influence of fluids released from the subducting slab during subduction.

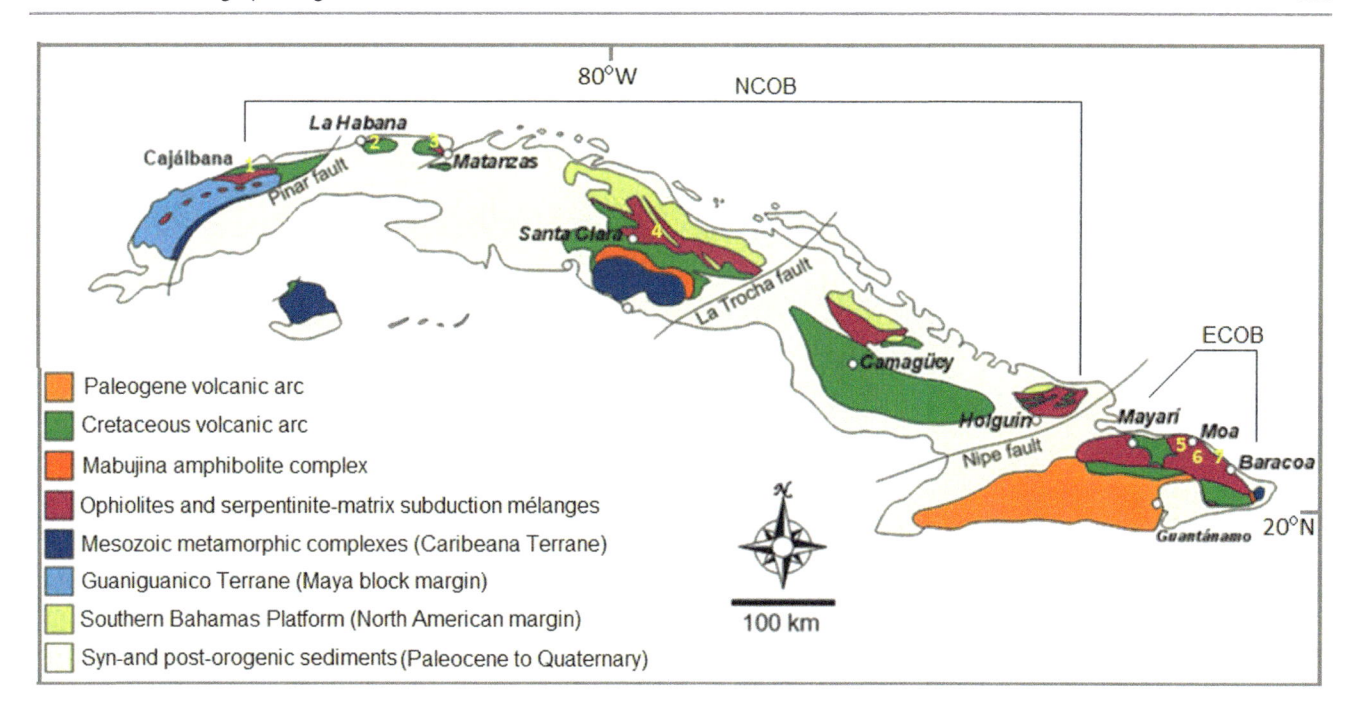

Fig. 6.1 Generalized geological map of Cuba (after Iturralde-Vinent 1998) with indication of the sampling sites. 1: ENC1, ENC2; 2: HAV5, SM1, SM2; 3: MGT1, MGT2, MT3, MT31; 4: SC1, SAG1; 5: CEN202; 6: MEL201; 7: M202

6.2 Materials and Methods

Our study has focused on mafic volcanic rocks which are closely related with ophiolite sequence from the NOCB: Western (Encrucijada Formation from Bahía Honda); Central (Habana basalts, Margot Formation basalts from Matanzas, Sagua la Chica Formation basalts from Villa Clara); and EOCB: Moa-Baracoa (Morel Formation, La Melba, and Centeno basalts). Except drilling well sample MT-31 (Margot Formation), the samples represent outcrop fragments. Major- and trace-element data of 15 samples were compiled from previous researches [20, 24, 25]. Geochemical data of sample SC-1 (Sagua la Chica Formation) is introduced in this study. Whole-rock major- and trace-element composition, geological setting, and location of the study samples are shown in Table 6.1. According to the data source of the samples, the analytical methods are described.

Whole-rock major and trace elements of sample SC-1 were analyzed at the certified Activation Laboratories Ltd. (http://www.actlabs.com/) in Ancaster. After powdering, major and trace elements were measured using lithium metaborate/tetraborate fusion with inductively coupled plasma optical emission spectrometry (ICP-EOS) and inductively coupled plasma mass spectrometry (ICP-MS) methods.

Whole-rock major element composition of samples SM-1, SM-2, MT-3, and MT-31 was analyzed at the Mineral Laboratory in La Habana using atomic absorption spectrometry (see [24] for further details). The trace-element determinations were done at the University of Granada (Spain) by means of X-ray fluorescence (XRF) and ICP-MS.

Whole-rock major and trace element of samples: ENC1, ENC2, HAV5, MGT1, MGT2, and SAG1 were analyzed at Leicester University using conventional techniques, mainly by XRF. The rare earth elements (REE), Th, Co, Sc, Ta, and Hf, were determined by Instrumental Neutron Activation Analysis (INAA) (see [20] for analytical details).

Whole-rock chemical composition of samples, M 202, MEL 201, CEN 202, was measured at the GeoAnalytical Lab facilities of the Washington State University (WSU). Whole-rock major and minor transition elements were analyzed by X-ray fluorescence (XRF) spectrometer. Whole-rock trace elements (Rb, Sr, Y, Zr, Nb, Cs, Ba, REE, Hf, Ta, Pb, Th, and U) were analyzed by inductively coupled plasma mass spectrometry (ICP-MS) (see [25] for more details of the analytical methods).

6.3 Results

The results are based on immobile element proxies for fractionation indices, alkalinity, mantle flow and subduction addition, and an immobile element-based TAS proxy diagram (from [27], after [40]) that allows proper classification even of altered rocks.

Table 6.1 Whole-rock major- and trace-element data for basalts from NCOB and ECOB

Sample	ENC1	ENC2	HAV5HAV5	SM1	SM2	MGT1	MGT2	MT3	MT31	55173	55176	SC1	SAG1	M202	MEL201	CEN202
(wt%)	NCOB														ECOB	
SiO$_2$	53.79	40.96	55.13	54.66	54.48	48.46	50.8	46.27	45	52.72	53.59	59.120	50.96	49.08	49.11	49.23
TiO$_2$	1.23	2.68	0.29	0.34	0.33	0.91	1.31	0.87	1.9	0.7	0.32	0.620	0.94	1.88	2.17	2.87
Al$_2$O$_3$	15.57	14.38	13.6	15.45	15.49	20.46	17.84	20.21	15.09	15.35	14.69	13.620	16.87	14.38	13.64	13.11
FeO*	–	–	–	5.02	8.46	–	–	5.65	11.41	8.57	5.86	9.590	–	11.04	13.19	14.31
Fe$_2$O$_3$**	10.74	15.42	8.45	–	–	6.67	6.15	–	–	–	–	–	8.95	–	–	–
MnO	0.18	0.35	0.18	0.08	0.13	0.11	0.12	0.1	0.18	0.41	0.21	0.170	0.15	0.22	0.26	0.28
MgO	5.66	9.62	11.71	6.68	6.78	5.99	7.09	5.14	4.5	4.69	7.43	2.160	7.49	8.16	7.2	6.9
CaO	7.12	14.49	8.27	9.47	8.71	12.65	12.15	13.99	9.63	4.12	6.15	6.010	10	9.77	9.09	8.12
Na$_2$O	5.45	0.85	1.74	4.35	2.32	3.78	4.04	2.63	4.16	5.72	5.03	2.700	4.11	2.82	3.79	2.83
K$_2$O	0.663	0.311	0.754	0.49	0.38	0.317	0.55	0.2	0.61	0.1	0.1	0.320	0.027	1.24	0.16	0.59
P$_2$O$_5$	0.104	0.307	0.028	0.04	0.04	0.105	0.148	0.13	0.23	–	–	–	0.097	0.18	0.2	0.29
Total†	100.61	99.36	100.15	96.58	97.12	99.45	100.2	95.19	92.71	92.38	93.38	94.31	99.6	98.77	98.81	98.64
LOI (ppm)	2.04	2.83	2.91	2.44	1.44	5.98	5.61	4.09	5.44	6.1	5.54	5.90	3.09	–	–	–
V	323	421	203	206	253	192	221	227	353	210	180	121.20	209	375	394	486
Cr	48	272	494	257	208	274	418	349	125	16	250	16.96	157	167	125	85
Ba	74	50	130	175.4	91.24	94	52	72	44	60	55	197.18	9	14	6.5	19
Nb	1.0	18.8	0.8	0.67	0.75	4.6	7.3	2.5	5.42	3	3	0.94	0.7	2.9	3.6	5.7
Zr	66	153	21	25.56	15.62	49	83	42	130	40	17	61.26	61	110	130	187
Y	27	39.8	10.8	10.2	14.92	18.4	24.9	16	41	16	11	22.08	21.9	42	48	64
Ta	–	1.17	0.04	0.05	0.06	0.3	0.44	0.28	0.4	–	–	0.08	0.07	0.23	0.27	0.43
Hf	–	4.37	0.65	0.9	0.72	1.54	2.28	0.89	3.7	–	–	2.05	1.59	3.1	3.7	5.2
Th	–	0.0.56	0.29	0.29	0.28	0.41	0.60	0.37	0.46	1.5	0.05	0.98	0.17	0.18	0.23	0.38
U	–	–	–	0.13	0.09	–	–	0.1	0.32	0.05	0.05	0.51	–	0.44	0.078	0.14
Ti	7380	16080	1740	2000	2000	5460	7860	5220	11400	4200	1900	3700	5640	11280	13020	17,220
Rb	5.9	1.8	4.6	2.988	2.741	5.2	6.3	14	18	1.5	1.5	4.80	0.6	11	0.77	2.7
Cs	–	–	–	0.036	0.023	–	–	0.12	0.33	–	–	0.35	–	0.36	0.013	0.031
Ni	28	110	197	161	103	92	141	111	92	20	100	8.94	59	59	52	42
Sr	143	47	55	337.7	177.5	490	324	288	171	47	54	356	41	107	133	122
La	2.0	12.3	1.3	1.84	2.02	3.7	5.3	4	6.45	–	–	7.24	2.3	3.8	4.9	7.4
Ce	9.6	32.2	2.5	3.71	4.9	9.2	13.5	10.11	17.3	–	–	15.94	6.8	11	14	21
Pr	–	–	–	0.53	0.77	–	–	1.38	2.86	–	–	2.42	–	1.9	2.3	3.3
Nd	8.6	22.6	1.4	2.67	3.9	6.7	10	6.1	15.87	–	–	11.82	6.2	11	13	18

(continued)

Table 6.1 (continued)

Sample	ENC1	ENC2	HAV5HAV5	SM1	SM2	MGT1	MGT2	MT3	MT31	55173	55176	SC1	SAG1	M202	MEL201	CEN202
Sm	–	5.81	0.59	0.94	1.27	2.01	3.03	2.18	5.19	–	–	3.18	2.12	4.3	5	6.8
Eu	–	2.12	0.23	0.36	0.45	0.87	1.23	0.88	1.71	–	–	0.99	0.93	1.5	1.7	2.3
Gd	–	N.D	N.D	1.22	1.63	N.D	N.D	2.59	6.5	–	–	3.59	–	5.8	6.7	8.9
Tb	–	1.27	N.D.	0.23	0.3	0.51	0.77	0.45	1.14	–	–	0.58	–	1.1	1.3	1.7
Dy	–	N.D	N.D	1.55	2.12	–	–	3.11	7.29	–	–	3.72	–	7.5	8.5	11
Ho	–	N.D	N.D	0.36	0.51	–	–	0.63	1.61	–	–	0.83	–	1.6	1.8	2.4
Er	–	N.D	N.D	1.06	1.56	–	–	1.81	4.32	–	–	2.37	–	4.4	5	6.7
Tm	–	N.D	N.D	0.18	0.37	–	–	0.24	0.67	–	–	0.38	–	0.63	0.73	0.96
Yb	–	3.78	1.11	1.15	1.74	1.77	2.37	1.8	4.37	–	–	2.46	2	3.9	4.5	5.9
Lu	–	0.58	0.17	0.18	0.28	0.26	0.35	0.22	0.62	–	–	0.38	0.28	0.62	0.7	0.93

*All Iron reported as FeO
**All Iron reported as Fe$_2$O$_3$
† Major element totals reported on a volatile free basis
# LOI—loss on ignition
ENC1, ENC2, HAV5, MGT1, MGT2, SAG1 (from [20]; SM1, SM2, MT3, MT31 (from [24]; M202, MEL201, CEN202 (from Marchesi et al. [25]

6.3.1 Rock-Type Signature

We used the Zr/Ti–Nb/Y projection [27]; Fig. 6.2) to recognize the rock type. This diagram provides a useful filter for basalt discrimination.

Most of samples plot on the basalt field, but some of them show an andesite and basaltic andesite signature. This feature could be caused by the fact that some volcanic arc basalts, commonly classified as intermediate rocks, due to their mantle sources are modified by subduction melts with high Zr/Ti ratios [27]. Thus, the case of the samples plotting in the intermediate rock field might be related to high mantle/fluid interaction.

6.3.2 Geochemical Fingerprint

We have used a combination of three discrimination diagrams (Fig. 6.3), following Dilek and Furnes [7] and Pearce [29]. In the Th/Yb–Nb/Yb diagram [28] (Fig. 6.3a), samples M202, MEL201, and CEN202 (Eastern Cuba) and most of Margot Formation basalts (Central Cuba) plot within the mantle array, whereas those of another group (Habana and Sagua la Chica Formation basalts, from Central Cuba) plot above and embrace oceanic and continental arc types. ENC2 sample (Encrucijada Formation, Western Cuba) plots near to the mantle array, close to E(enriched)-MORB (Mid-ocean ridge basalt) composition. In Fig. 6.3b (modified V-Ti diagram of [29], one can observe two main geochemical fingerprints, i.e. boninite-IAT (Habana volcanic rocks) versus MORB (Encrucijada Formation basalts (ENC1) Margot Formation and ECOB basalts).

In the TiO_2/Yb–Nb/Yb diagram, two samples from Margot Formation have E-MORB character, whereas another two of the Margot formation rocks and those from eastern Cuba are N (normal)-MORB in tectonic affinity (Fig. 6.3c). The Habana–Matanzas and Sagua la Chica Formation volcanic rocks plot close also to N-MORB field. ENC2 falls in the boundary OIB-MORB arrays (Fig. 6.3c).

In the MORB-normalized Th versus Nb diagram of Fig. 6.4, most samples from Habana and Margot Formations plot in the backarc field together with a sample from Sagua la Chica Formation. Other two samples from Sagua la Chica fall close and, in the field, of an intraoceanic arc, another one in the overlapping of backarc (BA)–continental arc (CA) fields. Most of the Margot Formation basalts and Moa-Baracoa volcanic rocks plot in MORBs fields. One sample from Encrucijada Formation (ENC2) falls in the E-MORB field.

6.3.3 Ophiolite Classification

Ophiolite-related basalts were first divided, following the classification procedure of Dilek and Furnes [6, 7] as well as representative geological features, into two main groups: subduction-unrelated and subduction-related. The subduction-unrelated group is further discriminated into rift/continental margin, MOR, and plume/MOR types, a distinction that is defined by their grouping in the TiO2/Yb-Nb/Yb diagram. MOR-type basalts plot within the MORB array while plume-related basalts plot within the OIB array and the rift/continental margin types fall into the alkaline field of the OIB array (Fig. 6.3a, b). The subduction-related basalts were subdivided into backarc, backarc to forearc, forearc, and volcanic arc types.

The backarc (BA), backarc to forearc (BA-FA), and forearc (FA) ophiolite types can be recognized when their compositions plot mainly in the oceanic arc field, as well as near to intraoceanic arc/continental arc field of the Th/Yb-Nb/Yb diagram (Fig. 6.3a), characteristic also for the VA ophiolite type [11] and in the MORB, MORB-IAT-boninite, and boninite fields in the V-Ti diagram (Fig. 6.3a, c). Further, the MORB-normalized Th and Nb diagram [35] allows distinction between basalts of subduction-related and subduction-unrelated backarc basins (Fig. 6.4). About half of the backarc field overlaps largely with those of N-MORB and E-MORB without subduction influence. This imply that those basalts classified as MOR type (N-MORB and E-MORB) may be part of a backarc system. A proposed classification for the studied basalts is summarized in Table 6.2.

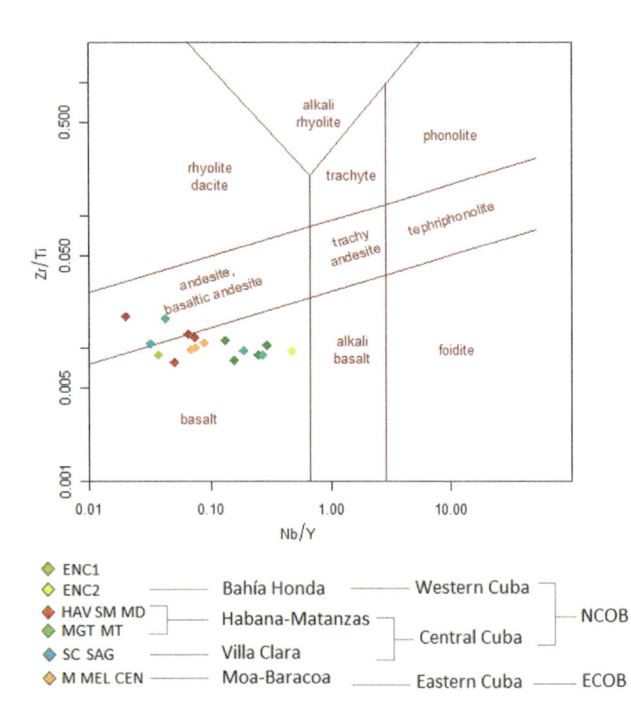

Fig. 6.2 Diagram of volcanic rock classification from Pearce [27]

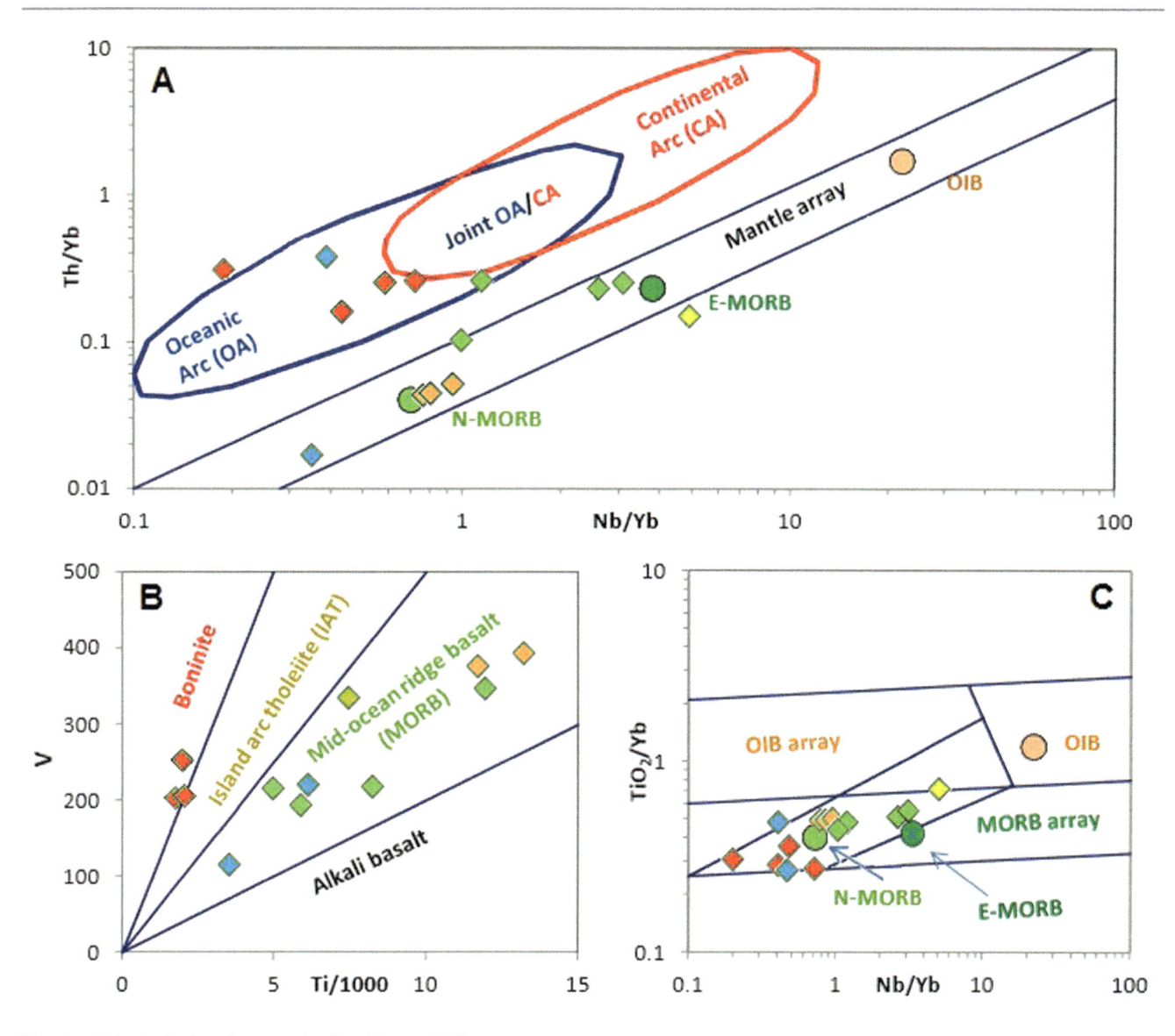

6 Geochemical Fingerprinting of Ancient Oceanic Basalts ...

Fig. 6.3 Discrimination diagrams (A-C) of Pearce [29]

Fig. 6.4 Discrimination diagram of Saccani [35]. The Th and Nb values are 0.12 ppm and 2.33 ppm, respectively (taken from [30])

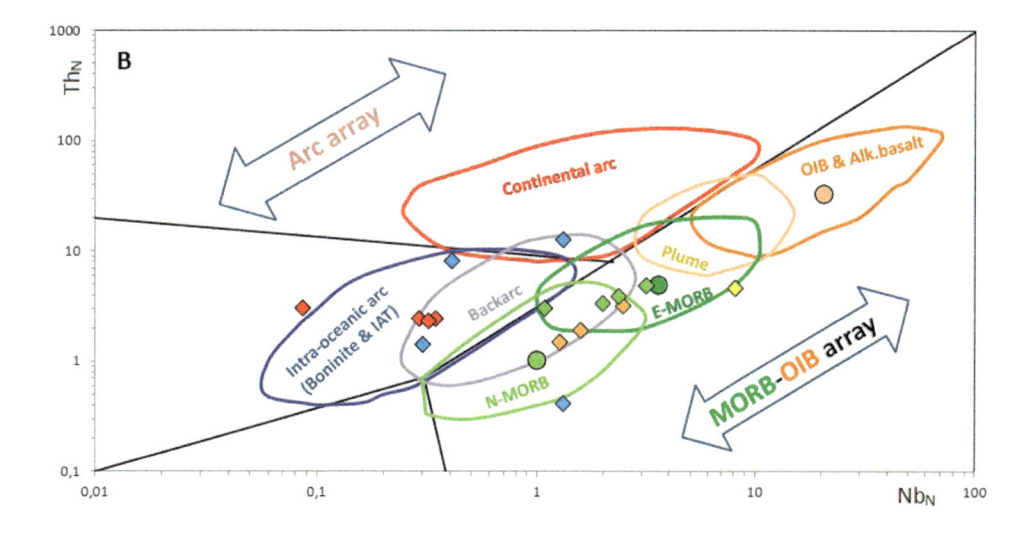

Table 6.2 Summary of studied basaltic rocks from NCOB and ECOB with their inferred tectonic setting of formation and proposed ophiolite classification

Sample	Ophiolite Belt	Geological setting	Th/Yb-Nb/Yb	V/Ti	TiO2/Yb-Nb/Yb	ThN-NbN	Ophiolite type
Subduction-unrelated							
EnC2	NCOB	Encrucijada Formation	E-MORB	MORB	OIB-MORB	E-MORB	MOR/Plume
Subduction-related							
ENC1	NCOB	Encrucijada Formation	No data to plot	MORB	No data to plot	No data to plot	MOR-BA
HAV5	NCOB	Habana ophiolite mélange	OA	Bon-IAT	NMORB	BA	FA
SM1	NCOB	Habana ophiolite mélange	OA	Bon-IAT	NMORB	BA	FA
SM2	NCOB	Habana ophiolite mélange	OA	Bon-IAT	NMORB	BA	FA
MGT1	NCOB	Margot Formation	EMORB	MORB	NMORB	EMORB-BA-NMORB	BA / C-MOR
MGT2	NCOB	Margot Formation	EMORB	MORB	NMORB	EMORB-BA-NMORB	BA / C-MOR
MT3	NCOB	Margot Formation	OA-CA	MORB	NMORB	EMORB-BA	BA / C-MOR
MT31	NCOB	Margot Formation	NMORB	MORB	NMORB	EMORB-BA-NMORB	BA / C-MOR
55,173	NCOB	Sagua la Chica Formation	No data to plot	IAT-MORB	No data to plot	BA	FA
55,176	NCOB	Sagua la Chica Formation	No data to plot	Bon-IAT	No data to plot	NMORB	FA
SC1	NCOB	Sagua la Chica Formation	OA	MORB	NMORB	BA-CA	BA
SAG1	NCOB	Sagua la Chica Formation	MA	MORB	NMORB	NMORB	BA
M202	ECOB	Moa-Baracoa ophiolite	MA	MORB	NMORB	BA-NMORB	BA
MEL201	ECOB	Moa-Baracoa ophiolite	MA	MORB	NMORB	BA-NMORB	BA
CEN202	ECOB	Moa-Baracoa ophiolite	MA	MORB	NMORB	BA-NMORB	BA

6.4 Discussion on the Tectonic Setting

Iturralde Vinent [15] proposed that peridotites and gabbros had to be from the late Triassic to the Jurassic, bearing in mind that they had to be formed at the same time as the breaking of Pangea and the formation of the Protocaribbean crust took place. Somin and Millán [38] obtained an age K-Ar of 160 ± 24 My in a body of anorthosites of the cumulative complex of Camagüey, which confirmed such opinion. Somin and Millán [38] obtained an age K-Ar of 160 ± 24 My in a body of anorthosites of the cumulative complex of Camagüey, which confirmed such opinion.

Nevertheless, from isolated fossiliferous samples recovered from volcanic-sedimentary sections, the tectonic episodes related to the ophiolite formation embrace a range age from Tithonian to Coniacian [1, 8, 10, 16, 17, 23, 26]; Pszczolkowski [19, 33]. On the other hand, Deschamps et al. [5] and Butjosa et al. [2] conclude that hydrated peridotites and serpentinites in the Villa Clara subduction mélange represent subducted and non-subducted mantle lithosphere of the Protocaribbean oceanic arm of the Central Atlantic.

The last conclusion is dealing with the seafloor spreading between North and South America since the late Jurassic after the break up of Pangea, and Caribbean (Pacific Carillon)-derived lithospheric domain of the Cretaceous forearc.

The geochemical studies during the last two decades have considered that most of basalts have been generated in a suprasubduction zone, formed in conditions of arc and backarc settings [3, 10, 18]; Cruz Gámez [4, 25, 31], or forearc [1, 22, 24, 32, 34]. Some basalts are thought to be of oceanic plateau type [6, 20], but also MORB (Simón et al. [9, 10, 37].

Here is present new considerations based on the whole-rock major and trace-geochemical signature of the ophiolite-related basalts from the NOCB and EOCB. Our research complements recent geochemical studies on the volcanic rocks in the Caribbean [12–14, 25, 32].

On the basis of the immobile element criteria revised in various diagrams (Figs. 7.3 and 7.4) as well as in characteristic geological features, we analyze the tectonic setting affinity of the ophiolite-related basalts. The majority of the studied samples were below subduction-related ophiolites (Table 6.2). The samples from Central Cuba (from La Habana and another two from Margot and Sagua la Chica Formations, respectively) have an intraoceanic arc affinity (Fig. 6.3a) and plot within the IAT and boninite fields in the V-Ti diagram (Fig. 6.3b). Therefore, we infer that these basalts with subduction-related fingerprints correspond most likely to FAB (Table 6.2). However, they plot in MORB array in the TiO$_2$/Yb-Nb/Yb diagram (Fig. 6.3c), probably dealing with a protoforearc basin. Then, the volcanic rocks (samples: HAV5, SC-1, from Central Cuba) which classified as andesite and basaltic andesite in the Zr/Ti-Nb/Y diagram (Fig. 6.2) with boninitic character (SiO2: 55–59 wt%, MgO: 7.49–11.71 wt%, Table 6.1) should represent Si- and Mg-rich boninitic magmas in the latest stages of subduction initiation.

Those basalts from Margot Formation and Moa-Baracoa that plot in the overlapping field MOR type (N-MORB and E-MORB)—backarc could be considered formed in a backarc system (Fig. 6.4). On the other hand, locally, the contribution of trapped continental sediments can lead to distinctive, C (contaminated)-MORB compositions, not unlike those found in subducted ridge settings [11, 29] as it is demonstrated by majority plots of Margot Formation samples and one sample from Moa-Baracoa (Fig. 6.4). Finally, sample ENC2 (Western Cuba) displays a geochemical composition from E-MORB to N-MORB and Ocean Island Basalt (OIB) representing then a Plume-Related Mid-Ocean Ridges ophiolite.

According to the field occurrence and the geochemical characteristics of basalts associated to the NCOB and ECOB, there are some variations on the petrogenesis of the Cuban ophiolites, although the studied basalts show a geochemical behavior mainly of the subduction-related ophiolites. Nevertheless, our research put some constraints to some previous opinions. For example, Kerr et al. [20] proposed that basalts of the Margot Formation be classed as oceanic intraplate lavas and could either have formed a part of the 90 My Caribbean oceanic plateau or could represent part of the Protocaribbean oceanic crust. The MORB-normalized values of Th and Nb (Fig. 6.4) show the distinction between basalts of subduction-related and subduction-unrelated backarc basins. About half of the backarc field overlaps mostly with those of N-MORB and E-MORB without subduction influence, where the Margot samples are plotted. Moreover, Margot volcanic-sedimentary section (Cenomanian-Turonian; [33] appears tectonically embedded within deformed serpentinites, and in some cases, imbricated with Cretaceous volcanic arc rocks. Then, it may be supposed that Margot volcanic-sedimentary sequence could have been originated in a backarc basin or in a subducted ridge setting. Thus, Margot Formation rocks could have come from the Protocaribbean oceanic floor. However, we have not enough data to suggest Margot basalts as a relic from Protocaribbean crust or from Pacific.

At the same time, following the projection for recognizing geochemical fingerprint of volcanic rocks analyzed by Pearce [29], Dilek and Furnes [6, 7], and Furnes and Dilek [11], it has confirmed the backarc setting for the origin of Moa-Baracoa basalts as previously suggested by Proenza et al. [31] and Marchesi et al. [25] which show also a similar tectonic affinity as Margot Fm. samples (Figs. 7.3 and 7.4). In the same way, the sample ENC2 (from Aptian–Albian Encrucijada Fm.) has E-MORB to N-MORB-OIB geochemical signature and may represent oceanic island basalt type, as also proposed by Kerr et al. [20] who observed a REE pattern which most closely resembles modern-day ocean island basalts.

The formation of the ancient oceanic lithosphere in the northern Cuban orogenic belt records a magmatic history from the Protocaribbean seafloor spreading to subduction initiation stage (from pre-Albian), including later a local plume-mantle event. The pending question is to reconstruct the event sequence to envisage these volcanic rocks from different sources (Caribbean mostly, and Protocaribbean) and formed in distinct tectonic setting, mainly subduction related (backarc and forearc) but some must be subduction unrelated (MOR, C-MOR, and/or plume/MOR).

6.5 Conclusions

The geochemical fingerprint of the Cuban ophiolite-related basalts, together with their field relationships, allows to propose that the NCOB and ECOB involve subduction-related ophiolites, mainly FA type, but also BA ophiolite type and/or alternatively, C-MOR. Some rare transitional lavas MORB-OIB from Western Cuba are probably related to a plume-type ophiolite (P-type).

Acknowledgements We thank the Instituto de Geología y Paleontología/Servicio Geológico de Cuba for covering the expenses of geochemical analyses of sample SC1 introduced in this study and for allowing in the framework of a research project to perform this scientific work. We express our sincere thanks to Dr. Manuel E. Pardo Echarte, Dr. Jorge Rabassa, and Springer Editorial for giving us the opportunity to contribute this paper to the special monograph on Geology of Cuba.

References

1. Andó J, Harangi S, Szkmány l, Dosztály l (1996) Petrología de la asociación ofiolítica de Holguín. In: Iturralde-Vinent MA (ed) Ofiolitas y arcos volcánicos de Cuba. International geological correlation program, Project 364, Special Contribution 1. IUGS/UNESCO, Miami, FL, pp 154–178
2. Butjosa L, Proenza JA, Aiglsperger T, Rebaza AM, Galindos M, García Casco A, Iturralde Vinent MA, Piñero Pérez E (2014) Layered gabbro-hosted Al-rich chromitites at Loma Iguana area, Camagüey ophiolitic massif, Cuba: originated by crustal recycling?. SGA Meeting, Nancy
3. Cruz Gámez EM (1993) Papel de vulcanismo básico en la formación de los yacimientos cupro-piríticos de la zona estructuro-facial Bahía Honda. Pinar del río. Tesis para la opción al grado de Doctor en Ciencias Geológicas. Universidad de Pinar del Río, Cuba, pp 1–110
4. Cruz Gámez EM, Simón Méndez A (1998) Principales rasgos del complejo de basaltos en la región de Bahía Honda, Pinar del Río. Minería y Geología, XIV 3:51–57
5. Deschamps F, Godard M, Guillot S, Chauvel C, Andreani M, Hattori K, Wunder B, France L (2012) Behavior of fluid-mobile elements in serpentines from abyssal to subduction environments: examples from Cuba and Dominican Republic. Chem Geol 312–313:93–117
6. Dilek Y, Furnes H (2011) Ophiolite genesis and global tectonics: Geochemical and tectonic fingerprinting of ancient oceanic lithosphere. Geol Soc Am Bull 123(3/4):387–411
7. Dilek Y, Furnes H (2014) Ophiolites and their origins. Elements 10:93–100
8. Fonseca E (1987) Características de la Asociación Ofiolítica de la Provincia de Pinar del Río. Universidad de Carolina. Praga, Tesis para la opción del grado de Doctor en Ciencias, p 241
9. Fonseca E, Zelepuguin VM, Heredia M (1984) Particularidades de la estructura de la asociación ofiolítica de Cuba. Ciencia de la Tierra y el Espacio 9:31–46
10. Fonseca E, González R, Delgado R (1989) Presencia de efusivos ofiolíticos y de boninitas en las provincias de La Habana y Matanzas. Boletín Técnico de Geología 1:1–9
11. Furnes H, Dilek Y (2017) Geochemical characterization and petrogenesis of intermediate to silicic rocks in ophiolites: A global synthesis. Earth Sci Rev 166:1–37
12. Geldmacher J, Hoernle KA, Bogaard PVD, Hauff F, Klügel A (2008) Age and geochemistry of the Central American forearc basement (DSDP Leg 67 and 84): Insights into Mesozoic arc volcanism and seamount accretion on the fringe of the Caribbean LIP. J Petrol 49:1781–1815
13. Hastie AR, Kerr A, Mitchell S, Millar I (2009) Geochemistry and tectonomagmatic significance of Lower Cretaceous island arc lavas from the Devil's Race Course Formation, eastern Jamaica. In: James K, Lorente MA, Pindell JL (eds). Origin and evolution of the Caribbean region. Geological Society of London, Special Publication, 328:3999–409
14. Hastie AR, Mitchell SF, Treloar PJ, Kerr AC, Neill L, Barfod DN (2013) Geochemical components in a Cretaceous island arc: The Th/La-(Ce/Ce*) $_{Nd}$ diagram and implications for subduction initiation in the inter-American region. Lithosphere 162–163:57–69
15. Iturralde Vinent MA (1981) Nuevo modelo interpretativo de la evolución geológica de Cuba. Rev. Ciencias de la Tierra y el Espacio 3:51–90
16. Iturralde Vinent MA (1986) Reconstrucción palinspática y paleogeografía del Cretácico Inferior de Cuba oriental y territorios vecinos. Revista Minería y Geología 1:1–4
17. Iturralde Vinent MA (1988) Composición y edad de los depósitos del fondo oceánico (asociación ofiolítica) del Mesozoico de Cuba, en el ejemplo de Camagüey. Revista Tecnológica, XVIII 3:13–24
18. Iturralde Vinent MA (1996) Geología de las ofiolitas de Cuba. En: Iturralde-Vinent, M.A. (Ed.). Ofiolitas y arcos volcánicos de Cuba. IGCP Project 364. Special Contribution 1:83–120
19. Iturralde Vinent MA, Díaz Otero C, Rodriguez Vega A, Díaz Martínez R (2006) Tectonic implications of paleontological dating of Cretaceous-Danian sections of Eastern Cuba. GeolActa 4(1–2):89–102
20. Kerr AC, Iturralde Vinent M, Saunders AD, Babbs TL, Tarney J (1999) A new plate tectonic model of the caribbean: implications from a geochemical reconnaissance of Cuban Mesozoic volcanic rocks. Geol Soc Am Bull 111:1581–1599
21. Khudoley KM, Meyerhoff AA (1971) Paleogeography and geological history of Greater Antilles. Geol Soc Am Mem 129:1–199
22. Lázaro C, Blanco Quintero IF, Proenza JA, Rojas Agramonte Y, Neubauer F, Núñez Cambra K, García Casco A (2016). Petrogenesis and 40Ar/39Ar dating of proto-forearc crust in the Early Cretaceous Caribbean arc: the La Tinta mélange (eastern Cuba) and its easterly correlation in Hispaniola. International Geology Review, pp 1019–1040
23. Llanes AI, García Delgado D, Meyerhoff D (1998) Hallazgo de fauna jurásica (Tithoniano) en ofiolitas de Cuba central. Memorias de Geología y Minería 98. Centro Nacional de Información Geológica (CNDIG), La Habana, Tomo 2:241–244
24. Llanes AI, Díaz de Villalvilla L, Despaigne AI, Ronneliah Sitall M, García Jiménez D (2015) Geoquímica de las rocas volcánicas máficas de edad Cretácica de la región de Habana-Matanzas (Cuba occidental): implicaciones paleotectónicas. *Ciencias de la Tierra y el Espacio*, 16 (2): 117–133. ISSN 1729-3790
25. Marchesi C, Garrido CJ, Bosch D, Proenza JA, Gervilla F, Monié P, Rodríguez Vega A (2007) Geochemistry of cretaceous magmatism in eastern Cuba: recycling of North American continental sediments and implications for subduction polarity in the Greater Antilles paleo-arc. J Petrol 48(9):1813–1840
26. Navarrete M, Andó J, Ríos Y y Kozak M (1989) Asociación ofiolítica de Holguín, particularidades petrólogo-geoquímicas. Primer Congreso Cubano de Geología 95
27. Pearce JA (1996) A User's Guide to Basalt Discrimination Diagrams. In: Wyman DA (ed) Trace element geochemistry of volcanic rocks: applications for massive sulphide exploration. Geological Association of Canada, Short Course Notes, 12:79–113
28. Pearce JA (2008) Geochemical fingerprinting of oceanic basalts with applications to ophiolite classification and the search for Archean oceanic crust. Lithosphere 100:14–48
29. Pearce JA (2014) Immobile element fingerprinting of ophiolites. Elements 10:101–108

30. Pearce JA and Parkinson IJ (1993) Trace element models for mantle melting: application to volcanic arc petrogenesis. In: Prichard HM, Alabaster T, Harris NBW, Neary CR (Eds.) Magmatic Processes and Plate Tectonics: Geological Society of London, Special Publication, 76:373–403
31. Proenza JA, Gervilla F, Melgarejo JC, Bodinier JL (1999) Al- and Cr-rich chromitites from the Mayarí-Baracoa Ophiolitic Belt (Eastern Cuba): consequence of interaction between volatile-rich melts and peridotites in suprasubduction mantle. Econ Geol 94:547–566
32. Proenza JA, García Casco A, Marchesi C, Rojas Agramonte Y, Lázaro C, Blanco Quintero I, Butjosa L, Llanes AI (2016) Petrology, geochemistry and tectonic setting of ophiolites in Cuba. GSA Annual Meeting, Denver, Colorado, USA
33. Pszczółkowski A (2002) The margot formation in Western Cuba, A volcanic and Sedimentary Sequence of Cenomanian–Turonian age, p 15
34. Rojas Agramonte Y, Kröner A, García Casco A, Kemp T, Hegner E, Pérez M, Barth M, Liu D, Fonseca Montero A (2010) Zircon ages, sr-nd-hf isotopic compositions, and geochemistry for granitoids associated with the northern ophiolite mélange of central Cuba: tectonic implication for Late Cretaceous magmatism in the northwestern Caribbean. Am J Sci 310:1453–1479
35. Saccani E (2015) A new method of discriminating different types of post-Archean ophiolitic basalts and their tectonic significance using Th-Nb and Ce-Dy-Yb systematics. Geosci Front 6:481–501
36. Shervais JW (1982) Ti-V plots and the petrogenesis of modern and ophiolitic lavas. Earth and Planetary Science Letters 32:114–120
37. Simón A, Rodriguez G e Izquierdo M (1983) Algunas consideraciones petrológicas sobre el magmatismo Mesozoico de Pinar del Río. Empresa Geólogo-Minera de Occidente
38. Somin M, Millán G (1981) Geologia de los complejos metamórficos de Cuba Moscú. Nauka, p 219
39. Stern RJ, Reagan M, Ishizuka O, Ohara Y, Whattam S (2012) To understand subduction initiation, study forearc crust: To understand forearc crust, study ophiolites. Lithosphere 4(6):469–483
40. Winchester JA, Floyd PA (1977) Geochemical discrimination of different magma series and their differentiation products using immobile elements. Chem Geol 16:325–343

Stratigraphy of the Quaternary Deposits in Cuba

Leandro L. Peñalver Hernández, Miguel Cabrera Castellanos, and Roberto Denis Valle

Abstract

In Cuba, the existence of an important group of geological formations and innominate deposits of Quaternary age is recognized. They encompass both carbonate and terrigenous sequences and currently twenty-one of these lithostratigraphic units are recognized, as well as five innominate deposits. Applying relative dating methods it has been possible to compile a stratigraphic subdivision scheme for these deposits that locates them specifically in each of the classic subdivisions that are established for the Pleistocene: Lower, Middle, and Upper. Six of the recognized units transited the Neogene/Quaternary limit, that is, they began to settle in the Upper Pliocene and concluded their deposition in the Lower Pleistocene. In the case of other units, it has been possible to better define their correspondence with the Middle Pleistocene, or with the Upper. Specifically with the Classical Upper Pleistocene corresponds one of the most extended formations of Cuba: the Jaimanitas Formation; and with the upper part of the Upper Pleistocene, in this case the Oxygen Isotopic Stage (OIS) 3, there are currently several units of local character, but very important for their paleoclimatic peculiarities. In its general conception, most formations correspond to marine transgressions that are closely related to the great sea level glacio-eustatic transgressions that occurred during the Pleistocene in a number not less than seven, according to the majority of the world's researchers. On the other hand, all the units that are recognized in the scheme reflect, in one way or another, a greater or lesser effect due to the chemical weathering processes that occurred in the tropical region where Cuba is located. In this sense, the most affected geological formations are located in the lower part of the Pleistocene of Western and Central Cuba and in the premontane border of the Sierra Maestra in Eastern Cuba. There, the terrigenous sequences are characterized by presenting profiles of kaolinite weathering, with a variegated coloration of the deposits, while the younger units do not have this type of profile and their colors vary to lighter shades. A type of innominate deposit of Quaternary age and four others of Holocene age are recognized, which due to their wide distribution are also described. Satellite images of the regions of the Guanahacabibes Peninsula, Zapata Peninsula, the south of the Isla de la Juventud, and the Cauto River Plain are presented, which are some of the places where the formations and deposits of the Quaternary in Cuba are best exposed.

Keywords

Quaternary in cuba • Pliocene • Pleistocene • Holocene • Glacio-eustatic transgressions • Kaolinite weathering profile

---

L. L. P. Hernández (✉) · M. C. Castellanos · R. D. Valle
Instituto de Geología y Paleontología-Servicio Geológico de Cuba, Vía Blanca No. 1002 y Carretera Central. San Miguel del Padrón, CP 11000 La Habana, Cuba
e-mail: leandro@igp.minem.cu

M. C. Castellanos
e-mail: miguel@igp.minem.cu

R. D. Valle
e-mail: denis@igp.minem.cu

List of Abbreviations

Fm	Formation
INQUA	Acronyms in English language of the International Union for Quaternary Research
OIS	Acronyms in English language of Oxygen Isotopic Stage
RGB	Red-Green-Blue

7.1 Introduction

Before the seventies of the twentieth century, there was practically no scheme of stratigraphic subdivision for the Quaternary of Cuba. Until that date only the Jaimanitas, Vedado, Casablanca, and Santa Fe formations had been described, the last three with a very local character, circumscribed to the area of the city La Habana. In 1976, Kartashov, Cherniajovski, and Peñalver, as part of the thematic research on the Quaternary in Cuba that they carried out for 10 years, published an article in which they describe, for the first time, the widely distributed Ávalos, Guane, Guevara, Villarroja, and Camacho formations in Western and Central Cuba. These authors published in 1981 the monograph "The Quaternary in Cuba", in which, for the first time, a scheme of stratigraphic subdivision of the Quaternary in Cuba was presented. This was the one used in the Geological Map of Cuba at a scale of 1: 250,000 [1], Cuban Geological Map at scale 1: 500,000 [2], and on the Map of the Quaternary deposits of Cuba at scale 1: 2,000,000, Peñalver [3] (in the New National Atlas of Cuba). This monograph, with a broad update, serves as the basis for this work.

On the other hand, Iturralde Vinent [4] elaborated a stratigraphic scheme for the Quaternary deposits of the provinces of Camagüey and Ciego de Ávila, although on a smaller territorial scale.

In the former province of Oriente, Cuban, and Hungarian specialists used a provincial scheme that recognizes some lithostratigraphic units typical of the Quaternary of eastern Cuba [5].

During the last 25 years, a group of Cuban researchers has continued to specify, correct, and expand, some of the ideas outlined above, publishing their scientific results in both Cuban and foreign journals. Examples of this are the works of Peñalver [3, 6, 7], Peñalver et al. [8–21], Cabrera and Peñalver [22, 23], Cabrera et al. [24]. Of particular importance has been the revision and updating of all the geological formations of the Quaternary System of Cuba that were recognized in the second version of the Léxico Estratigráfico de Cuba [25].

The present work will have the descriptions of the different lithostratigraphic units or innominate deposits that are recognized as mappable units and of which all the update referred to is provided.

7.2 Materials and Methods

The stratigraphic subdivision scheme presented is based on the same principles used by Kartashov et al. [26], but includes some units proposed by Iturralde Vinent [4] and other researchers who have worked on the Quaternary Geology in Cuba later.

As it is known, since there is not abundant fossil fauna in Cuba in the stratigraphic sequences, it is very difficult to establish a bio-stratigraphic basis for the division of the Quaternary.

The basis of the Quaternary could be quite well-fixed thanks to the study of some species of the microfauna of foraminifera and ostracods that allowed to date the rocks of the Vedado Formation as Lower Pliocene-Lower Pleistocene.

The multiple discoveries of vertebrate remains that have been made in Cuba could be used as complementary material for the dating of the Quaternary, but always taking into account the endemicity of most of this fauna and that most of the discoveries are related with caverns deposits.

These same authors pointed out that the detailed study of mollusks of the Pliocene–Pleistocene age in Cuba should allow the separation of complexes of different ages, whose differences could be used for stratigraphic purposes, although the variation of mollusk complexes in the tropical zone in the course of the Quaternary was, apparently, insignificant.

In fact, De la Torre and Kojumdgieva [27] were able to verify the existence of associations of mollusks corresponding to levels lower than the Upper Pleistocene, such as *Chlamys (Nodipecten) pittieri, C (N) arnoldi arnoldi, Lycyma (Linga) pensilvanica pegnalveri, Chioni cancelata francoi, Chamaa sp., Pecten sp.,* and *Cypraea cf. cervus*, among others. Those that correlated with the glacio-eustatic oscillations of the sea in Europe, the Caribbean, and North America.

Palynology, which is very useful in regions of high and medium latitudes, in the case of the Cuban Quaternary has had little effectiveness, due to the poor conditions of pollen conservation in sediments reworked by tropical weathering. It is important to carry out studies on the character of the Quaternary climatic oscillations in order to try to define to what extent these oscillations were reflected in the vegetation changes.

Taking into account the limitations related to fauna, the Quaternary Subdivision Scheme in Cuba was not elaborated on the basis of bio-stratigraphic data, but on geomorphological and lithological criteria. It reflects the genesis of the different sequences and the succession of their formation. The diagnostic signs of these sequences, that is, their lithological characteristics and their position in the cut and in the relief, allow making a quite safe differentiation between them.

At the same time, the lithological and geomorphological signs that served as the basis for the differentiation of the Quaternary, provided genetic and paleo-geographical information. It allows to relate these subdivisions with certain stages of geological development that were manifested not only in the territory of Cuba, but in a broader region, and in principle, obviously, can be correlated with such important events in the Quaternary history of the Earth as the glacial and interglacial ones.

Following the regulations of the International Union for Quaternary Research (INQUA) and the International Geological Congresses, the position of the lower limit of the Quaternary has been placed in the 2588 million years, including the Gelasian Stadium. This coincides with the transit between the Gauss and Matuyama epochs within the Paleomagnetic Scale and with the start of OIS 103 in the Isotope Oxygen Scale.

In addition, as a limit of first order was taken between the Pleistocene and Holocene, which corresponds to the age of 11,700 years.

In the course of the investigations carried out in Cuba, a very clear paleogeographic boundary was detected in the Pleistocene of Western and Central Cuba. The oldest sedimentary terrigenous formations are characterized by a profile of well-developed kaolinite weathering, and the youngest, starting with the Villarroja Formation, do not have this type of profile. In view of the fact that, at present, the intense chemical weathering with the destruction of smectite and its partial transformation into kaolinite occurs only in those regions of Cuba where the annual level of atmospheric precipitation exceeds 1800 mm per year. Besides, due the old variegated sequences with a profile of kaolinite weathering is also found in regions with a lower level of annual rainfall (up to 1100 mm), it is logical to assume that at the beginning of the Pleistocene the climate of Cuba was wetter than at the end. The extinction of the processes of chemical weathering in most of the territory of Cuba, related to the decrease of the humidity of the climate is precisely the limit that allows to divide the Pleistocene of Cuba into two parts: a lower one more humid, and one higher dry. This limit is on the Guevara Formation.

As stated by its authors, this subdivision of the Pleistocene conditioned by climate is involuntarily associated with the Pleistocene conception "preglacial" and "glacial" according to Selli 1967 (in Peñalver et al. [11]) and the division of the Anthropogen into Eopleistocene and Pleistocene [28]. In addition, Kaiser 1969 (in Peñalver et al. [11]) was of the opinion that the subdivisions within the Quaternary should be established by the climatic variations that are reflected in the sedimentary deposits and especially those belonging to the glaciations. A little more recently Bonifay 1980 (in Peñalver et al. [11]), in its chronological pictures distinguishes an old Pleistocene that goes between 1,85 and 0,9 million years and one Middle-Upper, from 0,9 millions of years.

For Kartashov and his collaborators [26], it was tempting to suppose that the boundary between the "Humid" and "Dry" Pleistocene of Cuba was related to a cardinal variation of the climate that determines the position of the boundary between the warm and cold Pleistocene of the high and medium latitudes. According to various authors this coincides with the paleomagnetic inversion of Brunnes/Matuyama, currently dated to 780,000 years.

7.3 Results

In the subdivision scheme presented (Table 7.1), six formations pass through the Neogene-Quaternary limit, and that is, they began to be deposited in the Upper Pliocene and concluded their deposition in the Lower Pleistocene. Other units can be correlated with the Lower Pleistocene. Until that last age comes the deposition and weathering of the sediments included in the Humid Pleistocene.

Later several units for the Middle Pleistocene are recognized and a greater number, that totalizes eleven, for the Upper Pleistocene. Specifically for the upper part of the Upper Pleistocene, correlated with the OIS 3, several formations are recognized, although they have a local character, from the paleoclimatic point of view they present their own peculiarities.

As innominate deposits of the Quaternary, that is to say, that have an age that includes both the Pleistocene and the Holocene and continue their development at present, the elluvial-colluvial-prolluvial deposits occur.

Finally, a Holocene formation and four unnamed deposits are exceptionally recognized.

Table 7.1 Scheme of the stratigraphic subdivision of the Quaternary of Cuba

Chronostratigraphic subdivision				Lithostratigraphic subdivision	
Chronostratigraphic subdivision (climatic-stratigraphic)				Carbonated formations	Terrigenous formations
QUATERNARY	Holocene			Los Pinos (lpi)	Marine deposits Alluvial deposits Marsh deposits Biogenic deposits Elluvial-colluvial-prolluvial deposits Siguanea (sgn) El Salado (sdo)
	Upper Pleistocene	Late	"Dry" Pleistocene	Playa Santa Fe (psf) La Cabaña (lcñ) Cocodrilo (ccl)	
		Early		Cayo Romano (cro) Jaimanitas (js)	Camacho (cmc) Jamaica (jmc) Cauto (ca)
Neogene	Medium Pleistocene		"Humid" Pleistocene	Guanabo (gnb) Versalles (vs)	Villarroja (vr)
	Lower Pleistocene			Calcarenitas López Orta (clo) Alegrías (alg) Río Maya (rm) Vedado (vd)	Guevara (gv) Bayamo (by) Dátil (dt) Guane (gne)
	Upper Pliocene				

7.4 Discussion

In the elaborated scheme, enriched in recent years and shown in Fig. 7.1, the carbonated and terrigenous formations of Western and Central Cuba can be correlated very well, which is more difficult when compared with those of Eastern Cuba.

It is noteworthy that in the course of the Late Cenozoic, the eastern regions of Cuba differed from the rest of the territory by the active tectonic regime. This is reflected very well in the constitution of the relief and in the lithological peculiarities of the Quaternary sediments, both terrigenous and carbonate.

In Eastern Cuba, there is the highest mountain system of the Cuban Archipelago: the Sierra Maestra, to which the deep Cayman valley is associated by the south. Planning surfaces that are considered as relicts of the Pliocene–Pleistocene marine abrasion platforms here reach the height of 800 m above sea level, while in Western and Central Cuba they do not exceed 200 m. Accordingly, the sand-clayey sediments of the Pleistocene of Western and Central Cuba are replaced in this region by sand-pebbled sediments, appearing in the carbonate sediments a considerable amount of gravel and fine pebbles of the magmatic and metamorphic rocks (Fig. 7.2).

It highlights a close relationship between the distribution of atmospheric precipitation and the lithological peculiarities of the terrigenous sediments, specifically their levels of weathering. Both the wettest and the driest regions of the island are located in Eastern Cuba. It is characteristic that in regions with an annual precipitation of less than 1100 mm, it is not possible to detect variegated terrigenous deposits with a well-developed kaolinite weathering profile, although the ancient terrigenous formations that correspond to that part of the Pleistocene are present in Western Cuba and Central was characterized by wet conditions.

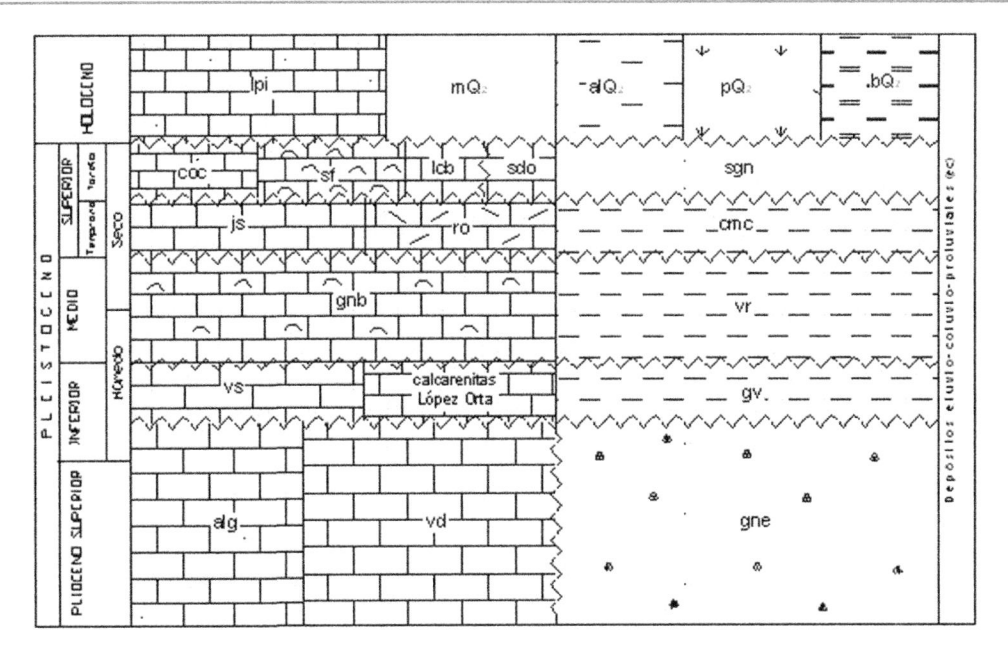

Fig. 7.1 Scheme of stratigraphic correlation of the Quaternary of Western and Central Cuba. (alg – Alegrías Formation; vd – Vedado Formation; gne – Guane Formation; vs – Versalles Formation; gv – Guevara Formation; gnb – Guanabo Formation; vr – Villarroja Formation; js – Jaimanitas Formation; ro – Cayo Romano Formation; cmc – Camacho Formation; coc – Cocodrilo Formation; sf – Playa Santa Fe Formation; lcb – La Cabaña Formation; sdo – El Salado Formation; sgn – Siguanea Formation; lpi – Los Pinos Formation; mQ_2 – marine deposits; alQ_2 – alluvial deposits; pQ_2 – marshy deposits; bQ_2 – biogenic deposits)

Fig. 7.2 Stratigraphic correlation scheme of the Quaternary of Eastern Cuba. (rm – Río Maya Formation; by – Bayamo Formation; dt – Dátil Formation; gv – Guevara Formation; vr – Villarroja Formation; js – Jaimanitas Formation; ca – Cauto Formation; jam – Jamaica Formation; lcb – La Cabaña Formation; sdo – El Salado Formation; mQ_2 – marine deposits; alQ_2 – alluvial deposits; pQ_2 – marshy deposits)

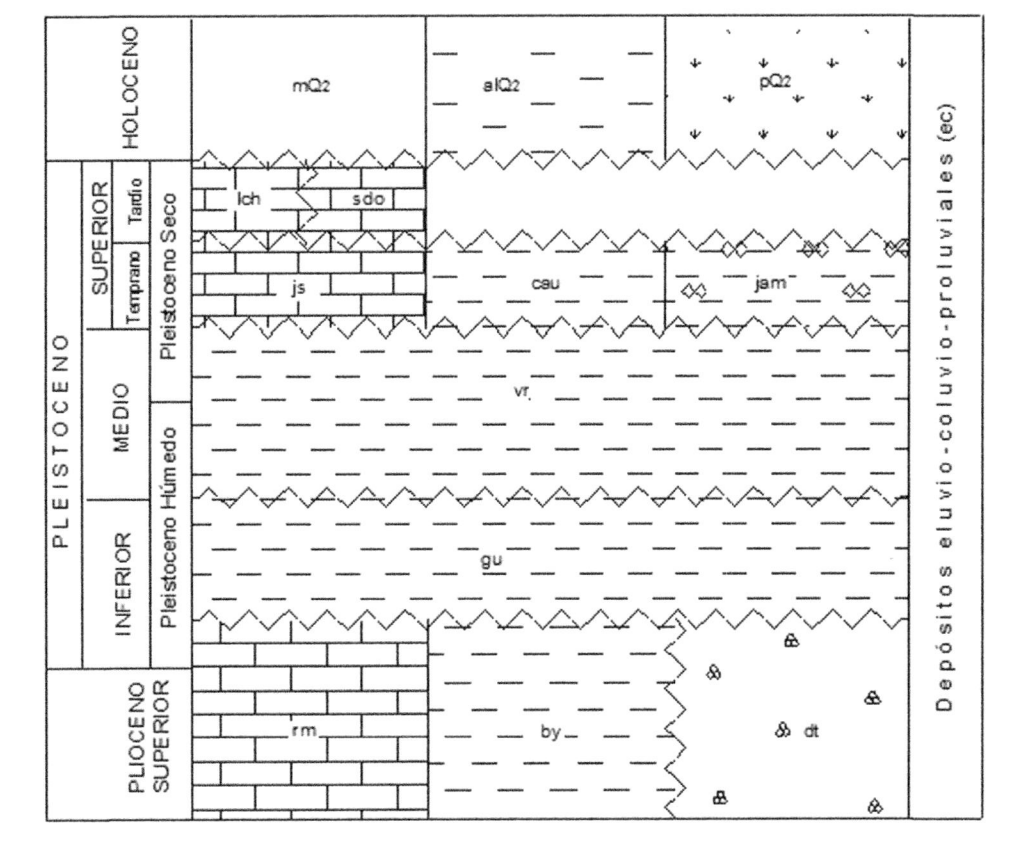

This can be explained by the fact that in the course of the humid Pleistocene the annual precipitation was greater than the current one in 600–700 mm, and in the drier regions the humidity was not enough for the development of the chemical weathering processes that led at emergence of motley coloring in terrigenous sequences. At the same time, the main features of the distribution of atmospheric precipitation in the territory of Cuba, apparently, did not undergo substantial variations in the course of the entire Quaternary. It can be assumed that the regions with the lowest level of atmospheric precipitation were located approximately in the same place as they are in the present. In addition, this assumption is not applicable to the periods of the great transgressions of the Quaternary, during which most of the Cuban territory was submerged under the waters of the ocean.

If it takes into account that all the terrigenous formations of the Pleistocene in Cuba have a marine origin as a result of the alternation of transgressions (accumulation of deposits) and regressions (weathering of deposits), it can be assumed that this alternation was conditioned by the Pleistocene glacio-eustatic oscillations of sea level. Considering that in Cuba these oscillations took place in a background of intense and differentiated neotectonic movements, it is difficult to admit an absolute synchrony of the transgressions and regressions in the whole island.

Therefore, there is the possibility that some of the terrigenous formations of Eastern Cuba do not coincide in the time of accumulation with any of the terrigenous formations of Western and Central Cuba.

The description appears structured according to the divisions proposed in the scheme, starting with the oldest and the carbonated ones, when there are carbonated and terrigenous on the same floor.

7.5 Upper Pliocene-Lower Pleistocene

As already noted, six formations are recognized in this interval.

The index fossils of the Vedado and Río Maya formations stand out, since they are among the few from the Quaternary of Cuba that present fauna.

In the other formations (Guane, Dátil, and Bayamo) no fossils have been reported to facilitate their dating, which is common for the rest of the Quaternary deposits of Cuba. In all probability, the Neogene/Quaternary limit transits in these units. The paleomagnetic studies carried out in the Vedado and Río Maya formations; confirm this criterion [10].

Vedado Formation:

It is constituted by coral-algal and biodetritic bio-thermal limestones, white, cream gray, yellowish and sometimes pink, massive or with local stratification unclear, hard, sometimes aporcelanated, partly porous and cavernous, recrystallized, containing corals in position of growth or its fragments, are often dolomitized. Its matrix can be micritical or micrite-arenitic. They contain, in general, numerous tubular impressions of the *Acropora prolifera* coral and are filled with a reddish-carbonaceous reddish material with goethite. Calcarenites lenses are observed.

The surface of the limestones is strongly karstified, with a very intense lapis, which sometimes reaches more than 1 m in elevation. The presence and aggressiveness of the karstic forms, which also reflect a greater humidity in the environment during its genesis, is confirmed by the existence of caves and caverns, of a much larger development than originally considered, which has been proven in works carried out in the North coast of Habana-Matanzas in recent years. In that region, as well as in the south of Central Cuba, these form the second and third level of the marine terrace and reach a height that ranges between 75 and 100 m above the current sea level.

The predominant mineral is calcite. Some portions of the rocks, undistinguished neither by their external appearance nor by the characteristics of their organogenic structure, are formed by magnesium calcite in association with proto-dolomite or dolomite. Most likely, they are epigenetic formations.

When diluting limes in 2% hydrochloric acid, generally the amount of the insoluble residue does not reach one percent and only in the pink limestones it increases up to 1–3%. In the composition of the insoluble remainder, sporadic quartz grains of fine sand or silt size are common, with much less frequent grains of serpentinites, effusive and metamorphic rocks of similar granulometry. The clay component is heterogeneous, because of its composition it can be correlated with the local peculiarities of the geological constitution. In the zone of distribution of volcanites of Cretaceous age in Central Cuba, the clay component consists mainly of smectites, chlorites and/or chlorite/smectites, and interstratified mica-smectites. Near the outcrops of the metamorphic complex and its weathering crust, a kaolinite-hydro micaceous association with one or another mixture of interlayered kaolinite-smectites represents the clay material of the limestones. Near the serpentinitic massifs is represented by smectites, talc, and serpentine.

The formation constitutes typical deposits of the reef complex. The following index fossils have been determined: ostracods: *Bairdia dimorfa, B. pillosa, Bairdoppilata triangularis, Cytherella dominicana, Perissocytheridea aff, P. bicelliformis, Quadracythere ex. gr, Q. bichensis, Radimella confragosa;* mollusks: *Calyptraea equestris, Chione wordwardi, C. elattocostatum, Cypraea cf., C. patrespatriae, Marginula depressa, Lucina cf, L. Podragoina, Ostrea frons;* Corals: *Acropora prolifera, Montastrea limbata, Pachyseris rugosa,* echinoids: *Brissus sagrae, Clypeaster cubensis, C. Dalli, Schizaster cubensis;* crustaceans: *Mithras hispida.*

Its upper cut is erosive or covered by the Jaimanitas, La Cabaña, and El Salado formations. In some places it interdigitates with the Guane Fm. It reaches a thickness of up to 196 m and extends as a discontinuous coastal strip in the western and central part of the country.

The inclusion of the deposits of the Punta del Este Fm., located in the southern part of the Isla de la Juventud, inside the Vedado Fm., is because both formations have the same age and have similar lithological characteristics. The exception is the oolites, which appear only in the limestones of the Punta del Este Fm.

Limestones similar to those of the Punta del Este Fm. had been described in neighboring territories to those of its type locality, such as Ávalos Fm. them by Kartashov et al. [29], as limestones of the Guanahacabibes Peninsula by Pszczolkowski et al. [30]. These proposals were not validated in the first version of the Léxico Estratigráfico de Cuba [31], because they were considered synonymous with the Vedado Fm. Something similar should have been done with the Punta del Este Fm., but it was not done. In the new version of the Léxico Estratigráfico de Cuba [25], it was concluded that these deposits are nothing more than a lithofacial variation within the Vedado Fm., that due to its small extension, non-affectability and similarity with it, it is not practical to segregate it as an independent unit.

In satellite images, this formation can be delimited with great precision. In the RGB color composition (Fig. 7.3) the deposits of this formation are observed with a Carmelite (brownish) color.

Río Maya Formation:

It is constituted by algal, coralline, and micritical biohermic limestones, very hard, with a micritical matrix, frequently aporcelanated, containing corals in growth position or their fragments, as well as subordinate mollusk molds and shells all recrystallized, being abundant the coral Acropora prolifera. Limestones are frequently dolomitized. The clay content is very variable.

It contains abundant clasts of terrigenous material, coming from the magmatic and metamorphic rocks of the neighboring emerged areas. Its granulometry varies between sands and ridges, which considerably differentiates it from the Vedado Fm. Sometimes there are intercalations of polimictic conglomerates of variable granulometry and calcareous cement. The surface of the limestones is strongly karstified. The predominant mineral is calcite. The color is white, yellowish, pink, or grayish. This formation includes deposits typical of the reef complex. It contains the following index fossils: corals: *Acropora prolifera, Diploria sarassotana, Montastrea cf. limbata*; mollusks: *Nodipecten ex gr. nunezi, Spondylus americanus cf. giant.*

It overlays prequaternary formations and its upper cut is erosive or is covered discordantly by the Jaimanitas, La Cabaña, or El Salado formations. Its thickness ranges between 30 and 80 m. It extends in the form of a discontinuous coastal belt in the eastern and southeastern part of

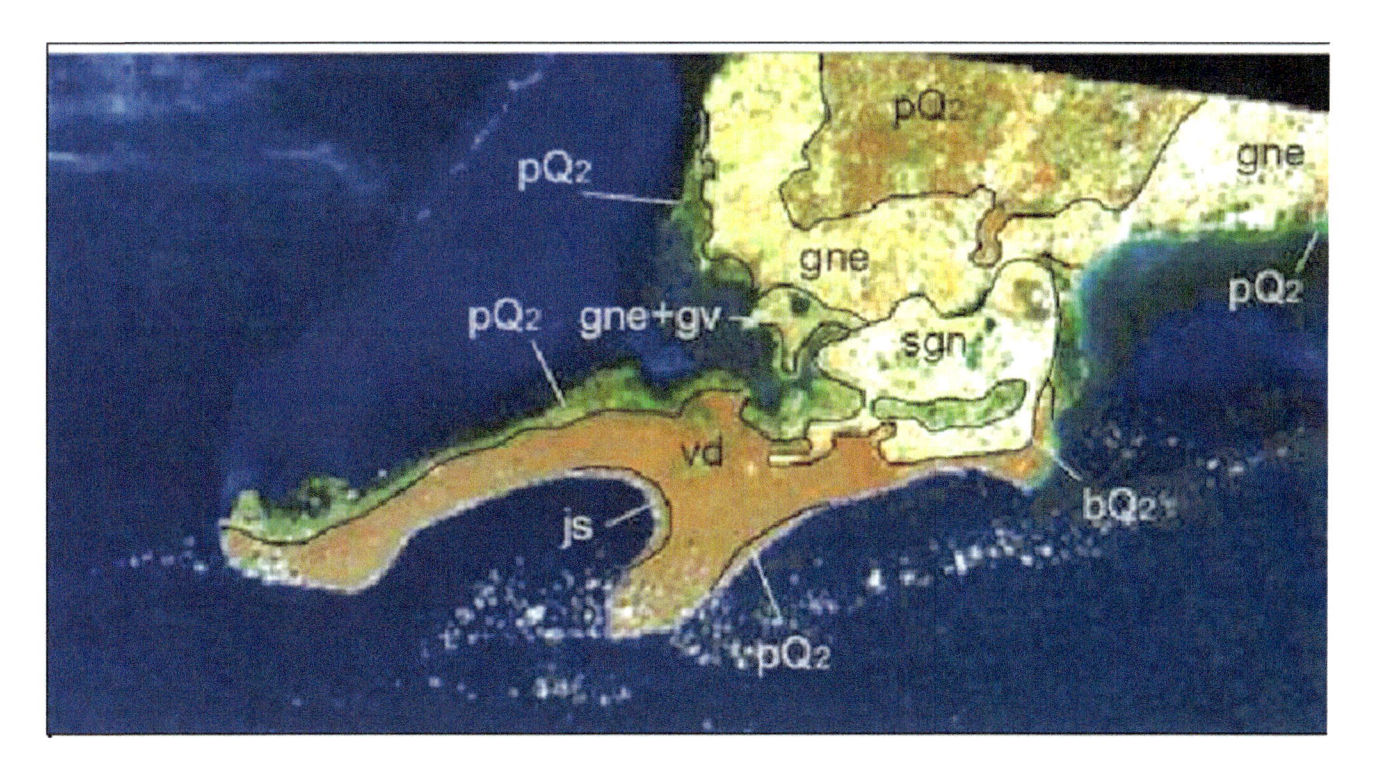

Fig. 7.3 Satellite image of the Guanahacabibes Peninsula and adjacent areas. (pQ$_2$ – marshy deposits; bQ$_2$ – biogenic deposits; gne – Guane Formation; sgn – Siguanea Formation; vd – Vedado Formation; js – Jaimanitas Formation; gv –Guevara Formation)

the country, where it frequently constitutes high levels of marine terraces, rising up to 300 m above present sea level, which is evidently linked to the great neotectonic ascents of this part of the island.

Alegrías Formation:

Its original reference is found in the explanatory text of the geological map of the Ciego de Ávila-Camagüey provinces at a scale of 1: 250,000 [4]. However, it was not taken into account in the first version of the Léxico Estratigráfico de Cuba [31] and no synonymy was established either.

In subsequent more detailed investigations [32], it was found that due to its extension, lithological and morphological peculiarities, it is feasible to consider it as a chronolithostratigraphic unit. It is composed of calcarenites and medium-grained, well-cemented, medium-sized, cream-colored calcarenites made up of very well-shaped remnants of calcareous algae, foraminifera and between 2 and 15% of quartz grains, plagioclase, and volcanic rocks. It appears recrystallized and chamfered on the surface, forming a cap of 1 to 3 m in thickness. This is a result of weathering, which also causes deeper disintegration, turning the rock into an earthy material. No sedimentary structures are observed. Its sedimentation environment corresponds to a cumulative sandy coast, coastal lowlands, beaches, bars, and coastal dunes, in communication with the territory emerged by means of tidal channels. The rocks that underlie it are not known. On the surface, it contacts laterally discordant with the Jaimanitas Fm.

It is possible that their lateral equivalents in depth are the rocks cut by drillings in neighboring zones and they are fine bioturbated and recrystallized calcarenites, biogenic-detrital limestones, and biogenic-oolitic limestones. Its visible thickness reaches 20 m. It is located in the highest elevations of the eastern part of the Sabana-Camagüey Archipelago. Its age has been assigned taking as a fundamental criterion its degree of recrystallization, which is similar to that of the limestones of the Vedado and Río Maya formations.

Guane Formation:

Conglomerates, gravels, sands, sandy clays, weakly cemented by clays, constitute it. These deposits have horizontal undefined stratification and more rarely crossed. The clastic material is angular, subangular, and has a variegated color.

Its composition in general is oligomictic (predominantly quartz, siliceous rocks, and sandstones); in the clasts of the conglomerates sometimes are also other local rocks very weathered.

The matrix appears in the weathered parts of the thicker cuts, up to where the weathering did not reach, with four facial types: mainly kaolinitic, hydro micaceous-kaolinitic, smectite-hydro micaceous-kaolinitic, and smectitic.

The deposits of the Guane Fm., with a mainly kaolinitic clay cement composition, were determined on the Isla de la Juventud, within the distribution limits of the Jurassic metamorphic schists. In addition to kaolinite, insignificant amounts of dioctahedral hydro mica, interstratified mica-smectite, and chlorite have been determined.

The hydro micaceous-kaolinitic associations gravitate spatially towards the outcrops of shales and sandstones of the Jurassic in the province of Pinar del Río. Here interstratified kaolinite-smectite with a small content of smectitic packages appears. In addition, considerable amounts of dioctahedral hydro micas and/or interbedded micaceous-smectite formations appear. The chlorites and smectites are found in subordinate amounts.

Smectite-hydro micaceous-kaolinitic clayey cement has a limited distribution within the development areas of the previous type, also in the province of Pinar del Río. The cut here can be divided into two parts. The smectites together with the imperfect kaolinites or the interlayered kaolinite-smectites are present in the lower horizons, mainly clayey. Further, up the cut, in the sandy-gravel layers, the smectites lose their predominant role, and the hydro micaceous-kaolinitic mineral association described above characterizes the clay cement from the deposits.

The smectite type of clay cement is characteristic of the regions where volcanoes emerge in Central Cuba (Province of Sancti Spíritus). In addition to the smectites, only an insignificant mixture of imperfect kaolinite appears.

The general color of the rocks, as already indicated, is mottled with predominance of red, gray, and yellowish, which resulted from the action of a strong weathering. Precisely during the process of weathering, redistribution of iron takes place in the form of roentgenomorphic hydroxides and goethite, and with that the appearance of the variegated colors that mask the primary stratification and other details of the constitution of the formation. Also occurs the accumulation of ferruginous concretions and lateritic armor, which is observed in the upper part of the cut. This formation was deposited in a sea with abundant alluvial contribution, being able to become estuarine. It can be interdigitated with the Vedado Fm., as it happens in the western end of the province of Pinar del Río. It overlays prequaternary formations and is covered discordantly by the Villarroja and Guevara formations. Reaches up to 50 m thick. It occupies several zones of Western and Central Cuba, constituting low hills and hilly plains.

Dátil Formation:

It is constituted by deposits of pebbles, blocks, and pebbles, generally of silicates originated by the hydrothermal alteration of volcanic rocks, which predominate in the pre-montane zone. There are also argillaceous sands of polimictic composition, in the form of intercalations and lenses, which come to predominate in areas far from the mountains. All the deposits are of variegated coloration, with a predominance of red, yellowish, and gray colors, which resulted from the action of a strong weathering. The fragmentary material is poorly rounded and weakly cemented. The stratification is horizontal and is manifested mainly by the alternating sediments of the different granulometry. The cementing material has a scaly-fibrous structure of heterogeneous composition. In the lower horizons, relatively little weathered, smectite predominates, but chlorite-smectite and hydroxides may be found. In the upper part of the cut, more weathered, the clay cement is more homogeneous, predominantly interbedded kaolinite-smectite. The sedimentation environment is alluvium-marine, predominantly the alluvial contribution, and may have the character of an estuary. This formation overlays prequaternary formations and is covered discordantly by the Cauto Fm. Reaches a thickness of 34 m. It forms discontinuous bands on the northern slope of the Sierra Maestra, in Eastern Cuba.

Bayamo Formation:

It is made up of gray and gray-yellow fine-grained sands, with sandstone lenses and conglomerates of fine pebbles and intercalations of sandy-gray clays of greenish gray color, with gray spots on the sands. In the sands, stratification can often be observed. The gravelly-sandy-silty fractions are of polimictic composition. Grains of metavolcanites, siliceous rocks, quartz, feldspars, chlorites, zeolites, and epidote grains predominate. The grains of coarse sand are well rounded, while the finer grains are subangular. The sediments are weakly bound by carbonaceous cement. The carbonaceous component of the cement is represented by calcite and in the clays, the smectites predominate and, subordinately appear, chlorite and probably interstratified chlorite-smectite. They were deposited in a marine environment, with some alluvial contribution.

This formation discordantly covers prequaternary rocks and is overlain by the Cauto Fm., being an important component of the Cauto river plain, where it sometimes appears or can be distinguished in the cuts. Its thickness ranges between 30 and 120 m.

7.6 Lower Pleistocene

With this interval, until now, the unnamed unit López Orta Calcarenites and the Guevara Fm. are related. However, one cannot rule out the possibility that the Versalles Fm., which appears in the Middle Pleistocene, could be included here.

On the other hand, the termination of this interval coincides with an important paleogeographic boundary. On the roof of the Guevara Fm. is the limit of the Humid Pleistocene of Cuba. From this interval the climate changed and became drier, which does not mean, in any way, that there have not been cycles in which the climate varied before and after the Guevara Fm., but always in the frames of a wetter or drier climate.

In fact, each time the area occupied by Cuba was subjected to the influence of glacio-eustatic transgressions and regressions of sea level, decreasing or increasing in size, the climate also varied. Thus, in the transgressions, when decreasing the emerged territory, the precipitations also diminished. On the other hand, in the regressions, when increasing the emerged area, the sediments that were deposited in the previous transgression were subjected to weathering (weathering), increasing precipitation and humidity. This happened in a general background more humid or drier, depending on the floor in which we are.

It can be noted that the most humid general bottom was prolonged until the deposition and subsequent weathering of the Guevara Fm. Therefore, the roof of this paleoclimatic interval is placed there.

López Orta Calcarenites:

It is an informal unit, constituted by calcarenites of grayish cream color, that present inclined oblique stratification. Its strata dip to the southeast at an angle of 32°. They overlap discordantly to the biocalcarenites of the Loma Triana Fm., of Late Miocene-Pliocene age and, frequently, to a clayey layer that can reach up to 1 m in thickness, constituted by calcarenites fragments of said formation whose dimensions generally reach 0.5–1.5 cm and they are surrounded by a thin cover of esparitic calcite (evaporites). The matrix is mainly composed of calcite crystals grouped in the form of microdrusas of different sizes and with inclusions of a yellow clay substance. This layer evidently formed in a continental phase that lasted a long time.

The López Orta calcarenites are developed between the towns of Martí and Corralillo, on the north coast of Central Cuba, and are a series of narrow ridges isolated from each other and with a parallel course, separating the low coastal plains. They are covered by clay sediments of the Camacho Formation, from the plain of Colón-Manacas, composed of gravel, clays, and sands, from the Guane, Guevara, and Villarroja formations.

In the López Orta Knoll, at 55 m altitude, there are obvious traces of transgression, which do not appear in other elevations. It is evident that this transgression took place after the fundamental features of the current landscape were formed on the surface of the calcarenites belonging to the Loma Triana Fm.

Diagonally stratified calcarenites could be deposited in the form of bars in a shallow marine environment, or they can represent eolianites. According to the specific geomorphological situation (top of elevation), the first interpretation seems to be the most probable.

The authors of this informal unit assigned it a Plio-Pleistocene or Early Pleistocene age, although De la Torre and Kojumdgieva [27] based on its content of gastropods, limited it to the Early Pleistocene. The thickness of the calcarenites does not exceed 3 m.

Guevara Formation:

Its authors described it as clayey and clayey sands, with gravels, sometimes with pebbles and even with boulders. It has a non-clear horizontal stratification masked by the weathering processes, which form blotches and stripes of whitish, greenish-gray, and red color. The weathering zone, within whose limits the deposits have variegated coloration, has a thickness of 2–4 m and, in most cases, covers all the formation and underlying rocks. Only in some cuts the lower layers of the formation are represented by unaltered deposits of monotone yellow-greenish gray color.

These deposits frequently contain iron pellets and pisolites, and even, below the humic-cumulative horizon of the upper soil, a 0.5–0.7 m thick layer is sometimes distinguished, represented by non-variegated deposits, impoverished in material clayey, containing ferruginous concretions (hardpan ferruginous).

The clastic material of the formation is oligomictic. Both in the fraction of pebbles and gravel, as in sand and silt, quartz predominates. In some sections, a considerable amount of siliceous rocks is also observed. In the form of isolated grains are potassium feldspars and dioctahedral micas. Frequently the pebbles and gravels in the formation correspond to concretions and ferruginous nodules, obtained from the denudated lateritic shells of the Guane Fm.

Two types of facies represent the primary-sedimentary associations of clay minerals in the cement of the deposits of this formation not altered by weathering: hydro micaceous-kaolinitic and smectitic. The first ones are spatially related to the outcrops of the Guane Fm. and the metamorphic schists. In this case, the imperfect kaolinite, the interlayered kaolinite-smectite, the dioctahedral micas, and the mica-smectites are characteristic. They also contain interstratified smectite chlorite. The chlorites and smectites are present in greater quantity than in the deposits of the Guane Fm.

The deposits of the Guevara Fm. with smectitic clay cement are distributed within the development limits of the carbonate and marly rocks of the Miocene and, to a lesser extent, of the volcanic rocks. The main component of cement is smectite. In some sections, based on thermographic analysis, the presence of aluminous-ferruginous smectite was established. Sometimes, in the clay cement also appears kaolinite-smectites, imperfect kaolinite, chlorites, palygorskite, and others. The smectites, in these cases, were obtained from the underlying rocks, which is attested by their presence in the insoluble residue of the Miocene limestones and marls.

The weathering of deposits with hydro micaceous-kaolinitic clay cement leads mainly to the destruction of a certain amount of roentgenomorphic and poorly crystallized components of the system, to the redistribution of the primary ferruginous pigment and to the formation of secondary variegated colorations.

In deposits with smectitic cement the effect of weathering is more noticeable. It can be noted that the kaolinization of the smectites is accompanied by the mobilization of certain amounts of aluminum and notably larger amounts of iron, which is not part of the structure of the kaolinite. As a result of it there are variegated colorations of ferruginous deposits and concretions.

Its authors [29], based on the lithological peculiarities of the deposits, the character of their lying (little potent and isometric layers), their close relationship with the coastal plains and the flat relief of their surface, supposed a marine origin for its deposits. They correlate it with one of the great glacio-eustatic transgressions of the Lower Pleistocene.

Overly discordant to the Guane Fm. and is found unconcealed by the Villarroja Fm. and by the Camacho Fm. It appears distributed in the South plain of Pinar del Río, in parts of the South Habana-Matanzas plain and in the plains of Central Cuba. Its thickness can exceed 30 m.

7.7 Middle Pleistocene

In line with what was proposed for the previous interval, a decrease in the humidity of the climate begins, which manifests itself in the lower alteration of the deposits caused by the subaerial chemical weathering. From here, the terrigenous sequences that are deposited with each transgression, do not develop profiles of kaolinite weathering, nor does the transformation of the smectites in kaolinites occur.

At the level of knowledge that is currently possessed, the Versalles, Guanabo, and Villarroja formations are being included in this floor.

Versalles Formation:

It is formed by biodetritic limestones of loamy-arenaceous matrix, massive, porous, recrystallized, or disaggregated, in part, with weak cementation, that contain alignments of small bioherms and intercalations of calcarenites and biocalcirudites in the form of irregular lenses. It also contains molds and shells of mollusks and other marine skeletal remains. They are traversed in the type locality by decapod galleries of the genus *Callianassa*. The color varies from cream to cream-brownish. Its environment of formation corresponds to the reef complex, with the growth of biohermes and episodes of destruction of the reefs. Its thickness is 20 m.

It lies discordantly on Neogene limestones. It is covered discordantly by the Jaimanitas Fm. It occupies two small areas in the City of Matanzas.

Guanabo Formation:

It is constituted by fine bioturbated biocalcarenites, with crossed laminar stratification (eolianites), which are disaggregated into slabs by the effect of weathering (weathering). According to report of De la Torre and Kojumdgieva [27], contains *Cerion tridentatun* and *Vanatta sp.*, Mollusk lung, coastal, living today in nearby locations.

The cut may appear interrupted by up to two generations of paleo soils. Brownish yellow color. Its sedimentation environment corresponds to coastal dunes, formed in a tropical climate by two well-defined seasons: one dry, with movement of sand and dune growth, and another wet, with proliferation of creeping vegetation and fixation of the dune. This seems to have occurred in at least two cycles, during the recess of the accumulation the paleo soils formed. Its thickness does not exceed 20 m. It is located in fossil dune fields on the northeast and northwest coasts of La Habana. In the northeast part, its upper limit is erosive, while in the northwest part it is partially covered in a discordant way by the Jaimanitas Fm. The underlying corresponds to Oligocene–Miocene limestones.

Villarroja Formation:

It is represented by clays, sandy clays and silty sands, clayey sands, and quartz sands of different granulometry, pigmented by iron hydroxides (goethite). It contains fine lenses and layers of gravel with clasts of varying size, frequently with good roundness and selection, constituted by quartz and more subordinately by fragments of ferruginous hardpan, as well as ferruginous concretions. They present variable tonalities, from yellowish red to violet red.

The authors of this formation [29] proposed that three varieties of facies stand out quite clearly in the sediments. In two regions, south of Guane (Western Cuba) and west of Cienfuegos (Central Cuba), are the sands and clay sands with red pebbles, with an absolute predominance of quartz. At the periphery of the basic and ultrabasic rock masses, the formation consists predominantly of clays and of heavy and loose ocher dark red and purple red with a mass of rolled fragments of ferruginous laterites, sometimes of siliceous rocks and of silicified serpentines. The fragments have the dimension of the pebble and gravel, sometimes even that of the rolled blocks. The most widespread is the third facial variety of the sediments of the formation: the clayey sands and the red clays with intercalations and lenses of sandy-gravel material, in which, together with the quartz, are always present. Sometimes predominate sandy ferruginous pisolites and ferruginous oolites, obtained from destroyed lateritic breastplates, which developed in older deposits. These sediments are found within the limits of most of the plains of Cuba. These same authors continue to point out that the structure and composition of the clay material of the Villarroja Fm. is kept in huge spaces. In the composition of the clayey material of the three varieties of facies, kaolinite-smectites predominate and the presence, ordinarily insignificant, of badly crystallized kaolinite and metahalloysite is recorded. The pigment, in addition to the goethite that was already mentioned, is represented by roentgenomorphic iron hydroxides. In the deposits that are distinguished by the predominance of quartz, the red clay material contains as insignificant impurities quantity of hydro micas and interstratified mica-smectite.

For the dark red clays and ochres the predominance of the roentgenomorphic iron hydroxides and the badly crystallized goethite is characteristic. As insignificant impurities in many regions of Cuba, it was possible to determine dispersed gibbsite, together with which scattered boehmite is rarely found.

The characteristic peculiarity of this formation consists in the absence of traces of vertical and lateral redistribution of the ferruginous pigment, as well as in the absence of any other sign of chemical weathering. Secondary variegated colorations or whitish spots are rarely found there. The configuration of the whitish spots, not typical for the weathering crusts, as well as the absence of the typical weathering profile in these epigenetically modified zones, testifies that in such cases, the local mobilization of iron by the groundwater, enriched with organic matter. This is confirmed by the preferential alteration of the sandier intercalations and lenses, which are between the argillaceous zones that conserve the primary red colors.

Its authors assumed for it a marine origin, considering that the deposits of it accumulated during the transgression that occurred in the Middle Pleistocene, which presented two unique characteristics for Cuba: a sequence of very varied red sediments and the coincidence of the beginning of this transgression with general conditions of greater dryness of the climate. In this interval the intensity of the tectonic movements also received a strong impulse and the relief was transformed abruptly.

However, authors such as Dzulynski et al. [33] consider that it is probable that the deposits of this formation correspond to redeposited rinds of weather generated from rocks of different composition and mobilized in conditions of emerged lands with a marine reelaboration in those areas of the territory that were later flooded by the waters from sea. This unit lies on a large number of prequaternary formations, as well as on the Guevara, Guane, and Vedado formations. Its upper limit is erosive. Its common thickness is 2–3 m, being able to increase up to 40 m in the karstic funnels. It is widespread in the plains of Western and Central Cuba.

7.8 Upper Pleistocene

In this interval, the geological formation of the Quaternary that was first described in Cuba—the Jaimanitas Fm.—[34]. Later, Ducloz [35] extended these deposits to the Ciénaga de Zapata and correlated them with those of the southeastern United States, which are linked to the Sangamon. Shantzer et al. [36] on the basis of radiometric determinations determined that it corresponds to the last Pleistocene interglacial (Sangamon), that is, the OIS 5e. These last authors added that there were also much younger deposits here, which correspond to Wisconsin (OIS 3).

In the Second Version of the Léxico Estratigráfico de Cuba [25] and in subsequent works, the following aspects have been specified:

1. Even though the Jaimanitas Fm. presents facial varieties, it is not practical to divide it into members, since there would be as many members as there are facies.
2. The two transgressions observed by Kartashov and his collaborators [26] in the Jaimanitas layers, can be separated into formations, keeping the name of Jaimanitas for the lower formation, which is the one with the highest occurrence and for which estimated an age of 82,000 ± 6000 years, based on the dating by the Io/U234 ratio [36] and even up to 100,000–130,000 years, according to the estimates of Ionin et al. [37]. For the younger, higher layers, whose age has been estimated between 26,000 and 34,600 years, based on determinations made by C14 [36, 37] the name of La Cabaña Formation is proposed, respecting a proposal made by Bronnimann and Rigassi [38], under another name, unable to use by the rules of the Léxico Estratigráfico de Cuba.
3. With the Jaimanitas Fm., which encircles a large part of the island of Cuba and forms the rocky socle of its marine platform, correlates the Cayo Romano, Camacho, Jamaica, and Cauto formations. These units are the oldest and have been dated as Early Upper Pleistocene.
4. From this interval, the climate in Cuba began to become drier, which does not imply that minor oscillations directly linked to the transgressions or regressions did not occur each time they occurred, but all linked to a general fund of the dryer climate.
5. The Cocodrilo, La Cabaña, Santa Fe, Siguanea, and El Salado formations, which overlap the Jaimanitas Fm. in different places, belong to the Late Upper Pleistocene. In this sense it is more convenient to correlate the same with the transgressive interstadial of the Middle Wisconsin. The radiometric ages noted above do not contradict this assumption.

7.9 Early Upper Pleistocene

Jaimanitas Formation:

It consists of massive biodetritic limestones, generally karstified, very fossiliferous, but without index fossils. Among the most abundant fauna are *Strombus gigas, Porites sp., Anadara notabilis, Argopecten nucleus, Lucina jamaicensis, L. squamosa, L. pensylvanica, Codakia orbicularis, C. costata, Chione cancellata, Astrea brevispina Chalamy (Feguipecten) museosus, Linga pensylvanica, Natica canrena, Politices lacteus guilding, Bulla occidentalis, Archaias angulatus, Ammonia beccari Jania and Amphiroa*, etc. Corals predominate, which may be forming bioherms. Between the karstic forms they are lapis and karstic pockets, occasionally filled by a fine carbonaceous-clayey-ferruginous mixture of brick red color. The limestones pass to biocalcarenites with variable or massive granulometry and stratification. The most frequent coloration is whitish, cream gray, pinkish, or yellowish. In the accumulation of these deposits beach, post-reef and reef facies predominate.

A broad description of this unit, based on hundreds of observations made throughout Cuba, indicates that the facies interrelation of this sequence is determined by the particularities of the structure of the marine platform, within the limits of which it was formed.

In the areas of a narrow marine platform, bounded abruptly from the seaside by great depths, coralline, bryozoan, and conchiferous limestones predominate. Calcarenite intercalations occupy a subordinate place in the cuts of these facies. There are lenses and intercalations of conglomerates with clasts, in most of the limestones of the Vedado Fm. and limestones coral-conchiferous with rounded clasts of rocks of aluminous-silicate composition of sand-gravel size.

The microscopic investigations show that the skeletal remains and fragments of carbonate rocks in the limestones of this facies are cemented with sand-silty and hairs-carbonate material. In the cement there are many remains of skeletons of foraminifera, spines of urchins, detritus of small shells, and pieces of bryozoans. Calcarenites are formed from this same material. The recrystallization

Fig. 7.4 Satellite image of the Ciénaga de Zapata region and adjacent territories. (pQ$_2$ – marshy deposits; mQ$_2$ – marine deposits; bQ$_2$ – biogenic deposits; vd – Vedado Formation; vr – Villaroja Formation; js – Jaimanitas Formation; preQ – prequaternary rocks)

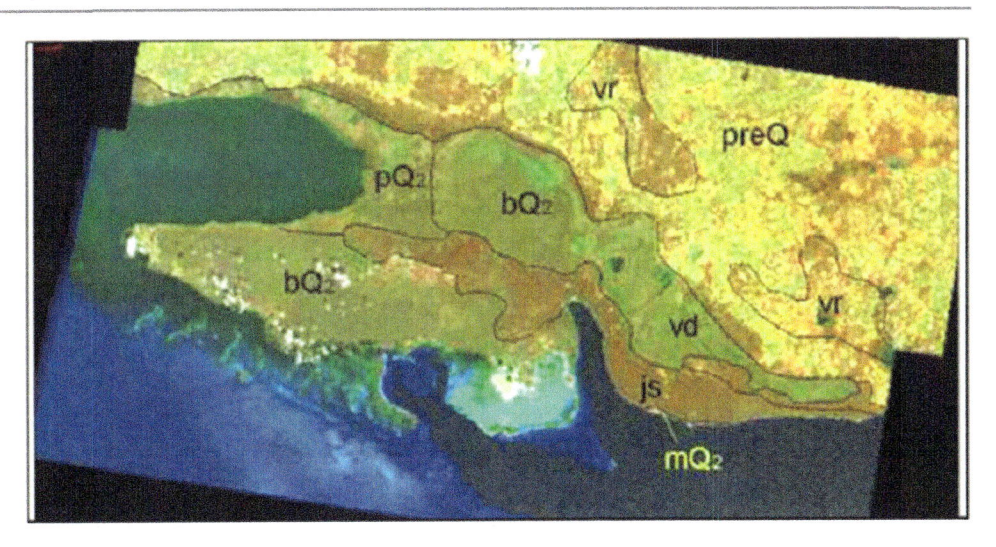

of the skeletal material is usually weak, sometimes almost non-existent. The corals and mollusk shells are made of fibrous carbonate material and the foraminifera testes are made of orange metacoloidal carbonate. The color of the mother-of-pearl layer is often observed in mollusk shells. The cement consists of a cryptocrystalline carbonaceous material with a lumpy or metacoloidal structure.

In the wide marine platforms, the limestone facies accumulated mainly in its outer part. In some regions of Cuba, when walking from the sea to the old coastline, you can see how the coral limestones with thick shells of mollusks that appear in the current coastal area, are replaced laterally by calcarenites.

The facades of calcarenites are represented by two subfacies, of fundamentally foraminifera and pelitic composition.

The determinations of the limestones and calcarenites showed that they are composed mainly of three components: aragonite, calcite magnesium, and calcite. It has been proved that the shells of the mollusks and conchiferous detritus, the foraminifera testes and the great majority of the pellets and ooids, are formed by metacoloidal aragonite.

In most cases the limestones and calcarenites of Jaimanitas contain an insoluble residue lower than 1% and in the remote cuts several kilometers of the old coastal line, they do not contain it at all.

The composition of the sand-siliceous fraction of the insoluble residue depends on the sources of local contribution, although frequently there are angular grains of quartz, hornblende, plagioclase, fragments of siliceous rocks, epidocites, etc.

The clayey material of the residue is also not homogeneous and among the clayey minerals kaolinite-smectite predominates and subordinately hydromica-smectite and chlorite-smectite.

In the case of the calcarenites, far from the old coastline, the main role is played by the roentgenomorphic combinations and in a subordinate way there are combinations of micaceous-smectite and interstratified chlorite-smectite.

The clay minerals characteristic of the residue are smectites, mica-smectites, chlorite-smectites, chlorites, and hydroxics. The thickness of the formation is about 10 m, although some authors have estimated it up to 30 m.

In the RGB color composition shown in Fig. 7.4, the deposits of this formation are observed with a Carmelite (brownish) color.

Cayo Romano Formation:

It is constituted by oolite-pisolitic limestones of dark reddish color. The oolites-pisolites vary between millimeters and up to 5 cm in diameter, presenting a very unique structure. The nucleus can be a gastropod, fragment of its shell or rock. The gastropods are filled with oolites and rock fragments cemented by a dark brown calcareous carbonate. Other nuclei are of blackish-colored calcium carbonate (dark brown in the thin section), but this same type of material forms a layer of 1–2 mm around the mollusk shells. On the outside of the dark layer there is a layer of reddish brown color 1–2 mm wide, which is the one that contacts the matrix of the rock. Said layer is formed by thin concentric sheets of calcium carbonate dyed red. The matrix is calcitic, detrital, or biodetritic with a reddish brownish color.

This unit juts out discordantly to the limestones of the Alegrías Fm. in Cayo Romano or rocks of the Upper Eocene in the Ballenatos Keys, in Central Cuba.

Its roof can be erosive or be covered by clayey sediments. Transition laterally with biocalcarenites and coral limestones of the Jaimanitas Fm.

It is correlated with the Cauto, Camacho, Jamaica, and Jaimanitas formations. Due to its sedimentation environment it seems to be constituted by elluvial-colluvial-prolluvial and marine chemogenic deposits, deposited in a very shallow sea of moderate waves that re-elaborated them. It has a thickness of 10 m and its age, based on its stratigraphic position, is estimated as possible Early Upper Pleistocene. As noted, these deposits are only distributed in the Sabana-Camagüey Archipelago in the northern portion of Central Cuba.

Jamaica Formation:

It is constituted by polimictic conglomerates of clay-carbonate matrix, with more or less rounded and medium-sized clasts, of gray and yellowish gray color, which also contain interspersed lenses and sand and loams of the same color.

The clasts correspond to rocks that emerge in the area: limestones, metavolcanites, silicites and ultramafites. In the sandy and silty fraction, in addition to these rocks, quartz, feldspars, amphibolites, pyroxenes, biotite, epidote, chlorite, etc. are present.

In the clay component of the cement smectite predominates and is also present mica-smectites, chlorites, and hydroxics. Serpentine was found in some samples. On this formation develops a gray-brown floor. The process of internal weathering of the soil is manifested mainly in the dissolution of the carbonated material within the limits of the superficial humic plant horizon, whose thickness is 0.3–0.4 m. Most of this material is deposited on the elluvial horizon, whose thickness is 1 m, in the form of pelitomorphic calcite and lublinite. No alteration is observed in the primary composition of clay minerals.

The deposits of this unit form thin coverings (1–3 m) on the flat peaks of the residual elevations of the coastal plain in southern regions of the province of Guantánamo, which apparently were preserved from erosion. They are also found in the low terraces of the rivers and in the intramontane depressions of the Tortuguilla-Imías region. The formation overlies different lithostratigraphic units and its uppercut is erosive. It represents correlatable marine littoral deposits with a small transgression that must have occurred in the Early Upper Pleistocene.

Camacho Formation:

Its authors [29] described it as clay clays and sands of greenish gray, dark gray, and brown, sometimes with gravels and small pebbles, which have a non-clear horizontal stratification and contain small ferrous manganic concretions, carbonate nodules dark gray. Sometimes they form thin horizons in the basal part of the sediments and in some regions crystals and druse of gypsum. The deposits are usually saline.

The clastic material of the deposits of the formation has polimictic composition. In the sandy-silty fraction, the presence of quartz, plagioclase, amphiboles, biotite, pyroxenes, epidote, fragments of siliceous rocks, metavolcanites, and serpentinites is detected, sometimes there are rounded ferruginous concretions the size of pebbles. The clayey material of the formation is distinguished by the embroiled fibrous texture. In its composition, smectites predominate. In some samples, the presence of kaolinite-smectite was determined in a negligible amount.

In these clays are constantly segregations of fine crystals of calcite. The secondary alterations of the deposits of the Camacho Fm., which are related to weathering and soil formation, lead to the formation of compact black soil. The transformations of the minerals are insignificant: in most cases, they do not even lead to the dissolution of the dispersed carbonated substance and the soil reacts with 2% hydrochloric acid directly from the surface. The thickness of the formation does not exceed 3 m. It is located in Central Cuba and partially in Western Cuba. They lie discordantly on prequaternary formations and on the Guane and Guevara formations. The upper limit is erosive.

Their authors estimate that their lithological characteristics, their forms of occurrence, and the close spatial relationship of their deposits with a characteristic shape of the relief, such as the coastal marine plain, the salinization of these deposits, and their high content of chlorine ions, clearly indicate the marine origin of this formation.

The geomorphological position of the unit, the absence of vestiges of weathering of its deposits, its close relationship with the current coast and with the Holocene marine, marshy and alluvial deposits, as well as other characteristics, place them in the Early Upper Pleistocene.

Cauto Formation:

It is constituted by clays, silts, sands, and polimictic gravels. The sands form layers with intercalations of sandy gravels, they are gray, brownish gray, of fine to medium grains, sometimes coarse. They have horizontal stratification and sometimes crossed, contain intercalations of sandy gravel with layers of 5–20 cm thick. The material of the gravels is porphyrite, tuff, tuffite, limestone, quartz, and silicite, presenting sub-rounded. The diameter varies from a few mm to 5 cm, with an average of 0.5 cm. They are usually covered by a thin black ferrous manganic bark. In the lower part of the formation, the gravels cemented with calcareous material frequently appear, while in the upper part, the amount of gravel decreases.

Fig. 7.5 Satellite image of the Cauto basin. (pQ$_2$ – marshy deposits; alQ$_2$ – alluvial deposits; ca – Cauto Formation; dt – Dátil Formation; by – Bayamo Formation; preQ – prequaternary rocks)

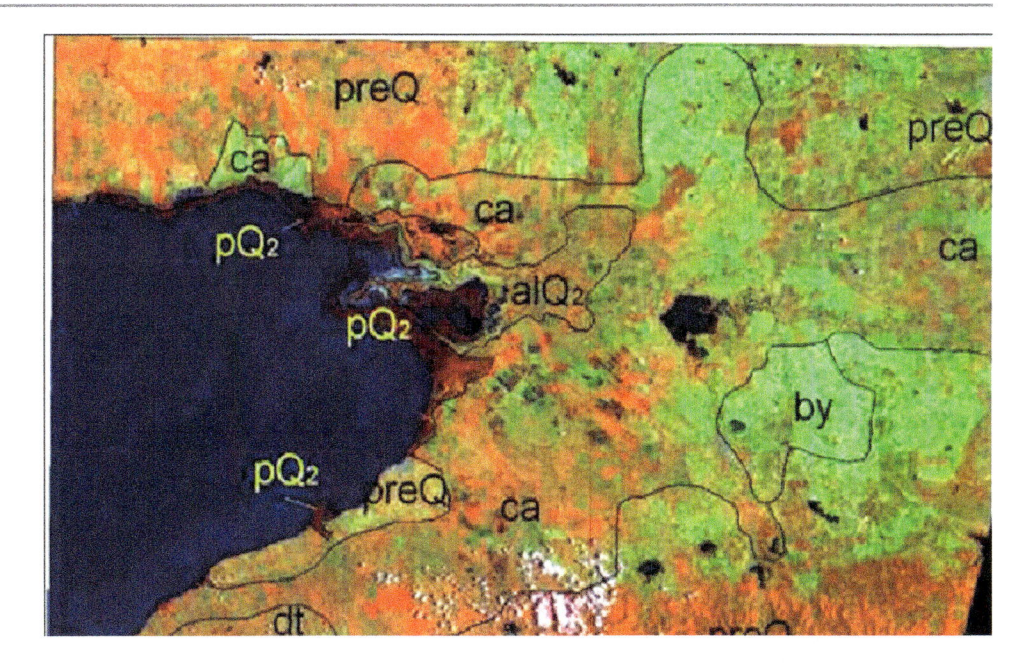

The thickness of the formation generally varies between 3 and 5 m, but sometimes it can reach up to 20–25 m. Overlays discordantly to different prequaternary formations, as well as to the Dátil and Bayamo formations. It is overlain by alluvial and marsh Holocene deposits, although in many places its upper limit is erosive.

Its distribution area is restricted exclusively to the Cauto river basin in Eastern Cuba.

On the basis of its stratigraphic position it is assigned an Early Upper Pleistocene age, presenting in its formation alluvial facies with fluvial-marine and marine influence closer to the mouth of the Cauto River.

In satellite images, this formation differs from those that surround it by presenting relatively clear tones (Fig. 7.5).

7.10 Late Upper Pleistocene

Cocodrilo Formation:

It consists of hard oolitic calcarenites of dark gray color, which in the coastal zone form bars with crossed laminar stratification. In addition to the oolites and pseudoolites, which constitute practically all of these deposits, fragments of corals and shells are sometimes located. The cementitious material is micritic and sparitic basal, sometimes very porous, with carbonate composition. The layers are millimetric and form packages of different thicknesses, oriented indistinctly. This orientation also varies in different localities, sometimes becoming less discernible.

As it advances towards the old coastal line, the calcarenites form a mantle, with successive stratification of thin, gently sloping layers, which generally dip in the direction of the current coastline.

The unit lies discordantly on the Jaimanitas Fm. in the regions of Isla de la Juventud and the western end of Pinar del Río, which are its distribution areas. Its upper limit is erosive although it is sometimes covered by storm ridges. It correlates with the El Salado, Siguanea, La Cabaña, and Playa Santa Fe formations.

In the satellite image shown in Fig. 7.6, the deposits of this formation are observed with a reddish brown color.

It has been assigned a Late Upper Pleistocene age. In radiometric determinations by C14 in deposits of a second terrace of the south of the Isla de la Juventud, an age of 28,000 years was reported, which seems to support this supposition.

The bars with layers of variable stratification in thickness, inclination, and direction seem to have accumulated in a coastal beach environment, under the action of a shallow sea of variable depth, with lateral currents of different directions, while the deposits with unidirectional layers they seem to be the product of a shallow sea of variable depth. The maximum thickness of this unit is 10 m, although the most common is 5 m.

La Cabaña Formation:

It is constituted by biocalcirudites of calcarenite marlish matrix, usually very cemented, formed mainly by nodules of

Fig. 7.6 Satellite image of the Isla de la Juventud. (pQ$_2$ – marshy deposits; mQ$_2$ – marine deposits; bQ$_2$ – biogenic deposits; gne – Guane Formation; gv – Guevara Formation; js – Jaimanitas Formation; ccl – Cocodrilo Formation; preQ – prequatenary rocks)

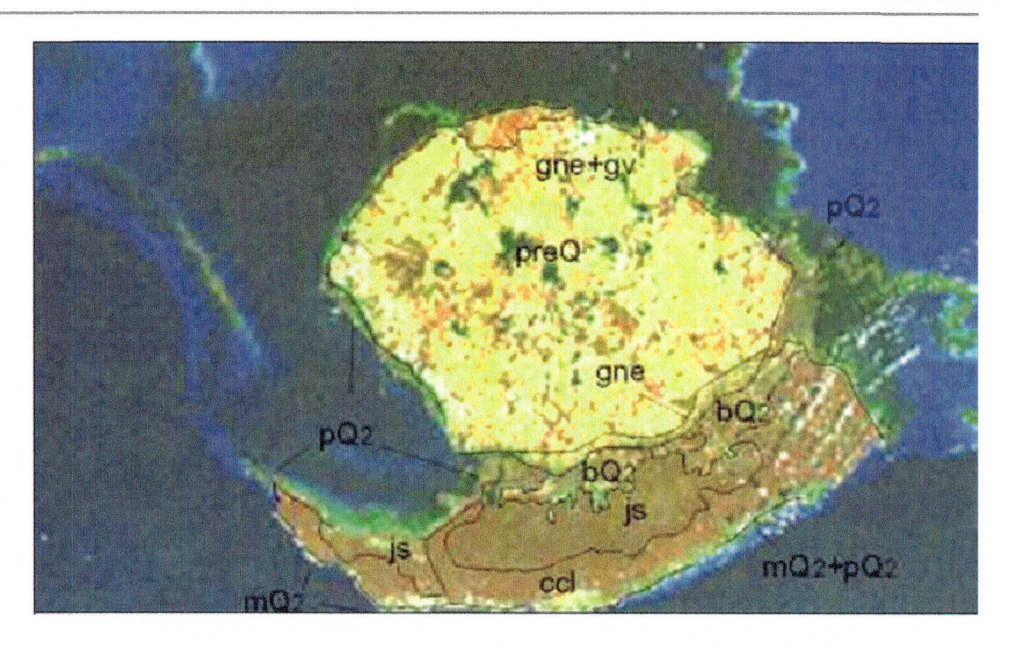

algae and fragments of mollusks and corals and sometimes biohermic limestones. Sometimes it breaks down, giving rise to an unctuous, powdery chalky material with the appearance of chalk. The calcirudites can happen to marlish calcarenites and sandy marls, as well as to limestones and conchiferous calcarenites less consolidated and with a weak filling of the chambers of the shells. The cement and the matrix are contact or filling.

It also includes lamellar biocalcarenites of lenticular and cross-stratification, with intercalations of polimictic gravelites (mainly volcanites), whose clastic material, well rounded, is likewise inside the calcarenites. They present white, whitish gray, and yellowish coloration. It has fossils with a wide range of age that includes the recent one. The foraminifera are abundant: *Amphisteginidae, Soritidae*; mollusks: *Argopecten gibbus, Bulla occidentalis, Chione paphia, Thachicardium muricatum, Trigonocardium medium*; corals: *Montastrea annularis, M. cavernosa, Porites porites, Siderastrea radians*.

This formation reaches a thickness of 10 m. It has been located in different parts of the coasts of Cuba and its keys. It lies discordantly on the Jaimanitas Fm. and its upper limit is erosive, although it is sometimes overlain by marine and marsh deposits.

This formation presents a high porosity, conditioned by the non-compact packing of the skeletal material and the weak filling of the chambers of different organisms. Only in some courts the conchiferous detritus, fragments of corals and bryozoans, foraminifera and other similar remains, preserve the metacoloidal and fibrous laminar structure formed by aragonite that they possessed in life. The rocks are cemented by pelitomorphic calcite or fine crystals.

The pink color of some varieties is conditioned by the presence of holomorphic clay substances colored with iron hydroxide. This substance forms incrustations in the walls of the pores or is in the form of globulites, disseminated in the rock.

On the banks of the Bacunayagua River in the deposits of this formation Rudnicki (in [39]) described a young and brief marine transgression, which started from a sea level lower than the current one and reached at least 10 m height above the current sea level. At the beginning of the transgression there were favorable conditions for the growth of the coral reefs, but during the development of the transgression there was a sudden increase in the terrigenous material transported by the Bacunayagua River, making the growth of the corals totally impossible.

Playa Santa Fe Formation:

It consists of fine to medium to medium-sized lamellar calcarenites with medium porosity and cross-stratification containing terrestrial mollusks. By alteration they are disaggregated into thin slabs. They present colors white and yellowish brown to yellowish gray. The eolianites of this formation are composed mainly of foraminifera testis and their fragments, conchiferous detritus, pellet-like grains, and fragments of calcite crystals, predominantly the foraminifera. This carbonated material has a great similarity with the material of the marine foraminifera calcarenites of the infrared Jaimanitas Fm.

Siliceous clastic material, including dark minerals, is also found in all eolianite samples. The amount of siliceous material varies in wide limits and can reach up to 30%.

The well-rounded siliceous and carbonated grains regularly have between 0.5 and 0.8 mm. The good selection of the material is characteristic; in each thin layer of eolianites carbonated and siliceous grains of similar size predominate.

The carbonated material of the Playa Santa Fe Fm. is represented by calcite magnesium and calcite.

The thickness of the formation does not exceed 10 m and is located in the form of dune fields on the north coast, east, and west of La Habana. It lies discordantly on deposits of the Jaimanitas Fm. through a paleo soil. Its upper limit is erosive. Contains Cerion shells. It was formed in conditions similar to those that formed the Guanabo Fm. Its correspondence to the Late Upper Pleistocene is undeniable when overcoming the Jaimanitas Formation.

El Salado Formation:

It consists of brownish red-carbonaceous clays, red clays, saturated with conchiferous detritus, fragments of corals, and fragments of the limestones of the formation. Underlying Jaimanitas and conchiferous limestones with a lateral transition to the La Cabaña Fm. At the eastern end of Cuba there are also red non-carbonated clays, clayey sands, and pebbles, with well-rounded quartz edges. Filling the karst cavities in the underlay limestones, the formation appears in the form of irregular elongated bodies, composed of hardened carbonated-clayey rocks of pink, red and reddish carmelite (brownish) colors, with marine and terrestrial mollusk shells and fragments of phosphatized bones.

In the cement, numerous pellets and ooids can be observed. The irregularly shaped part of the cement is composed of calcite and colofane. In the clastic material of the deposits carbonate rocks predominate, which are absent only in the Tortuguilla-Imías area, in variable quantities are quartz grains and fragments of ferruginous laterites, the clay material of the deposits is heterogeneous.

To the west of La Habana, in its type locality, the clay material colored intensely by red pigments enters the composition of the clay-carbonate cement of the deposits. The type of cement is contact or pore filling. Among the clay minerals, smectite predominates. In some samples, kaolinite-smectite appears. In other cuts appear impurities of chlorite and mica-smectite, more rarely chlorite-smectite. In the eastern end of Cuba, claystone minerals predominate interbedded kaolinite-smectite and smectite.

The existence of roentgenomorphic iron hydroxides conditions, in particular, the red color of the deposits of the formation. The thickness of the unit is 1–2 m. The appearance of these red sediments is related not to changes in the character of the sedimentation basins, but to an abrupt increase in the entrance of terrigenous material from the nearby land.

Cabrera and Peñalver [22] consider that the sedimentation environment corresponds to facies of low cumulative coasts, with intense alluvial contribution, developed in a stage of strong weathering and abrasion, in the presence of a hot and humid climate, probably similar to the current one.

The formation overlies discordantly the Jaimanitas and Río Maya formations. The upper cut is erosive or is covered discordantly by Holocene deposits. It can be correlated with the Cocodrilo, La Cabaña, Santa Fe, and Siguanea formations and also assigned to the Middle Wisconsin. It is located in different coastal portions of the country.

Siguanea Formation:

It is constituted by quartz sands of variable granulometry with intercalations of gravels of equal composition. Subordinately they contain fragments of hardpan and heavy minerals and not very potent intercalations of silty clays. Among the heavy minerals are distene, staurolite, tourmaline, ilmenite, rutile, leucoxene, zircon, and garnet. Presents whitish, yellowish, or gray colors, rarely reddish. The thickness does not exceed 10 m. Discordantly overlays diverse prequaternary formations, as well as the Guane and Guevara units. The upper limit is erosive.

The unit was included in the Late Upper Pleistocene, based on the existence of a low ridge, 6–7 m above sea level, formed by the waves that may have originated during the last transgression of the Pleistocene. It is correlated with the Cocodrilo, La Cabaña, Playa Santa Fe, and El Salado formations. It is only distributed on the Isla de la Juventud and on the western edge of Pinar del Río.

The lithological characteristics of the deposits of the formation, its close spatial relationship with the coastal plains, and the existence of a cumulative characteristic feature of the relief, such as the ridges originated by the above-mentioned waves, leave no doubt that the Siguanea Fm. has a marine origin. In many cases, their deposits approach directly to the sea. The current beaches that form on the western coast of the Isla de la Juventud arise at the expense of the re-laving of the sands of the Siguanea Fm. and the Holocene sands of these beaches do not lithologically differ from the oldest sands. In most cases, the formation is separated from the sea by a strip of mangroves. The materials that enter the composition of the sands have been derived from the weathering crusts of the neighboring metamorphic massifs.

7.11 Undifferentiated Quaternary

Elluvial-colluvial-prolluvial deposits:

Under this term are understood elluvial deposits developed "in situ", commonly associated with colluvial, and even prolluvial deposits, which originate at the expense of weathering crusts (eluvium). On the other hand, colluvial deposits formed by the detachment of large blocks, which accumulate in the foothills of the elevations, and which do not necessarily have to be associated with the weathering crusts, are also included in this category.

For this reason, this is a group of complex deposits in which the lithological characteristics vary greatly, as they pass quickly from clays, silts, and sands, to blocks of limestones, granodiorites, tuffs, etc. Something similar happens with the coloration, which is very variable.

In Cuba there are four types of rocks on which weathering crusts and their redepositions are developed. They are serpentinites, volcanic rocks, intrusive rocks, and metamorphic schists.

Some of the main characteristics of these deposits, in relation to the rocks that gave them origin, are offered below:

Serpentinites: Undoubtedly, the most important weathering crusts in Cuba are developed, not only because of the thickness they reach, but also because they are related to the ferronickel deposits widely distributed in the North of Eastern Cuba. A good example of this type of deposit is described in Pinares de Mayarí, where you can document the following profile from top to bottom:

1. Ochres "non-structural" of dark red color, with lenses of gravel, represented by ferruginous concretions. It has a thickness that reaches 10 m. Undoubtedly these are colluvial-prolluvial deposits formed at the expense of the eluvium itself.
2. "Structural" browns, reddish brown, light, friable, porous, cut by bodies of weak dams, composed of a greenish brown material and a network of veinlets and clay, whitish and greenish micro inclusions. They have a thickness that ranges between 5 and 10 m. They constitute true elluvial deposits, which conserve the structure of the parent rock.
3. Serpentinite dark green to black, in lighter places, with serpentine veins and bodies similar to dams, of medium composition, less frequently basic.

The serpentinites are composed of chrysotile, as well as smectite and magnetite, which are present in insignificant amounts. A mixture of chlorite, smectite chlorite, and serpentine has been determined in the veins.

Precisely the intertwining of chrysotile is the one that leads to the formation of structural ochres, composed of entanglements of iron hydroxides roentgenomorphic and goethite.

In some regions that have high humidity, the upper horizon, in addition to the dark red color, also has a yellowish red color and frequently contains a large number of ferruginous concretions, in them there are sand and pebble intercalations, represented by the same concretions and often cemented in a compact layer of ferruginous laterite. In addition to ocher, among these deposits are sandy clays, more compact, carmelite, and reddish brown.

The thickness of these deposits in the northeastern zone exceeds 20 m.

Volcanic and volcanic-sedimentary rocks: The elluvial-colluvial-prolluvial deposits developed on these rocks have a lower degree of weathering, when compared with serpentinites or granitoids.

For volcanic rocks the predominant sediments are fragments with some argillaceous matrix, which is conditioned by the hardness of the rock that gives rise to it. These deposits are scarce and in general have little alteration and outcrop.

In the case of volcanic-sedimentary rocks, the deposits are directly subordinated to the parent rock. For example, tuffs and tuffaceous sandstones give rise to medium and fine sandy deposits, with little content of clayey material, with thicknesses ranging from a few centimeters to a few meters. These deposits, in Central Cuba, are subordinated to the areas with low or little undulating relief, while in the Sierra Maestra the relief is of high and medium mountains.

The predominant colors of these deposits depend on the parent rock. They are black gray for porphyritic-basaltic lavas, dark-brown, which in depth turns cream-colored for xenolavas, reddish brown to brown-yellowish rocks for tectonized and even ocher-colored tuffs that deepen into Carmelite (brownish) gray, for greenish gray diabases. The thickness of this type of deposit rarely exceeds 10 m.

Intrusive rocks: They are represented by granodiorites, granosyenites, diorites, and other varieties and also develop elluvial-colluvial-prolluvial deposits. These are mainly represented by sands with granulometry from fine to coarse, depending on the degree of weathering of the rock and the degree of transport to which it was subjected.

Sometimes the destruction of granitoids gives rise to thicker sediments, such as gravels of different granulometry.

The composition of these sands is totally subordinated to the composition of the rock that gives rise to it and generally presents, to a greater or lesser degree, contents of clay material.

The colors vary from brownish yellow to grayish, sometimes gray-green, passing through yellowish beige, while the thickness is also variable and ranges from a few centimeters to tens of meters. They are located mainly in Central Cuba and partially in Eastern Cuba.

Metamorphic rocks: They include both the structural crusts and the redeposited ones according to Nikolaiev et al. 1981 (in Peñalver et al. [15]), which describe a profile located in the Sierra del Purial, Eastern Cuba, where the following are observed:

1. Non-structural clays, developed from metamorphic shales, uniformly dyed, yellowish brown, and reddish brown, with a thickness of 2.0–3.0 m.
2. Structural clays developed from metamorphic shales, uniformly dyed, of a yellow–brown color, with a thickness of 0.8–1.0 m.
3. Weathered metamorphic patches, stained with brownish yellow spots, with a thickness of 0.5–1.0 m.

In the Guamuhaya Massif, the Isla de la Juventud and the province of Pinar del Río, deposits of this type are also developed. In these regions the color that predominates is variegated, with abundance of reddish tones in the deposits.

In Central Cuba, possible elluvial-colluvial-prolluvial deposits have been recognized under the term "Arcillas Mayajigua", which are constituted by clayey rocks, gap-conglomerates with breccias, gravels and a bit of limestone and mixed carbonated clay rocks.

Two types of clays are differentiated: some are ocher and the others are gray to greenish.

Those of the first type are ocher yellow to reddish brown and reddish, the light green, gray and pinkish colorations are rarer occurring in the form of spots of different sizes, generally not related to stratification. Due to their texture, they are very thin, "greasy" pelitic. The structure is of non-clear stratification, predominantly massive. In general, they are homogeneous, soft and very clastic and are often distinguished from Fe–Mn pellets, decomposed to a different degree.

The clays of the second type are very thin, interspersed with CO_3Ca enriched with small gypsum crystals 7–8 mm in size. Decreases the amount of Fe pellets. These clays are very plastic.

The thickness of the elluvial-colluvial-prolluvial deposits, included in the Arcillas Mayajigua, is variable. The ocher and reddish clays with pellets exceed 40.0–50.0 m thick, the gray-green clays with plaster reach 80.0 m. The limestones fluctuate from 20.0 to 25.0 m.

As it has been verified, the elluvial-colluvial-prolluvial deposits are distributed in almost all the Cuban territory, although they are frequently not reflected in the geological maps at the average scales.

The age of these deposits is very variable. For many authors, their formation began from the Upper Oligocene and, consequently, they date as Neogene-Quaternary. Assuming this age, undoubtedly an important part of these deposits have been formed during the Quaternary and even continues to be formed today.

7.12 Holocene

The deposits of the Holocene, which form quite large areas of the insular territory and the marine platform of Cuba, are represented by a complex of deposits very varied by their genesis and their lithological peculiarities. Due to the reduced age range (10 thousand years), as well as the many facial variations they possess, it is not necessary to consider the facial, lithological, and genetic varieties of Holocene deposits as geological formations. In this sense an exception was made with the Los Pinos Fm., as will be explained later.

The Holocene in Cuba is represented by deposits of different lithological varieties, the main ones of which can be grouped into four genetic types, namely, marine, alluvial, marshy, and biogenic.

Marine deposits:

They are made up of sands of variable granulometry, gravels, pebbles, and even blocks distributed on beaches as well as on ridges of storms. The main varieties of the sands are carbonated. They are mainly of fine and medium grain, sometimes of coarse grain, as it happens in the south of Eastern Cuba. The grains are rounded or sub-rounded and are composed of calcareous algae, corals, mollusks, bryozoans, echinoids, foraminifera testes, ostracods, etc. Occasionally they may contain oolites and coprolites in varying proportions. The sands of the eastern beaches of La Habana and Matanzas are fine and medium-grained carbonates.

In addition to the carbonated sands there are polimictic sands. These minerals and fragments of local rocks are present in the most varied proportions (Pavlidis et al. 1976, in Ionic et al. [37]). The pebbles and boulders frequently with a perfect rounding are formed by magmatic and effusive rocks, and sometimes, limestones, which show a strong activity of the wave processes. A particular importance in Cuba are storm ridges, visible in the most varied coastal regions, where the accumulations of carbonated sands are observed, generally of medium grain, with an abundant quantity of blocks and corals belonging to the Jaimanitas Fm. They form elongated mounds several kilometers, with an average height of 3–5 m and an average width of about 40 m, moving away from the current coastline 20–50 m.

The marine deposits overlap discordantly to the Jaimanitas, Vedado, and Río Maya formations, as well as to another large group of prequaternary rocks. It has no upper stratigraphic limit. They constitute unconsolidated deposits, with little or no stratification. Morphologically they form different coastal types: plains, bars, dunes, beach ridges, etc., in cumulative or mixed coasts.

Normally they have thickness between 1.5 and 3 m, which can increase up to 8 m, according to drilling data.

Alluvial deposits:

They consist of gray silt and brownish gray, sandy silts, and sandy clays with rare intercalations of gravels and small pebbles.

The mineralogical characteristics of the Holocene alluvium are closely related to those of the power source. In the regions where there is no kaolinitic weathering bark, the alluvium has a polimictic composition, predominant in the clay material the dioctahedral smectite. In the regions of development of the weathering crusts, it is characterized by the oligomictic composition and in its clayey material are present the kaolinite, the kaolinite-smectite, hydronic dioctahedral, etc. In Eastern Cuba, blocks, boulders, gravels, silty sands, and clays derived from fluvial and regional erosion enter the composition of the deposits. In these cases, the material presents a different type of rounding. Nagy et al. [5] noted that sediments corresponding to relatively dry periods and those related to cyclonic disturbances are distinguished in the sequence.

These deposits overlap with all existing rocks in Cuba, prevailing over volcanic, intrusive, and sedimentary rocks. Its upper limit is erosive. They are present in all existing environments in Cuba.

Its general thickness is 2 m, although by drilling data it can reach up to 20 m. Its age is related to the Holocene, with the characteristic that they continue their formation.

Marshy deposits:

These sediments are formed in two essentially different environments and for this reason two facies can be distinguished: the mangrove swamps and the freshwater coastal swamps.

The mangrove swamps originate in the sea and their development represents a growth of the land at the expense of the sea. When bordering the coast, the coastline can only be traced conditionally. The composition of these swamps is dominated by plant residues and carbonated silts, since wave energy in the shallow coasts occupied by mangroves is insignificant and favorable conditions rarely arise for the entry of terrigenous material to its limits. However, among these deposits are also carbonate-clayey facies.

The most typical representatives of the terrigenous facies of the mangrove swamps are dark gray and dark brown clays heavily salinized and sandy clays with carbonized remains of trunks and mangrove roots. The main part of the deposits is represented by a dispersed thin, clayey substance with fluidal or cotton-like texture, formed by mica-smectite with packages of micas and smectites in different proportions, kaolinite-smectite and roentgenomorphic combinations. The dark color of the deposits is conditioned by the presence of organic material. The clay material contains numerous fragments of plant tissue, sometimes accumulations of angular quartz grains. Occasionally spicules of sponges formed by opals are found.

The carbonate facies of mangrove swamp deposits are represented by carbonated-organogenic silts with a variable amount of plant detritus. Here white silt-like silts are found, without vegetable detritus, gray silt with vegetable remains and carbonated peat. All these varieties contain terrestrial mollusk shells.

The carbonated material is represented by fine skeletal remains and a pelitomorphic carbonated cementitious mass.

The fine skeletal remains contain detritus of shells, ostracods shells and foraminifera testes, spines of urchins, etc. Also found is siliceous skeletal material, sponge spicules, etc. Sometimes the skeletal remains form almost 90% of the deposits, in other cases the pelitomorphic carbonates predominate.

All carbonated material from mangrove swamp deposits is represented by a mineral association of aragonite, calcite, magnesite, and calcite.

Coastal freshwater marshes originate in the land, in the discharge zones of groundwater, mainly karst, as a result of the containment of these waters by the sea. The development of these swamps does not change the boundary between land and sea, and the coastline, in the regions where it develops, has a clear configuration.

The deposits of coastal freshwater marshes do not penetrate the land more than a few kilometers and do not go beyond the limits of the period of periodic bogging, which originates in rainy seasons, at the same time, the entry of the sea into the land quickly can hardly facilitate the development of a broad strip of mangroves on shallow coasts and the prolonged accumulation of marsh deposits in the coastal strip of land.

Coastal freshwater marshes receive practically no terrigenous material and their deposits are mainly represented by vegetable waste and carbonated silt. The gray carbonated silts of these swamps are represented by two varieties.

One of them is represented by pelitomorphic carbonate material with a large quantity of terrestrial mollusk shells and freshwater and vegetal detritus with a different degree of conservation. In the structure of the other variety tabular threads of up to 0.2 mm long are observed formed by cryptocrystalline carbonate.

The carbonated material of both varieties of silt is represented by magnesite calcite.

The carbonate deposits of both types of marshes frequently contain a negligible amount of insoluble residue in 2% HCl (less than 1%). These are quartz-sized quartz grains and partially spicules of sponges and radiolarians.

The marshy deposits occupy the lowest position in the relief and are deposited in low coastal environments, inlets, cumulative coasts, river mouths, marshes, etc. They transgressively overlay the Jaimanitas Fm. as well as numerous prequaternary formations, both terrigenous and carbonate. Its upper limit is erosive.

The common thickness of these deposits reaches between 1 and 2 m, although some holes have been described up to 5 m.

Their age is Holocene and they continue to form today.

Biogenic Deposits:

They are constituted by semi-compost organic matter in different degrees, which can reach black consolidated peat. Sometimes carbonated peat or turbid carbonate of variegated color is observed, ranging from light gray to creamy white, plastic, with abundant shells and freshwater mollusk remains. Frequently they are of herbaceous origin, with predominance of the cortadera. There are also remains of jonquil, water arrow, lily, remnants of woody vegetation such as baga, júcaro, white mangrove, and guano prieto.

These deposits overlay different formations, among which we can mention Jaimanitas, Vedado, Río Maya, Güines, Paso Real, Bayamo, Cauto, Camacho, etc. Its upper limit is erosive. They accumulate in low places and inland, a few kilometers from the current coast, where conditions of excessive humidity and very little oxygenation predominate, which facilitate the decomposition of herbaceous and woody vegetation. They have a thickness of 1–3 m, although locally they can reach up to 8 m.

Its age is Holocene, oscillating between 2 and 10 thousand years. In a sample dated by a Dutch company, which worked in Cuba [40], an age of 10,700 years was obtained.

As is known, Cuba has an extensive current marine platform, covered almost entirely by Holocene deposits.

Some authors consider that silts and carbonate sands of biogenic origin (algae and peat, mainly) and chemogenic (oolitic) predominate in it. The terrigenous sediments are typical of the coastal areas and, therefore, there are no considerable amounts of heavy minerals in the sea, which is a consequence of several factors.

Los Pinos Formation:

As noted, it is the only exception that was made in the Holocene deposits when nominating a formation and is related to an area located in the Sabana-Camagüey Archipelago, which is part of the Cuban marine platform and can also be distinguished in some coasts of Cuba Central.

It consists of oolitic calcarenites and medium grain, white-cream-colored biocalcarenites; sometimes they present crossed lamination. The grains are rounded and well selected. It reaches thicknesses of up to 7 m. Overly discordant to the Jaimanitas Fm., being separated from this by a paleo soil. Sometimes it is covered by sands, which are the result of its own disintegration under the effect of weathering processes. It is also covered, in part, by turbid deposits.

The unit corresponds to a low cumulative coastal environment, represented by a complex of bars, probably formed during the Flandrian Transgression. It was separated as an independent unit due to its extension, its degree of lithification, and the clear definition of its morphological features, which have allowed its cartography in several of the keys of the aforementioned archipelago [4, 32].

7.13 Conclusions

- The stratigraphic division for the Cuban Quaternary proposed here fulfills all the requirements suggested by recognized researchers such as Flint 1965 (in Peñalver et al. [11]) when it affirmed that a flexible stratigraphic code is what allows the quaternary units to be based on in the two classical principles that divide the Quaternary: organic evolution and climate changes.

The scheme of stratigraphic division proposed, apart from dividing the Pleistocene from the paleoclimatic point of view in Humid and Dry, starting from the mineralogical characteristics that leave in the deposits the subaerial chemical weathering, also divides them from the chronological point of view, adjusting the Classical division of Lower, Middle, and Upper Pleistocene, as corresponds to each lithostratigraphic unit. In addition, on the basis of the new information, the Early Late Pleistocene has been separated from the Late.

Such considerations facilitate the process of geological cartography, as well as its correlation with other deposits both in the Cuban Archipelago and in neighboring territories.

- The idea of Kartashov et al. [26] when they considered the Guane, Dátil, and Guevara terrigenous formations as corresponding to the Humid Pleistocene, characterizing them because they have very well preserved profiles of kaolinitic weathering. Also with the Wet Pleistocene the Bayamo Fm., although this does not have well-developed weather profiles. This shows that the Cauto basin, its distribution area, was a much drier region than the rest of Cuba, and the humidity did not reach for the deposits to be completely transformed, and the primary smectites were converted into kaolinites and kaolinite-smectites. interstratified.

It should be noted that at present, the Cauto basin is still one of the driest regions of Cuba, and the premontane north edge of the Sierra Maestra, where the Dátil Fm., one of the rainiest.

In general, it can be stated that a large part of Western and Central Cuba had a higher humidity of the climate during the Humid Pleistocene, while in Eastern Cuba only the mountainous regions were more humid, including the northern slope of the Sierra Maestra.

During the Dry Pleistocene, those same zones of Western and Central Cuba, or a part of them, were characterized by a greater dryness of the climate, being observed that in the younger formations (Villarroja, Camacho, Siguanea, and El Salado), it was not developed a profile of kaolinitic weathering.

In Eastern Cuba and specifically in the Cauto basin, it is observed that the young deposits of the Upper Pleistocene that are distributed there (Camacho, Cauto, and Jamaica formations), present general characteristics that indicate a much greater dryness of the climate, when compared with their equivalent in age of Western and Central Cuba.

- Taking into account that the stratigraphic subdivision scheme for the Quaternary in Cuba previously analyzed was elaborated, mainly on the basis of lithological-geomorphological criteria and, therefore, with the use of relative dating methods, it is necessary to carry out specific investigations in some lithostratigraphic units, that allow to obtain data of more precise ages.

The following are indicated as the most perspective for the solution of this problem:

a. Investigations of the isotopic composition of oxygen in the sediments of the Vedado, Río Maya, and Versalles formations.
b. Paleomagnetic investigations of the sediments of the terrigenous formations Guane, Dátil, and Bayamo, as well as the carbonates Vedado, Río Maya, and Versalles.
c. Isotopic analysis by the 230 Th/234 U method of the marine terrace that constitutes the Jaimanitas Fm.

Especially interesting should be up to three levels of marine terraces linked to these rocks of Jaimanitas, raised to heights of up to 240 m, on the southeast coast of Cuba (in the region of Tortuguillas-Imias and possibly in the area of Cabo Cruz). These terraces could be the equivalent of Barbados III terraces (125,000 years BP), Barbados II (105,000 BP years), and Barbados I (82,000 BP), however, this may be speculative, since in the entire Caribbean levels Barbados II and I have only been found below sea level, forming underwater terraces (in Curacao, Jamaica, etc.). It should mean the extraordinary richness of corals in position of growth that was found at heights of up to 180 m above present sea level, in the profiles of the Tortuguilla-Imías region, in limestones that correlate with the Jaimanitas Fm.

It can be assumed that the corals that abound in the three levels of marine carbonate terraces, have largely escaped the effects of the diagenetic alteration in the phreatic subaerial environment, characterized by a low humidity of the climate, and therefore, retain even the aragonite as the prevailing mineral.

d. Palynological investigations in sediments of the Guevara, Villarroja, Camacho, Cauto, Jamaica, and Bayamo formations.
e. Continuation of the paleontological and micro-paleontological investigations of the sediments of the carbonate formations Vedado, Río Maya, Alegrías and Versalles, as well as the terrigenous Bayamo.

- The results presented here could serve as a comparative material with the studies that have been conducted on the Quaternary in neighboring countries of the Caribbean basin, or encourage those who do not yet have it to undertake such studies in a comprehensive and multilateral manner, at least in the stratigraphic aspect, which is the one that has been most abundant in this work.

Acknowledgements To Dr. Carlos Pérez Pérez, for his valuable contribution to the development of the research that has served as the basis of this chapter, by participating since the beginning of the present century in numerous field studies. Also to Geologist Technician Luisa Rodríguez González, who in recent years has supported the work of Quaternary specialists of the Instituto de Geología y Paleontología-Servicio Geológico de Cuba.

References

1. Mapa Geológico de Cuba a escala 1. 250000 (1989) Colectivo de autores. Editado en la URSS
2. Mapa Geológico de Cuba a escala 1: 500000 (1986) Colectivo de autores. Editado en la URSS

3. Peñalver LL (1989) Mapa de los depósitos Cuaternarios de Cuba a escala 1: 2000 000. En: Nuevo Atlas Nacional de Cuba. Editado en España
4. Iturralde Vinent MA (1981) Geología del Cuaternario, en: (Bermustakov I et al): Resultado de las investigaciones y levantamiento geológico a escala 1: 250 000 del territorio Camagüey-Las Tunas. Inst. Geol. Paleont. La Habana. (Inédito)
5. Nagy E, Bresznyansky K, Brito A, Coutín DP, Formell F, GL Franco, Gyarmaty P, Jakus P y Radocz Gy (1976) Texto explicativo al Mapa Geológico a escala 1:250 000 de la provincia de Oriente. Inédito Archivo del IGP. La Habana
6. Peñalver LL (1982) Correlación litoestratigráfica entre los depósitos cuaternarios de la plataforma noroccidental de Pinar del Río y las zonas emergidas próximas. Rev. Cien. Tierra y Espacio. No.5, pp 63–83. La Habana
7. Peñalver LL (2005) Sobre la existencia de la Formación La Cabaña en el tramo costero Canasí-Este de Bacunayagua. Memorias VI Congreso Cubano Geología y Minería
8. Peñalver LL, Oro JR y Barriento A (1982a) Las secuencias terrígenas del Plioceno Superior-Pleistoceno Húmedo del Occidente de Cuba. Rev. Cien. Tierra y Espacio. No.5, pp 43–61. La Habana
9. Peñalver LL, Oro JR y Barriento A (1982b) Las secuencias carbonatadas del Plioceno Superior-Pleistoceno Húmedo de Cuba Occidental. Rev. Cs. de la Tierra y Espacio. No. 5, pp 25–41. La Habana
10. Peñalver LL, Barriento A, Orbera L, Hernández CE, Estrada V, Méndez A, Pérez Lazo J y Fundora M (1997) Mapa de depósitos Cuaternarios de Cuba y su plataforma marina a escala 1: 500000. (Inédito) Arch. Inst. Geol. Paleont. La Habana
11. Peñalver LL, Cabrera M, Trujillo H, Pérez Lazo J, Fundora M, Molerio L, Morales H, Guerra M y Pedroso I (2001) Informe final sobre los cambios climáticos ocurridos en Cuba durante el Cuaternario sobre bases geológicas. (Inédito). Arch. Inst. Geol. y Paleont. La Habana
12. Peñalver LL, Alonso JA, Rodríguez A, Pérez Pupo R y Cruz Dávila F (2003) Sobre la existencia de la Formación Camacho en la llanura sur de Pinar del Río. Memorias V Congreso Cubano Geología y Minería.La Habana
13. Peñalver LL, Cabrera M, Cabrera M, Fundora M, Pérez Lazo J y Pedroso I (2005) La datación de los depósitos Cuaternarios en Cuba. Memorias VI Congreso Cubano Geol. y Minería. La Habana
14. Peñalver LL, Pérez C y Cabrera M (2007) Cambios climáticos en Cuba durante el Cuaternario y sus efectos en la paleogeografía. Memorias VII Congreso de Geología
15. Peñalver LL, Cabrera M y Pérez C (2009) Mapa digital de los depósitos cuaternarios de Cuba a escala 1:250000. Memorias III Convención Ciencias de la Tierra. La Habana
16. Peñalver LL, Pérez C, Denis R., Martín D y Rodríguez L (2011) La transgresión del Wisconsin Medio en Cuba. Memorias IX Congreso de Geología. La Habana
17. Peñalver LL, Denis R y Rodríguez L (2012) Nuevos datos sobre las terrazas marinas de Cabo Cruz, Granma, Cuba. Memorias IX Congreso Ciencias del Mar. La Habana
18. Peñalver LL, Denis R, Núñez A, Pérez C, Martín D y Rodríguez L (2013) Nuevos datos sobre los camellones de tormenta en Cuba. Memorias X Congreso de Geología. La Habana
19. Peñalver LL, Denis R, Cabrera M, Núñez A, Triff J, Pantaleón G, Rodríguez L y Martín D (2015) Caracterización geólogo-geomorfológica del Área Protegida Guanahacabibes. Memorias VI Convención Cubana de Ciencias de la Tierra - XI Congreso de Geología. La Habana
20. Peñalver LL, Denis R y Martín D (2016) Paleogeografía del Holoceno en Cuba Occidental y sus formas del relieve más características. Memorias Trópico 2016. La Habana
21. Peñalver LL, Díaz M y Denis R (2017) Precisiones de las edades de los depósitos del Wisconsin Medio en Cuba mediante dataciones radiométricas. Memorias 7ma Convención de Ciencias de la Tierra. La Habana
22. Cabrera M y Peñalver LL (2001) Contribución a la estratigrafía de los depósitos Cuaternarios de Cuba. Revista Cuaternario y Geomorfología Vol. 15 (3–4) p 37–49. Madrid
23. Cabrera M y Peñalver LL (2003) Contribución a la estratigrafía de la Formación Jaimanitas y su relación estratigráfica con las demás formaciones del Pleistoceno Superior. Memorias V Congreso Cubano Geología y Minería. La Habana
24. Cabrera M, Peñalver LL, Pérez CM, Ugalde C, Barriento A, Guerra R, Chávez ME, Nápoles E, De Huelbes J, Bouza O, Rivas L, Magaz A, Sosa M, Zúñiga A y Orbera L (1997) Estudio geológico-ambiental del Ecosistema Sabana-Camagüey. Manuscrito (Inédito). Inst. Geol. Paleont. La Habana, p 123
25. Léxico Estratigráfico de Cuba (2000) Inst. Geol. Paleont. Colectivo de autores. La Habana. (Inédito)
26. Kartashov IP, Cherniajovski AG y Peñalver LL (1981) El Cuaternario en Cuba. Edit. Nauka, Vol. 356, p 145, Moscú
27. De la Torre A y Kojumdgieva E (1985) Estudios sobre los moluscos del Neógeno-Cuaternario del Occidente de Cuba y sus implicaciones estratigráficas. Inst. Geol. Paleont. Acad. Cs. Cuba. Reporte de Investigación 5., pp 1–19
28. Krasnov II and Nikiforova KV (1975) Stratigraphic scheme of the Quaternary based on the materials of the last years. Roy. Soc. New. Zeal., p 58
29. Kartashov IP, Mayo MA, Cherniajovski AG y Peñalver LL (1976) Descripción de algunas formaciones geológicas del Sistema Cuaternario de Cuba, reconocidas recientemente. Inst. Geol. Paleont. Acad. Cien. Cuba. La Habana. Serie Geológica. N. 26, p. 1-6
30. Pszczolkowski A, Piotrowska K, Myczynski R, Piotrowski J, Skupinski J, Grodziccki A, Danilewski D y Haczewski G (1975) Texto explicativo al mapa geológico a escala 1: 250 000 de la provincia de Pinar del Río. Brigada Cubano-Polaca, Inst. Geol. Paleont. Minis. Ind. Básicas. La Habana. (Inédito)
31. Léxico Estratigráfico de Cuba (1991) Inst. Geol. Paleont. Colectivo de autores. La Habana (Inédito)
32. Iturralde Vinent MA y Cabrera M (1998) Estratigrafía de los cayos del Archipiélago Sabana-Camagüey entre Ciego de Ávila y Las Tunas. Memorias III Congreso Cubano de Geología y Minería. La Habana, pp 319–322
33. Dzulynski S, Pszczółkowski A y Rudnicki J (1984) Observaciones sobre la génesis de algunos sedimentos terrígenos cuaternarios del occidente de Cuba. Rev. Ciencias de la Tierra y del Espacio, No. 9, pp 75–90
34. Brödermann J (1940) Contribución al mapa geológico de la provincia de La Habana. ONRM. Min. Ind. Básicas. La Habana (Inédito)
35. Ducloz Ch (1963) Etude geomorphologique de la región de Matanzas, Cuba (avec une contribution a l'̈etude des deposit quaternaire de la zone Habana-Matanzas. Arch. Sci. Soc. Phys. Hist. Nat. Geneve, Paris 16:351–402
36. Shantzer EV, Petrov OM y Franco GL (1976) Sobre las formaciones costeras del Holoceno en Cuba, las terrazas pleistocénicas de la región Habana-Matanzas y los sedimentos vinculados con ellas. Inst. Geol. Paleont. Acad. Cs Cuba. Serie Geología. No. 2, pp 1–26. La Habana

37. Ionin AC, Pavlidis YA y Abello O (1977) Geología de la plataforma marina de Cuba. Edit. Nauka, Moscú, p 207
38. Bronnimann PD and Rigassi S (1963) Contribution to the geology and paleontology of the area of La Habana, Cuba, and its surrounding Eclogae Geol. Helvetia, 56 (1): p.193–480, 26 lams.
39. Piotrowska K, Pszczolkowski A, Piotrowski J, Rudnicki J, Myczynski R, Kuzniarski M, Peñalver LL, Franco GL y Pérez N. (1981) Texto explicativo del Mapa Geológico de la provincia de matanzas, a escala 1: 250 000. (Inédito). Arch. Inst. Geología y Paleont. La Habana
40. Wedman EJ and de Albear JF (1959) Reclamation of Cienaga de Zapata. Cuba. Proyecto de saneamiento y utilización de las zonas pantanosas. NEDECO. La Haya. Holanda. (Inédito)

Mineral Systems of Cuba: A Panoramic Vision of Cuban Metallogeny

Jorge Luis Torres Zafra and Xiomara Cazañas Díaz

Abstract

As a result of the preparation of the Metallogenic Map at scale 1:250,000 of the Republic of Cuba, forty-one models and eight sub-models of metallic mineral deposits were identified. These models, of descriptive—genetic type, together with the analysis of their spatial distribution and their relationship with the geology, allowed the identification and mapping of ten mineral systems, linked to the geodynamic environments present in the Cuban territory. The present work aims to provide the reader with an up-to-date image of Cuba's metallogeny, with the incorporation of new results obtained recently, which is presented through the description of each of the ten mineral systems identified. Special attention is given to the mafic-siliciclastic VMS-SEDEX-MVT mineral systems of the J-K_1 distensive basin, which covers the Guaniguanico and the Cangre structural-tectonic units, Pinos and Guamuhaya terranes, and the supergene environment, this represented in a large part of the Cuban territory. The first, hosts an important pyritic and polymetallic sulfide mineralization of Cu, Zn, Pb, Co, Au, and Ag, while the second contains the known Cuban lateritic deposits of Fe–Ni–Co, which also contain Au, PGE, Sc, and others metals. The text addresses some deposits that deserve attention due to their singularities, such as Matahambre, Lela, and El Cobre. The scattered and little abundant data on fluid inclusions and isotope geochemistry in metallic mineral deposits of Cuba have also been collected here. Finally, some general inferences about Cuba's metallogeny are presented and several aspects that require new works are pointed out.

Keywords

Cuba • Metallogeny • Mineral systems • Ore deposits • Matahambre • Lela • El cobre

---

J. L. T. Zafra (✉) · X. C. Díaz
Instituto de Geología y Paleontología-Servicio Geológico de Cuba, Vía Blanca No. 1002 y Carretera Central, San Miguel del Padrón, CP 11 000 La Habana, Cuba
e-mail: zafra@igp.minem.cu

X. C. Díaz
e-mail: xiomara@igp.minem.cu

© Springer Nature Switzerland AG 2021
M. E. Pardo Echarte (ed.), *Geology of Cuba*, Regional Geology Reviews,
https://doi.org/10.1007/978-3-030-67798-5_8

Abbreviations

CDT	Canyon Diablo troilite
CDM	Canyon Diablo meteorite troilite
Fm	Formation
IGP	Instituto de Geología y Paleontología (Institute of Geology and Paleontology)
IOCG	Iron Oxide Cupper Gold
IVA	Insular Volcanic Arc
MVT	Mississippi Valley Type
N-MORB	Normal-type mid-ocean ridge basalts
PGE	Platinum Group Elements
REE	Rare Earth Elements
SEDEX	Sedimentary Exhalative
SEM-EDS	Scanning Electron Microscopy/Energy Dispersive X-Ray Spectroscopy
TOC	Total Organic Carbon
TSU	Tectonic Stratigraphic Unit
VMS	Volcanogenic Massive Sulfide

8.1 Introduction

The Cuban territory is characterized by a varied and complex geology, in which several geodynamic environments are present. They are distensive continental margins, insular volcanic arcs, oceanic lithosphere (ophiolites), collision (orogenic), and intraplate. The presence of diverse depositional environments within the aforementioned geodynamic contexts contributes to the existence of a variety of types of metallic mineral deposits wider than what could be expected at first sight. Thus, in the continental margins there are siliciclastic and carbonated sequences; in the insular volcanic arcs there are sequences belonging to the tholeiitic, calcoalkaline, and alkaline series, as well as representative of the different sections of the volcanic arc (back arc and, to a lesser extent, fore arc). Within the ophiolites, the sequences belonging to the different levels of the ophiolitic section are distinguished, while the orogenic environment includes subvolcanic intrusive bodies—from mafic to felsic—and felsic intrusives, generated by partial melting of the continental crust. In the intraplate context appear the depositional environments related to the processes of sedimentation—chemical and mechanical—and those of supergene weathering.

8.2 Materials and Methods

For the preparation of this chapter, we started with the results of the Metallogenic Map at scale 1:250,000 of the Republic of Cuba [29]. This map contains 1181 metallic mineral occurrences, belonging to a total of 41 models and 8 sub-models of mineral deposits, georeferenced on a schematic structural-tectonic base that shows the folded basement (folded belt) of Cuba. These occurrences were also classified in deposits, prospects, and shows. According to the models of Law—Tonnage used, 156 deposits of the first two categories are classified, according to their size, into 17 large, 58 mediums, and 81 small.

The models of mineral deposits elaborated for the aforementioned metallogenic map are of descriptive—genetic type, which makes them more complete and detailed than those previously used, as well as more in keeping with the state of the art in the world on the subject [136]. The preparation of these models was made on the basis of the generalization and synthesis of the contributions and opinions published in international specialized literature by various authors. In this way it was tried to achieve that the models thus made were, without being excessively long and detailed, as universal and representative as possible, on one hand and, on the other, a reflection of the current state of the art on the subject.

In turn, within the models, when necessary, subtypes (sub-models) were distinguished, which differs from the main models to which they belong in a certain number of characteristics, which, however, are insufficient to make them independent separate models. An example of this is the deposit sub-model of alkaline low sulphidation epithermal Au–Ag, present within the main type of low sulphidation epithermal Au–Ag [136]. This type of model allows, in turn, a better identification and understanding of the mineral systems present in the different geodynamic environments and in the different regions of Cuba. For the identification of these systems, the concept of the mineral system was used.

The concept mineral system, originally enunciated by Wyborn et al. [142], can be described by Torres Zafra et al. [135] as "a set of spatially and temporally related geological processes and elements that lead to the formation of a mineral deposit or group of them". Its main parts are metal source, migration pathways, metal and other compound transport fluids, ore deposition site, and energy source (heat) of the system. Other components to consider are the source of binders and the discharge zone. A factor that, although not part of the mineral system, also influences the presence of mineral deposits in a given area is the degree of preservation, which in turn depends on tectonic, climatic, and geographic factors.

In addition to the metallogenic map and its associated books: the explanatory memory of metallogenic map [29] and descriptive—genetic model book [136], for the preparation of this chapter, data and unpublished results of investigations concluded after the publication of the aforementioned metallogenic map, and others currently in progress, as well as others that appeared in various publications.

In addition to the metallogeny of the different parts of the Cuban territory, a vision is offered about the particularities and problems of metallogeny in Cuba through the description of some mineral deposits or districts that deserve special attention because of their importance.

8.3 Results

In the Cuban territory, a total of 41 models and 8 descriptive—genetic sub-models of metallic deposits are present, identified with different degrees of certainty. To this is added the identification of another 9 models and 2 sub-models of deposits that could be represented in the national territory, although, until now, there are no known deposits representative of them Table 8.1. The genetic part of these models, together with the analysis of the spatial–temporal interrelation between the types of deposits located in the different geodynamic environments and tectonic-structural units of Cuba, allows to recognize the presence in the country of 10 mineral systems. By geodynamic environments, they are Fig. 8.1.

Environment of distensive continental margin:

1. Mafic-siliciclastic VMS-SEDEX-MVT of the distensive basin J-K$_1$ mineral system.
   Environment of insular volcanic arcs (IVA) of the Cretaceous:
2. Bimodal-mafic Ag–Au veins in volcanic sequences of K$_1$ mineral system.
3. Cu–Mo ± Au porphyry—skarn—veins of Ag–Au in volcanic sequences of K$_2$ mineral system.

4. Cu–Mo ± Au porphyry—skarn—Au–Ag epithermal K$_2$ mineral system.
   Collision environment:
5. Campanian—Maastrichtian orogenic mineral system.
6. Paleocene—Middle Eocene orogenic mineral system.
   Paleogene island volcanic arc environment:
7. VMS bimodal-mafic—veins of Ag–Au in volcanic sequences—Mn volcanogenic mineral system.
8. Cu–Mo ± Au porphyry—skarn—veins of Ag–Au in volcanic sequences mineral system.
   Intraplate environment:
9. Supergenic mineral system.
10. Sedimentary mineral system (placers of Au and Cr).

To these mineral systems belong the main deposits of Cu, Zn–Pb, Au–Ag, Mn, and W known in Cuba so far. In Mesozoic mafic–ultramafic sequences, representative of the oceanic crust environment, deposits of podiform chromitites and massive volcanogenic sulfides of the mafic–ultramafic type are present. However, although they are located in the same geodynamic environment, both are "loose models" that do not constitute a mineral system.

A panoramic view of the metallogeny of Cuba is given in Table 8.1, which shows the types of deposits known and possible to be found in the different geodynamic environments and tectonic units present in the Cuban territory.

8.4 Discussion

8.4.1 Continental Margins. The Mafic-Siliciclastic VMS–SEDEX–MVT of the Distensive Basin J-K$_1$ Mineral System

The continental margin sequences, generally rich in organic matter, are well represented in the Cuban basement Fig. 8.2. These sequences belong to the periphery of two different blocks of the North American Plate: the Maya Block and the Florida-Bahamas Block. To the first, belong the Guaniguanico Tectonic-Stratigraphic Unit (TSU), the Pinos and Guamuhaya (Escambray) terranes, as well as the discussed Arroyo Cangre structural-tectonic unit [29] or Arroyo Cangre Terrane [66]. The Cocos, Remedios, Camajuaní, and Placetas tectonic-stratigraphic units (TSUs) are part of the second. The ownership of the small Asunción structural-tectonic unit, located in the eastern end of Cuba, is the subject of controversy, because while Iturralde Vinent [66] considers it part of the terranes of Cuba SW and, therefore, part of the Maya Block, Cazañas et al. [29] consider it part of the Florida-Bahamas Block.

Table 8.1 Distribution of metallic mineral deposits models in Cuba for geodynamic environments and tectonic units

Type of mineral deposit	Subtype of mineral deposit	Metallic association	Examples of cuban deposits	Tectonic unit
Volcanogenic Mn Cuban type		Mn	Polaris, La Gloria	Back arc basin IVA Sierra Maestra–Cresta Caimán
Bimodal mafic VMS		Cu, Zn (Pb, Au, Ag, Cd)	There are no examples	
Bimodal mafic VMS		Cu, Zn (Pb, Au, Ag, Cd)	El Cobre, San Fernando, Antonio	Insular volcanic arc (IVA) Sierra Maestra—Cresta Cayman and IVA of the Hauterivian—Albian
Felsic VMS	Bimodal félsico sub-type	Cu, Zn, Pb (Au, Ag, Cd)	Infierno*	
Fe skarn		Fe	La Grande, Chiquita	
Fe–Cu skarn		Fe, Cu (Au)	Antoñica, Arroyo de La Poza	
Cu skarn		Cu, Au (Ag, Zn)	Marea del Portillo	
Ag–Au veins in volcanic sequences		Ag, Au, Pb, Zn (Cu, Mn)	Veta Rey, San Miguel	
Volcanogenic Mn Cuban type		Mn	Los Chivos, La Margarita, Barrancas—Ponupo IV—Sultana	
Barite veins		Ba	El Cedrón	
Cu–Mo ± Au porphyric		Cu (Mo, Au, Ag)	Buey Cabón	
C–Mo ± Au porphyric		Cu (Mo, Au, Ag)	Cobre Arimao, Guáimaro	Insular volcanic arc (IVA) of Middle Albian–Campanian
Cu–Mo ± Au porphyric	Cu–Au porphyric alkaline sub-type	Cu, Au	There are no examples	
Au–Ag epithermal high sulphidation		Au, Ag, Cu (As)	Golden Hill	
Au–Ag epithermal low sulphidation		Au, Ag	Florencia, Maclama	
Au–Ag epithermal low sulphidation	Au–Ag epithermal low sulphidation alkaline sub-type	Au, Ag	Jacinto	
Fe skarn		Fe	Maragabomba, Loma Alta	
Fe–Cu skarn		Fe, Cu (Au, Zn)	Guaos (Francisco Mayo)	
Cu skarn		Cu, Au (Ag, Zn)	Cañada Honda, Arimao Norte	
Au skarn		Au	Abucha, Jobabo I	
Th-REE veins		Th, REE, Cu, Au, Ag	Embarque, Tres Antenas	
Fe–P-REE Oxide Kiruna Type**		Fe (REE)	Palo Seco I	
volcanogenic Mn Cuban type		Mn	Naranjo	
Orogenic Au Mother Lode Type		Au, Ag (Cu)	Reina Victoria, Nuevo Potosí, Descanso	Collision Mesozoic Ophiolitic Association, Cretacic insular volcanic arcs Terrane
Orogenic Cu–Zn–Au–Ag		Cu, Zn (Au, Ag, Ti)	Santa María, Charco Prieto	
Orogenic Au hosted in siliciclastic sequences		Au, Ag (Sb)	Delita	Collision Pinos Terrane, Guamuhaya Terrane and Guaniguanico Tectonic-Stratigraphic Unit
Veins of five elements Ni–Co–As–Ag ± Bi, U		Ag, Ni, Co (Zn)	Loma del Viento	
Orogenic Au-EGP in black slates		Au (PGE)	There are no examples	
W veins		W (Mo)	Lela	Linked to felsic intrusive of collision. Pinos Terrane
Mo porphyric		Mo (W)	Lela	
Mo skarn		Mo (W, Cu, Pb, Zn, Sn, Bi, U, Au)	There are no examples	
Sn skarn		Sn (W, Zn, Fe)	There are no examples	
W skarn		W (Mo, Cu)	There are no examples	
Gold veins related to reduced intrusives		Au (Sb)	There are no examples	
Oceanic volcanogenic Mn		Mn	La Ligera	Oceanic lithosphere and supra-subduction Mesozoic Ophiolitic Association
Podiform chromitites		Cr (Au, PGE)	Merceditas, Camagüey I-II, Caledonia, Casimba	
Ultramafic mafic VMS		Cu (Au, Ag)	Júcaro, Buenavista, Río Negro	
Orogenic Cu–Ni ± Au, Co		Ni, Cu (Au, Co, PGE)	La Cruzada, Salomón	

(continued)

Table 8.1 (continued)

Type of mineral deposit	Subtype of mineral deposit	Metallic association	Examples of cuban deposits	Tectonic unit
Siliciclastic VMS mafic		Cu, Co (Au, Ag)	Unión-Juan Manuel, Hierro Mantua	Continental margin Guaniguanico Tectonic–Stratigraphic Unit, Pinos and Guamuhaya terranes
Zn–Pb–Ag exhalative sedimentary	Selwyn sub-type	Zn, Pb, Ag (Cu, Au, Ba)	Santa Lucía, Castellanos, Nieves	
Cu–Ag–Co-hosted in sediments		Cu (Ag)	Matahambre	
Stratiform barite		Ba	El Indio, Santa Gertrudis	
Zn–Pb MVT	Irish sub-type	Cu, Zn (Co, Ag, Au)	Carlota, Guachinango	Continental margin Pinos and Guamuhaya terranes; Guaniguanico Tectonic–Stratigraphic Unit
Pb–Zn–Ag polymetallic veins in siliciclastic sequences		Pb, Zn (Cu, Ag)	Lola	
Stratiform Mn		Mn	Mesa I y II	
sideritic Fe		Fe	La Ceiba	
Sedimentary Al		Al	Sierra Azul—Pan de Guajaibón, Manga Larga	
Polymetallic black shales	Black shales rich in Cu–Zn–Ni–Co ± EGP sub-type	Cu, Zn, Ni, Co (PGE)	There are no examples	
Broken Hill Type Pb–Zn–Ag ± Cu		Cu, Pb (Au, Ag)	Isabel Rosa	Continental margin. Arroyo Cangre Tectonic–Stratigraphic Unit
lateritic supergenic Fe–Ni–Co		Fe, Ni, Co (Au, Cr, Mn, Sc, PGE)	Yagrumaje Oeste, Pinares de Mayarí	Intraplate basins Eocene—quaternary cover
lateritic supergenic Fe–Ni–Co	Lateritic–pisolitic sub-type	Fe, Ni, Co (Au, Cr, Mn, Sc, PGE)	Luz Sur	
Fe–Ni–Co lateritic–saprolitic supergene	Lateritic–saprolitic serpentinitic sub-type	Fe, Ni, Co (Au, Cr, Mn, Sc, PGE)	Punta Gorda, Martí, Piloto	
Fe–Ni–Co lateritic–saprolitic supergene	Lateritic–saprolitic-clayey sub-type	Fe, Ni, Co (Au, Cr, Mn, Sc, PGE)	San Felipe	
Sedimentary Fe–Ni–Co	Goethitic sedimentary outcropping sub-type	Fe, Ni, Co (Au, Cr, Mn, Ti, Sc, PGE)	Levisa Norte, Playa La Vaca	
Sedimentary Fe–Ni–Co	Chamositic sedimentary confined sub-type	Fe, Ni, Co (Au, Cr, Mn, Ti, Sc, PGE)	There are no examples	
Supergenic REE		REE (Au)	There are no examples	
Supergenic Al		Al	Cayo Guam, Quemado del Negro	
Au–Ag gossan		Au, Ag	Loma de Hierro	
Supergenic Fe (elluvial-delluvial iron buckshots)		Fe	San Francisco	
Au placer		Au	Río Yabazón, Río Naranjo, Placer Maclama	
Cr placer		Cr	Placer Tau, Toa	

*This deposit contains more Pb than the Cuban mafic bimodal VMS, but less than that of a typical felsic VMS
**It is considered identified with a low degree of certainty
Notes The models and submodels without examples of Cuban deposits are those that may exist, they have not been found until now.
The metallic association is the one observed in Cuban deposits. In cases without examples, it is provided by the model or sub-model

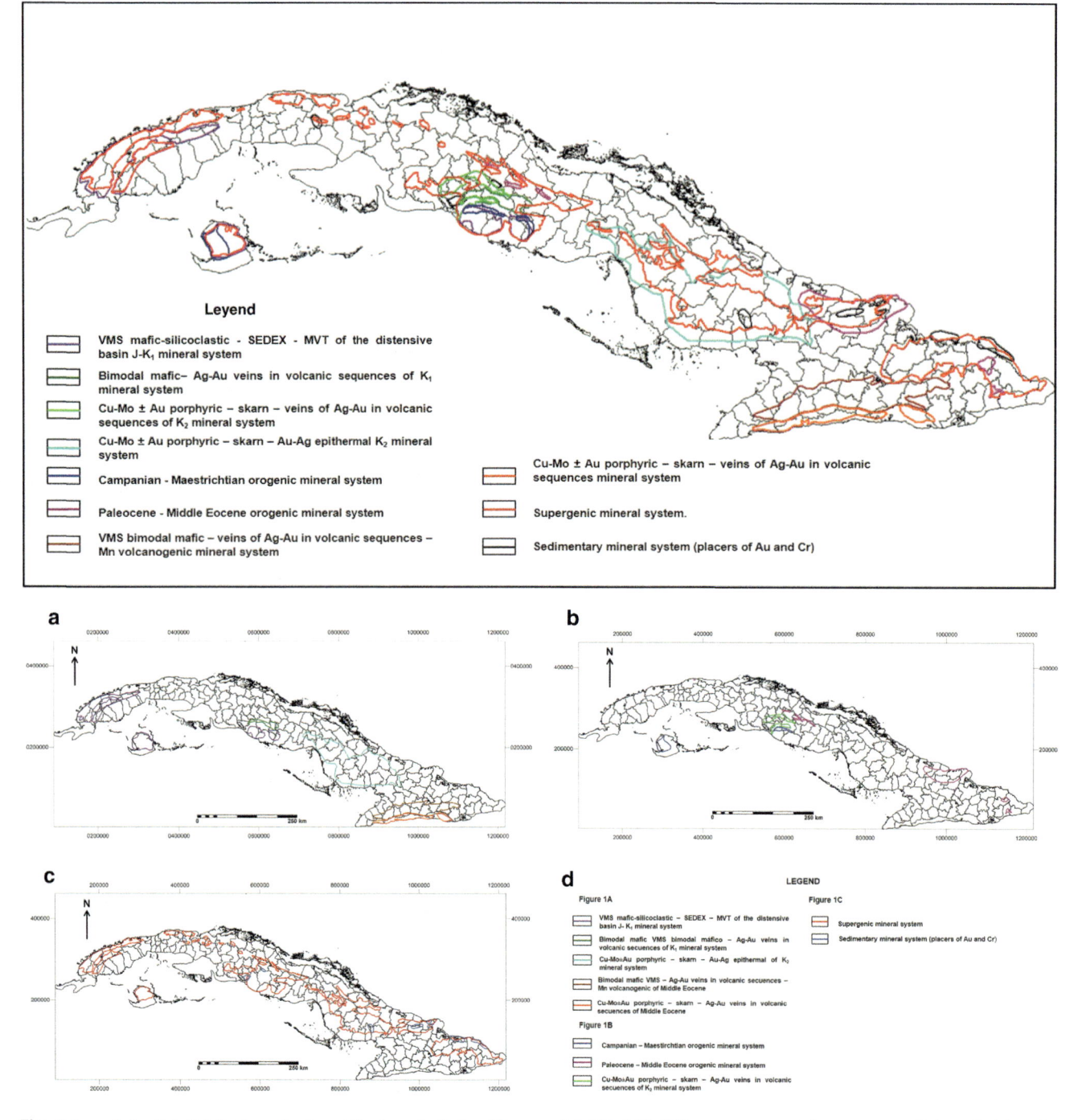

Fig. 8.1 a–d Spatial distribution of mineral Systems in Cuba. Maps and legend. The contours of the ten identified mineral systems, extracted from the Metallogenic Map of the Republic of Cuba at scale of 1:250 000 [29], are represented on a map of the administrative (municipal) political division of Cuba, shown in three separate maps to avoid overlapping and provide better visualization of them

The sequences belonging to the Florida-Bahamas Block are located in the northern part of the country, from La Habana province to Holguín Fig. 8.2, extending under the waters of the Bahamas Old Channel, the Florida Straits and, the easternmost part of the Mexico Gulf. They include important lithostratigraphic sequences that generate oil and gas, as well as the main deposits known in Cuba of this non-renewable natural resource.

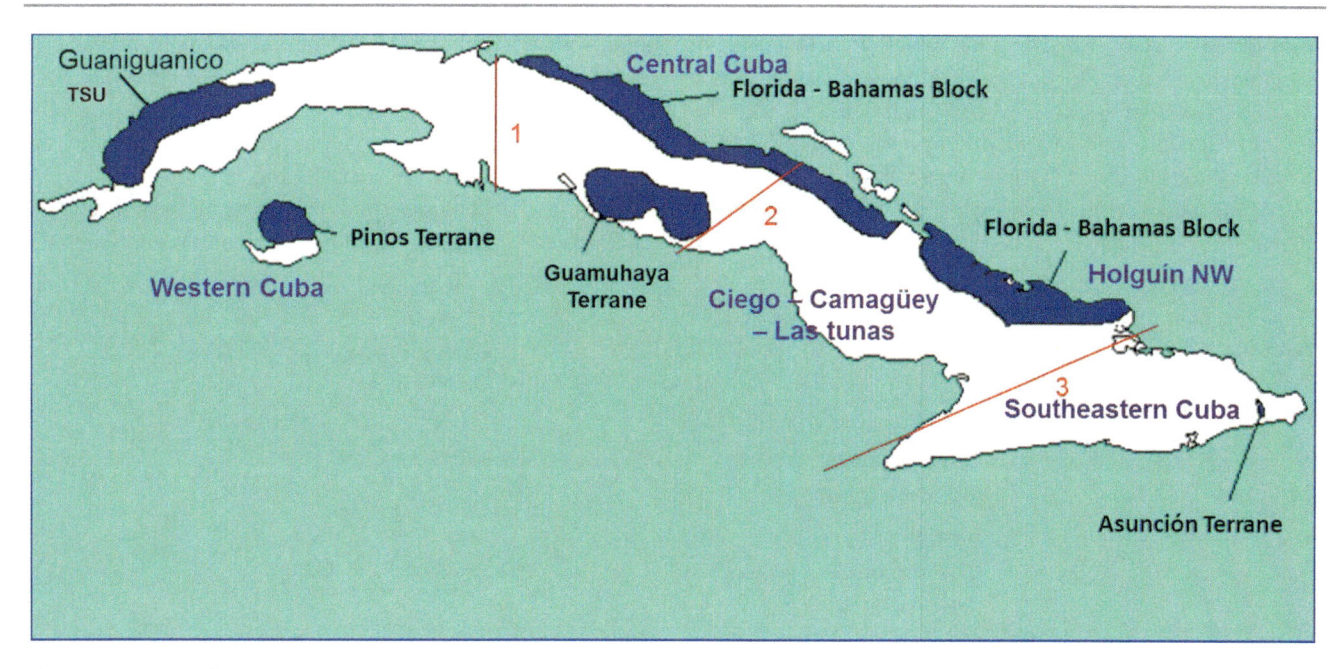

Fig. 8.2 Location of the outcrop areas of continental margin sequences in Cuba. In violet color the names of Cuban regions mentioned in the text. Numbers in red: 1. Yabre alignment, 2. La Trocha Fault and 3. Cauto—Nipe Fault

The sequences belonging to the Maya Block, on the other hand, are located in the western and central part of Cuba and are distinguished from those of the Florida-Bahamas Block by the shortage of evaporites and by hosting a significant number of metallic mineral deposits.

The mafic-siliciclastic VMS-SEDEX-MVT of the distensive basin J-K$_1$ mineral system was developed, as its name indicates, in a paleo-basin that existed and evolved during the Jurassic and Early Cretaceous in the northern part of Proto Caribbean, between Caribeana (Maya Block), located SW, the NE part of the Maya Block, located in the NW, and Florida-Bahamas Block, located NNE of the marine waters that separated them, according to Iturralde Vinent [66].

From the metallogenic point of view, this basin seems to have behaved in an asymmetric way, since the wealth of metallic minerals is concentrated in the units belonging to the Maya Block, very poor in evaporites, while those belonging to the Florida-Bahamas Block stand out for their large evaporitic deposits of salt and its oil deposits. This apparent asymmetry could be explained by the great depths to which the siliciclastic section, mainly Jurassic, is located in the Florida-Bahamas Block, and/or by certain paleogeographical peculiarities of the region.

During the evolution of the lower part of the rift distensive basins, a felsic magmatism usually appears first. Although this felsic magmatism has not been observed in the units belonging to the Maya Block, there is evidence of its presence in tectonically exhumed blocks within the Florida-Bahamas Block sequences: Jurassic granites in the Socorro Complex (border region between the Matanzas and Villa Clara provinces, near the north coast), with an U–Pb age of 172 million years [120], 139 and 150 million years according to Somin and Millán [130] and ortogneiss, of granodioritic origin, with an age K–Ar in muscovite of 196 million years, sampled in La Palma lithodemic unit, located in the NW of the province of Holguín [109].

This felsic magmatism is then replaced by a distensional tectonic mafic magmatism, which is present in the sequences belonging to the Maya Block in varying degrees (from moderate to absent). Magmatic rocks are represented here by gabbroid, diabases, and basalts, which generally appear in the form of intrusive or effusive concordant bodies and small sills, although dykes have also been reported [32, 111]. These bodies are generally of small size, with the exception of those present in Sábalo and the Olga-Jagua intrusive massif. These manifestations of continental margin magmatism appear in Arroyo Cangre (J_1–J_3^{ox2})—where they are metamorphosed—Esperanza (J_3^{th}–K_1^{val}), Francisco (J_3^{ox2}–J_3^{ox3}) and Jagua (J_3^{ox2}–J_3^{ox3}) formations, all located in the Pinar del Río province. This magmatism has also been reported in a locality in the Sierra de Camaján (Camagüey), where the base of the Placetas TSU is represented by alluvial basalts interspersed with thin layers of hyaloclastites, with an age of 146 ± 6 million years, i.e., Upper Jurassic Tithonian [65].

On the basis of the analyzes and studies carried out on Sábalo samples, Allibon et al. [6] conclude that the basalts of this unit are similar to the continental basalts of low content of Ti and P_2O_5. They also emphasize that they are derived from the mixture of a mantle source with an asthenosphere component, similar to the E-MORB type, with scarce quantity of markers of cortical contamination and assimilation of sediments, referring its origin to the rifting linked to the disintegration of the Pangea supercontinent.

This distensive rift basin, with the presence of large thickness of siliciclastic (and carbonate) sediments rich in organic matter Table 8.2, is a typical example of the presentation of all the geological, paleo-geographic, and paleo-climatic regional requirements necessary for the development of the metallogenic processes leading to the formation of SEDEX ore deposits [46, 58, 77].

With the metallogeny of the mineral system of the paleo-distensive basin J-K$_1$, the most important polymetallic

Table 8.2 Content of organic matter in the distensive paleo-basin of J-K$_1$

Lithology	Organic material*	Location, bibliographic sources (unpublished)
UTE Guaniguanico		
Crystalline limestones	Up to 30%	Unión II; Barrios et al. [13]
Dolomitized limestones	12%	
Black slates	Up to 30%	
Clays–carbonaceous slates	30%	Unión I; García et al. [53]
Recrystalized limestones	7%	
Schistose limestones	Up to 30%	
Dolomitized limestones	12%	
limestones enclosed at Juan Manuel ores	5–6%	Unión I-Juan Manuel; Shigariev et al. [126]
Dark limestones	1–10%	Hierro Mantua; Podkamenny y Guzmán [112]
Carbonaceous-carbonated slates	10–30%	
Dark limestones	1–15%	Hierro Mantua; Muliukov y Guzmán [94]
Limestones	1–10%	Santa Lucía Mineral District—Castellano, Vershigora et al. [139]
Carbonaceous slates	5–7%	Santa Lucía; Ovchinnikov et al. [103]*
Black limestones	Up to 25%	
Dolomitized limestones and dolostones	15–30%	
Quartz silt	5–10%	Baja, Veguita and La Vitrina sectors, W part of Santa Lucía Mineral District—Castellano; Lara et al. [73]
Quartz carbonaceous silt	Up to 15%	
Carbonaceous slates	Up to 30%	
Limestones	Up to 15%	
Argillites	10%	Castellano; Díaz et al. [40]
Carbonaceous limestones	10%	Santa Lucía Este; Sánchez y Ortega [122]
Black slates	Up to 35%	Santa Lucía Mineral District—Castellano; Vershigora et al. [139]
Carbonaceous-carbonated slates	10%	Dora—Francisco; Blanes et al. [15]
Quartz sandstones	5%	
Carbonaceous–carbonated slates	5%	Dora and Amistad; Novizky [100]
Carbonaceous silts	3–5%	Veguita II (Veguita Sector), Santa Lucía Mineral District—Castellano; Lara et al. [73]
Quartz silt–sandstones	1–2%	
Silt–shales	Up to 7%	
Slates	5–7%	
Limestones	7–25%	

(continued)

Table 8.2 (continued)

Lithology	Organic material*	Location, bibliographic sources (unpublished)
Limestones lens	Up to 20%	Pinar Ciego and La Estrella sectors; Puentes et al. [118]
Quartz silts	10–15%	
Black slate shales	23,86% of TOC**	Nieves Prospect; Mormil et al. [91]
Pinos Terrane		
Alternating shales, gneiss and quartzites	3–15% of graphite	Los Indios Sector in Cañada Fm.; Babushkin et al. [12]
Graphitic quartz marble	5% of graphite	La Ceiba Manifestation of Fe; Babushkin et al. [12]
Guamuhaya Terrane		
Marbled graphitic limestones	3–7% of graphite	Carlota Prospect; Bolotin [17]
Carbonate–graphite–quartz schists	4,87% a 33,63% of graphite	Arroyo Junco; Montano et al. [89]
Carbonate–quartz–graphite schists	17,6–24,21% of graphite	El Mamey A and B; Montano et al. [89]

*Expressed in percent of carbonaceous substance unless otherwise indicated
**TOC (Total Organic Carbon)

deposits of Cuba are listed: Hierro Mantua, Union-Juan Manuel (VMS mafic-siliciclastic), Castellanos, Santa Lucia, Nieves (SEDEX subtype Selwyn), Carlota and Guachinango (MVT subtype Irish). At the Cu ± Ag ± Co-hosted in sediments model belong the well-known Matahambre copper deposit, closed in 1997 without exhausting its ore resources, constituting an atypical case within the general model. The Broken Hill type, represented by the Isabel Rosa occurrence, is a product of the modification by regional metamorphism of deposits VMS siliciclastic-mafic type or Zn–Pb–Ag exhalative-sedimentary.

The petrography examination of the ore-bearing rocks of the Castellano, Santa Lucía, Matahambre, and Hierro Mantua ore deposits, as well as the Santa Lucia, Nieves, Unión-Juan Manuel, Esperanza, and Sur prospects, allow establishing that the host sequences of massive sulfides of the SEDEX and VMS mafic-siliciclastic types in the Guaniguanico TSU are shales and siltstones, both rich in organic matter and partially transformed into slates, finely stratified and with intercalations of limestones and dolomitized limestones, as well as calcareous sandstones, quartz-feldspathic and quartz. The shales and siltstones are composed of hydroxics, clays, quartz, feldspars, and carbonates, with constant presence of pyrite. Dolomite, calcite, and siderite are uniformly scattered, accompanied by chlorite and sericite. The shales and siltstones harbor pyritic-polymetallic mineralization, while the sandstones host basically copper mineralization.

The syngenetic nature of the ores of the VMS mafic-siliciclastic type deposits is evidenced by the presence of framboidal pyrite (Hierro Mantua and Unión-Juan Manuel) and globular (Unión-Juan Manuel), as well as of marcasite and organic matter in both deposits mentioned. The concordant disposition of the ores, folded together with the enclosing medium, the stratiform morphology of the mineral bodies, their distribution in several mineralized levels, and the diffuse character of the ore-rocks contacts, is other elements that indicate a primary syngeneic genesis for these deposits [53, 54, 72, 97].

In deposits of Zn–Pb–Ag exhalative-sedimentary also clearly distinguishes an early stage of mineralization, well studied in Castellano and Santa Lucía, characterized by the precipitation of iron di-sulfides (pyrite, framboidal, marcasite, and melnikovite), accompanied by other sulfides, mainly sphalerite and galena.

The acid character of the mineral solutions is also suggested by the existence of enargite and marcasite in Unión-Juan Manuel and Hierro Mantua ore deposits. Likewise, the abundance of organic matter also suggests the reduced nature of these, which is supported by the general shortage of barite. The maximum temperature during these processes seems to be linked, temporally and spatially, with the precipitation of the chalcopyrite in the second stage of mineralization, as suggested by the presence of cubanite associated with chalcopyrite. The existence of Au in the primary ores of the mentioned deposits, as well as in the Castellano and Santa Lucía ore deposits and in the Nieves prospect, is an indicator of the acid character of the mineral solutions. This is supported by the report of tetraedrite in the Malas Aguas I occurrence and tenantite-sanbergite and freibergite in the Sur prospect, as well as by the existence of marcasite, a mineral that is usually formed in low pH media [80, 139, 140].

The presence of Fe-rich sphalerite in the Santa Lucía ore deposit suggests, in turn, the presence of very low oxygen fugacity during the first stage of mineralization, given the absence in it of minerals and mineralogical forms of high temperature. Also, the presence of pyrrhotite in the Castellano ore deposit and the El Jíbaro prospect reinforce this suggestion. It is also supported by the general scarcity of barite mineralization during this stage, despite the fact that Ba is a relatively abundant element in the regional sequences. The decrease in Fe content in sphalerite II in relation to I in the Santa Lucía ore deposit suggests that this situation of highly reducing environment gave way, during the ore formation process, to more common reducing conditions. On the other hand, the presence of marcasite at this stage suggests rather low deposition temperatures (less than 160 °C) and acidity of the mineral fluids, [52, 80, 91, 103]. The abundance of organic matter in ores and enclosing rocks also supports the reducing nature of the metal-bearing solutions that gave rise to these deposits.

The values of $\delta^{34}S$ obtained in 8 samples of pyrite from the Unión-Juan Manuel prospect [54] range between +7.5 and +11.5‰ (CDM). This suggests that the dominant source of S was the inorganic reduction of marine SO_4, through the interaction of seawater with the basic effusive material at temperatures above 250 °C [54]. This does not rule out the presence of magmatic S, given that this study has not been performed for chalcopyrite. Regarding the source of metals, the presence of Co, generally housed in pyrite, and heavy impurities in ores (up to 400 mg/kg of Ni, up to 1000 mg/kg of Cr, up to 1000 mg/kg of Mn, up to 8000 mg/kg of Ti, up to 3000 mg/kg of Ba and up to 400 mg/kg of V), indicate the existence of a contribution of mafic magmatism to the metalliferous charge that originated these ores [53, 54, 97], in addition to the one derived from the leaching of oxidized rocks in the deep part of the paleo-basin.

On the other hand, the isotopic relationship of $\delta^{34}S$ in the ores of Santa Lucía, reported by García et al. [52], oscillates between −5.9 y −32.8‰ (CDM), what indicates an biogenic origin for present S in their ores. These results were corroborated by Valdés et al. [138], who reported a value of $\delta^{34}S$ in pyrite of −27.5‰ (CDM) in this deposit. The above-mentioned is in agreement with the presence in the same of abundant pyrite framboidal. On the other hand, the values of $\delta^{34}S$ determined in the year 1980 in samples coming from the Castellano ore deposit [9] showed a wide dispersion range, with prevalence of values of the order of −25‰ (CDM), indicative that also in this ore deposit the biogenic sulfur played an important role in the formation of the ores, although the presence of contribution of magmatic S is also revealed by the hydrothermal fluids.

The Cu Matahambre ore deposit, exploited between 1913 and 1997, during which some 13 million tons of copper ore were extracted, is a very interesting example of stratiform chalcopyrite mineralization in sandstones, slates, and shales rich in organic matter. The mine was closed in 1997, when level 45 was exploited at 1503 m depth and the opening of level 46 was prepared at 1545 m depth, without its resources being depleted, as the mineral body 44, the most important of all, it extends to greater depths, without any sign of reduction of the law of Cu in the ore. It consists of 4 copper mineral bodies (named 14, 19, 30, and 44, by the year of its discovery in the twentieth century), which are arranged from NW to SE. Its morphology is complex, counting with sectors of massive ore and others of disseminated chalcopyrite, reefs, and veinlets of ore with quartz. However, they always form continuous mineral bodies. These four bodies are located stepwise, with the upper ones being displaced toward the N and NW in relation to the lower ones, all dipping to the NW, concordant with the dip of the embedding rocks. Additionally, there is body 70, a fifth mineral body, parallel to the previous ones, but of polymetallic composition (Zn–Pb–Cu–Ag), which houses an additional 1.8 million tons of ore [138, 140].

In the ore deposit, from body 70 to bodies 44 and 14, chalcopyrite increases and pyrite and pyrrhotite decrease, as well as galena and sphalerite, indicating the existence of a chemical and mineralogical cross-sectional zonation on the whole ore deposit. The ore body 70, itself, in turn, has an internal zoning, with the pyritic ore at its base and polymetallic at the top, while it is enriched in pyrite on its NE flank and in pyrrhotite on its NW flank. Its ore consists mainly of sphalerite and galena, with subordinate amounts of chalcopyrite, pyrite, and pyrrhotite. This ore body presents at its base a small Stockwork zone, similar to that present in the deposits of SEDEX type.

The oxidation zone of Matahambre ore deposit varies between 12 and 63.6 m in thickness, averaging 23 m, with 1300 m long in North–South direction and more than 700 m wide. Inside it there are several gossans, located in the highest parts of the relief, where the presence of up to 0,85 g/t of Au has been detected. They are made up of Fe oxides with quartz, stained with Cu minerals.

In the shallow part of the ore deposit (first 150 m depth of bodies 14 and 19) was found massive chalcopyrite, also accompanied by malachite, azurite, bornite, and calcosine, appearing in great abundance native Cu. The Matahambre ore is perfectly concordant with the strata. The examination of the plans of the different levels of the mine indicates it [5, 82, 101]. The law of ore extracted at that time ranged between 1 and 18% Cu, with up to 1.015 Oz/t Ag, reaching up to 22% Cu in some unsorted starts [5, 101].

All this points to the fact that the deposit does not constitute a vein-shaped feeding zone as those commonly observed in proximal SEDEX type deposits, which, moreover, are crossing to the stratification of the embedding sequence. As for the presence of As (0.05–0.07%) and Te (mainly in the form of hessite) in the ore, this can be explained as a result of the concentrating effect that organic matter has on these elements, especially in the slates [110]. These latter authors also report the presence of selenides, arsenides, cobaltite, and arsenopyrite. On the other hand, the presence of Au in the ore and organic matter in the bedrock rocks suggests that, as in the SEDEX and VMS mafic-siliciclastic types deposits, the mineral fluids were of reduced and acidic nature. Regarding the thermal regime, Lavandero et al. [75] proposed, on the basis of their mineralogical and textural observations, a temperature range of 200–400 °C for the formation of this deposit. Unfortunately, there are no studies of fluid inclusions that allow us to specify this range.

The displacement experienced by the ores along the planes of faults that cut them have caused their dislocation and fracturing, which is reflected in the presence of fractured pyrite grains and ore fragments cemented by a more plastic clay-carbonaceous material [91]. It has also been observed the presence of pressure twins in chalcopyrite and blastic and sub graphic textures, which indicates that the deposit was subjected to intense post-mineral stresses and deformations, probably during the Cuban orogeny Figs. 8.3 and 8.4.

When comparing this ore deposit with the Cu ± Ag ± Co-hosted in sediments model, we observe, as a fundamental difference, the non-existence in Matahambre ore deposit of any reddish sequence or clearly identifiable redox boundary. The whole enclosing sequence is rich in organic matter in variable degree and presents black and gray colors more or less dark, this feature that brings it closer to the subtype Cu ± Ag ± Co Kupferschiefer (also known as Cu in reduced facies). However, the marked mineralogical

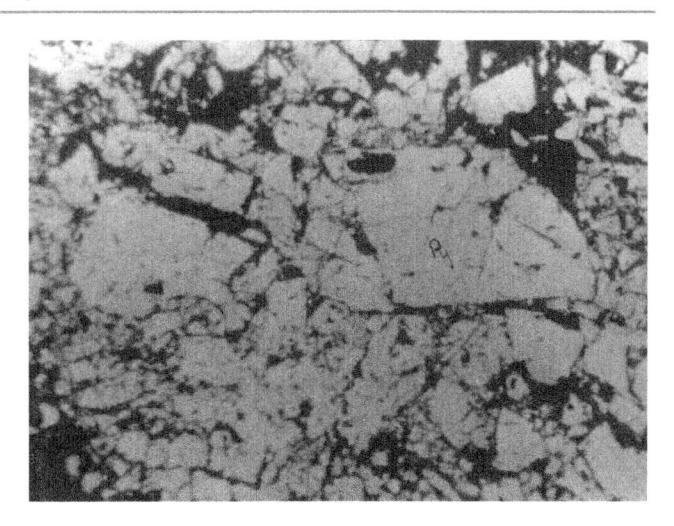

Fig. 8.4 Matahambre ore deposit. Pyrite (Py) with cataclastic texture with cracks filled with chalcopyrite and non-metallic minerals (20X)

zoning of this deposit subtype is not present in it either. On the other hand, the composition of the embedding rocks (quartz-feldspathic sandstones with intercalations of siltstones, shales, and shales) brings it closer to the Revett subtype of the mentioned model. Another notorious feature of the Matahambre site is that the chalcopyrite constitutes the main mineral of ore, something that does not happen in any of the four subtypes of the deposits that the model groups, although its presence is important in all of them. On the other hand, only a partial coincidence is observed between the textural features of the ores and those indicated in the model, which could be due, at least in part, to the aforementioned modifying influence of the Cuban orogeny, which occurred long after formed the deposit. Regarding the thermal range of ore formation, the one proposed for the Matahambre deposit, from 200 to 400 °C, is higher than the range, generally of low temperatures, estimated for the sediment-hosted strata bound copper [61].

In relation to this last aspect, it should be noted that the presence of Co, Ag, Pb, and Zn in some deposits of this type in various parts of the world and their absence in others suggests the existence in the former of exhalative-sedimentary processes of mineralization, in which deep fluids from basins with high temperature were transferred along the faults to the ore deposition sites, place where they over primed the existing mineralization, or were mixed with brines rich in Cu [36]. In this sense, it should be noted that the copper ores of the Matahambre ore deposit contain up to 0.7% of Zn, up to 0.12% of Pb, and 0.001–0.003% of Co, which is reflected in the subordinate presence in them of sphalerite and galena, as well as rare cobaltite and linnaeite. The presence of native gold is also highlighted [91].

In general, the reduced and acid nature of the hydrothermal mineral fluids, both for the SEDEX type deposits and for the siliciclastic-mafic type VMS, makes it

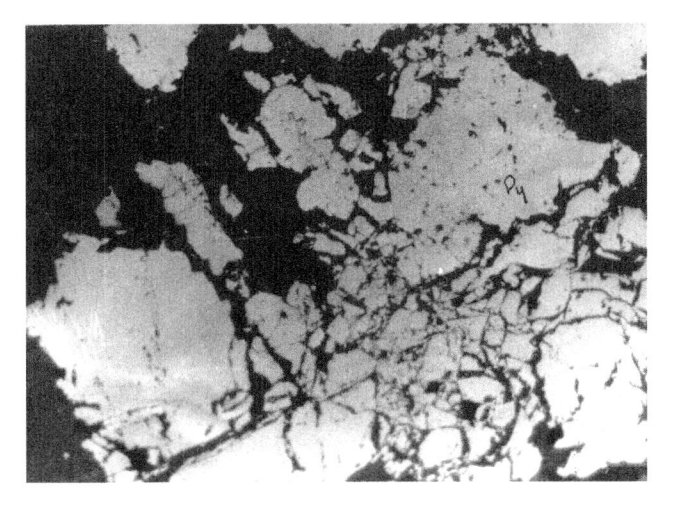

Fig. 8.3 Matahambre ore deposit. Fragmented pyrite (Py), allotriomorphic to hip idiomorphic aggregates (10x)

possible to establish, first of all, that the deposits of Zn–Pb–Ag of the SEDEX type present in the Guaniguanico TSU are from the Selwyn subtype [124]. Secondly, it allows to characterize the mafic-siliciclastic VMS—SEDEX—MVT Irish subtype of the paleo-distensive basin J-K$_1$ mineral system as a mineral system of the Selwyn Basin type.

The Arroyo Cangre unit is composed of siliciclastic rocks with intercalations of diabases, all affected by a regional metamorphism of high pressure and low temperature, between the transition from the greenschist facies-amphibolites to that of glaucofanic schists [37]. The presence of pyrite and chalcopyrite disseminated in these mafic bodies, representative of the continental margin magmatism, suggests the existence of a paragenetic link between them and mineralization, represented by several polymetallic manifestations (Isabel Rosa, La Guabina, El Guayabo), characterized for the presence of a mineralogy of ore and a varied and complex gangue. Although there is no quantitative data on the content of organic matter present in this tectonic-stratigraphic unit or terrain, this is considerable, particularly in the black graphitic schist layers that have been observed.

In this unit is the only deposit (Isabel Rosa occurrence) identified as belonging to the model Pb–Zn–Ag ± Cu Broken Hill type, considered, from the genetic point of view, a metamorphosed version of the VMS mafic-siliciclastic, or Zn–Pb-exhalative Ag-sedimentary [132]. In addition to this model, the presence of Pb–Zn–Ag polymetallic veins in siliciclastic sequences is likely here. However, the small size of the outcrop area of this unit limits its potential for mineral resources.

In geological and paleogeographical terms, some authors (including Iturralde Vinent [66]) consider the Pinos terrain as the Cuban unit of continental margin closest to the core of the Maya block. It covers most of the Isla de la Juventud, including areas of the Batabanó Gulf near the island mentioned by the W, N, and E.

As part of the paleo-distensive basin of the J-K$_1$, as in the Guaniguanico TSU, the stratigraphic section consists of a siliciclastic lower part, corresponding to the synrift, and a carbonated upper part, representative of the postrift. However, it is distinguished from this and the other continental margin units of western and central Cuba by the absence of bodies of serpentinite mélange and ophiolites [88]. This suggests, in the opinion of Iturralde Vinent [66], that their sequences may not have become detached from its foundation. Another important feature of this terrane is the presence of anatectic granite bodies, related to regional metamorphism [130], a trait that, as will be seen below, has great importance for the metallogeny of this part of the Cuban territory (see the section dedicated later to the Campanian–Maastrichtian orogenic mineral system).

The organic matter present in the sequences of the Pinos Terrane, as a result of the action of regional metamorphism, generally appears in the form of amorphous graphite, which usually occurs in schist and graphitic marbles in concentrations between 3 and 10%, reaching up to 25%. The highest contents of organic material, in the form of carbonaceous substance and/or graphite, are found in the siliciclastic sequences, particularly those belonging to the Cañada (J$_{1-2}$) and Agua Santa (J$_{2-3}^{ox}$) formations, representative of the stage of synrift. However, graphite is not the only way organic matter appears in this terrane, since it also occurs in the form of a carbonaceous substance. Thus, the sample for ATD 14/29, taken from well 14 at 68.50–68.56 m depth, reports well-crystallized kaolinite and amorphous carbon of the gray stone coal variety. The sample for ATD 14/31, taken in this same well at 69.0–69.66 m depth, reports kaolinite, amorphous carbon of the anthracite type, and amorphous carbon of the gray coal type [12]. This suggests a subaerial depositional environment for these siliciclastic sequences.

In this terrane the continental margin magmatism in the form of amphibolites is also present, whose protolith is of the E-MORB or N-MORB series [36], formed in conditions similar to the basalts of Sábalo (Guaniguanico TSU).

The mineralization belonging to the mafic-siliciclastic VMS—SEDEX—MVT of the paleo-basin of the J-K$_1$ mineral system is represented here by deposits of siderite stratiform Fe (La Ceiba occurrence) and polymetallic veins of Pb–Zn–Ag in siliciclastic sequences (Maria Josefa occurrence).

La Ceiba occurrence is of interest for itself and for to demonstrate the existence of potential for polymetallic mineralization of SEDEX type in this terrane. It consists, as far as it has been investigated, in a tabular body 750 m long, 116 m wide, and up to 20 m thick, located at the contact of Cañada (J$_{1-2}$) Fm. with Agua Santa (J$_{2-3}^{ox}$) Fm. Its average content of total Fe is 21.5%, varying between 18 and 56%. The ore minerals reported in this occurrence, little studied, are siderite, sphalerite, and pyrite, accompanied by quartz, actinolite, and phosphates as gangue minerals [12]. It is noteworthy that this association between Fe and phosphate minerals usually constitutes the distal lateral facies of the host horizons of SEDEX type polymetallic mineralization [58].

Similar to the two large units previously described, the Guamuhaya Terrane is composed of a lower section of siliciclastic sequences and an upper section of carbonate rocks. These sequences were deposited during the Jurassic and Early Cretaceous and formed in shallow water environments, which progressively became deep water and open sea [43, 68], dominating the environmental conditions reducing during the Jurassic and the lower part of the Lower Cretaceous. The continental margin magmatism, of sub

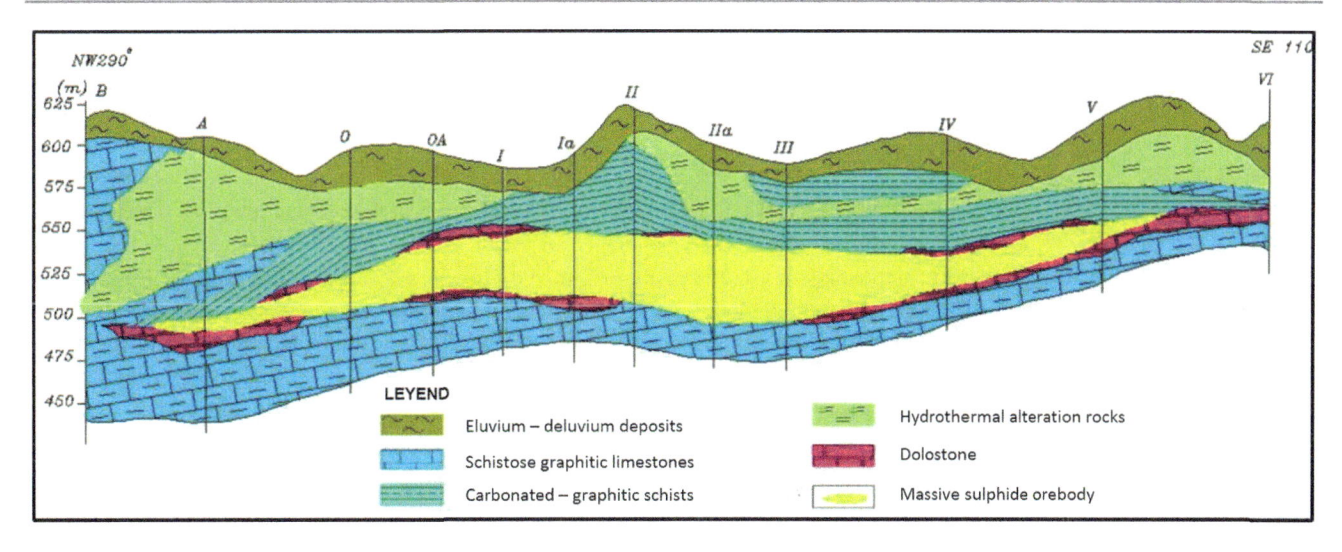

Fig. 8.5 Longitudinal profile of the Guachinango prospect. As can be seen, the body is always concordant with the gently folded lay of the embedding rocks. Taken from Bolotin [16]

alkaline basic composition [133], is represented by metamorphosed gabbros and diabases and appears well manifested in the La Chispa (J_1–J_3^{ox2}?), Cobrito (J_3), Boquerones (J_3), and Los Cedros (K_1) formations, mainly in the form of concordant lenticular or tabular bodies [64].

In the Guamuhaya Terrane, graphitized rocks have very broad representation, being the siliciclastic protolith units the most favorable for it. The normal content of graphite in graphitized rocks oscillates between 1 and 7%, existing localities in which it exceeds 10% (see Table 8.2).

In the metallogeny of the Guamuhaya Terrane there are two outstanding features: (1) Presence of Cu–Zn–Pb ± Co, Au, and Ag massive sulfide deposits of the MVT subtype Irish type, hosted in metamorphosed carbonated sequences. (2) Spatial superposition of the Campanian–Maastrichtian orogenic mineral system on the mafic-siliciclastic VMS—SEDEX—MVT of the distensive basin J-K_1, present in the deposits of the Mineral District Carlota-Guachinango.

The pre-metamorphic origin—and prior to the tectonic deformations—of these massive sulfides deposits is supported by the following facts, collected in the well-studied prospects Carlota, Guachinango, and Victoria of the mentioned mineral district [14–18, 21, 47, 59, 75, 125, 134, 144]:

1. Presence of lepido granoblastic, porphyro granoblastic, and other similar textures, typical of the action of regional metamorphism on the minerals of hydrothermally altered rocks. Particularly, the presence of a grain texture in the dolomite directly associated with the mineral bodies and the presence of sulfide mineralization disseminated in these rocks is noted.
2. Ores have high concentrations of heavy impurities, characteristic of mafic magmas (Ti, Co, Ni, Cr, and others). In addition, there is almost total absence of Li, F, Be, and W, as well as other elements related to granitoid magmas. This strongly suggests the relationship of the mineral solutions with the metamorphosed mafic intrusive bodies.
3. The content of Co in the ores varies between 0.042 and 0.093%. This supports the existence of a relationship between hydrothermal solutions and mafic igneous bodies. On the other hand, is reported the existence of high concentrations of Co in the MVT deposits, formed in non-metamorphosed carbonates [45, 77].
4. The presence of holomorphic, globular, and meta-colloidal relict texture in the ores indicates their formation from colloidal solutions, which had relatively low temperatures.
5. Most authors agree that the mineral bodies were deformed together with the sequences that contain them. Also that these are associated with soft folds and flexure zones. This is well observed in the geological profiles of the explored deposits Fig. 8.5.
6. Existence of micro-fracturing and micro-folding zones, as well as plans and sliding grooves in the Guachinango, Carlota, and Victoria prospects.
7. Report of crushed pyrite grains in the core samples of the well PT-3A, drilled in the mineral zone No 1 of the Carlota prospect.

The elements of judgment exposed above indicate, beyond any doubt, that the ores were affected by regional metamorphism. According to the existing data, the high pressure metamorphism took place in the Upper Cretaceous, when the Guamuhaya Terrane occurred under the Cretaceous Volcanic Arc Terrane, with the consequent suture of the subduction zone associated with the insular volcanic arc of the Upper Cretaceous. Consequently, these deposits are prior to this event.

The prospects and shows of the Mineral District Carlota-Guachinango are localized in the Los Cedros (K_1), Cobrito (J_3), and San Juan (J_3^{ox2}–K_1^{ba}?) formations. The widespread and abundant presence of graphite, as well as the ubiquitous authigenic pyritic dissemination in these carbonate rocks, indicates that they were deposited in euxinic conditions, accumulating large amounts of organic matter together with carbonates. These sedimentation conditions can correspond to both restricted basins and open sea [137]. The high content of S is evidenced by the existence of sectors in which the rocks, when crushed, give off an odor of H2S [18, 81] and by the presence of marbleized silty graphitic limestones [56, 104]. This also suggests euxinic conditions in this part of the paleo-distensive basin of J-K_1.

The mineral aqueous solutions transported to the sites of ore formation fundamentally Fe, being very poor in base metals. This is proven by the same composition of the sedimentary pyrite ores, since sulfides of base metals have a lower solubility product than pyrite, so sulfides, if present in mineral fluids, can accumulate by breaking the pre-existing pyrite [129].

During the regional metamorphism of the greenschist facies, occurred in a collision environment of continental margin-insular volcanic arc type, the second mineral event (pyrrhotite-polymetallic) that affected the deposits took place. This mineralization event consisted essentially of a mineralogical modification of the pre-existing ore and its local remobilization under the action of hydrothermal metamorphogenic aqueous solutions, originated by the important water release of the rocks occurred during the development of regional metamorphism [43].

Associated with this collision, the deformations and metamorphism modified the morphology of the mineral bodies and erased—to a large extent—the structures and primary textures of the ores. The mineral bodies were also folded and sheared, as well as divided into several lenticular segments, with the thickest sections associated with the anticlinal folds and flexures of the embedding rocks. As a result of this process, the mineral bodies were distributed in a staggered or echelon way. The ores were also recrystallized, breached, and locally remobilized.

With respect to the thermal parameters of this second mineral event, Tolkunov et al. [134] determined decrepitating and homogenization temperatures between 340 and 200 °C in pyrite and between 250 and 110 °C in chalcopyrite, sphalerite, galena, ankerite, and calcite. As already stated, regional metamorphism erased the geothermal indications of the thermal regime of pre-existing sedimentary mineralization.

The ^{34}S (CDT) measured in pyrites from the Carlota deposit and in pyrite and chalcopyrite from the Guachinango deposit varies between −3.9 and 2.8‰ [23].

In sharp contrast to the remaining continental margin units discussed so far, the Florida-Bahamas Block presents a very underdeveloped metallogeny, characterized by the presence of small Mn stratiform occurrences, which appear mainly in Mata (K_1^{al}–K_2^{cm}), Santa Teresa (K_1^{ap}–K_2^{cm}) and Carmita ($K2^{cm-st}$) formations of the Placetas Tectonic Unit-Stratigraphic. Among them, the Maria Victoria occurrence stands out, with up to 43.2% of MnO, composed of pirolusite, psilomelane, and wad. In addition, some geochemical anomalies of certain Cu, Pb, Zn, Ag, Au, Co, Ba, and Cd occur, associated with exhalatites with Mn in the contact between the Carmita Fm. (carbonated) and the Santa Teresa Fm (silicite—jasperoid).

8.4.2 Island Volcanic Arcs

The Cretaceous sequences of the insular volcanic arc that extend throughout Cuba correspond to two distinct island volcanic arcs [66]: an insular tholeitic volcanic arc K_1, possibly from the Hauterivian to the Middle Albian, and an insular volcanic arc tholeitic-calcoalkaline-alkaline younger, of Upper Albian-Middle Campanian age [41, 69]. It is noteworthy that in Cazañas et al. [29] different age ranges are assigned to both arcs: Aptian?-Albian for the oldest and Cenomanian—Upper Campanian the youngest. This shows the still existing insufficiency of data on the age of these units, which gives rise to the existence of disagreements between different authors about this topic.

Considering the differences in evolution, ages, rocks, and types of mineralization, Yanev et al. [143] distinguished in the cretaceous sequences of insular volcanic arc two zones with different metallogenic characteristics: (1) Central zone (Cienfuegos, Villa Clara, and Sancti Spíritus provinces) and (2) Central-eastern zone (Ciego of Ávila, Camagüey and Las Tunas provinces). In relation to this, Cazañas et al. [29] established that, due to its metallogenic behavior, the Cretaceous Volcanic Arcs Terrane can be divided into five sections:

1. Western Cuba (tholeitic to calcoalkaline), up to the Yabre lineament (see Fig. 8.2).
2. Central Cuba (tholeitic and tholeitic-calcoalkaline), between the Yabre lineament and La Trocha fault.
3. Ciego-Camagüey-Las Tunas (normal to calcoalkaline and alkaline), between the La Trocha fault and Cauto-Nipe fault.

4. Holguín (calcalkaline), extended by the eastern end of the Las Tunas province and the NW part of the Holguín province, north of the Cauto-Nipe fault.
5. Southeastern Cuba (tholeitic), to the E of this last fault.

The metallogenic variations between the different mentioned segments of the Cretaceous Volcanic Arcs Terrane are related to the changes in the chemism of the intrusive and volcanic rocks observed between them. It should be noted, incidentally, that the last two mentioned faults, of slip by the course, extend toward the WSW beyond the Cuban territory, also subdividing the Yucatan Basin, located in the Caribbean Sea [20].

So far there are no known metallic mineral deposits of any importance in the section of the Cretaceous Insular Volcanic Arcs of Western Cuba. On the other hand, it is interesting, the presence of an occurrence of B: Mogote Simón. It consists of a body of hornblendic breccia of diorites with a system of veinlets of datolite whose individual thicknesses go from some millimeters until 2,5 cm and sometimes more. They are composed of datolite (96–97%), with a little quartz and carbonate (possibly ankerite) accompanying. The veinlets have different directions, which is observed in the whole area covered by the diorite body. Although the presence of datolite has been reported up to 15 m deep [14], this occurrence, being part of an olistolith, lacks economic value.

The bimodal VMS—Ag–Au veins in volcanic sequences mineral system, located in the Insular Volcanic Arcs Terrane in central Cuba, belongs to the insular volcanic arc of the Lower Cretaceous (Hauterivian- Albian). It is located mainly in the Los Pasos Fm. (K_1^{val-b}), characterized by a bimodal basaltic-rhyolitic volcanism, although it is also present in the Mataguá (K_1^{ap}–K_2^{t}) y Cabaiguán (K_1al) formations. The main deposits are San Fernando and Antonio (Cu–Zn–Ag $\pm$ Au bimodal-mafic VMS) and the Las Breñas occurrence of Cu, Au, and Ag (veins of Ag–Au in volcanic sequences).

In Los Pasos Fm. (K_1^{val-b}) there are two basic types of hydrothermal alterations: epidote-quartz-chlorite (sometimes only epidote-quartz or quartz-chlorite) and quartz-sericite. The first is the most widespread, usually corresponds to mafic photoliths and metallic mineralization is scarce (magnetite and pyrite). The second type of alteration is usually very intense, not allowing to identify the original rock, and it is associated to the areas with metallic mineralization. There is also the presence of exhalative levels, consisting of silicite and pyrite, closely related to the mineralization in the San Fernando, Los Cerros, and Antonio deposits.

Regarding the formation conditions of the VMS deposits within this lithostratigraphic unit, in the Antonio prospect, the isotopic composition of S was studied in 20 samples of pyrite and chalcopyrite, obtaining a narrow range of variation of $\delta^{34}S$ (CDT), between −3 and 5.8‰, with a concentration of values between 0 ± 3‰ (14 samples). This indicates a predominantly magmatic source of S, with a certain contribution of S derived from the thermochemical reduction of marine sulfate.

Iturralde Vinent [66] considers that the boundary between the Mataguá and Los Pasos formations is concordant. This is a point that is worth confirming, because if so, there would be possibilities to find the Los Pasos Fm. below this and, consequently, would raise the prospects to find non-surfacing bimodal VMS deposits.

Finally, it should be noted that there is no known epithermal Au–Ag mineralization in these sequences, although there are Ag–Au veins in volcanic sequences, which is due to its submarine character. Likewise, porphyritic mineralization presents a limited development, which is in accordance with its tholeiitic nature.

The Cu–Mo $\pm$ Au porphyry-skarn-veins of Ag–Au in volcanic sequences mineral system, also located in the terrain section of Cretaceous insular volcanic arcs corresponding to central Cuba, belongs to the insular volcanic arc of K_2 and is associated with its main magmatic axis. This mineral system is well represented by the Arimao and Macagua porphyry Cu–Mo $\pm$ Au deposits, accompanied by Cu skarn deposits (Arimao Norte, Guaos, and Los Cedros occurrences), Fe skarn (Loma Alta occurrence) and by Ag–Au veins in volcanic sequences (La Vega occurrence). This system has as a relevant feature the presence of Au in all types of deposits that are part of it.

The Ciego-Camagüey-Las Tunas section of the Cretaceous Volcanic Arcs terrain, unlike that of central Cuba, is notable for the wide predominance of the Upper Cretaceous sequences. Another notable feature is the abundance of vulcanite of alkaline potassium tendency in the Camujiro Formation (considered the Albian-Cenomanian or Turonian) and, to a lesser extent, Caonao Formation (considered Coniacian-Campanian). Another prominent feature is the presence of abundant polyphasic intrinsic felsic bodies.

The Cu–Mo $\pm$ Au porphyry—skarn— Ag–Au epithermal mineral system of K_2 of this section of the Cretaceous Insular Volcanic Arcs Terrane is represented by Guáimaro, Palo Seco, Tres Casas I (Cu–Mo $\pm$ Au porphyry); Abucha (skarn of Au); Cañada Honda and Asientos de Tamarindo (skarn of Cu); El Pilar and Golden Hill (Au–Ag epithermal of high sulphidation); Florencia and Maclama (Au–Ag epithermal of low sulphidation); and Jacinto (Au-Epithermal Ag of low sulphidation of alkaline subtype) deposits.

The Jacinto prospect covers an area of about 5 km^2, where, to date, 10 gold veins have been identified, with a maximum length of up to 1500 m. The fluid inclusions in quartz reveal the participation of a boiling process in the formation of the gold veins, with homogenization temperatures between 200 and 320 °C, and fluid salinity of less than 1% by weight of Na Cl equivalent. Regarding their age, in two samples of adularia, coming from quartz veins, they were determined, using the $^{40}Ar/^{39}Ar$ method, absolute ages of 73.3 million years and 69,8 million years, respectively [127]. This suggests that the epithermal gold mineralization present in this part of Cuba was formed in the Campanian and the Maastrichtian.

The Jacinto prospect is found in a sequence of rocks of the alkaline series, formed by basalts, lava-breccias, lavas, clastic-lavas, and tuff of dacitic and shonshonitic composition. In addition, there are also rocks of andesitic-dacitic composition, with intercalations of sandy and detrital limestones, as well as rhyolites and subvolcanic-intrusive bodies [127]. Alkaline leucocratic granite bodies also appear in the area.

On the other hand, the Maclama prospect, classified as low sulphidation epithermal [29], is found in basaltic and basaltic shonshonític andesites and basalts, as well as basaltic, andesite-basaltic, and andesitic calcoalkaline tuffs. These rocks host micro dispersed gold mineralization, isolated lenses of weakly skarnified limestones and jasperoid bodies. All this sequence is crossed by dykes of andesitic hornblendic-plagioclastic porphyry and by diorite and monzosienitic stocks [3]. Likewise, the presence of tetradymite, ricardite, petzite, altaite, melonite, tellurium-bismutite, calaverite, chlorite, epidote, calcite, quartz, and very pure native Au [7] is reported in the ores.

The Florencia prospect is another Au–Ag low sulphidation epithermal deposit, consisting of two mineral zones. The North zone is divided into three bodies, in which the pyritic-chalcopyrite mineralization predominates. In the South zone also appears polymetallic mineralization (lead, zinc, silver), which suggests that it represents a portion of the highest part of deposit in the section. This deposit draws attention especially by the abundance of tellurides: calaverite, silvanite, petzite, hessite, volinskite, and altaite. Of all of them, petzite is the most abundant. This tellurides abundance is, within epithermal deposits of low sulphidation, a typical feature of the alkaline subtype. This feature of the Florencia prospect suggests, once again, the intervention of alkaline rocks in the mineralization process that gave rise to a part of the low sulphidation epithermal deposits of the Ciego-Camagüey-Las Tunas region, reason why these can be considered Au–Ag epithermal of low sulphidation of alkaline subtype.

For this deposit an isotope ratio of $\delta^{34}S$ between −2.53 and 5.26‰ was determined. The $\delta^{13}C$ in the calcite is −13.06‰. In early quartz the isotope ratio of $\delta^{18}O$ is 5.56‰ and in the late quartz it is 14.87‰. Finally, in the hydrothermal fluid it is 4.41‰. From the thermodynamic analysis of the mineral associations in the Au–Ag–Te system, [19], obtained formation temperatures for the deposit between 70 and 280 °C.

As already noted, the prospects Jacinto, Maclama, and Florencia testify to the intervention, to varying degrees, of alkaline rocks in the mineralization process and the presence, in this part of the Cuban territory, of the epithermal low sulphidation alkaline subtype. Another element that elevates the perspectives of this region for Au and Ag is the possible existence of a second magmatic axis, not outcropping, south of the main magmatic axis, with predominance of this type of lithology, as suggested by the important propagation of the volcanic sequences of alkaline tendency observed to the South of the main magmatic axis. In favor of this testifies the interpretation of large buried volcanic structures through the coincidence of magnetic axes and annular drainage, as well as the presence, suggested by the interpretation of TM-5 images, of the presence of saprolite typical of the arc on intermediate-felsic magmatic rocks in the center of one of the mentioned structures [24]. Likewise, its presence is reflected in the aero-gamma-spectrometric map (K-channel) of the territory. From the above it is inferred that the territory where this presumed second axis is located has a significant level of favorability for the presence of the Cu–Mo ± Au porphyry, skarn, and Au–Ag epithermal deposits, by a higher degree of preservation of the upper part of the geologic section.

The skarn deposits of Au and Cu are well represented in the territory, particularly to the E of the Camagüey fault. In the province of the same name, the auriferous skarn predominate, and more to the E, in the Las Tunas province, the copper skarn predominates. However, this type of mineralization does not have a significant contribution to the Cu and Au potential of the territory so far and its current level of study is low.

The Cu–Mo ± Au porphyry mineralization of the territory, in general, has been little studied despite appealing interesting contents of Au. Thus, the Guáimaro prospect presents up to 3 g/t, with 0.1–0.54% Cu and 0.02–0.096% Mo; Tres Casas I presents 0.2–0.6 g/t of Au, with 0.1–0.9% of Cu and 0.02–0.07% of Mo. Starting from the structure of the Cu–Mo ± Au porphyry—skarn—Au–Ag epithermal mineral system can expect the existence of other porphyry deposits, hitherto unknown, genetically and spatially associated with the epithermal and skarn deposits present in the territory. This, together with the broad development in the area of the calcoalkaline and alkaline intrusive bodies, argues its high favorability for the presence of this type of mineralization.

The wide development of alkaline sequences and intrusive bodies of alkaline tendency, together with the polyphasic character of the igneous activity, of mafic to felsic and alkaline composition, argues the possible presence of other types of deposits in the Ciego-Camagüey-Las Tunas region that would expand the territory's perspectives regarding its mineral wealth. In this sense the Embarque, Tres Antenas, and Palo Seco I occurrences are interesting. The first two are examples of rare earth mineralization and the third of magnetite Fe mineralization hosted in metamorphosed mafic rocks.

The Embarque occurrence is located in the periphery (0.5 km away) of an intrusive alkaline complex, made up of syenites of varied composition (hornblende, biotite, leucocratic, quartziferous, aplitic, and porphyritic), monzonites and monzodiorites, embedded in andesitic lavas-basalt, cut by diorite porphyritic dikes and by subvolcanic necks of andesitic porphyrites. The occurrence consists a mineralized zones system of ferruginous carbonates (ankerite, siderite), with 100–280 m in length by the course and thickness between 2 and 5 m, in which quartz and quartz-carbonated veins develop from 40 to 100 m in length by the course and from 1 to 3.7 m thick. In this veins there is 0.1–0.6% La and 0.1–0.5% Ce, also containing up to 2 g/t Au and 0.6–1% Cu [102].

The Tres Antenas occurrence presents two areas with intense aero-gamma spectrometric anomalies of Th, spatially associated to a zone of hydrothermal alteration, located inside intrusive syenites rocks and probably associated to a system of faults. Both anomalous zones have a linear shape, with similar characteristics and intensities. There are also geochemical anomalies of Y, La, and Au in soil and La, Ce, Sm, Eu, and Yb in rocks. Ten wells were also drilled, which located auriferous mineralization of interest in skarn and syenites (contents between 0.2 g/t and 0.9 g/t Au), which reaches 6 g/t Au and 23 g/t Ag in a quartz vein 2.3 m thick, also intercepted by the perforation.

The characteristics of both manifestations allow to classify them as deposits of Th-REE veins type.

The Palo Seco I occurrence was detected by magnetometry, as it is associated with a very intense magnetic anomaly, up to 16,000 nT [44]. The interest of this occurrence lies in the atypical features that it presents with respect to the other mineral deposits of magnetite Fe known up to now in Cuba, as well as in its considerable dimensions. It is hosted in plagioclase-biotitic amphibolites, cut by some dykes of hornblendic-biotitic diorite, affected by a regional metamorphism of, apparently, moderate intensity. In the very nearby Palo Seco locality, large-grain, often porphyritic, biotitic-hornblende syenites were described. From the petrochemical point of view, these rocks belong to the alkaline series with a high K content, with K_2O predominating over Na_2O [87].

The mineralized body is, apparently, tabular in shape. The mineralization occurs in massive and disseminated form, observing the presence of alterations such as potassium (biotitization), sericitization, and argilitization. The Fe is presented in the form of magnetite (main mineral ore), accompanied by pyrite. Four samples collected on the surface [90] from the area where this occurrence is located, allowed to establish the presence of hydrothermally altered basalts (albitization and silicification), phlogopitic-tremolithic rocks (composed of phlogopite and light green mica). There are 7% of magnetite disseminated in the form of metallic powder, as well as quartz—micaceous metasomatites (bluishgreen, micas, quartz, and topaz, the latter very affected by internal kaolinization.

All this points to the possible belonging of this occurrence to the Oxide of Fe–P-ETR Kiruna type, although the data so far available are insufficient to make this classification safe.

A different situation presents the sequences of insular volcanic arcs of the Cretaceous in the Sierra del Purial, in the easternmost section of the Cretaceous Insular Volcanic Arcs Terrane, where they appear metamorphosed, generally in the greenschist facies, reaching locally the schist blue high pressure. The protoliths were volcanogenic-sedimentary successions, with tholeitic and calcoalkaline vulcanite, estimated to be from the volcanic arc of the Upper Cretaceous [66], although the truth is that there are no direct data of age, paleontological or radiometric, to sustain this opinion.

The presence of high graphite content, especially in its middle sequence, is a feature that distinguishes this part from the rest of the Cretaceous Volcanic Arcs Terrane, which presupposes that here the magmatic-volcanic activity of the insular volcanic arc coexisted with an abundant supply of siliciclastic material and with a high degree of preservation of organic matter, perhaps related to the presence of some of the global anoxic events that occurred between the Albian and the Campanian [67, 124].

According to Popov et al. [113], the metamorphic rocks of the Sierra del Purial Fm. are subdivided, according to their protoliths, into three parts or sequences:

- Lower: basaltic and andesitic—basaltic, with rare limestone lenses and siliciclastic material.
- Intermediate: andesitic rocks, with intercalations of diorites, rhyolite-dacite, and carbonaceous siliciclastic rocks rich in organic matter.
- Upper: volcanic-sedimentary section of varied composition, with an andesitic lava-basalt horizon.

Most of the mineral occurrences known in the Sierra del Purial are located in the middle sequence of vulcanite and siliciclastic rocks rich in organic matter. Of all the lithologies present, the Cu–Zn–Au–Ag mineralization host rocks are the different types of graphitic schist, which present a varied mineralogical composition. This composition changes in the following ranges: graphite: 20–25%, calcite: 10–50%, sericite: 15–20%, quartz: 5–20%, chlorite: 5–30%, pyrite: 2–10% [113].

Silicification processes usually affect the aforementioned intermediate sequence. The dykes that cut it, as well as the igneous bodies of little thickness concordant with it also present a schistosity similar to that of the enclosing rocks, which shows that both were affected by regional metamorphism, which occurred after its penetration into this sequence. In it, a Cu–Zn–Au–Ag mineralization was developed (The Asientos, El Ñame, Quibiján, and Aníbal occurrences, of the metamorphosed siliciclastic-mafic VMS type). A feature of these manifestations is the scarce association of the mineral zones with the igneous bodies, reduced to dikes and thin concordant bodies of diorite porphyrites in those cases in which they occur. On the other hand, if the relationship with tectonism is common, with the existence of a lithological and tectonic control of the mineralization clearly expressed.

The controlling and conductive role of the mineralized fluids played by the faults is evidenced by the fact that the law of useful components and the density of the mineralization (which becomes massive) increases with the proximity to the fault zones [99, 113], which is explained, firstly, by the increase in the permeability produced in rocks by tectonic fracturing. This facilitated the circulation of hydrothermal fluids and their interaction with rocks. This geological picture suggests that in addition to the magmatic sources of S and metals, in the formation of the deposits they also participated, to a greater or lesser extent, S reduced from marine sulfate and leached metals contributed by the enclosing sequence.

In the Sierra Maestra-Cayman Ridge insular volcanic arc, of Paleocene-Middle Eocene age, there exists a bimodal-mafic VMS—Ag–Au veins in volcanic sequences—volcanogenic Mn mineral system of Middle Eocene. The first of the models is represented by the well-studied El Cobre ore deposit (from, [5, 10, 39] and to [28, 25]), exploited at various times from 1535 to 2001. Although in that last date the mine was closed, the deposit still contains significant resources of chalcopyrite ores, to which is added the presence of polymetallic and auriferous ores. The mineralization in the deposit is located on the El Cobre Fault, in an area of approximately 1200 m long and 140 m wide. The mineralization has been contoured at depths between 200 and 500 m, with a Cu law of 0.7%. Below this depth range the resources are not delimited. In general, about one million tons of ore have been extracted from El Cobre with a Cu law between 14 and 18% and more than 2 million tons of ore with less than 2% Cu [28].

El Cobre deposit consists of four main types of ores:

1. Stratiform of anhydrite and barite sulfates.
2. Strata bound of sulfides of Cu–Zn with Au
3. Vetiform of Cu and Stockwork of quartz with Cu
4. Sulfates with Cu, Zn ± Pb.

The stratiform ores of sulfates (anhydrite, barite, and cryptomelane) were formed by the discharge of hydrothermal fluids into the seabed. The vetiform mineralization of Cu, on the other hand, constitutes the root zone of the stratiform bodies, with morphology of Stockwork in the superficial part and of subvertical veins in depth. It is dominated by pyrite and chalcopyrite in association with quartz, except in some shallow veins, where the sphalerite associated with anhydrite predominates. In superficial veins, Au may also occur in association with tellurium. The related alterations are chloritization and silicification in the deepest sectors and sericitization and kaolinitization in the most superficial ones.

Strata bound mineralization of Cu–Zn sulfides with Au is contemporaneous with vetiform Cu mineralization and was produced by the replacement of several favorable levels of the embedding sequence: limestones in the lower part and the stratiform of sulfates in the upper part, carriers of most of the existing Au in the deposit (usually in the form of free Au).

This deposit and others similar to it, such as La Cristina, La Mañana, and La Nicolasa, are distinguished by their very low Pb content, which is in accordance with the tholeiitic nature of the enclosing volcanic sequence.

The study of fluid inclusions revealed that the fluids that formed the majority of the mineralization in the El Cobre deposit have salinities between 2.3 and 5.7% by weight of equivalent NaCl, with a formation temperature between 177.3 and 293 °C. These results suggest the contribution of at least three different sources of fluids, whose contribution varied over time: seawater (fluid I), a salinity fluid lower than that of seawater (fluid II) and a saline fluid, with salinity greater than that of seawater (fluid III).

The isotopic composition of sulfur (CDT) in the sulfides of the El Cobre deposit presents a range of values of $\delta^{34}S$ between −1.4 and 6.7‰. These values can be explained by assuming a mixed source of S, represented mainly by S from the inorganic reduction of marine sulfate at high temperatures, with S of magmatic origin, which could be incorporated by leaching from the volcanic sealant and also by direct contribution of magmatic volatiles. In the case of the sulfate from the El Cobre deposit, the isotopic composition $\delta^{34}S$ shows a range of general variation between 16.3 and 20.9‰. These values identify the sulfate of the seawater as the main source of S. On the other hand, the $^{87}Sr/^{86}Sr$ ratio in the sulfate levels is close to that of the seawater, so these mineralizations were predominantly, but not exclusively, formed with the participation of heated seawater. In the case of the lower body of barite, where the ratio $^{87}Sr/^{86}Sr$ is much lower than that of seawater, a mixture of fluids is assumed, in which the marine fluid provided the sulfate and a deep solution provided the Ba and good part of Sr [28].

Perhaps the most relevant aspect of the geology of the El Cobre deposit is the relationship observed on its flank W (Melgarejo sector) between sulfur ore and Mn ore. Here appears a mineralized zone with manganese oxides, which has a total thickness of around 4 m. On its lying side this mineral body is lodged in altered tobaceous rocks, with the presence of nontronite, montmorillonite, and zeolites (stellerite, chabasite, and clinoptilolite) as alteration minerals. At the base of the Mn ores appear bodies of jasper of decimetric thickness, with thin intercalations of iron oxides and manganese oxides. The celadonite alteration, associated with hematite and goethite, develops near the contact of the Mn oxide bodies with the jaspers. The mineral bodies of Mn oxides have a brecciated texture in the lower part, formed by fragments of enclosing rocks cemented by manganese oxides. There are also ores with nodular texture, composed of macfallite and oxides of Fe and Mn. Higher up, these bodies are formed by massive botryoidal ore of Mn oxides, constituted by cryptomelane cemented by very fine aggregates of manganite and grouthite [28]. All this set is cut by veins, a few mm thick, of macfallite and quartz, intercepted by even later veins of calcite. This structure is very similar to that of the most representative manganese deposits of the Cuban-type volcanogenic Mn model (El Quinto, Los Chivos La Margarita, Barrancas-Ponupo IV-Sultana), housed in the upper part of the geologic section in the Sierra Maestra–Cayman Ridge Insular Volcanic Arc.

In the El Cobre deposit, the mineralization of Mn occupies a distal position with respect to that of sulfides, with an evident spatial association, which can also be considered genetic, between both types of mineralization. The latter is suggested by the presence of laumontite veins, which cut off the underlying celadonite alteration. Another element in favor of this is the presence of high contents of Mn in the chlorites and epidotes present in the anhydrite-epidotic veins of the upper part of this deposit. These anomalous contents, which in the epidotes is of the order of 1%, sometimes more, give testimony of the probable enrichment in Mn of the mineralized fluid during its interaction with the volcanic rocks through which it circulated. On the other hand, the deposit of Mn La Margarita reports the presence of scarce grains of chalcopyrite and pyrite, of submicroscopic size, scattered among the crystals of celadonite. Likewise, the presence of anomalous contents of Ba, Sr, and Zn is reported there [27].

This spatial and genetic association between sulfide and Mn mineralization within the same volcanogenic mineral system has been recognized in other parts of the world. A good example of this is the Vani deposit (Greece), hosted in a volcanogenic calcoalkaline sedimentary sequence, accumulated in a shallow submarine environment and locally in subaerial conditions [5, 71, 96].

The vein-shaped deposits of Ag–Au (±Cu) in volcanic sequences of the Sierra Maestra were exploited on a small scale, due to their high content of Ag, between 1920 and 1955 and are very little studied. It is necessary to point out that although in a series of aspects these deposits have a behavior similar to the epithermal intermediate sulphidation deposits, they differ from these by forming in underwater depositional environments and by their scarcity of Pb and Zn.

Deposits of Cuban-type volcanogenic Mn, already mentioned when talking about the El Cobre deposit, represent the highest level of this mineral system. The high degree of study reached by these deposits in Cuba before 1959, [106, 107, 128, 131, 141], made them be used as a world reference for their type [92]. Later studies [1, 27], Cazañas and Melgarejo [26], Cazañas [28] led to the discovery of new deposits and the expansion of knowledge about their characteristics. The distribution area of these deposits covers the upper part of the axial zone of the insular volcanic arc Sierra Maestra-Cayman Ridge and the back arc basin located to the north. A large number of deposits have been discovered here, with masses of ore ranging from 300 tons to more than 4 million tons (La Margarita, Barrancas-Ponupo IV-Sultana). This mineralization, as the name of the model indicates, is genetically associated with the volcanism (submarine) and is hosted in volcanic and volcanoclastic rocks, preferably of intermediate-felsic or intermediate-mafic composition, with a tendency to form groups of deposits. In spite of the intense exploitation developed during the first half of the twentieth century, the deposits of Mn in the eastern region of Cuba are home to a significant amount of resources, most of them discovered and evaluated during the second half of the last century.

A second mineral system present in the Sierra Maestra-Cayman Ridge insular volcanic arc is the Cu–Mo ± Au porphyry—skarn—Ag–Au veins in volcanic sequences mineral system. The porphyry Cu–Mo ± Au type mineralization, represented by the Buey Cabón deposit (0.808% Cu average in an intrusive complex of diorite-hornblende porphyrites and fine-grained diorites) is underdeveloped. This is due both to its masking by the depth of deposit deposition and the tholeiitic character of this insular volcanic arc [31], Cazañas and Melgarejo [26].

The skarn mineralization (of Fe and Fe–Cu) has as main representatives the deposits of the Hierro Santiago Mineral District, located in the southern slope of the Sierra de La Gran Piedra, to the ESE of Santiago De Cuba City. Usually this district has appeared named in the literature published as a Daiquiri or Firmeza. It is therefore necessary to clarify that Daiquiri is geographically the name of a beach and, in the geology of Cuba, is the name of one of the two intrusive complexes (Daiquiri and El Norte) with which the magnetite iron mineralization of this mineral district is related. On the other hand, Firmeza is the name of the wharf—and associated village—disappeared between 1946 and 1950 when the mining works ceased, where the extracted iron ore from the quarries of La Grande, Chiquita, Abundancia (El Este and El Oeste quarries) and Arroyo de la Poza was exported. It is also convenient to clarify that it is not a single deposit, but a set of them, which as indicated, is named in the Cuban geological literature after 1959 Mineral Field or Mineral District Hierro Santiago. It covers the deposits La Grande, Chiquita, Concordia, Yuca, El Norte, Chicharrones, Folia, Vinent, Alfredo, Abundancia, Providencia (Fe skarn), as well as Antoñica, Arroyo de la Poza and Descanso (Fe–Cu skarn). Throughout the period of its exploitation (from 1884 to 1947) of the skarn deposits of Hierro Santiago Mineral District, about 22 million tons of rich ores were extracted, with total Fe between 58 and 65%. The important magnetitic ore resources still existing here have an average law of 40.25% of total Fe.

This set of deposits is located mainly in the exocontact of differentiated mafic-acid intrusives, belonging to the aforementioned Daiquiri and El Norte intrusive complexes, with the volcanogenic-sedimentary sequences of the Paleocene-Middle Eocene (El Cobre Fm.). The skarns are mainly of the calcic type, with varieties determined mainly by the different proportions in which the main minerals are found (garnets, pyroxene, epidote, magnetite, and amphiboles), being the garnet-magnetite and garnet-epidotic-magnetite the most abundant. Associated to them there is a wide development of the clarification (albitization) of the igneous host rocks.

In the Mineral District Hierro Santiago and in other parts of the Sierra Maestra, skarn deposits appear that belong to the Fe–Cu Skarn model [124]. However, this is not the first time that we speak of Fe–Cu skarn, since Smirnov [129], when referring to copper skarns, states that in several regions they are intimately associated with the magnetite ores, forming deposits of magnetite and copper complex skarn. On the other hand, Kesler [70] describes skarn with Cu and Fe in Meme, Haiti, while Harnish and Brown [60] point out the existence of skarn deposits of this type in this same country (Casseus and Meme).

Similarity to the IOCG, the Fe–Cu skarns contain Fe oxides, Cu and Au. A distinctive feature related to this is the presence of significant concentrations of sulfides in ferrous magnetite ± hematite ores. Although they present albitization and other features that resemble them, these skarns are clearly distinguishable from the IOCG deposits [98, 119, 136, 95].

The Fe-Cu skarns of the Hierro Santiago Mineral District are characterized to contain from 36.75 to 41.66% of Fe, between 0.5 and 8.9% of Cu; 0.03 g/t to 2 g/t Au, from 0.7 to 1.40% Zn and 2.2 g/t to 60 g/t Ag. The main ores minerals are magnetite, hematite, pyrite, and chalcopyrite, which are accompanied by martite, limonite, goethite, and maghemite, by sulfides and secondary minerals such as sphalerite, galena, bornite, pyrrhotite, covelline, cuprite, chalcocite, malachite, cubanite, azurite, and chrysocolla. Electrum and native gold may also appear. Outside this mineral district, in the western part of the Sierra Maestra, there are some deposits of Fe and Fe–Cu skarn of some importance, such as Sexta and Camaroncito.

Deposits of the type Ag–Au veins in volcanic sequences belonging to this mineral system are, until now, of little interest.

8.4.3 Collision Mineral Systems

The deposits associated with the collision processes are one of the most complex problems in the metallogeny of Cuba. In the Cuban territory two events of this type are recognized:

1. Campanian–Maastrichtian: Subduction-collision of the southern distensive continental margin, corresponding to the Maya Block, with the Cretaceous Insular Volcanic Arcs Terrane.
2. Paleocene-Middle Eocene: Collision of Insular Volcanic Arcs Terrane, of magmatism extinguished by the previous collision, with the Florida-Bahamas Block, a process that also involved the foreland basins.

Although it also occurs in the Guamuhaya Terrane, the Campanian–Maastrichtian orogenic mineral system has its best representation in the Pinos Terrane, where the regional metamorphism associated with a collision process continental margin—insular volcanic arc type caused the appearance of generated felsic intrusive bodies by partial melting of continental crust. With this collision magmatism are linked the mineralizations of Au-Sb and W, Mo, Cu, and Au present in the western half of the Isle of Youth. The Delita deposit (Au orogenic hosted in siliciclastic sequences) represents the first mineralization type and the second type by the Lela deposit (veins of W). These deposits are related to the presence of felsic igneous dykes and are located within Jurassic siliciclastic sequences of the synrift, affected by a regional metamorphism of low grade to high grade and relatively high pressure [51]. Ar/Ar data from micas and amphiboles yield about 72 million years, which allows us to consider that the high pressure metamorphism is from the Upper Cretaceous. The Ar/Ar dates in biotite and muscovite, obtained in medium to high grade phyllites, consistently yield 68 ± 2 million years, which is interpreted as the age of cooling [66].

The predominant magmatism present in the Pinos Terrane is of felsic composition and, possibly, highly evolved. This affirmation is supported by the physical properties (low density and magnetic susceptibility) deduced from the behavior of physical fields, by the close spatial relationship of some of the bodies of igneous rocks interpreted by geophysics with felsic magmatism outcrops, by the presence of uvite Fig. 8.6, a relatively rare mineral related to magmas more evolved and rich in volatiles, due to the absence of a relationship between these intrusive bodies and geochemical anomalies of siderophile elements (Cr, Ni, Co), due to the clear spatial association of the mineralization of Mo and W with these bodies and the outgrowth of the magmatism, as well as the presence in the rhyolites sampled of high contents of the radioelements U and Th [29]. This is supported, in the case of the intrusive body of Lela, due to the existence of radioactive anomalies associated with riodacites, as well as to quartz-tourmaline veins, detected in wells depths greater than 100 m, [12].

According to the results of the study of fluid inclusions, the deposition of the minerals of the first stage occurred at temperatures of 270–400 °C and a pressure of 1.5 Kbar. The second stage of mineralization occurred at temperatures between 210 and 320 °C, at a pressure of 900–980 bars [19, 81].

In the main mineral body, the cataclastic texture in arsenopyrite is observed Fig. 8.7, which could be indicating a certain degree of tectonic affectation of the ore after its formation, produced by some compressive tectonic event. Such type of texture has also been described in quartz-tourmaline veins in Lela and may have an explanation similar to this.

Although Delita is, until now, the most important metallic deposit known in the Isla de la Juventud, the Cu–W–Mo ± Au, Ag Lela prospect is undoubtedly the most interesting and complex. Also located in the Cañada Fm. (J_{1-2}), the presence of W was apparently, discovered around the year 1900. In 1940, a plant was built to process from 200 to 250 tons of ore per day. The deposit geology was studied by Page and Allister [105] when it was in exploitation, which continued until 1945.

This prospect is housed in graphite-micaceous-quartz schists that rest in concordant contact under sericite, biotite, and bimicaceous quartzite as well as schists of similar composition, all belonging to the aforementioned Cañada Formation. It is located within a large area of greisenization of about 10 km long and 2 km wide, outcropping only in a third of its total extension and associated with the Siguanea Tectonic Zone, of NNE heading. To this system there is associated a high number of rhyolitic dikes and some of

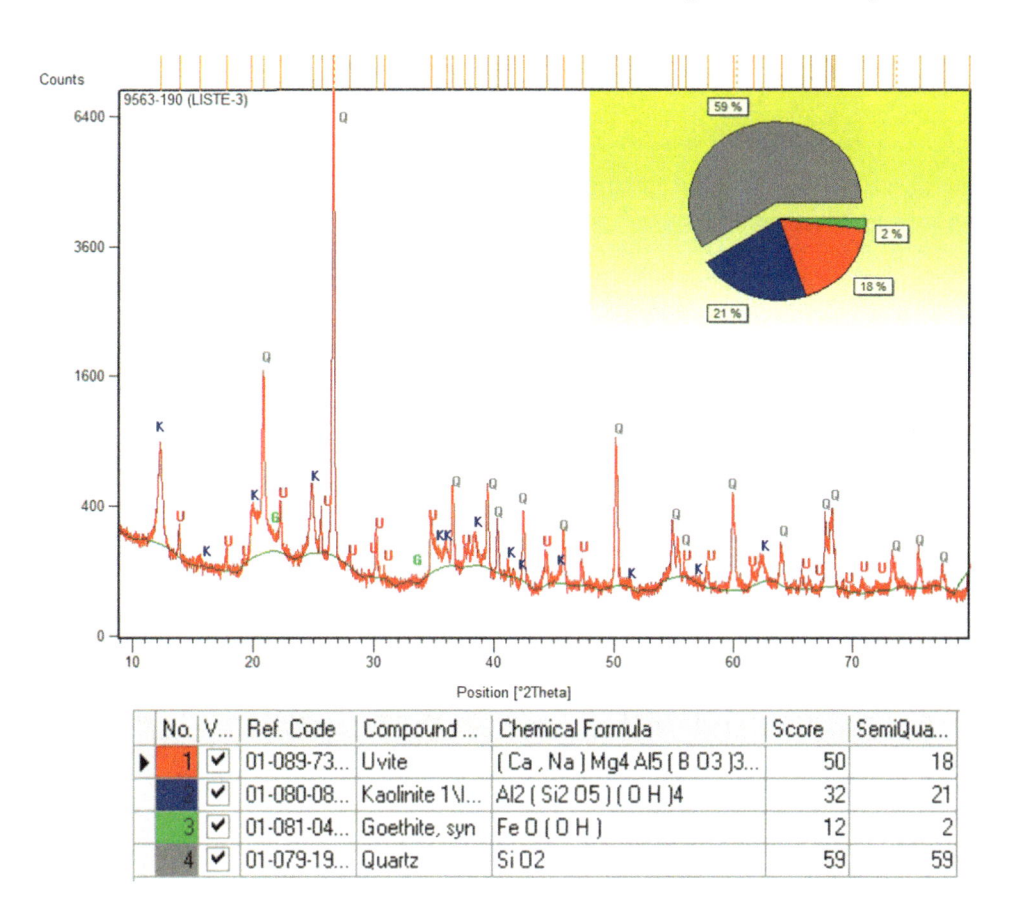

Fig. 8.6 Diffraction of RX of sample LISTE-3, taken in a greisen outcrop in Santa Elena, NW Isla de la Juventud. Symbols used: 1—U-uvite, 2—K-kaolinite, 3—G-goethite, 4—Q-quartz. The goethite here appears in small quantity. There is a relatively abundant phase of uvite (member of the solid solution dravite-uvite) of the tourmaline group, which is a relatively rare mineral and its presence is frequently related to evolved, volatile rich magmas and other characteristic elements of these

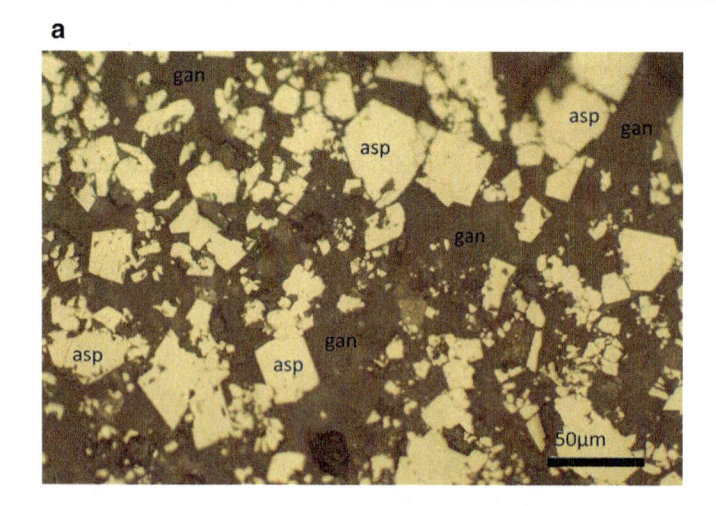

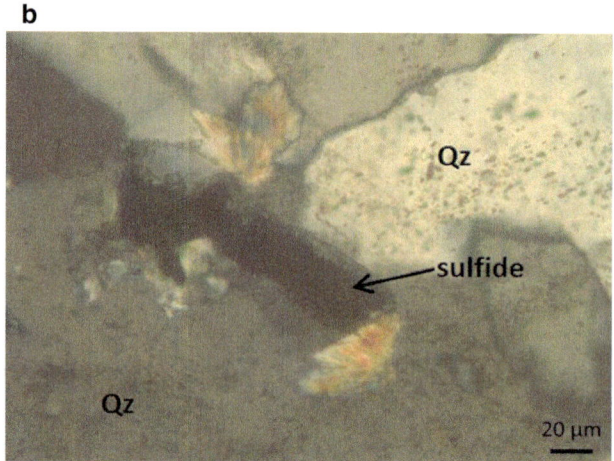

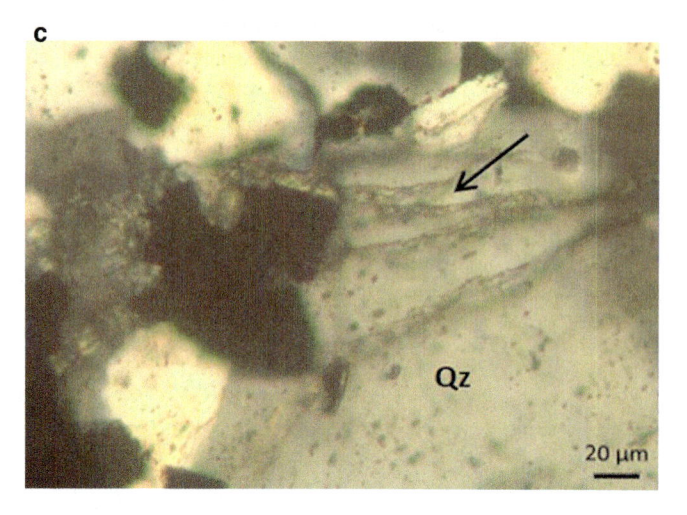

Fig. 8.7 a–c Polished sections of LID-4 sample. **a** Euhedral and idiomorphic crystals of arsenopyrite (asp) with presence of cataclastic in gangue minerals (gan) in section polished. Parallel Nicoles. **b** Thin section with parallel Nicoles. Quartz with abundant fluid inclusions and sulfide mineralization bordered by scaly mineral. **c** Thin section with crossed Nicoles. Quartz (Qz) with abundant fluid inclusions and sulfur bordered by scaly mineral, this also in veinlets

lamprophiids, in its majority extended with courses between the N and the NE. Five to ten percent of these dikes are younger than the rest, showing directions from NW to N. The dating by the K–Ar method of a sample taken from one of the riodacites dikes gave an age of 60–68 Ma [22], which is considered the approximate age of intrusive bodies spatially and genetically associated with mineralization.

Investigations with electronic microscopy (SEM–EDS) in samples from the main sinkhole of Lela, preserved in the Mineralogy and Petrology Archive of the IGP (Geology and Paleontology Institute, by initials in Spanish), allowed to verify that the ferberite is the main mineral carrier of W in Lela [29]. This mineral occurs in nests from 1 to 2 mm up to several centimeters, distributed in an irregular way in the veins. Other less frequent minerals are wolframite and scheelite. In addition to the minerals of W, pyrite, arsenopyrite, marcasite, pyrrhotite, and chalcopyrite appear in the central part of the deposit. Galena, sphalerite, bornite, stannite, magnetite, ilmenite, and rutile, as well as pilsenite, tetradimite, alexite, cosalite, bismutite, and native gold are also rarely found.

The most important finding made in this prospect after 1959 is the discovery of the presence of Cu mineralization, mainly toward its NE flank and that of Mo, located toward its SW flank [11, 12, 49, 123]. Detected in several parts (Garapko et al. [50], Sánchez et al. [148]; Avdeev et al. [11], Fernandez et al. [48]), gold mineralization has never been the subject of a systematic study at the scale of all the deposit. Although there is no firm quantitative relationship between the contents of Au and As, if it is well established that when the As increases, the Au content increases. It should be noted that the content of As here is several times lower than in the previously exposed Delita prospect.

The analysis of the results obtained by the group of authors cited shows that the Lela prospect is not a simple deposit of W, but a Mo, W, and Cu ± Au, Ag complex deposit, with three types of mineralization: porphyritic Mo, veins of W (and Cu–W) and concordant copper. It is characterized by a clear horizontal and vertical geochemical and mineralogical zoning, where the mineralization of Mo is the most proximal to the intrusive and that of Au the most distal.

As can be seen in Fig. 8.8, toward the NW part of the area there is an auriferous mineralization (accompanied by As in the form of arsenopyrite). This mineralization has been studied in more detail in the Aeropuerto sector but is by no means limited to the same. It also appears in other parts of the deposit superimposed on the mineralization of W. As can be seen in the aforementioned figure, in addition to the well-studied area of the prospect (named Lela in the figure), there is another mineralization core of W in the Siguanea zone, with a much lower degree of study. Between both cores extends, from NW to SE, an area (NE Lela) where copper mineralization predominates, which unlike that of W and Cu-W, is concordant, being better studied in its SE portion, located at the ENE of the best-studied area of Lela. The fact that Cu is hosted in different lithologies (quartzite, sericite graphite, and micaceous graphitic schists), not infrequently cracked and breached, suggests the possibility that it is a mineralization by lateral replacement along inter-strata faults.

Finally, and not least important, to the SW of Lela three wells (two drilled in cross and one vertical) intercepted a mineralization in porphyry Mo Stockwork hosted in rhyolites and metamorphic rocks, associated with the presence of greisenization, tourmalinization, and silicification, as well as with quartz-tourmaline veins. As seen in Fig. 8.8, these wells are located on the NW flank of the main intrusive body mapped by geophysics. One of them (vertical well P-2) reached 463 m depth and in its last 60 m (from 403 to 463 m) it advanced in silicified, sericitized, and carbonated riodacites porphyries, with abundant presence of quartz-tourmaline veinlets. This suggesting that it may have reached an apophasis of the mentioned felsic intrusive body, probable source of metals, and heat of the hydrothermal forming system of at least a part of the deposit, although the geophysical interpretation indicates a deeper lying.

A fact of interest shown in Fig. 8.8 is that the main intrusive body mapping by geophysics is located at the S and SE of the investigated mineralized zone. Another possible intrusive body, also delimited by geophysics, located on the E flank of the same, not studied until now,where it appears a best-developed system of faults oriented in the same direction of those that control the mineralization known in Lela and its surroundings.

Another well-established fact in the studied part of the Lela prospect is the presence of a marked geochemical zonality that from SW to NE is Mo–W–Cu–As–Bi, which has been interpreted as a reflection of the spatial and genetic relationship between mineralization and existing hypabyssal and subvolcanic magmatic activity [11]. However, in Fig. 8.8 this zonality is not clearly observed, except in the case of the Au (and the As, not shown), which appear in areas clearly far from the intrusive delimited in it. This has two causes: (1) Differences in the level of erosive cut between the different tectonic blocks in which the deposit is divided and (2) "Telescopic" effect produced during the cooling of the igneous bodies and the hydrothermal system associated with them.

El Descanso locality, on the other hand, is relevant for the finding of wolframite grains crushed and oriented in the direction of the schistosity, which could constitute an evidence of the existence in the Pinos terrain of pre-metamorphic or synmetamorphic, mineralization of W Fig. 8.9 in addition to that associated with the intrusive bodies after metamorphism.

In part W of the Isla de la Juventud (Pinos Terrane), within the rich in organic matter sequences of the Cañada Fm. (J_{1-2}) and Agua Santa Fm. (J_{2-3}^{ox}), there is a close relationship between gold mineralization and the presence of pyrite and arsenopyrite, very well expressed in the area of the Au-Sb Delita prospect, the largest gold deposit in Cuba, and to a lesser extent, in the W-Mo-Cu ± Au Lela prospect. The authors are of the opinion that in this case is applied the genetic model of formation of the orogenic Au deposits at the basin scale, divided into two stages, proposed by Large et al. [74]. In this model, the source of Au and As are siliciclastic sedimentary rocks rich in organic matter.

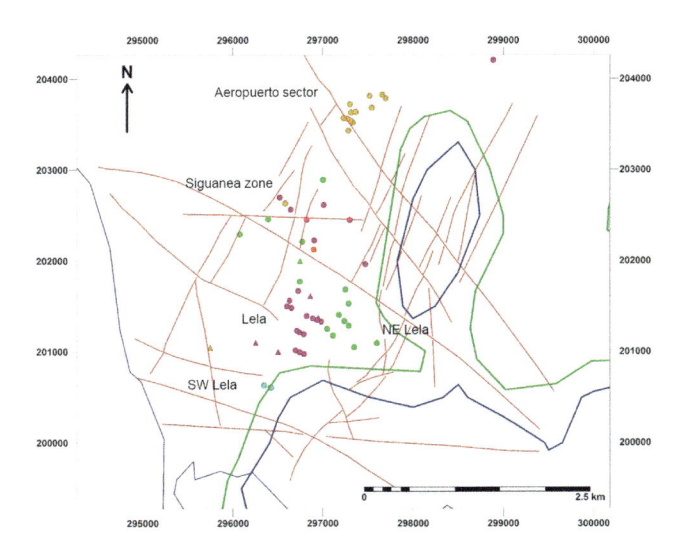

Fig. 8.8 Distribution Map of useful components in Lela. Legend. Circles: wells. In golden color Au; in light green Cu; in blue cyan Mo and in magenta W. Triangles: outcropping mineralization; coloring for Au, Cu, Mo, and W as in the case of wells. Red lines: faults. Dark green thick outline: Possible contour of intrusive body at 600–800 m depth. Dark blue thick contour: Possible contour of intrusive body at 400–600 m depth (in both cases deduced from the interpretation of the gravimetric data). The scale of the map does not allow to show the abundant faults present in the central area of the deposit

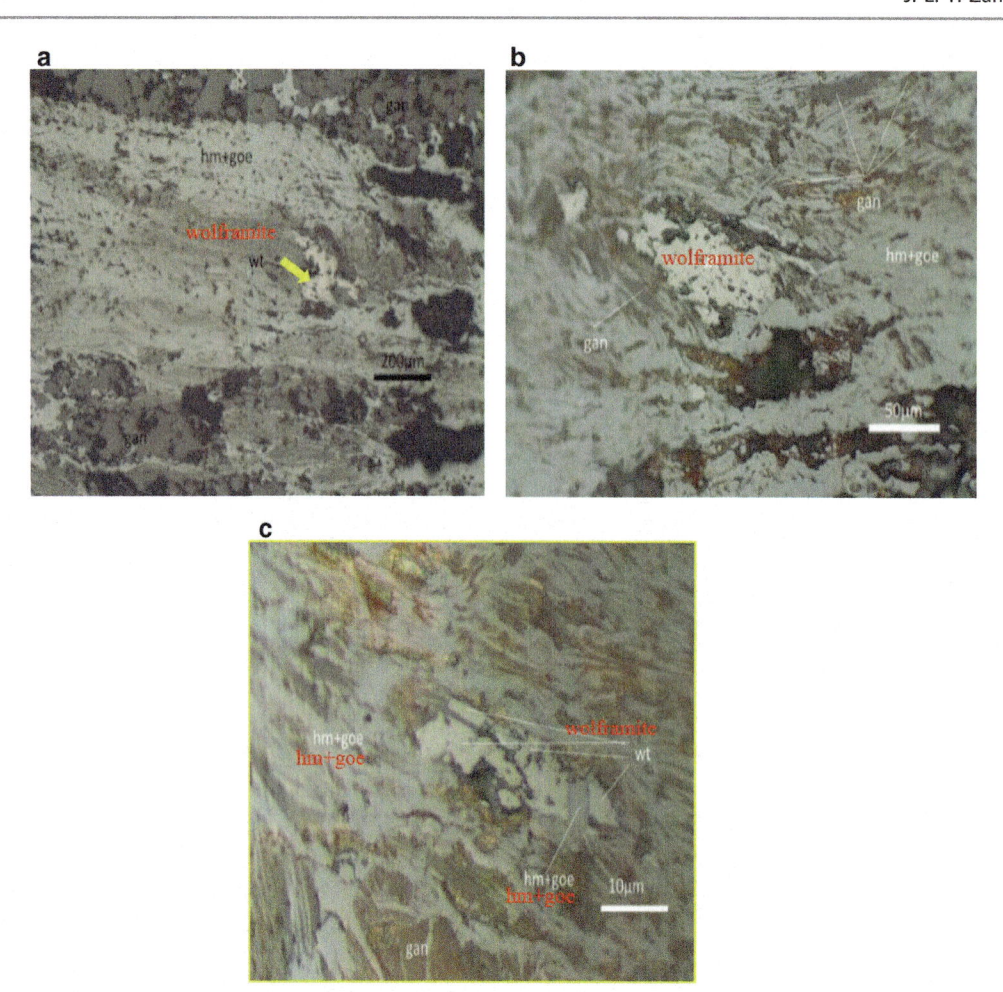

Fig. 8.9 Results of the description of the polished section corresponding to a sample of ferricrete by schists of the. Cañada Fm. (J$_{1-2}$). **a** and **b** Hematite + goethite (hm + goe) with texture banded according to the direction of schistosity and Wolframite (wt) in gangue minerals (gan). Parallel Nicoles. **c** Grains of hematite and goethite intercepting wolframite. Parallel Nicoles

With respect to the pyrrhotite, in the analyzed territory there are only three wells outside the known mineral deposits that report a significant presence of pyrrhotite. This suggests that the main horizons of conversion of pyrite to pyrrhotite are located at more or less considerable depths. In relation to this, it has to be in the Bendigo deposit—a classic example of an orogenic Au hosted in siliciclastic sequences, a deposit model to which the Delita prospect belongs—the carbonaceous schists that exhibit the conversion of diagenetic pyrite to metamorphic pyrrhotite are developed at 300–1000 m below the mineral bodies [74].

Unlike the other segments of the Cretaceous Volcanic Arcs Terrane, the geodynamic environment of collision dominates the NW of Holguín province. It is characterized by the presence of subvolcanic bodies, younger and tectonically aligned, which cut all the elements of the over thrust sequence (volcanic-sedimentary rocks, ophiolites, and tectonic zones) intensely scaled and imbricated. Related to the above, here is presented the Paleocene-Middle Eocene orogenic mineral system, which contains two models of deposits, characterized by a mixed magmatic-tectonic control, and more than one mineral event may be present.

This mineral system responds to a continental margin-insular volcanic arc collision type, which involves the cretaceous sequences of insular volcanic arc and those of the continental margin of the Florida-Bahamas Block. This process that also affected the foreland sequences located between them and whose consequence was the location in allochthonous position of considerable volumes of mafic and ultramafic rocks, as well as the development of a subvolcanic magmatism of mafic, intermediate, and felsic composition. This mineral system, although it has representation throughout the entire mafic–ultramafic Cuban belt, reaches its highest development in the northwest of the Holguín province, where deposits of Mother Lode type and Cu–Zn–Au–Ag orogenic are located. Also associated with this mineral system, in the provinces of Guantánamo and Havana, some manifestations of Cu–Ni orogenic appear.

As already explained, the fundamental element in the geological environment in which these deposits are developed is the presence of younger subvolcanic bodies that cut the pre-existing overthrust sequences. In favor of the younger age of said bodies, Pentelenyi et al. [109] point out the following facts found by them during the field study of the Las Cuevas prospect:

- The subvolcanic body has an almost ideal circular shape (1300 m × 1400 m), which contrasts with the intense tectonic deformation of the environment in which it is located.
- The subvolcanic body discordantly cuts the contact between the bands of serpentinites and volcanic-sedimentary rocks of insular volcanic arc.
- There is active igneous contact between the subvolcanic rhyolitic body and the serpentinite in its northern part.
- By geophysics (magnetometry and resistivity) it is verified that this rhyolitic body is continuous up to depths greater than 150 m, which suggests that it does not constitute a block inside the mélange.

Other arguments in favor of the posterior location of the subvolcanic bodies in the overthrust sequence are the following:

- In the outcrops and wells made in the quarry of the Reina Victoria deposit it is observed that the hydrothermal alterations affected the already cataclastic enclosing rocks, reason why they are later to the tectonic location of the ophiolites.
- The places where subvolcanic bodies appear are well distinguished in the magnetic and gravitational fields, both regionally and locally. This suggests that they are not mega blocks without roots, but apical bodies belonging to a deeper focus.

As an additional element it can be pointed out that the intrusive bodies of mafic and intermediate composition existing in the Nuevo Potosí prospect contain numerous serpentinite xenoliths. This also supports the thesis that the subvolcanic bodies penetrated after the overthrust of the ophiolites and volcanic-sedimentary Cretaceous rocks or toward their final phase. Likewise, erosion witnesses Colorado and Cerro Boston hills, linked to the presence of volcanic necks, seem to testify to the cutting nature of the bodies and their subvertical position.

In general, Mother Lode type Au orogenic deposits present in the northwest of Holguin (Nuevo Potosí, Reina Victoria, Agrupada, Holguinera, Cuatro Palmas, Floro Pérez and others) are characterized by Chaliy and Dovnia [30], Pentelenyi et al. [109], Costafreda et al. [33, 34, 35]:

1. Presence of ophiolitic, subvolcanic-intrusive, and volcanic-sedimentary lithologies. Generally, the host rocks of the mineralization are of the first two classes.
2. Spatial and genetic association of the mineralization with the subvolcanic bodies (from basaltic to rhyolitic) and the tectonic zones, usually corresponding to transcurrent and/or overthrust faults.
3. Two sets of hydrothermal alterations are clearly distinguished. One for the subvolcanic bodies and the other for the ophiolites and tectonic zones (faults and shear zones). The difference between both sets increases with the silica content of the subvolcanic rocks.
4. The auriferous mineral bodies of high and medium grade are formed by quartz and quartz-carbonated veins, sometimes surrounded by an extensive and diffuse zone of poor disseminated mineralization.
5. The mineralization is predominantly auriferous and may be accompanied by Cu, Zn, and Ag.
6. Mixed tectonic–magmatic control of the mineralization, exercised mainly by the faults and subvolcanic bodies. The observed picture can be additionally complicated by the lateral displacement of mineral zones and bodies under the action of younger transverse faults.
7. There is usually no clear delimitation between the mineralized rocks and the sterile embedding rocks, because the thin disseminated gold mineralization of low grade is hosted in all the lithologies present in a more or less indiscriminate way.
8. Pyritization behaves in a similar way to disseminated gold mineralization, becoming intense in some places.
9. It is common that the host lithologies of the mineral deposits are metamorphosed in the zeolite and prehnite-pumpellite facies. In some places they acquire the appearance of green schists.
10. Gold occurs in a native form, electrum, and amalgam, as well as associated with arsenopyrite.
11. Presence of a generally complex geochemical signal, with Au, Ag, Cu, Zn, Pb, Ba, and Ga as main indicator elements, to which are added Ti, Co, and Ni, related to the ophiolites.

These deposits present a wide mineralogy, in which the main minerals of ore are pyrite, chalcopyrite arsenopyrite and native gold, accompanied by pyrrhotite, marcasite, sphalerite, galena, gersdorffite, electrum, amalgam, cinnabar, pentlandite, tellurium-bismuthine, melnikovite, mispiquel, maquivanite, bournonite, chromite, ilmenite, rutile, magnetite, leucoxene, limonite, malachite, azurite, rhodochrosite, chrysocolla, ankerite chalcocite, and covelline. The main gangue minerals are quartz, calcite, talc, and chlorite, accompanied by epidote, garnet, vesuviana, chrysotile, magnesite, and chalcedony. Equally varied are the present hydrothermal alterations: silicification, carbonization, chloritization, talcitization, epidotization, dolomitization, albitization, argilitization, sericitization, secondary serpentinization, amphibolitization, and skarnification.

The microprobe analyses performed on 16 samples of the gold mineralization of the Nuevo Potosí prospect Lopez Kramer [81] show the existence, in all cases, of an intimate association of Au and Ag with Hg, whose content ranges from 0.14 to 0.94% in native gold samples and from 7.57 to 12.02% in amalgam samples. As shown by Pein [108], the precipitation of Hg occurs as the acidity of the hydrothermal solution that transports it increases. This allows to suppose that during the formation process of the gold ore, the pH of the mineralizing solutions tended to decrease.

An interesting fact related to the Nuevo Potosí prospect is the identification and description [93] in high purity gold grains (90–95% by weight of gold), based on scanning electron microscopy, of the presence of cubes and octahedrons of 5 Mk of native silicon. According to Lopez Kramer et al. [83], these data give rise to the new mineral and in MinDat they define the Nuevo Potosí deposit as the type locality of the native silicon. A visit to this website (mindat.org URL: http://www.mindat.org/min-3659.html) in March 2015 allowed the authors to confirm that the Nuevo Potosí deposit appears as the first location where registered this mineral, also described in localities of China, Russia, USA, and Turkey.

In relation to the thermal regime of ore formation, the presence of minerals, such as chalcedony (Las Cuevas), opal (Holguinera), amalgam, and electrum (Nuevo Potosí), indicates formation temperatures between 200 °C and less than 100 °C. The first two minerals mentioned also indicate the acid character of mineral fluids. Lopez Kramer et al. [7, 8] used the calcite-dolomite pair (a single sample), to determine the ore formation temperature of Nuevo Potosí: 330 °C. All of the above suggests that at the beginning of the mineralization process the temperature of the hydrothermal fluids was relatively high, possibly higher than 300 °C. At the end of the process, the temperatures were of the order of 100 °C and below.

In this same region of the NW of Holguín, along the longitudinal tectonic zones, there are also deposits of massive sulfides with significant Cu and Zn contents, accompanied by Au and Ag, sometimes with Ti and rarely Co. These deposits (Santa Maria, Mayabe, Charco Prieto, and Las Margaritas, among others) have been assembled in an independent deposit model: that of Cu–Zn ± Au–Ag orogenic [136]. They are characterized by a marked structural control and by appearing not infrequently at the border between contrasting units due to their lithological composition. The mineralization is veinlet-disseminated dense and massive, developed in mineral bodies of lens, tabular, or filonian form, hosted in zones formed by breccias and mylonites, composed by the lithologies present on both sides of the tectonic zone.

Its mineralogy is wide and complex. Pyrite, chalcopyrite, and marcasite appear as major minerals. Subordinate and rare minerals are bornite, covelline, chalcocite, azurite, malachite, limonite, hematite, goethite, barite, native copper, native gold, amalgam, sphalerite, antimonite, titan magnetite, rutile, leucoxene, magnetite, cinnabar cassiterite, and chromite. The main gangue minerals are quartz and the serpentine group, accompanied by calcite, gypsum, epidote, talc, chlorite, biotite, kaolinite, montmorillonite, plagioclase, pyroxene, tremolite-actinolite, and sericite. The present hydrothermal alterations depend, largely, on the lithological composition of the affected enclosing rocks, being the silicification, chloritization, talcitization, carbonatización, argilitization, kaolinitization, epidotization, albitization, and amphibolitization the most frequent. This mineralogical variety is reflected in its complex geochemical signature: Cu, Zn, Ni, Au, Ag, Pb, Ba, Mo, Ti, V, Mn, Cr, and Co.

Some of these deposits (Mayabe and Santa Maria) have features that suggest that they were originally volcanogenic massive sulfide deposits of the mafic–ultramafic type. They are.

- High sulfur content.
- Pyrite-chalcopyrite mineralization accompanied by Au and Ag.
- Presence of massive sulfides
- Predominance of ophiolites in the composition of carrier breccias and mylonites.

In these cases, considerable parts of the ores of these deposits would have been crushed and relocated tectonically during the collision, to then experience the action of new hydrothermal fluids, especially in the final phase of the event, when the compressive stress began to diminish, favoring the opening of cracks and fissures in fractured rocks. Monte Rojo is the deposit that best preserves the characteristics of the VMS mafic–ultramafic mineralization, so that in Cazañas et al. [29] it is classified as such, despite of having also some features of the deposits of Cu–Zn ± Au–Ag orogenic, as it is the presence of a subvolcanic rhyolitic body in the western part of the deposit.

In other cases (Picos Verdes, Las Margaritas, Charco Prieto, Vista Hermosa), the polymetallic sulfide mineralization is accompanied by titanium mineralization, sometimes cobaltiferous and chromite, of ophiolitic origin, which speaks in favor of leaching by hydrothermal fluids from the mafic and ultramafic rocks during the mineralization process.

8.4.4 Metallic Mineralization in Ophiolites

Although both are housed in mafic and ultramafic rocks, representative of the oceanic lithosphere tectonically located on the continental margins, the podiform chromitites and the mafic–ultramafic VMS do not constitute parts of the same mineral system, remaining as "loose models" in the general picture of the Cuban metallogeny. An important and influential feature in the metallogeny of the ophiolites of Cuba is that their formation took place in an environment of supra subduction [8, 62, 63, 79, 84, 114, 121].

The orthomagmatic mineralization of chromitites of the podiform chromitites type (Merceditas, Casimba, Cayo Guam, Mamina, Ferrolana, Camagüey I–II, Victoria I–II) is associated with the harzburgite tectonites and the mantle-crust transition zone. The mineral bodies are predominantly tabular and lent form. Their size is variable and they have a wide range of chemical composition, ranging from chromite rich in Cr (metallurgical type) to chromite rich in Al (refractory type). The most important deposits of chromitites in Cuba are located in the Mayarí, Moa-Baracoa and Camagüey ophiolitic massifs. In the second of the mentioned massifs, the largest podiform chromitites deposit in Cuba and the American continent (Merceditas) is located, with original total resources of more than 5 million tons [115]. On the other hand, in the third of the mentioned massifs there are more than 300 deposits and occurrences of chromitites, ranging from a few tons to more than 0.5 million tons.

On the presence of PGE, Distler et al. [42] obtained platinum and palladium results in samples of chromitites from several chromium deposits in the eastern region of Cuba:

- Merceditas 0.05 g/t of Pt; 0.01 g/t Pd.
- Amores 0.004 g/t–0.06 g/t Pt, 0.002 g/t–0.4 g/t of Pd
- Potosi 0.004 g/t Pt; 0.001 g/t–0.003 g/t Pd
- Caledonia 0.003 g/t–0.006 g/t of Pt; 0.001 g/t–0.07 g/t Pd
- Blas 0.004 g/t Pt; 0.03 g/t–0.04 g/t Pd

Additionally, they established the presence of the minerals of the laurite-erlichmanite group in the chromitites of the deposits Merceditas, Amores, Potosí, Cayo Guam, Blas, Albertina, Caledonia, and the Cajálbana occurrences, the latter located in the Pinar del Río province.

Also González Jiménez et al. [57] established the following platiniferous minerals in the Sagua de Tánamo chromitites: laurite, irarzite-hollinworthite, alloys, and oxides of Ru–Os–Ir–Fe–Ni–(Rh) and iridium (Ir).

Proenza et al. [114] and Gervilla et al. [55] report PGE contents in deposits of chromitites rich in Cr, representative of the Mayarí Mineral District and part of the Sagua de Tánamo Mineral District (up to 3.7 mg/kg of total content). They also suggest that of all the deposits studied, the chromitites that show the highest PGE contents are those from the Caridad deposit (Sagua de Tánamo Mineral District), which systematically present total PGE contents >1 mg/kg. They also point out that sulfide-rich chromitites from the Potosí deposit in the Moa-Baracoa Mineral District show EGP contents of up to 1.1 mg/kg. On the other hand, in the NE of the province of Camagüey, Victoria I–II and Camagüey I–II (outcrops and tailings) were reported Au contents between 0.7 and 1.3 g/t.

The results of all these studies of the mineralogy and the content of PGE in chromitites show that only chromitites rich in Cr are enriched in minerals of platinum group elements, while the chromitites rich in Al do not contain them. The textures observed and their composition indicate that all these minerals found are of magmatic origin, formed during the crystallization of chromite. Likewise, with high Cr chromitites, high Au contents are associated, reporting in them from 0.5 g/t to 5.1 g/t Au, with up to 1.5 g/t Ag.

The metallogeny of the mafic–ultramafic sequences is completed with the presence of mafic–ultramafic VMS deposits type (Júcaro in Pinar Del Río, Río Negro in Camagüey and Monte Rojo in Holguín) and small dimensions Au–Ag veins of the Mother Lode type. The set of present models is completed with the deposits of Cu-Ni orogenic. There is also presence of deposits of Mn of oceanic type (La Ligera occurrence, Bahía Honda municipality, Artemisa province).

8.4.5 Intraplate Environment

The supergene mineral system, developed in the Neogene and the Quaternary, is of great metallogenic importance for Cuba, since it is related to its known Fe–Ni–Co lateritic deposits. They are represented by the Pinares de Mayarí Oeste, Luz Sur, and Camarioca Norte deposits (Fe–Ni–Co supergenic lateritic type), by Punta Gorda, Moa, Marti, San Felipe (Fe–Ni–Co supergenic type lateritic-saprolite), and Levisa Norte and Punta Gorda Norte (Fe–Ni–Co type sedimentary).

The laterites cover in Cuba 25–30% of the outcrop surface of the mafic–ultramafic massifs, with an average thickness of 10 m. They are characterized by a vertical zonal lithological structure that varies in a horizontal sense, with the regular omission of some (s) of the lateritic or saprolite lithological zones [136]. The first two mentioned models possess practically all of the Fe, Ni, and Co resources of the weathering mantles developed on ultramafic rocks in Cuban territory. They have an essentially elluvial genesis and form typical supergene deposits. In these deposits, the cumulative chemical elements in the supergene environment produce industrial and potentially industrial metallic concentrations of Fe, Ni, Co, Sc, Al, Cr, Mn, Ti, Au, and PGE.

The presence of PGE in Cuban laterites is explained by their existence in dispersed form in the parent rocks (ultramafic and mafic rocks) from which they are formed. The PGE is concentrated chemically and mechanically in the weathering mantle, along with other heavy elements such as the Au. The laterites also receive the contribution of the deposits of chromitites rich in Au and PGE that become part of them when their enclosing rocks and ores are weathered.

The presence and behavior of PGE and Au in ferruginous ores and in leached serpentinites of Nicaro and Moa has been demonstrated in the works of Distler et al. [42], Proenza y otros [116]; Lazarenkov et al. [76]; Proenza [117] and Aiglsperger et al. [2]. The last mentioned authors suggest that the concentration of PGE in the deposits of Moa Mineral District varies between 27 and 61 ng/kg in saprolite and from 109 to 239 ng/kg in the limonite-rich crust. They also point out that the concentration of Sc varies between 8 and 17 ppm in the saprolite and from 70 to 98 ppm in the limonite-rich weathering mantle.

Subsequently, 14 samples from the Camarioca Este (eight samples), Yagrumaje Sur (two samples) and Yagrumaje Norte (four samples) deposits were sent to Canada, where they were analyzed by neutron activation to determine Au and PGE; REE and Ag were determined by ICP-MS) and ICP-OES [78]. Although for the small number of samples taken cannot be considered fully representative of the average contents of PGE and scandium present in these deposits, they are sufficient to argue the perspectives for this type of mineralization in the lateritic ores of Cuba.

Of the 14 samples taken, 6 reported PGE contents greater than 0.2 g/t, with values of more than 0.3 g/t being reported in two of them [78], which indicates that they are ores with potentially economic PGE contents. The average contents obtained, of 0.202 g/t for Camarioca Este, 0.298 g/t for Yagrumaje Norte; 0.178 g/t for Yagrumaje Sur, with a general average for the 14 samples of 0.226 g/t, agree well with the averages presented (109 to 239 ng/kg) by Aiglsperger et al. [2], which reinforces the reliability of the results obtained by Cuban researchers.

Besides the lateritic mantles, in Cuba there are bauxite and other types, developed on vulcanite and intrusives of diverse compositions, which include granitoids and alkaline rocks. There are also weathering mantles on metamorphic rocks, particularly those of siliciclastic protolith, in which concentrations of rare and scattered elements of interest could be located. This last type of weathering mantle occurs in the Pinos Terrane, where supergene deposits of Fe appear formed by dense elluvial-delluvial surface accumulations of limonitic pellets. The content of Fe in these deposits ranges between 18 and 44%, sometimes accompanied by up to 9,5% of Mn. A good example of this type of mineralization is the San Francisco deposit.

This mineral system is completed with the presence of Au–Ag gossans, developed as a result of the supergene oxidation of sulfides ores, as in the case of Hierro Mantua, Loma de Hierro, San Fernando, and the Castellano and Santa Lucía deposits.

The sedimentary mineral system covers the Au placer deposits—and Au paleo-placer—(Río Yabazón, Río Naranjo) and Cr (Tau, Toa). Since the hydrographic network is characterized by the small length and generally limited flow of the rivers, these deposits are small in size and of little significance. In addition to the alluvial placers, there are also alluvial-coastal ones, also of small size.

8.4.6 General Facts of Cuban Metallogeny

The general examination of what has been exposed so far allows for a set of general inferences about the metallogeny of Cuba. In the first place, it is possible to distinguish five metallogenic epochs:

1. $J-K_1$: associated with the rift distensive tectonic regime and the development of continental margins.
2. K_1-K_2: related to the development of subduction zones and associated island volcanic arcs.
3. K_2 Campanian-Eocene: associated the successive collisions of the southern distensive margin (Maya Block, including Caribeana), represented by the Pinos, Guamuhaya, and possibly Cangre) terranes with the Cretaceous Insular Volcanic Arcs and after this with the northern distensive margin continental (Florida-Bahamas Block).
4. Paleocene-Lower-middle Eocene, associated with the development of the Sierra Maestra-Cayman Ridge insular volcanic arc.
5. Upper Eocene-Quaternary, related to the development of weathering mantles and supergene and sedimentary deposits in the tectonic regime of intraplate basins.

The metallogeny of continental units of western and central Cuba belonging to Maya Block has as main features the presence of large thicknesses of sedimentary sequences more or less rich in organic matter, of massive sulfide deposits hosted in sediments and the reduced and acidic nature of the hydrothermal fluids that gave them origin. This explains the presence of Au in their ores and bases the high favorability of these units to find Au in the pyritic-polymetallic ores of Cu, Zn, Pb, Ag, and Co.

In the Cretaceous Insular Volcanic Arcs, five regions or sections are distinguished differentiated by their petrological and metallogenic characteristics. Of these, the one of least favorability for the presence of shallow mineralization is that of Western Cuba, due to having little outcrops of the volcanic and intrusive sequences. The one from central Cuba stands out for its favorability for Cu, Zn, and Mo with Au and Ag on bimodal—mafic VMS and porphyry type deposits with subordinated skarn mineralization. The Ciego-Camagüey-Las Tunas section stands out for the great development Au–Ag epithermal mineralization, accompanied by Cu–Mo ± Au, Ag porphyry mineralization, integrated in a porphyritic—skarn—epithermal mineral system. The presence of a significant volume of alkaline magmatic rocks gives this region of the country a metallogenic favorability of great interest for Au, Ag, and other mineralizations very little studied up to now such as Th-REE and magnetite iron with P in metamorphosed intrusive bodies. The Holguín section is characterized by the predominance of an Au and Ag and base metals orogenic mineral system, sometimes accompanied by Ti and Co, with a high favorability for the finding of new deposits. The southeastern Cuba section is characterized by having bimodal-mafic VMS deposits, metamorphosed or not, and orogenic Cu-Ni, whose degree of study, in both cases, is low.

The insular volcanic arc Sierra Maestra-Cayman Ridge is the type region of occurrence of the mineralization of world reference by volcanogenic Mn Cuban type. Here there are important identified resources. This tectonic unit has a high degree of favorability for locating new Mn deposits of this type, as well as base metal deposits of the VMS type, which in some cases may be spatially and genetically associated with those of Mn.

The sequences of mafic and ultramafic rocks of Cuba are characterized by the development of podiform chromitites, sometimes carriers of Au and PGE, and Cu and Zn mafic–ultramafic VMS, as well as by a hitherto weak mineralization of Au and Ag. These sequences also have great importance for being the source of the supergene mineralization of Fe–Ni–Co laterites, enriched by accompanying elements such as Au, PGE, Cr, Mn and rare and dispersed elements such as Sc, which could be of interest economic.

The presence of high contents of elements of the PGE in Cuban laterites can be considered confirmed, at least as regards the deposits of the Moa Mineral District, and revealed the presence of significantly elevated contents of Sc, also of potential economic importance.

8.4.7 Knowledge Gaps

In spite of the studies carried out to date it is necessary to expand the geochemistry knowledge about the mafic and ultramafic rocks of Cuba and it's age. This would allow a more adequate identification and separation between the ophiolites sensu *strict*, the mafic rocks related to the Cretaceous insular volcanic arcs, and the mafic intrusive bodies corresponding to the collision magmatism.

Although several authors have advanced opinions on the geochemical affinity of the felsic igneous bodies of the Pinos Terrane, until now no geochemical research has been carried out to adequately argue their belonging to the ilmenite or magnetite series, as well as other aspects of its geochemical affinity and degree of evolution of this magmatism. Likewise, it is necessary to carry out absolute dating of these igneous rocks and ores, which allow to specify their age and location sequence. All this will allow a better understanding of the metallogeny of this part of Cuba, particularly the genesis of the Au-Sb and W-Mo deposits of the territory.

Likewise, it is necessary to carry out petrological studies and absolute dates that allow deepening the knowledge of magmatism in the Ciego-Camagüey-Las Tunas section, with a view to achieving a better differentiation and delimitation of the different intrusive phases, particularly with regard to the distribution and characteristics of alkaline igneous and volcanic rocks. This is important because of its implications for the metallogeny of this territory.

There is a significant deficit of studies of fluid inclusions, isotope geochemistry, and mineral chemistry of Cuban endogenous ores, required for a better understanding of their genesis, for the correct classification of the deposits according to the models and for a better identification of the regularities of their distribution.

The study of the accompanying metals (Au, PGE, Cr, Mn, Sc, and others) in the weathering mantles, particularly in the Fe–Ni–Co lateritic deposits and in the anthropogenic deposits created by the nickel industry, is insufficient and it is only in its beginnings. Something similar happens with the knowledge of the content of rare and dispersed elements in the endogenous mineralizations, particularly in the mafic-siliciclastic VMS, SEDEX, and W veins.

The adequate argumentation of belonging to one or another model of interesting deposits, but with a very low degree of study, could expand the spectrum of deposit models represented in Cuba, which is still a pending issue in the Cuban metallogeny.

8.5 Conclusions

- The metallogeny of Cuba is wide and varied, characterized by diversity of geodynamic environments and host lithologies. This, together with a complex geological history, is the cause of the wide variety of types and styles of mineralization observed in the Cuban national territory.
- The most important mineral system, due to its extension and economic significance, is the supergene. Other mineral systems that stand out for their geological interest and economic importance are (1) Mafic-siliciclastic VMS—SEDEX—MVT of the J-K$_1$ distensive basin. (2) The two orogenic mineral systems. (3) Cu–Mo ± Au porphyry—skarn—Au–Ag epithermal in the Ciego-Camagüey-Las Tunas region. (4) Bimodal-mafic VMS—veins of Ag–Au in volcanic sequences—volcanogenic Mn, present in the Paleogene insular volcanic arc of Sierra Maestra-Cayman Ridge.

- The orogenic mineral system of the Campanian–Maastrichtian, which includes Delita, the largest Au deposit known in Cuba so far, is of great interest for its metallogeny of Au, Ag, W, Mo, Sb, and Cu.
- In a series of aspects (petrology, mineral chemistry, fluid inclusions, isotopic studies, and radiometric dates), the degree of metallogenic study of Cuban deposits is insufficient and requires new work.

Acknowledgements The authors wish to express their gratitude to all those who taught us how to do science and to M. Pardo Echarte for his contribution to the final revision of this chapter.

References

1. Adamovich A, Chejovich V (1969) Búsqueda de yacimientos de manganeso en la zona de Guisa-Los Negros, provincia de Oriente. Revista Tecnológica VII (1):24–37
2. Aiglsperger Th, Proenza JA, Lewis JF, Labrador M, Svojtka M, Rojas Purón A, Longo F, Durisova FJ (2016) Critical metals (REE, Sc, PGE) in Ni laterites from Cuba and the Dominican Republic, Ore Geology Reviews 73:127–147
3. Alemán I, De la Torre A, Barroso AM, Lamas P, Pérez J, García M, Escobar A, Rodríguez RE (1993) Informe sobre la prospección detallada de oro a escala 1:50 000 en el sector Jobabo, Las Tunas. Yacimientos Maclama, Georgina, Iron Hill, Abucha: Inédito. Archivo Oficina Nacional de Recursos Minerales. La Habana
4. Alfieris D, Voudouris P (2005) Ore mineralogy of transitional submarine to subaerial Magmatic hydrothermal deposits in Western Milos, Greece. Au-Ag-Te-Se deposits IGCP Project 486, 2005 Field Workshop, Kiten, Bulgaria: 14–19. September 2005. Geochemistry, Mineralogy and Petrology 43, 6 pp. Academia de Ciencias de Bulgaria. Sofia
5. Allende R (1927) Yacimientos Minerales de la República de Cuba. Boletín de Minas 11:1–70
6. Allibon J, Lapierre H, Bussy F, Tardy M, Cruz Gámez EM, Senebier F (2008) Late Jurassic continental flood basalt doleritic dykes in northwestern Cuba: remnants of the Gulf of Mexico opening. Bulletin Society Geology of France 179(5):445–452
7. Alonso JL, González V, Pérez J, González CJ, Padrón M, Lugo R (2004) T. T. P. Reevaluación de la información geológica del Campo Mineral Maclama y su introducción en una base de datos: Inédito. Archivo de la Oficina Nacional de Recursos Minerales. La Habana
8. Andó J, Harangi S, Szkmány B and Dosztály L (1996) Petrología de la asociación ofiolítica de Holguín. En: Iturralde Vinent, M.A. (ed.). Ofiolitas y arcos volcánicos de Cuba. Miami, USA, IGCP Project 364, Special Contribution n.1:154–178
9. Aniatov I, Lavandero RM (1983) Capacidad Menífera del complejo carbonatado – terrígeno del Jurásico - Cretácico Inferior de Cuba Occidental. Serie Geológica, No 2, 1983:19–47. La Habana
10. Ansted D (1856) The copper of Santiago de Cuba. In: Proceedings of the geological society of London, pp 145–153. London
11. Avdeev S, Ferreiro M, Machado A, Fernández MA, Horta J y Bosch M (1986) Búsqueda Evaluativa para Au y W Lela y sus Flancos. Inédito, archivo ONRM, La Habana
12. Babushkin V, Tseimakh E, Akilvekov S, Sverov V, Kurtigueshev V y Orlov N (1990) Informe de los trabajos de levantamiento geológico – geofísico a escala 1:50 000 y búsquedas acompañantes en el municipio especial Isla de la Juventud en colaboración con la URSS (CAME III). Inédito, archivo ONRM, La Habana
13. Barrios F, Ávila A, García C, Pérez M, Cabrera H, Padrón C, Rodríguez E, Díaz A (1988) Informe búsqueda evaluativa Flanco Este Unión II, Inédito. Archivo Oficina Nacional de Recursos Minerales, La Habana
14. Biriukov B, Messina V, Ponce N, Navarro N (1969) Informe sobre los resultados de los trabajos de búsqueda y levantamiento a escala 1:50 000 realizados en los años 1967–1968 en la parte oriental de la provincia Pinar del Río (región La Palma): Inédito. Archivo de la Oficina Nacional de Recursos Minerales. La Habana
15. Blanes JA, Valdés EL, Chávez O, Chang JA (1991) Informe de los resultados del trabajo temático – productivo revisión de las minas antiguas Dora y Francisco. Inédito. Archivo Oficina Nacional de Recursos Minerales. La Habana
16. Bolotin YA (1968) Informe de los trabajos de exploración geológica realizados en el prospecto de Guachinango en 1965–67 con el cálculo de las reservas de las menas piríticas: Inédito. Oficina Nacional de Recursos Minerales, La Habana
17. Bolotin YA (1969) Informe con el cálculo de reservas de las menas piríticas de los prospectos Carlota y Victoria, según los trabajos geológicos de exploración realizados en 1966–8: Inédito. Oficina Nacional de Recursos Minerales, La Habana
18. Bolotin YA, Yidkov AY, Maximov AA, Sosa R (1970) Prospectos de minerales sulfurosos de la serie metamórfica Escambray en la parte noroeste del macizo montañoso del mismo nombre: Revista Tecnológica, VIII (2):35–48. La Habana
19. Bortnikov NS, Kramer JL, Guenkin AD, Krapiva LY y Santa Cruz M (1988) Paragenesis of gold and silver tellurides in the Florencia deposit, Cuba. International Geology Review 30:294–306
20. Boschman LM, van Hinsbergen DJ, Torsvik TH, Spakman W and Pindell JL (2014) Kinematic reconstruction of the Caribbean region since the Early Jurassic. Earth-Science Reviews 138:102–136. Editorial Elsevier, journal homepage: www.elsevier.com/locate/earscirev
21. Brovin M (1966) Informe sobre los trabajos de búsqueda de los minerales cupríferos realizados en la zona del yacimiento Carlota en los años 1964–65: Inédito
22. Buguelskiy YY, and others (1985) Ore deposits of Cuba. UDC 553.042/061. Editorial Nauka. Moscú. In Russian and English
23. Cabrera R (1986) Geología y regularidades de la distribución de los yacimientos de cobre y oro de la región mineral de Las Villas. Instituto de Geología y Paleontología, Academia de Ciencias de Cuba, p 130
24. Capote C, Santa Cruz M, González D, Altarriba I, Bravo F, De la Nuez D, Carrillo DJ y Cazañas X (2002) Evaluación del potencial de metales preciosos y base del arco cretácico en el territorio Ciego - Camagüey - Las Tunas: Inédito. Instituto de Geología y Paleontología, La Habana
25. Cazañas X, Pura A, Melgarejo JC, Proenza JA, Fallick AE (2008) Geology, fluid inclusions and sulphur isotope characteristics of the El Cobre VMS deposit, Southern Cuba. Miner Deposita 43:805–824
26. Cazañas X, Melgarejo JC (1998) Introducción a la metalogenia del Mn en Cuba. Acta Geologica Hispanica, 33 (1–4):215–237
27. Cazañas X, Melgarejo JC, Alfonso P, Escusa A y Cuba S (1998) Un modelo de depósito vulcanogénico de manganeso del arco volcánico Paleógeno de Cuba: el ejemplo de la región Cristo-Ponupo-Los Chivos. Acta Geológica Hispánica, 1998, 33, 1–4:239–276. ISBN/ISSN: 0567–7505
28. Cazañas X (2000) Depósitos volcanogénicos del Arco Paleógeno de la Sierra Maestra. El ejemplo del yacimiento El Cobre. Tesis de Doctorado en Geología. Departamento de Cristalografía,

Mineralogía i Depòsits Minerals. Facultat de Geologia. Universitat de Barcelona

29. Cazañas X, Torres Zafra JL, Lavaut Copa W, Cobiella Reguera JL, Capote C, González V, López Kramer JL, Bravo F, Llanes AI, González D, Ríos Y, Ortega Y, Yasmany R, Correa A, Pantaleón G, Torres M, Figueroa D, Martin D, Rivada R y Núñez A (2017) Mapa Metalogénico de la República de Cuba a escala 1:250 000

30. Chaliy D, Dovbnia V (1966) Informe acerca de los trabajos de búsqueda y exploración realizados durante los años 1963–65 en la zona de Holguín: Inédito. Archivo de la Oficina Nacional de Recursos Minerales. La Habana

31. Cobiella Reguera JL (1988) El vulcanismo paleogénico cubano. Apuntes para un nuevo enfoque: Revista Tecnológica XVIII (4):25–32

32. Cobiella Reguera JL (1996) El magmatismo jurásico (Calloviano? - Oxfordiano) de Cuba occidental: ambiente de formación e implicaciones regionales: Revista de la Asociación Geológica Argentina 51(1):15–28

33. Costafreda JL, Correa B, Guerra M, Mesa G (1993) Informe prospección detallada oro Aguas Claras, provincia de Holguín, cancelado por el Período Especial en tiempo de paz: Inédito. Archivo de la Oficina Nacional de Recursos Minerales. La Habana

34. Costafreda JL, Quiñones CL, Recouso Y, Rubio MD, Ge Bartelemi P, Correa B y Del Toro A (1994) Informe exploración orientativa y detallada oro Reina Victoria (cancelado): Inédito. Archivo de la Oficina Nacional de Recursos Minerales. La Habana

35. Costafreda JL (1996) Secuencia histórica de los trabajos mineros realizados en los cotos mineros Guajabales y Aguas Claras: Inédito. Archivo de la Oficina Nacional de Recursos Minerales. La Habana

36. Cox DP, Lindsey DA, Singer DA, Moring BC, Diggle MF (2007) Sediment-Hosted Copper Deposits of the World: Deposit Models and Database. Open-File Report 03–107. Versión 1.3. Disponible online en https://pubs.usgs.gov/of/2003/of03-107/

37. Cruz Gamez EM, Maresch W, Cáceres D, Balcázar N, Martín K (2003) La Faja Cangre y sus rasgos metamórficos. Pinar del Río. Cuba. V Congreso Cubano de Geología y Minería. Memorias Trabajos y Resúmenes. Centro Nacional de Información Geológica. La Habana. ISBN 959–7117–11–8

38. Cruz Gámez EM, Velasco Tapia F, García Casco A, Despaigne Díaz AI, Lastra Rivero JF y Cáceres Govea D (2016) Geoquímica del magmatismo mesozoico asociado al Margen Continental Pasivo en el occidente y centro de Cuba. Boletín de la Sociedad Geológica Mexicana

39. De la Sagra R (1842) Historia Física, Política y Natural de la Isla de Cuba. Capítulo VIII. Edición Española. París

40. Díaz A, Prieto A, Bárzana A, Padrón C, Borges E, Salinas A, Fernández R, Cabrera F, Guerra J, Izquierdo M (1993) Informe evaluación orientativa de cobre Castellano. Inédito. Archivo de la Oficina Nacional de Recursos Minerales. La Habana

41. Díaz de Villalvilla L (1988) Caracterización geológica y petrológica de las asociaciones vulcanógenas del arco insular cretácico en Cuba central: Instituto de Geología y Paleontología, resumen de la tesis presentada en opción al grado científico de candidato a doctor en ciencias, 1–28

42. Distler VV, Kriasko VV, Sansdomerskaya SM, Botova MM, Sviechkova VV, Nikolskaya NM, Eschenova ZA, Bielovsov GE, Falcón J, Campos M, Muñoz N, Rodríguez J, Guardado R, Rodríguez A, Cabrera R, Fonseca E, Rodríguez D, Guerra I, López I y Ávila E (1989) Información de la presencia de EGP en lateritas y cromitas de Cuba. Inédito. Instituto de Geología de los yacimientos minerales, Petrografía, Mineralogía y Geoquímica, Academia de Ciencias de la URSS, Rusia, Instituto Superior Minero Metalúrgico de Moa, Cuba, Instituto de Geología y Paleontología, La Habana, Cuba pag. 45

43. Dublan L, Álvarez H, Lledías P, Svestka J, Vázquez C, Marsal W, González E, Micoch B, Molak B, De los Santos E, Soucek J, Pérez M, Mihalikova A, Bernal L, Zoubek J, Ordóñez M, Lavandero R, Marousek J, Manour J, Pérez R y Rodríguez R (1986) Informe final del levantamiento geológico y evaluación de los minerales útiles en escala 1:50000 del polígono CAME I Zona Centro: Inédito. Oficina Nacional de Recursos Minerales, La Habana

44. Durañona D, Rodríguez A, González CJ y Pimentel H (1990) Informe final de prospección preliminar de polimetálicos en los sectores Guáimaro - Palo Seco y otros: Inédito. Archivo de la Oficina Nacional de Recursos Minerales. La Habana

45. Edwards R, Atkinson K (1986) Ore deposit geology and its influence on mineral exploration: Editorial Chapman and Hall, London, New York, p 466

46. Emsbo P (2009) Geologic criteria for the assessment of sedimentary exhalative (SEDEX) Zn-Pb-Ag deposits: U.S. Geological Survey Open-File Report 2009−1209, p 21

47. Espinosa A (1985) Informe geológico sobre la toma de muestras tecnológicas Carlota – Guachinango – Victoria: Inédito

48. Fernández MA, Díaz M, Pardo Echarte M (2000) TTP Reprocesamiento de la información geológica de las áreas revertidas por la AE Matlock Delita SA. Inédito. Oficina Nacional de Recursos Minerales, La Habana

49. Ferreiro M, Fernández MA, Díaz M, Rodríguez N y Vázquez A (1992) Prospección preliminar metales raros Colony. Inédito. Oficina Nacional de Recursos Minerales, La Habana

50. Garapko I, Yurov L, Chulga A, Sorokin B, Eguipko O (1974) Constitución geológica y minerales útiles de la Isla de Pinos. Informe sobre el levantamiento geológico y las búsquedas a escala 1:100 000 realizados en los años 1971 – 1974. Inédito. Archivo de la Oficina Nacional de Recursos Minerales, La Habana

51. García Casco A, Iturralde Vinent M, Pindell J (2008) Latest cretaceous collision/accretion between the caribbean plate and caribeana: origin of metamorphic terranes in the greater antilles. International Geology Review 50(9):781–862

52. García L, Escalona N y Seisdedos G (1979) Exploración detallada del flanco NE del yacimiento Santa Lucía, Pinar del Río. Inédito. Oficina Nacional de Recursos Minerales, La Habana

53. García CA, Ávila A, Hung A, Pérez M, Kindelán R y Basulto M (1988) Ejecución de los trabajos del proyecto de exploración orientativa del yacimiento de piedra y pirita cuprífera Unión II. Inédito. Oficina Nacional de Recursos Minerales, La Habana

54. García CA, Ávila A, Kindelán R, Pérez M, Valdivia M, Castañeda JA, Díaz A, Hernández G e Izquierdo M (1990) Informe de los trabajos de Exploración Orientativa para menas sulfurosas-cupríferas del Yacimiento Juan Manuel y el complemento de la Exploración Orientativa del yacimiento Unión

55. Gervilla F, Proenza JA, Frei R, González Jiménez JM, Garrido CJ, Melgarejo JC, Meibom A, Díaz Martínez R, Lavaut W (2005) Distribution of platinum-group elements and Os isotopes in chromite ores from Mayarí-Baracoa Ophiolite Belt (eastern Cuba). Contrib Miner Petrol 150:589–607

56. González H y Sotolongo R (1998) Reporte trabajos de reconocimiento en la concesión Trinidad. Inédito

57. González Jiménez JM, Proenza JA, Gervilla F, Melgarejo JC, Blanco Moreno JA, Ruiz Sánchez R and Griffin WL (2011) High-Cr and high-Al chromitites from the Sagua de Tánamo district, Mayarí-Cristal ophiolitic massif (eastern Cuba): Constraints on their origin from mineralogy and geochemistry of chromium spinel and platinum-group elements. LITHOS-02396, p 21

58. Goodfellow WD and Lydon JW (2007) Sedimentary exhalative (SEDEX) deposits: in Goodfellow, W. D. (Ed.). Mineral Deposits of Canada, A Synthesis of Major Deposit Types, District Metallogeny, the Evolution of Geological Provinces, and Exploration Methods, Geological Association of Canada, Mineral Deposits Division, Special Publication No. 5: 163–184
59. Gorielov VE, Gorielova VG y Stareva M (1965) Informe del prospecto Carlota en la provincia de Las Villas, con el cálculo de reservas del mineral pirita según los trabajos efectuados en el año 1963. Inédito. Oficina Nacional de Recursos Minerales, La Habana
60. Harnish DE and Brown PE (1986) Petrogenesis of the Casseus Cu - Fe skarn, Terre Neuve district, Haiti. Economic Geology 81 (7):1801–1807
61. Hayes TS, Cox DP, Piatak NM and Seal RR II (2015) Sediment-hosted stratabound copper deposit model: U.S. Geological Survey Scientific Investigations Report 2010–5070–M, p 147
62. Iturralde Vinent MA (1986) Reconstrucción palinspática y paleogeografía del Cretácico Inferior de Cuba oriental y territorios vecinos. Revista Minería y Geología 1:1–4
63. Iturralde Vinent MA (1989) Rol de las ofiolitas en la constitución geológica de Cuba (russian/english). Geotectonics 4:63–76
64. Iturralde Vinent MA (1995) Implicaciones tectónicas de magmatismo de margen continental pasivo en Cuba. En A. C. Ricardii y M. P. Iglesias Llanos (Editores). Jurásico de Cuba y América del Sur. Proyecto UNESCO/IUGS PICG 322: 14–30. Buenos Aires
65. Iturralde Vinent MA (1996) Magmatismo de margen continental en Cuba. En: Iturralde-Vinent, M., (ed.), Ofiolitas y arcos volcánicos de Cuba. International Geological Correlation Program, Project 364. Geological Correlation of Ophiolites and volcanic arcs in the Circumcaribbean Realm: 121–130. Miami, Florida
66. Iturralde Vinent MA Editor (2011) Compendio de Geología de Cuba y del Caribe. Primera Edición. DVD-ROM. Editorial CITMATEL, La Habana, Cuba
67. Jones ChE, Jenkyns HC (2001) Seawater strontium isotopes, oceanic anoxic events, and seafloor hydrothermal activity in the Jurassic and Cretaceous. Am J Sci 301:112–149
68. Kantchev I, Boyanov I, Popov N, Cabrera R, Goranov A, Iolkoev N, Kanazirski M y Stancheva M (1978) Informe Geología de la provincia de Las Villas. Resultados de las investigaciones geológicas y levantamiento geológico a escala1:250000 realizados durante el período 1969 - 1975. Inédito. Oficina Nacional de Recursos Minerales, La Habana
69. Kerr A, Iturralde Vinent MA, Saunders A, Babbs T, Tarney J (1999) A new plate tectonic model of the Caribbean: Implications from a geochemical reconnaissance of Cuban Mesozoic volcanic rocks. Geol Soc Am Bull 111(11):1–20
70. Kesler SE (1968) Contact-localized ore formation at the Meme mine, Haiti. Econ Geol 63:541–552
71. Kilias SP, Detsi K, Godelitsas A, Typas M, Naden J and Marantos Y (2007) Evidence of Mn-oxide biomineralization, Vani Mn deposit, Milos, Greece, in C.J. Andrew et al eds., Proceedings of the Ninth Biennial SGA Meeting, Dublin 2007: 1069–1072
72. Kozulin VA, Antoneev IV y Shulzhenko VM (1973) Complemento al informe yacimiento Hierro, confeccionado como resultado de los trabajos de exploración geológica realizados en este yacimiento en los años 1971 - 73 y del cálculo de reservas el 1 de Marzo de 1973. Inédito. Oficina Nacional de Recursos Minerales, La Habana
73. Lara J, Izquierdo M, Padrón C, Martínez N, Córdova R (1989) Informe sobre los resultados de la búsqueda evaluativa de sulfuros en los sectores Baja – Veguita – La Vitrina, dentro del campo mineral Santa Lucía – Castellano. Inédito. Archivo Oficina Nacional de Recursos Minerales, La Habana
74. Large RR, Bull SW, Maslennikov VV (2011) A Carbonaceous Sedimentary Source-Rock Model for Carlin-Type and Orogenic Gold Deposits. Econ Geol 106(3):331–335
75. Lavandero RM, Estrugo M, Santa Cruz Pacheco M, Bravo F, Melnikova A, Cabrera R, Trofimov VA, Romero J, Altarriba I, Álvarez P, Aniatov II, Badanchivin B, Barishev AN, Carrillo DJ, Cazañas X, Cuellar N, Dovbnia AV, Formell F, García M, González D, Gue GG, Janchivin A, Krapiva LJ, López J, Lozanov I, Montenegro J, Pantaleón G, Estefanov N, Vázquez A, Zagoskin AM y Zhidkov AYa (1985) Sistematización y generalización de los Yacimientos Minerales Metálicos de Cuba. Informe inédito. Oficina Nacional de Recursos Minerales, La Habana
76. Lazarenkov VG, Tikhomirov I, Zhidkov AYa, Talovina IV (2005). Platinum Group Metals and Gold in Supergene Nickel Ores of the Moa and Nicaro Deposits (Cuba). Lithology and Mineral Resources, 40 (6): 521–527. Translated from Litologiya i PoleznyeI skopaemye, No. 6, 2005, pp 600–608. Original Russian Text Copyright © 2005 by Lazarenkov, Tikhomirov, Zhidkov, Talovina
77. Leach DL, Taylor RD, Fey DL, Diehl SF and Saltus RW (2010) A deposit model for Mississippi Valley-Type lead-zinc ores, chap. A of Mineral deposit models for resource assessment: U.S. Geological Survey Scientific Investigations Report 2010–5070–A. p 52. Available from: https://www.usgs.gov/pubprod.
78. Legrá García I, Palanco CS, y Capote Flores N (2018) Caracterización geoquímica de los perfiles lateríticos de los yacimientos Yagrumaje y Camarioca Este. DG P 114. Memorias MINEMETAL 2018. IV Congreso de Minería y Metalurgia. Resúmenes y Trabajos. ISBN 978 – 959 – 7117 – 97 – 2. Centro Nacional de Información Geológica, Instituto de Geología y Paleontología – Servicio Geológico de Cuba. La Habana
79. Llanes AI, Díaz de Villalvilla L, Despaigne AI, Ronneliah Sitali M y García Jiménez D (2015) Geoquímica de las rocas volcánicas máficas de edad Cretácica de la región de Habana-Matanzas (Cuba occidental): implicaciones paleotectónicas. Ciencias de la Tierra y el Espacio, 16(2):117–133. ISSN 1729-3790
80. Lobanov P, Zhidkov A, Estrugo M, Vavilov G, Shelagurov V, Ruizhkov J (1976) Informe sobre la exploración preliminar del yacimiento pirito - polimetálico Castellano durante los años 1972-74 con el cálculo de reservas según estado el 1 - 1 - 75. Inédito. Oficina Nacional de Recursos Minerales, La Habana
81. López Kramer JM (1988) Composición sustancial y asociaciones mineralógicas de los yacimientos auríferos hidrotermales de Cuba. Tesis para la obtención del grado científico de Dr. C. Geólogo - Mineralógicas. Instituto de Geología de los yacimientos minerales, mineralogía, petrografía y geoquímica. IGEM. ACC URSS. In Russian
82. López Kramer JL, Poznaikin VV, Morales A, Echevarría BT y otros (1990) Informe final del tema 409 – 09 Fundamentación de los trabajos de búsqueda y exploración de oro en el territorio de la República de Cuba, con la evaluación de los recursos pronósticos a escala 1:500 000: Inédito. Archivo de la Oficina Nacional de Recursos Minerales. La Habana
83. López Kramer JM, Pimentel H, Redwood S, Gandarillas Hevia J y Pérez Vázquez RG (2008) Depósitos primarios de oro y plata del Archipiélago Cubano. Ciencias de la Tierra y el Espacio, 9:49–61, ISSN 1729-3790
84. Marchesi C, Garrido CJ, Bosch D, Proenza JA, Gervilla F, Monié P, Rodriguez Vega A (2007) Geochemistry of cretaceous magmatism in eastern Cuba: recycling of North American continental sediments and implications for subduction polarity in the Greater Antilles paleo-arc. J Petrol 48(9):1813–1840

85. Maximov A, Grachev G y Sosa R (1968) Geología y minerales útiles de las pendientes nor – occidentales del sistema montañoso Escambray. Informe sobre los trabajos de búsqueda – levantamiento a escala 1:50 000 realizados en la parte sur de la provincia Las Villas: Inédito
86. Mc Cormick DF (1918) Minas de Matahambre (Parte I). Boletín de Minas No 4:61–68
87. Milia González I, Pérez M, Torres M (2013) Acerca de la presencia de Arfvedsonita y Baddeleyita en las rocas sieníticas de Camagüey, Cuba. V Convención Cubana de Ciencias de la Tierra, Geociencias` 2013. Memorias en CD-ROM, La Habana, 1 al 5 de abril de 2013. ISSN 2307–499X
88. Millán G (1997) Geología del macizo metamórfico Isla de la Juventud. In: G. F. Furrazola Bermúdez, K. E. Núñez Cambra, (eds.). Estudios sobre Geología de Cuba: 259–270. La Habana, Cuba, Centro Nacional de Información Geológica
89. Montano JL, Torres JL, Suárez A, Altarriba I, Lavandero RM, Moreira J, Pardo M, González D, Cazañas X, Bravo F, Puentes G (2001) Reevaluación metalogénica de los recursos minerales de oro, plata y polimetálicos asociados en esquistos negros (secuencias ricas en materia orgánica) Escambray. Inédito. Instituto de Geología y Paleontología, La Habana
90. Moreira J, Torres Zafra JL, Montano JL, Morales A, Altarriba I, Bravo F, Suarez A, Echevarría BT, Carrillo DJ, Chang JL, y González D, (1999) Reevaluación metalogénica del potencial de recursos minerales de metales preciosos y bases en Cuba Oriental: Inédito. Archivo del Instituto de Geología y Paleontología, La Habana
91. Mormil S, Norman A, Varvarov A, Skosiriev V, Linares E y Vergara F. (1980) Geología y Metalogenia de la provincia de Pinar del Río. Inédito
92. Mosier DL and Page NJ (1988) Descriptive and grade – tonnage models of volcanogenic manganese deposit in oceanic environments–A modification. US Geological Survey Bulletin 1811. 28 págs
93. Mozgova MI, Boronijin BA, Generalov ME, Lopez Kramer JL (1989) Doklady Akademii Nauk, Soviet Union. Tom, 309 (5):1181–1186
94. Muliukov E M, Guzmán I V (1969) Informe de los trabajos geológicos realizados en el yacimiento Hierro con el cálculo de reservas según su estado para el 1 - VII- 1969. Inédito. Archivo Oficina Nacional de Recursos Minerales, La Habana
95. Murray WH, Oreskes N, Einaudi MT (1992) Geological characteristics and tectonic setting of proterozoic iron oxide (Cu-U-Au-REE) deposits. Precambr Res 58 (1–4):241–287
96. Naden J, Kilias SP, Derbyshire DBF (2005) Active geothermal systems with entrained seawater as analogues for transitional continental magmato hydrothermal and volcanic-hosted massive sulfide mineralization-the example of Milos island, Greece. Geology 33:541–544
97. Nanian B, Sedov V, Shulzhenko V, Andreev S, Fernández R, Estrugo M, Efimova Z, Semionov I y Mezentsev A (1972) Informe con el cálculo de reservas de las menas pirítico - cupríferas del yacimiento Unión I en base a los trabajos de exploración geológica ejecutados en los años 1967 - 1972. Inédito. Oficina Nacional de Recursos Minerales, La Habana
98. Niiranen T, Poutiainen M, Mänttäri I (2007) Geology, geochemistry, fluid inclusion characteristics, and U-Pb age studies on iron oxide–Cu–Au deposits in the Kolari region, Northern Finland. Ore Geol Rev 30:75–105
99. Nikolaev A, Núñez A, Sánchez R, Cordovés R, Reborido J y Rosales C (1981) Trabajos de levantamiento a escala 1:100 000 y resultados de las búsquedas a escala 1:50 000 y 1:25 000 ejecutados en la parte este de la provincia Guantánamo. Inédito. Archivo de la Oficina Nacional de Recursos Minerales. La Habana
100. Novizky V (1964) Informe sobre los trabajos de exploración geológica efectuados en 1962 –1963 en la zona de Mina Dora y el cálculo de reservas de los yacimientos Dora y Amistad en la zona Matahambre, provincia de Pinar del río. Inédito. Archivo Oficina Nacional de Recursos Minerales. La Habana
101. Ortega P (1916) Ojeada retrospectiva y reseña sobre el estado actual de la minería en Cuba. Boletín de Minas No 1:2–29
102. Ovchinnikov V, Balbis C, Biriolin V, Volcov V, Bolotov R, Díaz G, Lugo R, Pérez F, Laverov Y y Pak G (1982) Informe sobre los trabajos geológicos de búsqueda orientativa a escala 1:25 000 y detallada 1:10 000 para oro, cobre y molibdeno en la región Martí – Bartle - Las Tunas en los años 1976–1981. Inédito. Archivo de la Oficina Nacional de Recursos Minerales. La Habana
103. Ovchinnikov V, Robaina M, Hernández A, Valdivia M, Rodríguez A, Salinas A, Fernández R, Estrada N y Martínez A (1993) Informe sobre los resultados de la exploración detallada del yacimiento pirítico - polimetálico Santa Lucía en la provincia Pinar del Río, realizados durante los años 1982 - 1987 con el cálculo de reservas hasta el 1 de Diciembre de 1993. Inédito. Oficina Nacional de Recursos Minerales, La Habana
104. Padilla I, Lufriú L, Leal A, Millán G, Corbea L y Prieto F (1994) Resultados del levantamiento aerogeofísico complejo en el territorio de las provincias Cienfuegos, Villa Clara y Sancti Spiritus (Sector Escambray): Inédito. Oficina Nacional de Recursos Minerales, La Habana
105. Page LR and McAllister JF (1944) Tungsten deposits Isla de Pinos, Cuba. U.S. Geological Survey Bulletin 935-D:177–246
106. Park, C. F. (1942) Manganese deposits of Cuba. Geologic investigations in the American republics 1941 - 42. U. S. Geological Survey Bulletin 935 - B:75–97
107. Park CF and Cox M (1944) Manganese deposit in part of the Sierra Maestra, Cuba. Geologic investigations in the American republics. U. S. Geological Survey Bulletin 935 - F:307–355
108. Pein BJ (1997) The role of mercury – organic interactions in the hydrothermal transport of mercury: Economic Geology, 91(1): 20–28
109. Pentelenyi I, Foldessy J, García EM y Velázquez J (1990) Informe sobre los resultados del levantamiento geológico complejo polígono IV CAME Holguín: Inédito. Oficina Nacional de Recursos Minerales, La Habana
110. Pérez RG y Melgarejo JC (1998) El yacimiento Matahambre (Pinar del Río, Cuba): estructura y mineralogía. , Geología y Metalogenia de Cuba: una introducción. J. C. Melgarejo y J. A. Proenza Eds. Acta Geológica Hispánica, Vol. 33 (1–4):133–152
111. Piotrowski J (1977) First manifestations of volcanism in the Cuban geosyncline. Academie Polonaise des Sciences Bulletin, Sere des Sciences de la Terre 24 (3–4):227–234, Warsaw
112. Podkamenny A, Guzmán L (1971) Informe de los resultados de los trabajos de búsqueda y búsqueda – exploración en la región del yacimiento Hierro. Inédito. Oficina Nacional de Recursos Minerales. La Habana
113. Popov M, Cordovés R, Urgellés J y Reborido J (1984) Informe final sobre la búsqueda orientativa y detallada en escala 1:25 000 y 1:10 000 del sector La Cruzada, en las montañas de la Sierra del Purial: Inédito. Archivo de la Oficina Nacional de Recursos Minerales. La Habana
114. Proenza JA, Gervilla F, Melgarejo JC, Bodinier JL (1999) Al- and Cr-rich chromitites from the Mayarí-Baracoa Ophiolitic Belt (Eastern Cuba): consequence of interaction between volatile-rich melts and peridotites in suprasubduction mantle. Econ Geol 94:547–566
115. Proenza JA, Melgarejo JC (1998) Una introducción a la metalogenia de Cuba bajo la perspectiva de la tectónica de placas. En Acta Geológica Hispánica 33(1):89–131

116. Proenza JA et al. (2004). Distribución de elementos del grupo del platino (EGP) y Au en la faja ofiolítica Mayarí-Baracoa (Cuba oriental). En: Pereira, E., Castroviejo R., Ortiz F., (Editores), Complejos ofiolíticos en Iberoamérica: guías de prospección para metales preciosos: 309–336
117. Proenza JA (2015) Mineralogía y geoquímica de Ni, Co, EGP, Sc, REE en Yacimientos Lateríticos. Macla No 20. Junio15. Revista de la Sociedad Española de Mineralogía. 10p. Disponible en https://www.ehu.eus/sem/macla_pdf/macla20/Macla20_003-9.pdf.
118. Puentes R, Pedraza JV, Barrios AM, Castañeda JA, Valido A, Salinas A, Rodríguez E, González R, Díaz D, Guerra J, Solier S, Barrios R, Hernández N, Hernández JL, González JL (1991) Informe prospección preliminar de Cu en las Alturas Pizarrosas del Norte. Inédito. Oficina Nacional de Recursos Minerales, La Habana
119. Ray GE and Lefebure DV (1999) A Synopsis of Iron Oxide ± Cu ± Au ± P ± REE Deposits of the Candelaria-Kiruna-Olympic Dam Family. Geological Fieldwork 1999, Paper 2000–1. British Columbia Geological Survey: 267–271
120. Renne P, Mattinson IM, Hatten CW, Somin MI, Onstott TS, Millán G, Linares E (1989) ^{40}Ar/^{39}Ar and U/Pb evidence for Late Proterozoic (Grenville age) continental crust in North Central Cuba and regional tectonic implications. Precambr Res 42:325–341
121. Rojas Agramonte Y, Kröner A, García Casco A, Kemp T, Hegner E, Pérez M, Barth M, Liu D, Fonseca Montero A (2010) Zircon ages, Sr-Nd-Hf isotopic compositions, and geochemistry for granitoids associated with the northern ophiolite mélange of central Cuba: tectonic implication for Late Cretaceous magmatism in the northwestern Caribbean. Am J Sci 310:1453–1479
122. Sánchez, JL Ortega, JA (1988) Informe búsqueda detallada para barita en los sectores Este y Baritina. Inédito. Archivo Oficina Nacional de Recursos Minerales. La Habana
123. Sánchez J, Nuñez R, y Fernandez MA (1983) Informe final de Búsqueda Orientativa de Au, W y otros en Siguanea, Isla de la Juventud. Inédito. Oficina Nacional de Recursos Minerales, La Habana
124. Schlangert SO, Jenkyns HC (1976) Cretaceous oceanic anoxic events: causes and consequences. Geologie en Mijnbow 55(3–4):179–184
125. Shevchenko IA, Stepanov VM y Montes A (1967) Informe sobre los resultados de los trabajos de búsqueda – revisión para minerales cupríferos, realizados en 1965-66 en la parte oeste de la provincia Las Villas: Inédito. Oficina Nacional de Recursos Minerales, La Habana
126. Shigariev V, Blazheiev M, Kozulin V, Vavilov G, Shelagurov V, Fajardo M, Sinobas H (1975) Informe sobre los resultados de los trabajos de búsqueda compleja a escala 1:10 000 en el campo mineral Unión y en el flanco NW del yacimiento Hierro realizados en los años1972 – 74. Inédito. Archivo Oficina Nacional de Recursos Minerales, La Habana
127. Simon G, Kesler SE, Russell N, Hall CM, Bell D and Piñero E (1999) Epitermal gold mineralization in an old volcanic arc: the Jacinto deposit, Camagüey District, Cuba. Economic Geology, 94:487–506
128. Simons FS and Straczek JA (1958) A geology of the manganese deposits of Cuba. U.S. Geological Survey Bulletin 1057:1–289
129. Smirnov VI (1982) Geología de yacimientos minerales: Editorial Mir, pp 360–384; 510–586
130. Somin M y Millán G (1981) Geologia de los complejos metamórficos de Cuba. Nauka, Moscú 219 pp. In Russian
131. Spencer AC (1903) Manganese deposits of Santiago de Cuba. US Geol Surv Bull 213:251–255
132. Spry PG, Marshall B and Vokes FM eds. (2000) Metamorphosed and Metamorphogenic Ore Deposits. Review Economic Geology. 11:310
133. Stanik E, Manour J, Ching R y Vázquez C (1981) Informe de los levantamientos geológico, geoquímicos y trabajos geofísicos realizados en la parte sur de Cuba central en las provincias Cienfuegos, Sancti Spiritus y Villa Clara: Inédito. Oficina Nacional de Recursos Minerales, La Habana
134. Tolkunov AE, Bolotin YA, Cabrera R, Maximov AA y Zarianov DP (1974) Regularidades de la distribución y condiciones de formación de los yacimientos tipo "lentes piritosos" en el Anticlinorio de Trinidad: Geología de los minerales útiles de Cuba. Publicación Especial No 3:62–81.Academia de Ciencias de Cuba – Instituto de Geología y Paleontología. La Habana
135. Torres Zafra JL, Lavandero RM, Moreira J y Montano JL (2005) Aproximación a una evaluación del potencial metalífero y petrolífero del Terreno Guaniguanico. Primera Convención Cubana de Ciencias de la Tierra. Memorias Trabajos y Resúmenes. Centro Nacional de Información Geológica. La Habana. ISBN 959-7117–03-7
136. Torres Zafra JL, Lavaut Copa W y Cazañas X (2017) Modelos descriptivo – genéticos de depósitos metálicos para el mapa metalogénico A escala 1:250 000 de la República de Cuba
137. Tourtelot H (1979) Black shales – Its deposition and diagenesis: Clays and clays minerals, 27(5):313–321
138. Valdés EL, Diaz A, Davies JF, Whitehead RE and Fonseca L (1993) Cogenectic Sedex Zn-Pb and stockwork copper ores, Western Cuba. Exploration Mining and Geology, 2(4):297–305
139. Vershigora N, Prianjin A, Fernández NA, Karpov Y, Fernández R, Oubiña L, Castañeda J, Ayata T, Valdivia M, Rosas JR, Klimov A, Kranin L, Escalona N, Santamaría L, Kumalakov A y Konoplin D (1983) Informe de búsqueda a escala 1:10000 realizada en el campo mineral Castellano – Santa Lucía. Inédito. Oficina Nacional de Recursos Minerales, La Habana
140. Whitehead RE, Davies JF, Valdés EL and Díaz A (1996) Mineralogical and chemical variations, Castellanos shale – hosted Zn – Pb – Ba deposit, Northwestern Cuba. Economic Geology, 91(4):713–722
141. Woodring WP and Davies SN (1944) Geology and manganese deposits of Guisa-Los Negros area. Oriente province, Cuba. Geologic investigations in the American republics 1941–43 US Geological Survey Bulletin 935 - G:357–386
142. Wyborn LAI, Heinrich CA, Jaques AL (1994) Australian Proterozoic mineral systems: essential ingredients and mapable criteria. AusIMM Publication Series 5(94):109–115
143. Yanev SN, Tchounev DC y. Tzankov TV (1993) El complejo volcano-sedimentario cretácico de Cuba Central, (Unificación litoestratigráfica y ambiente paleo geodinámico). Doc. Lab. Géol. Lyon. (125):223–240
144. Zelenenko V, Feubell M, Varvin I, Hernández H, Pelicier M y Rivas MC (1986) Informe técnico No 42. Beneficio de las menas de los yacimientos Carlota – Guachinango – Victoria: Inédito. Oficina Nacional de Recursos Minerales, La Habana

The Minerageny of Cuba

Waldo Lavaut Copa and Rolando Batista González

Abstract

The complex and prolific geological scenario of Cuba has led to geological-genetic conditions favorable for the formation of varied deposits, manifestations, and mineralized points of various industrial raw materials, decorative and for multiple practical uses, known worldwide as industrial rocks and minerals, or simply as industrial minerals. Thus, Cuba has large deposits of limestone, loam, dolomite, kaolin, gypsum and anhydrite, rock salt, marbles, sands and clays of different types, zeolites, peat, therapeutic peloids, and many more. There are manifestations of decorative and precious rocks such as jasper, jadeite, different varieties of quartz, and even xylopals. The degree of study of the different varieties of these rocks and industrial minerals is still low in the face of their great diversity and possible uses in different industrial branches of industry, medicine, and in the field of research. Its origin can be summarized in seven genetic groups: magmatic, sedimentary, pyroclastic-sedimentary, metamorphic, skarn, hydrothermal, and supergenic (residual). The different genetic groups are not governed, in general, by the same or similar geotectonic situations. They appear in different geotechnical areas and geological contexts, such as limestone rocks, granitoids, and zeolites. There are also tendencies, in some genetic types, to be located in specific environments such as peat, in the biogenic zones of the Miocene–Quaternary sedimentary cover and, the kaolin, in the Mesozoic accreted terrane.

W. L. Copa (✉)
Instituto de Geología y Paleontología (IGP)-Servicio Geológico de Cuba, Calle Vía Blanca no. 1002, San Miguel del Padrón, La Habana, CP 11000, Cuba
e-mail: waldo@igp.minem.cu

R. B. González
ENIA/INVESTCONS-Dirección General, La Habana, Cuba
e-mail: rolando@enia.co.cu

© Springer Nature Switzerland AG 2021
M. E. Pardo Echarte (ed.), *Geology of Cuba*, Regional Geology Reviews,
https://doi.org/10.1007/978-3-030-67798-5_9

Keywords

Cuba • Minerageny • Industrial rocks and minerals • Accreted terrane • Genetic type • Mineragenic map

9.1 Introduction

The present chapter seeks to offer a more complete and more up-to-date synthesis of the *Minerageny of Cuba*. It is supported in the current knowledge on the geologic complex structure of this Caribbean archipelago of hardly 111 thousand km^2, especially of its bigger island, Cuba. It occupies the 94.7% of that surface [1] and it contains the biggest diversity of geologic atmospheres. Seated in them, there are big resources of rocks and industrial minerals with a total amount that overcomes the 10,900 million tons in more than 40 types of no-metallic natural useful materials.

During most of the twentieth century, mainly in their second half, Cuba substantially developed the study and exploitation of its resources of rocks and industrial minerals. This also carried the escalation of the geologic, technological, and economic study, for what was necessary to introduce environmental and legislative regulations for the rational use, control, and care of them and of the Cuban Nature. The results of such investigations were taken into account for this work.

However, there are still resources of industrial rocks and minerals little studied and necessary to complete the material base of the country. Among them are the precious stones, including jades, jaspers, xylopals, as well as multiple varieties of quartz, opals, chalcedonies, etc. Larger facilities exist for dimensioned rock (micritical limestones, granitoids, dolomites, etc.) and the development of other existing varieties in the country: white, gray, black calcitic marbles. In these cases, the mineralization characterization is still limited, which prevents a better illustration in this chapter.

9.2 Theoretical Frame

The actual theoretical framework includes the precision of some lexicographical and conceptual aspects used in the present work, due to the existence of different approaches and certain lexicographical diversity of nomenclatures used in the world during the treatment of the no-metallic mineral resources and its geology.

9.2.1 The Mineragenic Science

The **Minerageny** is the branch of the geologic sciences in charge of the study of the regional geologic regularities of the formation and localization of deposits of useful rocks and minerals. It studies the genetic-space relationships of these in connection with the geodynamics and geologic-tectonic scenarios that generate them.

As a science, it had a long development with the main concepts introduced at the beginning of the six decade of twentieth century. In this way, Ray [22] subdivides the Minerageny into two branches: *Metallogeny* to treat only the metallic deposits and *Minerageny* for the no-metallic ones. Also do Bauman and Tishendorph [6]. Other metalogenists, as Lingren, Bilibin and Tijomirov [7, 18, 24], also united both disciplines into only "*Minerageny*," although when treating the metallic deposits they used "*Metallogeny.*"

For the effects of the present work, Ray's article conceptualization will be used representing under the concept of **Minerageny** only the no-metallic deposits of Cuba, since the metallic ones are exposed separated in the precedent chapter of this book.

9.2.2 Cuban Geologic Scenarios

The territory of the Cuban archipelago includes a diversity of geologic scenarios as a consequence of a long evolution conditioned by the movements of geotectonic plates, the generation and development of several volcanic arcs, the collision with the North American platform, and the development of multiple sedimentary basins of different geologic ages.

To the moment, three main episodes are considered in the geologic evolution of Cuba, according to Iturralde Vinent et al. [14]:

- Rupture of Pangea in the Jurassic with an associated passive margin with oceanic sedimentation and magmatic evolution.
- The sedimentary, magmatic, and metamorphic evolution of an intraoceanic ophiolite-arc complex of Cretaceous–Paleogene age.
- "Soft collision" and transfer of the Northwestern Caribbean plate (with Cuba) toward the North American plate in the Paleogene.

These regional events have conditioned that two main geo-structural units characterize the Cuban archipelago, now: (a) Fold belt or Basement and (b) Neoautochthon or Sedimentary Cover [10, 12].

The folded basement includes the scenarios of continental margins, allochthon oceanic units, volcanic arcs, and a series of sedimentary basins classified as foreland basins, post-volcanic and transported (piggyback) basins, and first and second cycle's basins. The Neoautochthon or sedimentary covers composed of Cenozoic sedimentary rocks with varied lithologic composition.

The *continental margins* include the following geologic scenarios:

- **Block of Florida Strait** composed of rocks of Lower Jurassic to Lower Cretaceous age.
- **Mega platform of Florida–Bahamas** with sedimentary rocks and evaporites of varied lithologies (J_{1-2}-K_2m).
- **Platform of Bahamas** with sedimentary rocks of Lower to Upper Cretaceous ages, in some cases metamorphosed in variable grade. It is subdivided into the following zones:
  - Cayo Coco zone (intraplatformic channel).
  - Remedios zone (external platform).
  - Camajuaní zone (continental slope). Asunción Massif (Terrane).
- **Southwestern Terranes:** (a) *Guaniguanico*: Los Órganos and Cangre Zones. Zones of Northern Rosario and Southern Rosario. Quiñones–Guajaibón Zone; (b) *Escambray*: Metasedimentary sections. Metavolcanic sections. Metaophiolites of Yayabo; (c) *Pinos*: Sedimentary rocks metamorphosed in variable grade (J_1-K_2cn).

The *allochthones oceanic units* comprise the following scenarios: (a) Placetas Terrane (ProtoCaribe Basin) is composed of sedimentary rocks metamorphosed in variable grade (J_1-K_2cn); (b) ophiolites of the northern belt which are mafic and ultramafic rocks (K_1).

The **volcanic arcs** encompass volcanic terranes of different geologic ages and geochemical characteristics, that is, (a) *boninitic* (with possible pre-Albian age); (b) *primitive*, with rocks of pre-Albian age of the PIA series; (c) *Albian–Campanian volcanic arc*, with its volcanic-sedimentary complex divided into three basins (backarc, axial, and frontal basins of the arc) and the plutonic and metamorphic complexes; (d) *basement of the PIA and Albian–Campanian volcanic arc*, composed of metamorphosed volcanic rocks (pre-Neocomian Cretaceous); (e) *post-volcanic and transported (piggyback) Cretaceous basins*, containing sedimentary rocks of varied lithologies ($K_2cp\text{-}m\text{-}E_1^1$); and (f) *Paleogenic volcanic arc* with its volcanic-sedimentary complex divided mainly into the backarc and axial basins, and with the plutonic and metamorphic complexes that include volcanic rocks ($E_1\text{-}E_2^2$), granitoids (E_2^2), and volcano-sedimentary rocks.

Other sedimentary basins that cover the fold belt are the following ones: (a) *foreland basins*, among which are the Paleocene–Upper Eocene Basin of North of Cuba and the Upper Paleocene–Eocene of Guaniguanico; (b) *post-volcanic and transported (piggyback) basins*; (c) *the Upper Campanian–Lower Maastrichtian–Danian first cycle basins*; and (d) *the Upper Danian–Upper Eocene second cycle basins*.

The Neoautochthon or cover constitutes a practically sedimentary geologic scenario that can be divided into transitional complexes with Lower Eocene–Lower Miocene ages and several stages of geological ages from Eocene up to Quaternary.

9.2.3 Cuban Advances in Mineragenic Science

The knowledge and exploitation of industrial minerals and rocks in Cuba go back to so old times as to the indigenous native communities (eighth century of our era), a lot before the arrival of the Spanish conquerors in 1492. Already the first inhabitants, mainly of the Island of Cuba, had the lithic industry and the ceramic among the main ones [13, 23], using the flint, the clay, the dimension stone, as the coralline and other, as well as the micro-lithicand coloring stone industries.

In the time of the colony, the Spaniards used the "Jaimanita" stone in form of blocks for the construction of strengths as El Morro, La Fuerza, La Punta, La Cabaña, and others in strategic localities of the island of Cuba, as well as the clay for bricks and other no-metallic natural materials. Later on, in the Neo-colony, the "Capellanía" limestone was used for the construction of the National Capitol and Presidential Palace. Big quantities of sands and construction gravels were assimilated for the development of the civil constructions, as much inside the cities as in the rural ones with the construction of highways, railroads, and roads.

Special relevance has the geologic investigations starting from the sixty decade in the twentieth century with the appearance in the country of the Instituto Cubano de Recursos Minerales. In the mark of it, in the subsequent decade, intense prospecting works and geologic exploration were developed. These included important technological studies of the nonmetallic materials for industrial and constructive diverse aims, as well as applications in the medicine, jewellery, and for the agricultural sphere.

The preponderance of the deposits of calcareous rocks (limestone, calcareous limestone, marl, calcareous dolomite, dolomite, marble and marble-like limestone) was detected, as well as of the volcanic rocks. Among them, of more importance, were found tuff, zeolitic tuff, basalts, and andesites. A second order of importance was established for clays, bentonite, polimictic sand, and quartzose sand.

The main perspectives, because of their wide distribution, quality, and other facilities, have the deposits of the industrial rock limestone and calcareous dolomites, clays, polimictic sands, andesites, asphaltites and bitumen, zeolitic tuff, and quartzose sand.

Also are present deposits of kaolin, turf, feldspar, phosphorite, barite, quartzite, granitoids, gypsum, calcarenite, gabbro, volcanic glass, sandstone, quartzite, magnesite, micas, palygorskite, rock salt, chert, wollastonite, chrysotilic and amphibolic asbestos, opals, garnet, jadeitite, graphite, kyanite, secondary quartzite, dacite, diatomite, amphibolite, and diorite.

Parallel with the development of the knowledge of the facilities of industrial minerals and rocks of the country, outstanding advances were achieved in the geologic cartography. This included the deep study of the sedimentary basins, as well as for the forecasting of no-metallic minerals resources, and for the metallic ones and fuels. In this way, it was possible to establish the existence of more than 2000 deposits of nonmetallic useful materials of some 75 varieties, systematizing and generalizing their main structure and formational regularities [5].

In this respect, there was established and studied

- more than 100 deposits of limestone, dolomitic limestone, and marl-limestone with more than 1300 million m³ of calculated resources;
- more than 30 deposits of calcarenite and sandstone with more than 150 million m³ of calculated resources; more than 30 locations of andesite, basalts, gabbros, and granitoids with more than 400 million m³ of calculated resources;
- more than 100 deposits of sands and gravels with 300 million m³ of calculated resources; more than 25 locations of ornamental rocks with calculated resources of about 35 million m³, mainly of marble-like limestone of different colors and textures, what give them a high decoration possibility;
- more than 110 deposits of clays and marl with resources calculated for more than 220 million m³; more than 25 deposits of limestone, marl, and marbles for carbonated component of cement with resources calculated for more than 4800 million tons;

- more than 35 deposits of clays, kaolin, sandstone, sands, feldspar, and marl for silica-aluminous component of cement with resources calculated for more than 900 million tons; 8 locations of ferruginous clays (mining tailings of laterites and Cu-Zn-Pb sulfides are included) with more than 40 million tons of calculated resources;
- more than 10 pozzolan (vitreous and zeolitized tuffs) deposits with resources calculated for more than 80 million tons; more than 100 manifestations of rock gem (rock glass, chalcedony, agates, jaspers, chert, opals, amethysts, and jade).

At the moment, the study and amplification of the resources of white, gray, and black marbles with more than 2388.7 thousand m^3 were developed, settling down that their geologic scenarios of formation are related with contact metamorphism of limestones with arc granitoids [25]. The marbles or metamorphosed carbonated rocks for ornamental uses, as well as for source of calcium carbonate and crashed arid in Cuba are three deposits with more than 4 million m^3 of calculated resources.

9.3 Materials and Methods

The present chapter is based on materials published in Cuba and abroad, as well as data and geologic reports of unpublished files, those with the biggest upgrade and general possible acceptance. Several quarries of nonmetallic rocks (limestone, clays, kaolin, sands) were visited and inspected, including some no longer active. The methods of the inquiry, direct observation, induction, deduction, and generalization were used.

The initial information comes from specialists of recognized authority and sources with enough publicity. The unpublished used filed materials were made by collection of authors of the geologic prospecting/exploitation, as well as from the mining that constitutes official sources approved by the competent organisms of the geologic Cuban branch.

9.3.1 Classification of Rocks and Industrial Minerals

For the work, the denomination of *industrial minerals and rocks (IMR)* instead of nonmetallic minerals was adopted, and therefore deposits (locations, prospects, etc.) of industrial minerals and rocks were used instead of nonmetallic deposits (locations, prospects, etc.), although this last denomination appears effective still in the present century, as it seen in Luz and Lins [19].

Majorities of the classification systems were developed in the twentieth century, being used different approaches, mainly the practical use of these raw materials ([2, 11]; and others), economic approaches ([20, 26]; and others), as well as geologic (genetic) approaches combined with the practical uses ([3, 15]; and others). Some classification systems also include inside IMR the mineralized and underground waters [9].

In the case of Cuba and for this chapter, it has been used with emphasis the geologic approaches combining genesis with geodynamic scenarios, since it is a more advanced and more concordant nomenclature with the metallogenic theories of the present decade ([21, 27]; and others).

In this way, the industrial minerals and rocks of Cuba here are grouped into 6 groups and 14 genetic types:

- Magmatic (Intrusive/Extrusive; and the related Skarn/Metasomatic);
- Metamorphic (Metamorphized, Related Pegmatitic);
- Hydrothermal;
- Piroclastic (Diagenetic);
- Sedimentary (Clastic; Turbiditic; Biochemical; Evaporitic; Caustobiolithic); and
- Supergenic (Residual).

The genetic tipification is completed with the scenarios or geodynamic/geotectonic context where each IMR is located, in the most precise way inside the different terranes that compose the territory of Cuba, i.e., ancient accreted terranes, terranes of the Caribbean Plate, and Sedimentary Cover. The classification includes the geologic age and the name of the rock or industrial mineral (generally coincident with their petrographic name).

The productive reach of each genetic type within the geotectonic scenarios for IMR is exposed in form of total tonnage for the related deposits of more relevance, see Table 9.1.

9.3.2 The Cuban Industrial Minerals Database

Cuba has important databases of industrial minerals and rocks, located, maintained, and protected by the institutions of the geologic branch of the country (ONRM, IGP, GEOMINERA, MICONS, MINEM, and others). These databases almost totally, except for operative works, are in digital support as well as in printed form (some in manuscript).

Among the main databases of IMR of Cuba count the following:

- National Balance of Mineral Resources and Reserves at ONRM that contains the up-to-date information of the national movement in quantity and quality of them.
- INFOYAC_Nonmetallic Raw Materials, by the IGP-Servicio Geológico de Cuba.

Table 9.1 Mineragenic characteristics of Cuban industrial rocks and minerals

Geotectonic settings	IMR commodities	Genetic types	Ages	Bulk resources (million tons)	Examples of deposits
Magmatic genetic group					
Cretaceous granite-granodioritic pluton of CVA-terrane	Granitoide	Intrusive	K_2	23.4	Piedrecitas
Upper cretaceous volcanic neck of CVA-terrane	Andesite/Granitoide	Intrusive	K_2	242.3/83.3	Arriete; El Cuero; Ciego Alonso/Peña Blanca
Lower cretaceous volcanic arc terrane	Andesite	Extrusive	K_1	9.1	Rebacadero
Upper cretaceous volcanic neck of CVA-terrane	Andesite	Intrusive	K_2	27.5	Jicotea—Jose San Mateo; La Mulata
Upper cretaceous volcanic arc terrane	Andesite/Basalt	Extrusive	K_2	208.7/24.6	El Pilón; Flor de Mayo/San José
Pyggyback basin of PVA-terrane	Felspar	Metasomatic	K_2	8.1	El Purnio
Succession of volcanites and piroclastics of axial PVA-terrane	Andesite	Extrusive	E_1-E_2 (e_5)	167.8	Botija; Los Guaos II
Shallow gabbroic intrusives of meridional ophiolite suite	Gabbro	Intrusive	K_2	412.1	Serones II-La Prudencia
Metamorphic genetic group					
Mesozoic metamorfic accreted Pinus terrane	Marble/Mica/Cyanite	Metamorphic	J_3-K_1/J_2-J_3	13/0.72/1.6	Punta Colombo; Piedra La Fe/El Aleman/Las Nuevas
Cretaceous-subducted metasediments and metaophiolite Escambray terrane	Quartz/Marble/Garnet-Mica/Graphite	Metamorphic	J_1-K_1	0.04/2.8/17.4/3.4	Cacahual/El Mirador/La Belleza/Algarrobo
Cretaceous-napped Jurassic terrigenous sediments of Guaniguanico terrane	Quartzite	Metamorphosed	J_1-J_3	4.8	Bejuquera; Ceja del Negro; Llano de Manacas
Chaotic high pressure metamorphic-serpentinitic mélange	Quartz	Pegmatitic	J-K_1	0.005	La Corea
Skarn-hydrothermal genetic group					
Upper cretaceous sedimentary cover of CVA-terrane	Wallastonite-Garnet	Skarn	K_2	16.6	Arimao Norte
Green schist complex of Escambray metamorphic terrane	Talc	Hydrothermal	K_2	0.38	Crucesita-La Sabina; Higuanojo
Upper cretaceous volcanic arc terrane	Kaolin	Hydrothermal	K_2	7.5	Pontezuela
Pyroclastic genetic group					
Upper cretaceous volcanic arc terrane	Zeolite	Diagenetic	K_2	47.7	Las Pulgas; El Chorrillo
Pyggyback basin of PVA-terrane	Zeolite	Diagenetic	E_2 (e_{4-5})	20.4	San Cayetano
Upper cretaceous sedimentary cover of CVA-terrane	Zeolite	Diagenetic	K_2	2.3	San Andrés-Loma Blanca
Upper cretaceous volcanic arc terrane	Zeolite/Tuff	Diagenetic	K_2	139.48/20.8	Carolina II; Orozco I; Tasajera/La Victoria
Pyggyback basin of PVA-terrane	Zeolite/Vitreous tuff	Diagenetic	E_1^1-E_2 (e_{4-5})	66.31/102.6	Palmarito de Cauto; Seboruco; Caimanes; Palenque/El Lirial; Amansaguapo; Ají de la Caldera
Sedimentary genetic group					
Miocene-Quaternary sedimentary cover	Turf/Terapeutic pelites	Caustobiolithic	Q_2	79.25/1.8	Ciénagas de Lanier-Zapata; Majaguillar; Playa Guanímar/Maracayero; El Soldado; Monte Gordo; Jutía

(continued)

Table 9.1 (continued)

Geotectonic settings	IMR commodities	Genetic types	Ages	Bulk resources (million tons)	Examples of deposits
Miocene-Quaternary sedimentary cover	Sand/Quartzose sand	Clastic	Q/Q_1	201.01/90.4	Cayo Inés de Soto; Arimao; Pilotos; Imías; Río Bayamo; Río Arimao; Tacre-Cajobabo; Indios Norte; Río Guantánamo; El Angel/Siguanea; Panzacola I
Cretaceous-Napped Jurassic evaporitic sediments of Bahamas platform terrane (Diapires)	Gypsum/Rock salt	Evaporitic	J_1-J_2	198/22.24	Corral Nuevo-Canasí; Punta Alegre-Máximo Gómez; Punta Alegre-Mamon/Punta Alegre
Oligo-Miocenic sedimentary cover	Phosphoric rock	Biochemic	E_3-N_1	37.95	Trinidad de Guedes; Meseta Roja; Loma Candela Norte; Cañada Honda
Miocene-Quaternary sedimentary cover Los Palacios basin	Clay	Clastic	N_2^2-Q_1^1	102.58	La Conchita; Ovas I; Kessel; Moja Hueva; Bermejales; Club de Cazadores
Miocene-Quaternary sedimentary cover Cauto Valley	Clay	Clastic	N_2^2-Q_1	124.1	Ullao; Bayamo; San Rafael; Aguadores; Contramaestre; Caney del Sitio
Oligo-Miocenic sedimentary cover	Clay	Clastic	N_1-Q	187	Rodas—Abreu; Coloradas II; Rio Caunao; Castano - Siguaney; La Fortuna; Sancti Spiritus Zona III A; Charco Soto; La Pachanga; Ciego de Avila; Las Tunas zona XII; La Naza
Distal turbidites of sinorogenic basin of PVA-terrane	Clay	Turbiditic	E_2^2-E_2 (e_3)	4.52	Novaliche; Platanillo; Maceiras Carretera
Distal turbidites of sinorogenic basin of PVA-terrane	Clay	Clastic	Q_1^3-Q_2	2.7	El Quinto; Maceiras Vaquería
Oligo-Miocenic sedimentary cover	Bentonite	Clastic	E_3 (e_9)-N_1^3	21.52	Santiago de Las Vegas; Managua; Bauta; Redención
Miocene-Quaternary sedimentary cover	Bentonite	Clastic	Q_1^3-Q_2	21.29	Chiqui Gómez; Vado del Yeso
Lower Paleogenic olistostromes of foreland basin	Limestone	Biochemic	E_1^2-E_2 (e_4)	350.14	La Lata—Teté Contino; El Chivo—Quiñones; Quiñones—El Llano; La Molina; Senado
Maastrichtian Upper Middle Eocene piggyback Cabaiguán basin	Limestone	Biochemic	E_2 (e_4)-E_2^2	248.87	Seboruco; Nieves Morejón
Mesozoic metamorfic accreted metacalcareous Cangre terrane	Limestone	Biochemic	J_3-K_1	75.46	Lagunitas—Km 13 Viñales
Cretaceous-napped Mesozoic calcareous clastic sediments of Bahamas platform slope	Limestone	Biochemic	J_3-K_1	1071.58	Los Azores—Zulueta; El Purio; Jumagua—Cantera Lazaro Penton; Jiquí; Sierra de Cubitas; Los Caliches—Cantera 200 mil; Gibara—Los Caliches; La Yaya—Gibara—Velazco
Cretaceous-napped Jurassic terrigenous sediments of Guaniguanico terrane	Limestone	Biochemic	J_3	30	Sitio Pena
Cretaceous-subducted metasediments of Escambray terrane	Limestone	Biochemic	J_3	35.72	Trinidad Zona II
Cretaceous thin skin napped_J3-K2 terrigenous sediments of Guaniguanico terrane	Limestone	Biochemic	J_3-K_2	163.93	La Reforma; La Muralla
Lower-Middle Eocene basal sediments of Eocene-Quaternary cover	Limestone	Biochemic	E_2^2-E_2 (e_7)	22.08	La Caoba—Caraballo

(continued)

Table 9.1 (continued)

Geotectonic settings	IMR commodities	Genetic types	Ages	Bulk resources (million tons)	Examples of deposits
Oligo-Miocenic sedimentary cover in Vegas-La Broa-Mercedes-Guacanayabo-Nipe basins	Limestone	Biochemic	E_3 (e_9)-N_1^1	782.61	Sierra Anafe—La Cruz; Santa Teresa—Cantera II; Capellanía; Cantera Blanca; San Jose Sur—Camoa—La Moderna; Beluca; La Colina; La Union; Boca de Jaruco; Loma de Cura; Planta Libertad; San Antonio de Cabezas; Coliseo; Cantel-Zona I; Calera José Martí; Pilon; Levisa
Miocene-Quaternary sedimentary cover	Limestone	Biochemic	N_2^2-Q_1^3	900.42	Jaimanitas-Zona Guajaibón; Rio Mosquito; Jaimanitas Matanzas; Monte Alto; Carbonato Cayo Piedra; Aguadores Este; Cabo Lucrecia
Upper Cretaceous sedimentary cover of CVA-terrane	Limestone	Biochemic	K_2 (k_5-k_6)	125.79	Real Campiña; La Matilde; Rodolfo Rodriguez; Guayacán
Eocenic sedimentary cover of PVA-terrane	Limestone	Biochemic	E_2 (e_2-e_5)	681.83	Rio Frio; La Inagua; La Aurora—San Rafael; Farallones de Moa; Julio A. Mella; Mucaral—La Maya; Rosa Aurora; El Cacao
Upper Eocene-Low Miocene transitional sedimentary cover	Limestone	Biochemic	E_3 (e_9)-N_1^1	61.15	San Antonio del Sur
Backarc Eocenic sediments of PVA-terrane	Limestone	Biochemic	E_1^1-E_2 (e_5)	25.92	Mogote San Nicolas
Miocene-Quaternary sedimentary cover	Marl	Biochemic	N_1^1-N_1^3	2288.24	Matanzas—Bacunayagua; Aguadores Oeste
Oligo-Miocenic sedimentary cover in Mercedes basin	Marl	Biochemic	E_3 (e_9)-N_1^3	3.38	Simpatía-II
Upper Maastrichtian-Middle Eocene piggyback Santa Clara basin	Marl	Biochemic	K_2-E_2 (e_5)	70.71	Las Pilas
Upper Cretaceous sedimentary cover of CVA-terrane	Marl	Biochemic	K_2 (k_6)	205.19	Loma Cantabria
Lower-Middle Eocene basal sediments of Eocene-Quaternary cover	Marl	Biochemic	E_2 (e_5)	5.18	Trinidad—Santa Rosa
Eocenic-Oligocenic sedimentary cover	Marl	Biochemic	E_2 (e_7)	142.63	Pastelillo
Miocene-Quaternary sedimentary cover	Dolomite	Biochemic	N_1^1-N_1^3	1.50	Vedado
Oligo-Miocenic sedimentary cover in La Broa-Mercedes basins	Dolomite	Biochemic	N_1^1-N_1^3	119.26	Puerto Escondido; Alacranes; Perico; La Montana
Miocene-Quaternary sedimentary cover	Calcareous mud	Biochemic	Q_2	30.89	Cayo Moa Grande
Upper Cretaceous sedimentary cover of CVA-terrane	Calcarenite	Clastic	K_2 (Upper k_6)	257.10	La Fe—La Colorada—La Concordia; Victoria II-III-IV; Arenisca Matanzas-Navarro; Arango-Miraflores; La Guayaba Sector II; Peñalver; San Luis II
Supergenic genetic group					
Mesozoic metamorfic accreted Pinus terrane	Kaolin	Residual	N_2-Q	47.5	Río del Callejón; Siguanea
Cretaceous-napped Jurassic terrigenous sediments of Guaniguanico terrane	Barite	Residual	N_2-Q	0.24	Francisco

(continued)

Table 9.1 (continued)

Geotectonic settings	IMR commodities	Genetic types	Ages	Bulk resources (million tons)	Examples of deposits
Cretaceous thin skin napped_J3-K1 terrigenous sediments of Guaniguanico terrane	Kaolin	Residual	N_2-Q	5.7	Malcasado
Cretaceous thin skin napped_J3-K1 terrigenous sediments of Guaniguanico terrane	Ferruginous clay	Residual	N_2-Q	0.29	Olga
Mesozoic metamorfic accreted glaucophanic Cangre terrane	Clay	Residual	N_2-Q	15.9	Esquistos Paso Viejo
Obducted oceanic litosphere_J3-K1 Cajálbana ophiolite mélange	Ferruginous clay	Residual	N_2-Q	11.3	Cajálbana
Miocene-Quaternary sedimentary cover Los Palacios basin	Clay	Residual	Q_1-Q_2	10.7	La Conchita
Oligo-Miocenic sedimentary cover Vegas basin	Clay	Residual	Q_1-Q_2	0.4	El Jardín
Oligo-Miocenic sedimentary cover La Broa basin	Clay	Residual	Q_1-Q_2	0.2	Sabanilla
Upper Cretaceous sedimentary cover of CVA-terrane	Ferruginous clay	Residual	Q_1-Q_2	5	Palanquete
Chaotic polymictic serpentinitic mélange	Magnesite	Residual	Q_1-Q_2	0.01	Las Minas
Cretaceous granite-granodioritic pluton of CVA-terrane	Felspar-quartzose sand	Residual	N_1-Q	48.5	Cumanayagua-El Canal
Cretaceous plagiogranitic intrusive of CVA-terrane	Felspar-quartzose sand	Residual	N_1-Q	22.3	La Macagua-El Salto
Cretaceous granite-granodioritic pluton of CVA-terrane	Clay	Residual	N_1-Q	9.6	Carranchola-La Moza
Cretaceous granite-granodioritic pluton of CVA-terrane	Sand and gravel	Residual	N_1-Q	14.3	Jíquima-Alfonso
Cretaceous granite-granodioritic pluton of CVA-terrane	Sand	Residual	N_1-Q	66.2	Granodiorita—Sancti Spiritus—Aeropuerto
Cretaceous dioritic intrusive of CVA-terrane	Sand	Residual	N_1-Q	34.6	Venegas
Cretaceous Calc-alkalic Arc Basalts	Clay	Residual	N_2-Q	16.1	Sabana Grande
Cretaceous granite-granodioritic pluton of CVA-terrane	Sand	Residual	N_1-Q	6.2	Corojo
Eocenic-Oligocenic sedimentary cover	Ferruginous clay	Residual	Q_1-Q_2	17.5	El Cebadero
Cretaceous granite-granodioritic pluton of CVA-terrane	Felspar-quartzose sand and gravel	Residual	N_1-Q	110.25	La Mota; La Conchita; Cascorro; Las Tunas II; Galbis; Las Canoas-Dolores
Upper Cretaceous volcanic arc terrane	Ferruginous clay	Residual	N_1-Q	0.095	Golden Hill
Succession of volcanites and piroclastics of axial PVA-terrane	Clay	Residual	$E_2(e_5)$-Q	98.1	Guamuta; La Sierrita; Vega Grande; Sabicú
Shallow gabbro-plagiogranitic intrusives of PVA-terrane	Clay	Residual	N_2-Q	0.01	Pinar de las Canas
Succession of volcanites and piroclastics of axial PVA-terrane	Kaolin	Residual	N_2-Q	0.05	El Cobre
Middle Eocenic tonalite-granodioritic pluton of PVA-terrane	Sand	Residual	N_2-Q	118.4	Daiquirí; Juraguá; Las Guásimas; El Refugio; Sector III; La Punta; Damajayabo II

(continued)

Table 9.1 (continued)

Geotectonic settings	IMR commodities	Genetic types	Ages	Bulk resources (million tons)	Examples of deposits
Cretaceous tonalite-granodioritic pluton of PVA-terrane	Clay	Residual	N_2-Q	16.4	Santa Isabel
Distal turbidites of sinorogenic basin of PVA-terrane	Bentonite	Residual	N_2-Q	4.3	La Caoba
Succession of alcaline volcanites and sediments of Upper CVA	Clay	Residual	N_2-Q	0.325	Arroyo del Medio
Upper Eocene-Lower Miocene transient sedimentary cover	Clay	Residual	N_2-Q	2.5	La Clarita
Mafic gabbroic cumulates of meridional ophiolite suite	Clay	Residual	N_2-Q	2.5	Nuevo Mundo

- Geologic database of the Servicio Geológico de Cuba (GeoDato).
- Databases of geologic and mining companies of the country (GEOMINERA, Grupo Empresarial Industrial de la Construcción, GECEM, and others).
- Multimedia of the Potential of IMR of Cuba for municipalities (IGP 2011).
- Databases of the geologic universities of the MES (ISMMM; CUJAE; UPR; UCLV-CIDEM).

For the aims of the present chapter, it was carried out a compilation, revision, and generalization of the main databases of industrial minerals and rocks of Cuba.

9.3.3 The Cuban Industrial Minerals Study Grade

The industrial minerals and rocks of Cuba have a high grade of geologic study, except for the gem stones where most of the manifestations are still for prospecting. Important geologists, Cuban institutions, and foreigners during decades contributed to the study grade reached to the moment.

The actions that made possible the achievements in the study grade can be summarized in the following aspects:

- Prospecting and geologic exploration of different raw nonmetallic materials in diverse regions of the country.
- The systematizing and generalization studies for nonmetallic materials with reports and maps of locations and manifestations for the national territory by the Centro de Investigaciones Geológicas (CIG) of MINBAS (1980–1985).
- Making of the map of deposits and manifestations of nonmetallic materials and fuels of the Republic of Cuba in scale 1:500000 by the CIG MINBAS (1988).
- Mineragenic investigations with forecast of raw nonmetallic materials by the Instituto de Geología y Paleontología in scale 1:500000 (1990–1996).
- Realization of thesis of diplomas, masters, doctorates, and publications (articles, scientific and promotional books, reports, etc.).

Among the main geologic supports that contain and expose the study grade of industrial minerals and rocks of Cuba are the following ones:

- *Maps and reports for regions by the systematizing and generalization of present raw nonmetallic materials in the national territory.* Department of Nonmetallic Deposits (CIG, 1980–1985). Authors: Coutín DP, Martínez J, González Z, Lippstreu L, and others.
- *Map of Minerageny and Forecast for raw nonmetallic materials 1:500000* (IGP, 1985–1990). Authors: Coutín DP, Martínez J, and others.
- *Map of deposits and manifestations of nonmetallic materials and fuels of the Republic of Cuba in scale 1:500000* (IGP 1988). Authors: Coutín DP, Martínez J, and others.
- *Map of Rocks and Industrial Minerals of Cuba*, scale 1:100000, CD-ROM (IGP 2011), [5].
- *Informative system for the Mineral Resources of the Republic Cuba (INFOYAC)*—Database (IGP 2016), [4].
- *Deposits of Rocks and Industrial Minerals of Cuba, their main structure-formational regularities and perspectives.* Memoirs of the 1st Cuban Convention of Sciences of the Earth, CD-ROM, La Habana 2005. Authors: Batista González R, and others.
- *National Balance of Mineral Resources and Reserves* (ONRM, La Habana, Cuba). Authors: Collective of specialists.
- *Geologic and technological reports of the Prospecting and Exploration* of the industrial minerals and rocks of

Cuba (ONRM, until 2018). Authors: Several Cuban and foreigner geologists and specialists of companies, mines and universities, among the Cubans outstanding are Linchenat A, Vidal P, Rizo R, Rojas JC, Rams J, Sosa L, Regueiferos M, Leyva C, Ramirez J, Rabilero A, and many others.
- *Ornamental Rocks of Cuba. Geology, Localization and Properties* (in Russian). AC USSR—AC of Cuba, Edit. "NAUKA", Moscow 1981. Authors: Belikov BP, Coutin DP, Lizarev MTo, Skaris J.
- *Zeolitites of Cuba: Recount of 45 year dedicated to their study*. IGP (2017). Authors: Brito A, Coutin DP.
- *Geologic-descriptive model of El Canal sand deposit*. Cienfuegos. Thesis of Master. ISMMM-IGP (2018). Author: Shales S.

As can be appreciated, the study grade of industrial minerals and rocks of Cuba, in general, overcomes 80%.

9.3.4 Applicability and Perspectives of Cuban Industrial Minerals

The industrial minerals and rocks of Cuba constitute a material and economic fundamental resource of the country. On the basis of them, high levels of civil constructions as never before seen were achieved in the second half of the twentieth century in the country.

Cuba has the fundamental types of industrial minerals and rocks for the normal operation of its cement industries, sugar and nickel plants, the chemical and agricultural industries, and others. So that industrial minerals and rockshave applicability and development perspectives.

This way, constantly the study is continued for new applications of nonmetallic materials in the country, as, for example,

- the investigations carried out by CIDEM (year 2015) on kaolinitic clays, limestones and clays for the production of cement of low carbon (LCC);
- the studies carried out by the Cuban enterprise for marbles, especially in 2018 on white marbles and other kind of marbles of Cuba; and
- those carried out with zeolites for natural fertilizers, and recently other developments on raw zeolites by CIPIMM.

Also with the auspice of the Servicio Geológico de Cuba now are carried out works on other kinds of mineral raw materials for the construction, the agriculture and other applications, like on jade, talc, sheridanite, diatomite, carbonate, phosphoric rocks, natural fertilizers, and sands. The application of clays is developed for bricks and dimension rocks to a more economic construction styles as the use of vaults and arches.

w?>Wide applicability has been found in the Medicine in Cuba in the pelites and other nonmetallic materials that can also be used in other countries. On the natural zeolitites, the country has working plants and commercialization abroad; in this respect, the possibilities of Cuba in resources and reserves are enormous. As example of zeolite, for massive uses are ready among 1–10 million tons, for special uses between 30 thousand and 500 thousand tons, for pozzolanic additives from 3 to more than 30 million tons [8].

Among the recent applications in Cuba of national zeolite count: (a) for diets of birds; (b) elevation of the efficiency of the nitro-fertilizers; (c) as additive in anti-fungus paintings; (d) filtration of the water; (e) production of economic paintings; (f) pozzolanic additive to the cement; (g) preservation of floors of dairies and for organ-mineral fertilizers; (h) membranes for biogas production; (i) recovery of nutrients in bio-refineries and environmental management; and so many others.

Also recently, other Cuban industrial minerals and rocks have been applied in very diverse utilities as the levigated kaolin (kaolinite); the low tenor phosphoric rocks for direct fertilizers; pharmaceutic calcium carbonate precipitate; elevation of the quality of the natrificated bentonite with sodium carbonate; application of natural micronized alum silicates for fungicides; antibacterial and drying purposes; carbonate of calcium for drilling; obtaining aluminum sulfate kaolin; magnesite for refractory; bentonite for the regeneration of used oil; etc.

According to Kuzvart [15], the industrial minerals and rocks (IMRs) will be the typical raw materials for the second industrial revolution, the raw materials of the third millennium. Cuba also stands for this.

9.4 Results

The mineralogenic characteristics, distribution regularities, geologic ages, and relationship with the paleotectonic structures and complex of rocks, where the occurrences of industrial minerals and rocks of Cuba were generated (or where they are now), are shown in the map of Fig. 9.1 and with more details the main ones in Table 9.1. The corresponding descriptions are exposed in the subsequent epigraphs, see Fig. 9.1 and Table 9.1.

9.4.1 Distribution of Rocks and Industrial Minerals in the Different Geo-Structural Contexts

The industrial minerals and rocks of Cuba for a better systematizing of their distribution have been grouped in correspondence with the geotectonic terranes that compose

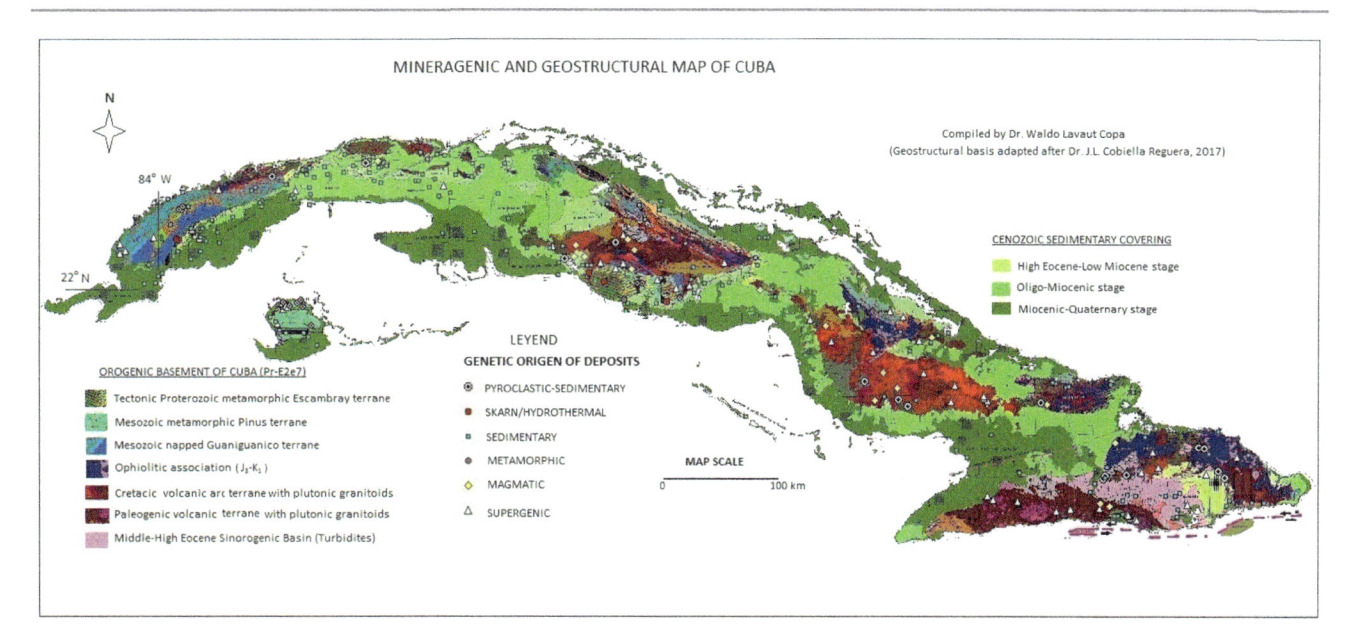

Fig. 9.1 Mineragenic and geo-structural map of Cuba

mega structures and that occupy significant territories in the country. These lithotectonic structures or geotectonic scenarios are treated as ancient accreted terranes, Caribbean terranes, and Cenozoic sedimentary cover with which are related the deposits of IMR and its specificities, as it is detailed next.

9.4.1.1 Accreted Ancient Terranes

Include the Guaniguanico terrane, the Florida–Bahamas plataformic terranes of continental margin, and the metamorphic Caribbean terranes (Cangre, Pinos, and Escambray).

Main IMRs of Guaniguanico are the following ones: **quartzite** deposits, as Bejuquera, Ceja del Negro, and Llano de Manacas; those of biochemical **limestone** as Sitio Pena, La Reforma, and La Muralla; and supergenic (residual) in weathering regolites of **barite** (Francisco), **kaolin** (Malcasado), and of **ferruginous clay** (Olga).

In the Florida–Bahamas plataformic terranes are biochemical **limestone**, which composed of more than a dozens of deposits, such as Los Azores–Zulueta, El Purio, Jiquí, Sierra de Cubitas, Los Caliches, Gibara, La Yaya, and others. Also the Jurassic evaporitic **gypsum** and **rock salt** deposits of diapirs of intraplataformic channel (Cayo Coco Unit) that compose the big deposits of Corral Nuevo-Canasí, Punta Alegre-Máximo Gómez, and Punta Alegre-Mamon y Punta Alegre.

In the metamorphic Caribbean terranes, the distribution of IMR is the following one, for part:

- Cangre terrane: biochemical **limestones** (Lagunitas y Km 13 Viñales) and **residual clay** (Esquistos Paso Viejo).

- Pinos terrane: **marble** (Punta Colombo y Piedra La Fe), **mica** (El Alemán), **kyanite** (Las Nuevas), **residual kaolin** (Río del Callejón, Siguanea, and others), see Fig. 9.2.

- Escambray terrane: **quartz** (Cacahual, Veta II), **marble** (El Mirador), **garnet and mica** (La Belleza), **graphite** (Algarrobo), **talc** (Crucesita-La Sabina y Higuanojo), and biochemical **limestones** (Trinidad-Zona II).

9.4.1.2 Caribbean Plate Terranes

Iturralde Vinent et al. consider these terranes [14] allochthonous fragments of the lithosphere of the Pacific Ocean that migrated until their current position in Caribbean during Cretaceous. Comprise several miner geodynamic scenarios: Ophiolitic Association, complex of the Cretaceous Arc, Meta-Arc Complex (Mabujina), Paleogenic Arc Complex, and a series of Synorogenic Basins.

The Ophiolitic Association contains a mineralogeny that raises in intraoceanic geodynamic and supra-subduction scenarios, as well as the mineralogeny of serpentinitic mélanges, also the one carried out by weathering under conditions of neoplatform. With them associate the following IMR of Cuba: **gabbro** (Serones II-La Prudencia), **ferruginous (lateritic) residual clay** on ultramafic rocks (Cajálbana) and **aluminic residual clay** on gabbros (Nuevo Mundo), and residual **magnesite** (Las Minas).

The complexes of the Cretaceous Arc extend for the whole country, with more development in the center-oriental parts of this. It contains the following IMR of Cuba: **granitoids** (Piedrecitas, Peña Blanca), **andesite** (Jicotea-José, San Mateo, La Mulata, El Pilón, Flor de Mayo), **Tuff** (La Victoria), **zeolite** (Las Pulgas, El Chorrillo, San Andrés-Loma

Fig. 9.2 Frontal view of Río del Callejón residual kaolin deposit

Blanca, Carolina II, Orozco I, Tasajera), **kaolin** (Pontezuela), **ferruginous residual clay** (Palanquete, Carranchola-La Moza, Sabana Grande), **feldspar–quartzose sand** (Cumanayagua-El Canal, La Macagua-El Salto), **sand and gravel** (Jíquima-Alfonso), **sand** (Granodiorita-Sancti Spiritus, Aeropuerto, Venegas, Corojo), **feldspar–quartzose sand and gravel** La Mota, La Conchita, Cascorro, Las Tunas II, Galbis, Las Canoas-Dolores), **ferruginous residual clay** (Golden Hill), and **residual clay** (Arroyo del Medio, Sabana Grande), see Fig. 9.3.

The Meta-Arc Mabujina Complex contains **amphibolites** intruded by small pegmatitic bodies and **quartz** veins with **mica** and post-metamorphic **bastard quartz** [10].

The Synorogenic Sedimentary Basin terranes were developed from Late Campanian until Late Eocene and they constitute an extensive foreland basin (from NW Pinar del Río until Gibara in Holguín) with olistostromic sequences. Also, the K_2 sedimentary cover of the Cretaceous Arcand partially of Cuban Ophiolite, as well as a series of small piggyback basins, developed mainly in early Paleogene, to the back of the over thrusts, and that contain besides silty and calcareous deposits, also epiclastic material and, in Eastern Cuba, pyroclastics, sometimes lavas.

With the Cretaceous Sedimentary Cover following IMR of Cuba is related: **zeolite** (San Andrés-Loma Blanca), **limestones** (Real Campiña, La Matilde, Rodolfo Rodriguez, Guayacán), **marl** (Loma Cantabria), **calcarenite** (La Cooperativa, La Fe—La Colorada—La Concordia, Victoria II-III-IV, Arenisca Matanzas-Navarro, Arango-Miraflores, La Guayaba Sector II, Peñalver, San Luis II), **ferruginous residual clay** (Palanquete), and **wollastonite–garnet** of skarns (Arimao Norte).

With the Olistostromic Sequences of Lower Paleogene in the Foreland Basin are related **limestone** deposits (La Lata—Teté Contino, El Chivo—Quiñones, Quiñones—El Llano, La Molina, Senado), **arcilla** (La Naza).

In the case of the piggyback basins, the IMR of Cuba is related fundamentally with those inherited that were subducting from final of Cretaceous, and with an Upper Maastrichtian–Middle Eocene age, developed in the central part of the country: Cienfuegos, Villa Clara, Ciego de Avila, and Camagüey Provinces, where deposited **marls** (Las Pailas), **limestones** (Nieves Morejón, Seboruco), and **clays** (Colorada II). With the Lower Danian–Eocene piggyback basins is related the **clay** deposit Ciego de Avila.

The terranes of the Paleogene Volcanic Arc were developed in the oriental part of the country: Sierra Maestra until the Cayman Crest. In the terrestrial portion, away from the waters of Caribbean Sea, they are prodigal in diverse mineralizations, those metallic as nonmetallic ones, embracing the axial region and the backarc basin mainly.

With the axial territory of arc, the following IMR is related: **andesite** (Botija, Los Guaos), **kaolin** (El Cobre), several **residual sand** regolitic deposits (Daiquirí, Juraguá, Las Guásimas, El Refugio, Sector III, La Punta, Damajayabo II), and several of **residual clay** (Guamuta, La Sierrita, Vega Grande, Sabicú, El Marañonal).

With the calcareous Eocenic sedimentary cover of the arc are related **limestone** deposits (Rosa Aurora, El Cacao, Río Frío, La Imagua, Aurora-San Rafael, Mucaral-La Maya,

Fig. 9.3 View of the El Canal residual feldspar sand deposit

Julio Antonio Mella, Mogote, San Nicolás, Farallones de Moa).

Besides the sedimentary covering, other basins that are related with the Paleogenic Arc contain IMR. In this way, in the backarc basin, there are **zeolite** deposits (Palmarito de Cauto, Seboruco, Caimanes, and Palenque), **vitreous tuff** (Amansaguapo, El Lirial, Ají de la Caldera), **tuff** (Palmarito de Cauto), and **bentonite** (La Caoba).

Other capacities of IMR are in the synorogenic basins of Middle and Upper Eocene that comprise a vast intramountainic territory in the south of eastern Cuba. These basins have a turbiditic terrigenous nature with clasts of the erosion of the rocks of the Paleogene Arc and of its calcareous cover, being classified in nearer and distal turbidites [10]. The distal turbidites originated **clay** deposits (Platanillo, Maceira, Carretera, Novaliche), also spread to a series of deposits of **alluvial clays** (El Quinto, Maceira, Vaquería, Ullao, and others), and coming from the nearer turbidites were formed **alluvial sands** (Imías), located in the equal spot name.

9.4.1.3 Cenozoic Sedimentary Cover

The terrane of the Sedimentary Cenozoic Cover, also denominated Neoautochthon, constitutes the upper structural stage of the Cuban Orogen. It is deposited in different moments of Eocene in most of Cuba: from Pinar del Río until Northwestern Holguín, after the ceasing of the Cuban Orogeny and, in Eastern Cuba, after the ceasing of the magmatism. This terrane is also subdivided in different geo-structural stages starting from Lower–Middle Eocene until the Quaternary, and it is outstanding as for the generation of deposits of sedimentary IMR as well as for some supergenic (residual) regolites.

The structural stages of the covering with smaller mineragenic productivity contain the deposits of older IMR of the Neoautochthon that arises under conditions of tectonic instability and related with basal sediments of age Lower–Middle Eocene of **marls** (Trinidad-Santa Rosa), and with transitional depositions of Upper Eocene–Lower Miocene **limestones** (San Antonio del Sur), as well as **residual clays** (La Clarita). In the Eocenic–Oligocenic structural stage, also of little productivity in IMR, were developed **residual ferruginous clays** (El Cebadero).

The biggest mineragenic productivity is related with the two structural more superficial stages of the cover; these are those of Oligo–Miocene and Miocene–Quaternary ages. They have generated a great number of deposits and a variety of IMR.

In this way, the structural Oligo–Miocene stage contains **limestone** deposits (Sierra Anafe-La Cruz, Santa Teresa-Cantera II, Capellanía, Cantera Blanca, San Jose Sur-Camoa-La Moderna, Beluca, La Colina, La Unión, Boca de Jaruco, Loma de Cura, Planta Libertad, San Antonio de

Cabezas, Coliseo, Cantel-Zona I, Calera José Martí, Pilón, Levisa), located in several basins (Vegas, Broa, Mercedes, Guacanayabo, Nipe); **dolomite** (Puerto Escondido, Alacranes, Perico, La Montana) in Broa and Mercedes basins; **marl** (Simpatía II) in the basin Mercedes; **phosphoric rock** (Trinidad de Guedes, Meseta Roja, Loma Candela Norte, Cañada Honda); **clay** (Rodas–Abreu, Coloradas II, Rio Caunao, Castano-Siguaney, La Fortuna, Sancti Spiritus Zona III A, Charco Soto, La Pachanga, Ciego de Ávila, Las Tunas zona XII, La Naza); **bentonite** (Santiago de Las Vegas, Managua, Bauta, Redención); and **residual clay** (El Jardín in Vegas Basin and Sabanilla in Broa Basin).

With the structural Miocene–Quaternary stage, the most superficial, there are related the following IMR: **limestone** (Jaimanitas–Zona Guajaibón, Rio Mosquito, Jaimanitas Matanzas, Monte Alto, Carbonato Cayo Piedra, Aguadores Este, Cabo Lucrecia); **marls** (Matanzas–Bacunayagua, Aguadores Oeste); **dolomite** (Vedado); **sand** (Cayo Inés de Soto, Arimao, Pilotos, Imías, Río Bayamo, Río Arimao, Tacre-Cajobabo, Indios Norte, Río Guantánamo, El Ángel); **quartzose sand** (Siguanea, Panzacola I); **calcareous mud** (Cayo Moa Grande); **clay** (La Conchita, Ovas I, Kessel, Moja Hueva, Bermejales, Club de Cazadores, en la cuenca Los Palacios; Ullao, Bayamo, San Rafael, Aguadores, Contramaestre y Caney del Sitio, en la cuenca Valle del Cauto); **bentonite** (Chiqui Gómez, Vado del Yeso); **turf** (Ciénagas de Lanier-Zapata, Majaguillar, Playa Guanímar); **therapeutic pelites** (Maracayero, El Soldado, Monte Gordo, Jutía).

Besides the occurrences of related IMR in the epigraphs of this section of the text, it is necessary to point out that Cuba counts with more than 100 manifestations of gem stones, of which jaspers, agates, rock glass, chalcedonies, flint, opals, amethyst and jade mainly stand out that are related with different terranes and genetic sources. Nevertheless, due to their low study grade here won't be approached in details.

9.4.2 Genetic Description of Cuban Rocks and Industrial Minerals

The mineral deposits present in the Cuban territory are framed inside the following genetic main groups: **magmatic**, **metamorphic**, **hydrothermal**, **pyroclastic**, **sedimentary**, and **supergenic**. Inside each genetic group, up to where it was possible, they were defined genetic types of deposits, what constitutes a refinement for the genesis inside each group. In this way, for the magmatic group were defined the *intrusive*, *extrusive*, and *metasomatic* genetic types; in the metamorphic group were included those *metamorphosed* and the *pegmatitic* one; in the pyroclastics, the *diagenetical* was indicated; in the supergenic group are the *residual* ones, and lasting the skarn-hydrothermal group like such.

A special group is composed of the sedimentary deposits, with which are related several types: *clastic*, *evaporitic*, *caustobiolithic*, *biochemical*, and *turbiditic*. This is due to the great diversity of genetic phenomena inside the sedimentary process.

Next are described the mentioned main genetic groups separately.

9.4.2.1 The Pyroclastic-Sedimentary Group

The main mineral deposits associated to this genesis in Cuba are those of tuffs that go from vitreous tuffs to zeolitized tuffs or zeolitites.

Tuffs: They are of the type diagenetic and they are represented by tuffs of intermediate-basic composition, in layers of different lithologic types and with variable thicknesses, with fine to coarse grain, fresh ones of gray clear to greenish colors with different tonalities. They lie jointly with tuffaceous sandstones, clays, and shales. In general, they are very affected by weathering processes and they show up with ocherus to brown colors. Their geologic more frequent age is Upper Cretaceous.

Vitreous tuffs: They are of the diagenetic type, of acid to intermediate composition, gray colors to white, very porous and slight, those of fine and middle grain prevail, in occasions something altered to montmorillonites, carbonated, or silicified, and they can be interlayered with other rocks. Their frequent age is Paleocene–Eocene.

Zeolitic tuffs: They are of the diagenetic type and they have predominant mordenitic–clinoptilolitic composition with variable contains of celadonite, montmorillonite, and quartz, prevailing in some regions of the mordenite and in others of the clinoptilolite. They are vitroclastic and massive crystal vitroclastic rocks, slight, porous, of clear colors, showing up in the profile without inclusions with other rocks or with alternation of sandstones, tuffites, and conglomerates or overlapped by lavas. The zeolitization is irregular averaging out 70%. Their bodies show up in layers of until hundreds of meters of thickness reaching territorial great development. They are used in the agricultural sector, cement, and demanded for multiple uses. Their geologic ages go from Upper Cretaceous to Middle Eocene.

9.4.2.2 The Skarn-Hydrothermal Group

With this genetic group associate the following IMR:

Barite: It is composed of small deposits of the genetic hydrothermal type. The mineralization is of barite, quartz–barite, or barite–polymetallic, associating to tectonized and cracked areas in shales, quartzose sandstones, quartzite, schists, and volcano-sedimentary rocks. They form veins, nests, lenses, and pockets with variable dimensions. When together with the polymetallics they show up occasionally forming independent bodies. Their geologic age is in dependence of its location and mineral genetic context (K_1-Q).

Garnet: They are of the skarn type and in the way of small garnet-epidotic deposits accompanied by magnetite, garnet–actinolite, piroxenite, and wollastonite. It is composed of 20–75% of garnet (grossular, almandine, rarely andradite types) and mica muscovite. It is associated to the rocks of the Cretaceous and Paleogenic volcanic arcs, and their geologic age is related with that of them.

Mica: They are related with greisenized mica–quartzose schists, product of plagiogranitic intrusions in schists, as well as in amphibolic schists and schistose serpentines. In the continental margin (J_1-K_2cn), they form very mineralized layers or quartz-micaveins in the proximities of the aplite, where the mica forms grains with average size of 2–5 cm. Several positive points of micas of the aplite–pegmatite genetic type exist associated to the feldspar-rich pegmatites in the Cretaceous granitoids and to the pegmatitic veins inside Jurassic (J_1) amphibolites.

9.4.2.3 The Magmatic Group

The intrusive genetic type comprises bodies of granitoids, gabbro and troctolite, andesite, basalt, dacite, dunite, and diabase that are described next.

Granitoids: Several bodies linked to the intrusives of the Cretaceous and Paleogenic volcanic arcs exist, represented by granites, granodiorites, diorites, and syenites. Their general composition is quartz, plagioclases, and biotite, while the syenitic types also contain feldspars and amphiboles. Their extensive bodies form plains with soft tectonized hills, solid mountainous massif, or they are as exotic blocks included in serpentinites by way of isolated hills. They can also be granodioritic dikes among syenites or of porfidic character among the granitic massifs, or crossing layers of lavas and tuffs. They present good physic-mechanic properties and useful enough thickness to be used as crashed arid and ornamental rocks.

Gabbros and troctolites: They are in the ophiolitic ultramafic massifs (K_1). Their mineral composition is plagioclase (Labrador), pyroxenes, and amphiboles. They are big bodies forming promontories of boulders in peniplanized regions being relics of erosion, or small and lengthened bodies among serpentinitic masses. They are in variable grade tectonized and serpentinized rocks, sometimes very weathered. The useful thickness with good physic-mechanic properties is considered in more than 10 m.

The *dunite* (olivine) exists evaluated in a small location in Pinares de Mayarí but with big potentialities for the metallurgic industry and source of magnesium.

Andesites: They are linked to the Cretaceous and Paleogenic volcanic arcs and are composed of plagioclases (andesine and oligoclase), amphiboles, and epidote. They form mantles of massive lavas, lava-breccias, subvolcanic bodies, or dikes, many times with complicated relationships, with tuffs and agglomerates, in flat areas and hills, tectonized and altered with imprints of pelitization, chloritization, albitization, and others. They have enough industrial thickness and mechanical properties for arid of construction.

Basalt: They associate to the formations of the Cretaceous and Paleogenic volcanic arcs, appearing among the younger sandstones. They are hard subvolcanic bodies of middle grain that form big massifs or small bodies with smoothly wavy to hill relief and tectonized, cracked and weathered textures, with a loamy irregular covering, but being even enough useful thickness of fresh rock. Their mechanical properties are acceptable for arid of construction.

Dacite: Is found in the Cretaceous volcanic arch (J_3ox-K_2cp), being related to subvolcanic bodies and with andesite–dacitic lava extrusions. They form variable bodies of big volumes, but with very scarce height on the land. The rock is tectonized and weathered until more than 10 m of depth.

Diabase: Forming part of the ophiolitic ultramafic (K_1) as subvolcanic bodies of reduced dimensions and limited volume they are porfidic, compact, massive, of middle grain diabases that are affected in variable grade by the cracking and the weathering. Their thickness and mechanical properties are favorable to the industry of the construction.

Besides the industrial rocks characterized above, several feldspar and plagioclase bodies of aplite-pegmatitic type exist, related with Jurassic (J_1) amphibolites and with Cretaceous and Eocenic granitoids. The composition is quartz–feldspar (albite–oligoclase or microcline–orthoclase). They form dikes, veins, or sufficiently big lenses fitted in amphibolites and granitoids. There stays a tectonic clear dependence and relationship with the Cretaceous granitoids, associating mainly to their exocontact.

Rhyolites and liparitic-dacitic lavas with feldspar and plagioclase bodies of the Paleogenic volcanic arc represent the extrusive genetic type.

The *rhyolites* form small bodies of compact porfidic, massive, hard lavas in high of the land amid a plain, with thicknesses of more than 10 m, and with good physic-mechanical properties, but they are tectonized and weathered in variable grade.

In the extrusive ***liparitic-dacitic*** bodies, the feldspar rock is softened, fragile, with hydrothermal alterations, mixed with clays and kaolin. The alkali is sodium and variable in the cut, but capable for the white ceramic.

Lastly, also produce ***feldspar and plagioclase*** the granitoids with leucocratic magmatic rocks, hornblendic syenite and sequences of pyrophillite, kaolinite, quartz and mica–quartz–kaolinite of the Cretaceous granitoids, and as well hydrothermally in metasomatites with clear gray color of the volcanic Cretaceous Arc terranes (J_3ox-K_2cp).

The prospected granitoids, basalts, andesites, and gabbros are used in the industry of the construction like arid or attaches to manufacture concrete and for ornamental rock in smaller scale ("Mantilla" in Villa Clara and "Las Guásimas" in La Gran Piedra, Santiago de Cuba). The feldspar of the aplito-pegmatitic and liparitic-dacitic kaolinized rocks (these last of the "El Purnio" in Holguín) is exploited for the ceramic industry.

In general, the magmatic rocks are employed for crashed arid and in the case of some granitoids and gabbro-troctolites like ornamental rock in blocks or foils.

9.4.2.4 The Sedimentary Group

The group of the sedimentary deposits is one of the more prolific and productive including several genetic types: clastic, biochemical, caustobiolithic, and turbiditic. The industrial minerals and rocks of Cuba related with these genetic types have geologic ages from Upper Cretaceous until the Quaternary, and they are the following ones.

Clastic genetic type:

Calcarenite: It is in several deposits of conglomerates, calcareous gravel stone, and sandstones linked to the volcanic Cretaceous Arc and sedimentary cover of Middle Paleogene–Late Neogene that form thick layers with some veinlets of interstratified calcite with other sterile lithologies. The whole group is tectonized in variable grade. It is exploded mainly as arid for the construction.

Another type is the biocalcarenites, porous, with pallid yellow color, compact, massive, thick grain, with presence of karst, suitable for filters of water when taken form Pliocene–Quaternary rocks of the sedimentary cover.

Marl: They are derived rocks of the calcite mixture and montmorillonite, such as argillaceous marls, calcareous marls, and marly limestones. They form massive layers or stratified with their own lithologic types or other sterile ones. They have a variable plasticity and they can present dolomitization. Used mainly for the Portland cement production and red ceramic.

Magnesite: They are linked to the clayey series where they form horizontal big and potent irregular masses that lie to variable depth. They contain loamy mixtures and they present a general conglomeratic or breccious aspect. They occupy fluvial wide and flat basins. They are used for fertilizers and they are studied for metallurgy.

Clay: They form significant bedded deposits, sometimes lenses, in the sedimentary sequences of the covers from the Oligocene one until the Quaternary one. The clayey layers are stratified with marls, clayey limestones, karstic organogenic limestones, and clayey sandy deposits, sometimes with conglomerates and breccias. In occasions, the clays contain moderately (5–15%) concretions of carbonates or iron oxide-hydroxides. They have wide employment in ceramic, cement, etc.

Bentonite: They are montmorillonitic, bentonitic clays, plastic, and very plastic of several colors: gray, grizzly, yellowish, grizzlygreen, and even motley, sometimes containing ferruginous concretions. The bentonites have high plasticity, they are soapy in water, they are low refractory, and in occasions their chemical composition can be something calcic. They are also developed by tuffs altered to bentonitic clays of cream color and with veinlets of iron hydroxides for the cracks. The same as the clays, the bentonites associates to the sequences of the sedimentary cover with Oligocene until the Quaternary ages.

Sand: They are clastic deposits generated in the Quaternary age and they are part of the Sedimentary Miocene–Quaternary Cover. They are represented by sands of biogenic marine origin (remains of algae, mollusks, corals, etc.) of little thickness average (2.5 m), varied grain (fine to coarse), carbonated, like in Cayo Inés de Soto; and also alluvial sands in form of lenses in the terraces of big rivers like Arimao, Bayamo, Guantánamo, and others. These alluvial ones associate to other lithologies like cobble, pebble, gravel, and clays.

Quartzose sand: They are quartz-containing sands of fine-middle grain with different colors: gray (41%), yellow (33%), white (33%), and black, as those of Siguanea, generally with high content of sand (more than 80%) and with low clay containing, as the deposits Panzacola in Santa Cruz del Sur (province Camagüey). They were generated by clastic-sedimentary phenomena and they are part of the Sedimentary Miocene–Quaternary Cover.

Biochemical genetic type:

Limestones and dolomitic limestones: They are located in carbonated multiple deposits that participate in the sedimentary structures, mainly of age Oligocene–Miocene and Miocenic–Quaternary, and much smaller in the Sedimentary Transitional Cover of Upper Eocene–Lower Miocene, including certain proportion in the volcano-sedimentary depositions. They associate to basins, massifs, terraces, mogotes, and olistoliths. Their composition is calcitic with variable parts of dolomites. They represent several lithologic types such as organogenic, biodetritic, pelitomorphic, marly, tuffaceous, and breccias among others. They form massive layers in formations with carbonated wide development, but in other regions they can be stratified with varied rhythm and thickness next to other lithologies. They are affected in variable grade by weathering (karst), cracking, tectonic, recrystallization, and dolomitization.

Dolomites and calcareous dolomites: Their dolomitic composition is predominant in some regions, but it varies in others due to the irregular dolomitization process. They form massive, thick bodies, in extensive layers, generally with karst and cracks. The calcareous dolomites form flat lands to hilly, in layers of variable thickness with sterile interstratifications of soft rocks or of another composition, affected by the karst and the supergenic alteration. The biggest productivity corresponds to the sedimentary structures of age Oligocene–Miocene.

Marl: They are clayey marls and calcareous marls, soft, of white, white-yellowish, gray, graygreenish colorations, sometimes compact, sandy, in carbonated parts and that they can transition to calcarenites and even to limestones. The biggest accumulations were deposited in the sedimentary covers of age Eocene–Miocene.

Phosphoric rock: They are calcareous granular phosphorites, phosphatic limestones, and phosphatic clays. The granular phosphorite is located in calcarenites with carbonated matrix of fine to coarse grain, very porous, and in grizzly clays, very plastic, very phosphatic, and associated to the terrigenous matrix of the chaotic policomponent deposits. The phosphorites associated to phosphatic limestones, calcarenites, and until with conglomerates, they are characterized by a variable phosphatization of the granular phosphorites (1–19%). These form big deposits and small and irregular bodies within organogenic limestones of Oligo–Miocene cover. They are used as fertilizers.

Carbonated muds: They are constituted by carbonated silts with quantities of biogenic sands formed from detritus of mollusks, foraminifera in the environment of calcareous keys, like in front of the Cayo Grande de Moa. They are characterized by a fine grain and coloration gray white. In Cuba, they are exploited for industrial chemical applications, like in the HPAL plant Moa Nickel S.A. (Cuba-Sherritt).

Caustobiolithic genetic type:

Turf: They associate to the paludal-swampy deposits of the sedimentary of Miocene–Quaternary cover. They constitute not consolidated deposits and developed in lagoon and coastal sites. They have varied texture, in general, with territorial great development, and thicknesses of few meters. They are blended with turfy clays, slimes, sapropel, sands, and others. The underlying flour generally is karstic carbonated rocks. It is used mainly in the agriculture.

Therapeutic pelites: They are very fine, soft, unctuous, creamy mires, adhesives, with sulfurous scent. They have a color of black to gray dark, sometimes greenish and mechanical with very few impurities. In Cuba, they are used thoroughly with therapeutic ends and cosmetics.

Turbiditic genetic type:

Clay: They were generated under conditions of synorogenic basins bound to the distal turbidites of the Paleogene volcanic arc terrane, and they are represented by plastic clays of greenish color, less plastic clays of yellow-greenish color, with segregations of sand with many fragments of gritty. The clays are very fine; they have high plasticity; bake temperature of 950°; and a yellow-greenish, greenish coloration, grizzly, and until brown dark. They are used for light elements, blocks, and solid brick.

9.4.2.5 The Metamorphic Group

Metamorphic genetic type:

Marble: They form alternation of marble layers and varied schists in a karstic relief, with quartzite lenses in the breccia areas. The deposits of this rock are characterized by a variable composition: from calcitic to dolomitic. The coloration goes from gray until gray dark and gray dark white, with banding of varied tonalities. The grain is middle with massive to banded texture, and sometimes schistose. They form mogotes jointly with some schists, but in other regions they are only lenses inserted in layers of limestones or sandstones.

Mica: Their main mineral is the muscovite associated to mica schists of Lower Jurassic age. This muscovite is characterized locally by segregations of no more than 2 mm, locally some appears of 30 cm^2, and an abundance of 24.2%. The mineralization is in form of bands with the participation of granites (50%), migmatites (35%), and gneisses (15%). It is used for paintings, as refractory, in the construction and the electronics.

Garnet with mica: The garnets with mica and rutile are in schists of half-coarse grain, in the low part rocks green eclogites of Lower Jurassic–Lower Cretaceous, constituted by bodies of garnet pegmatite of white-yellowish color. The garnet is of the almandine (red) and espesartite (red and black) types, accompanied by mica and rutile. The mineral composition of the deposit is garnet (6.78–13.35%), mica (6.4–11.87%), and rutile (0.5–7.5%). The physical properties of the garnet are absorption = 0.5%, volumetric weight = 3.14 g/cm^3, and micro-hardness = 1274 kgt/mm^2. These matters are capable of the electronic industry, cosmetics, paintings, and construction.

Kyanite: They are impregnating the J_1-K_2cn age schists, formed by a high such grade of metamorphism as quartz–kyanitic, chloritic–kyanitic, or purely kyanitic, graphitized. The bodies only formed by kyanite are strange and of little thickness. The crystals are of 3–4 cm long, but commonly they are of 0.5–0.7 mm. Their use is proposed for special and fine ceramic.

Quartz: It is of vein type, white-milky, of massive texture, with structure of middle (1–2 cm) grain and big (2–10 cm) massive grains, and in occasions it presents several padded cracks of iron oxide.

Graphite: It forms lenses of enough pure graphite inside quartz–muscovite schists, strongly impregnated of amorphous graphite, inside of metaterrigenous sequences, with thickness of 10–35 cm and low mineral content. It is also observed as bands in gneisses and lenses in schists contacting with serpentinites, or in magnetite-rich schists. The useful thickness oscillates of 1–12 m. The graphite is of microcrystalline type and its chemical composition is carbon = 3.78%, ash = 90.4%, and sulfur = 1.3%. The age of the mineralization is J_1-K_1. They are useful for metallurgy and several (refractory, pigments, lubricant) industries.

Metamorphosed genetic type:

Quartzite: It is a fractured quartzite, cracked, breccious, tectonized, of white and yellow coloration, fit in clays and gravel stones; the percentage quantity of clays increases in depth. The chemical composition of these quartzites is Fe_2O_3 1.3%, SiO_2 96.94% and the physical–mechanical properties are the following ones: resistance to the compression 480–720 kg/cm^2, volumetric weight 2.2–2.28 g/cm^3, and natural humidity 2.6–8.9%.

With this genetic type other IMR of Cuba is also related, little studied, although of practical potentialities, as the serpentines, amphibolites, and jadeitites, which will be pointed out here shortly by their importance.

The *serpentines* associate to the solid extensive Cretaceous ultramafic rocks, as well as to bodies of variable size of pre-Jurassic serpentinites of the Northern ophiolitic belt of Cuba. The serpentinites are of the antigoritic type with a fibrous-scaly texture or crisscross and they form blocks of enormous to medium sizes inside lizardite–chrysotilic serpentinitic bodies in tectonized areas, or sub-concordant bodies in metaterrigenous or calcareous schists, where they show up as compact, solid rocks, and of schistose or breccious texture. They form isolated hills or lengthened hills. These serpentines are used with ornamental ends.

Of same nature, also exist Cretaceous *amphibolites*, composed of amphibole and few plagioclases. They are derived rocks of the regional metamorphism of volcanic basic rocks. They form hills by way of isolated bodies of considerable territorial distribution and great thickness, being able to also find as regal packages or intercalations in crystalline schists or gneisses. For their mechanical good properties, it is intended to be used as arid and some sectors for ornamental aims.

And, lastly, it is necessary to mention to the *jadeite* associated to jadeititic rocks of intrusive origin that lie with marbles and carbonated schistose rocks of continental margin (J_1-K_2cn). This jadeite forms a monomineral mass of radial attachés; it is compact, massive, and of fine grain. Also, in areas of mélange of the oceanic Lower Cretaceous rocks, associated to metamorphosed mafic and ultramafic rocks, have been reported jadeititic rocks of green clear color to dark. The veins and nests of the green dark variety are related with white mica.

9.4.2.6 The Supergenic Group

The climatic conditions of Cuba are favorable to weathering of the rocks, what has been conditioned by their geographical position and the character of the surrounded sea island, where the influence of oceanic and eolic currents favors the changes of temperature and precipitations. In this way, the annual received solar radiation, as an average is of 80 kkal/cm^2; the temperature of the air has an annual mean of 24.7–26 °C in most of the country (plains), and annual abundant precipitations (1000–1300 mm in the plains, 1600–1800 mm in the piedmonts and more than 2000–3000 mm in the mountains) [1].

These factors, together with a favorable geomorphology, have conditioned the formation of excellent regolites and their intrinsic lithologies of weathering, among which are *alterites* (**saprock and saprolites**) that conserve the fabric of the parental rocks or protoliths; *lateritites* (**laterites, bauxites, and sialitites**) that have lost the fabric of their protoliths due to high degradation, as well as *duricretes* (mainly **ferricretes, alcretes, magnesicretes, and silcretes**) [16, 17].

These rocks of the weathering house a wide variety of IMR deposits of supergenic or residual genesis, among the main ones are counted: kaolin, magnesite, barite, clay and ferruginous clay, sand, sand and gravel, feldspar–quartzose sand and bentonite. In turn, these different IMRs arose under conditions of tectonic stability that was essentially from Miocene, for what they had their biggest development during Pliocene–Holocene, and evidently some continue being formed until the present time, since the climatic and other conditions have not changed substantially. Next, the description of main weathering or supergenic/residual IMR is exposed.

Kaolin: *Kaolin* is formed inside elluvial regolites with variable and typical thicknesses up to 20–30 m average that are developed by intense weathering during the Pliocene–Holocene of metaterrigenous schists of Jurassic until Neocomian age (J_3ox-K_1). They show up as fine aluminic saprolites and sialitites, composed fundamentally of variable contents of quartz, montmorillonite, iron impregnations, and mainly of kaolinitic clays of different colors: white, reddish white, motley white, yellowish white, gray, dark gray, black, these last due to organic contained matter. The chemical typical composition is the following one (in%): Al_2O_3 = 21.14, SiO_2 = 64.48, TiO_2 = 0.80, CaO = 0.56, Fe_2O_3 = 1.46, K_2O = 1.62, MgO = 0.54, Na_2O = 0.12, LOI = 8.44. The most significant locations, including the one bigger in Cuba: Río del Callejón, they are in the Isla de la Juventud and Pinar del Río. They are used in white ceramic and white cement in spite of their variable parameters of quality.

Clay: They are represented by sialitites and fine saprolites formed by the weathering of different types of rocks: quartz–biotitic schists (Isla de la Juventud), loamy and calcareous schists and loamy (Pinar del Río) loams, as well as starting from tuffs and hydrothermally altered lavas (Sierra Maestra), although these last give smaller deposits. The clays starting from schists are characterized by a plasticity 72.91–85.76% and a chemical composition (in %) $SiO_2 = 57.40$–69.49; $Al_2O_3 = 14.59$–19.73; $Fe_2O_3 = 6.53$–8.40. The clays formed starting from argillaceous–carbonated rocks have a plasticity of 48–78% and a chemical composition (in%) of $SiO_2 = 42.22$–91.06; $Al_2O_3 = 7.6$–18.8; $Fe_2O_3 = 0.75$–13.1. Lastly, the clays starting from pyroclastites and rhyolitic lavas contain fragments of these altered rocks, and the clays are kaolinitic, very plastic, and they have a yellow and motley color. Their main physical–mechanic and compositional parameters are the following: plasticity 64.4%, $Al_2O_3 = 11.4$–18.01%; $Fe_2O_3 = 10.36$–15.06%; $SiO_2 = 46.82$–60.68%. All these clays have application mainly for red ceramic, and alternatively for cement.

Sand and feldspar–quartzose sand: They are in the weathering lithologies of the fine and coarse saprolites and sandy typical sialitites, in extensive elluvial mantel over granite–granodioritic and plagiogranitic Cretaceous and tonalite–granodioritic Eocenic plutonic rocks of the volcanic arc terranes, being the parental protoliths more frequent the quartzose diorites and granodiorites, minor the tonalites and plagiogranites. These sands are something clayed and they have a variable polimictic composition with prevalence of feldspars, plagioclases, quartz, and tiny relictic fragments of the source rocks. The clay content is of 20–40% and of sand it is 60–70%, which contains 12–30% of quartz and 40–50% of feldspars. The chemical main composition of the sands is (in %) $SiO_2 = 50.72$–64.97; $Al_2O_3 = 15.7$–16.4; $Fe_2O_3 = 3.20$–7.42; $K_2O = 1.40$; and $Na_2O = 3.5$. After having washed the sands they are used as arid in the construction, being able to be good for the glass production and cement. Of same way, the gravel content separated from the coarse sandy saprolites is used as arid for the construction.

Ferruginous clay: They are laterites and ferruginous and alumoferruginoussialitites, nodular and no-nodular, formed on igneous (plutonic rocks and lavas) of the ophiolite and arcs, the metamorphites, and over carbonated rocks (limestones and marls). They form mantels and lenses with 2–7 m thickness of materials of mostly ferruginous loamy composition, and in smaller proportion lateritic (typical laterites, clayey laterites, bauxites, and little bauxitic clays). The chemical fundamental composition depends on the parental rock (protolith) and of the intensity of the weathering, in general, it is (in %): for igneous basic and ultrabasic protoliths $SiO_2 = 6.38$–37.61; $Al_2O_3 = 9.12$–31.60; $Fe_2O_3 = 15.16$–68.21; for pyroclastic protoliths of intermediate and acid compositions: $SiO_2 = 41.25$–56.53; $Al_2O_3 = 10.18$–16.07; $Fe_2O_3 = 6.05$–8.60; and those developed over calcic protoliths (limestones and marls) in %: $SiO_2 = 10.22$–34.66; $Al_2O_3 = 15.17$–26.69; $Fe_2O_3 = 17.56$–44.37. All these highly ferruginous clays are used as cement corrector, and for their fine grain and composition they can be used as natural pigments.

Besides these main IMR, occurrences of smaller range and study grade take place, but they constitute perspective resources. Among them we have the *elluvial phosphorites* of the silicic-clayey type with variables contents of fluorapatite by weathering of chaotic policomponent phosphatized deposits that form layers with more than 40 m of thickness of a regolith with amorphous mass, blended with clays and phosphatic material in concentration up to 25%, due to the destruction of the terrigenous matrix. Alternatively, these phosphatic regolites can develop to expenses of detritus-fragmentary limestones that produce a reddish soft ocherous regolite ("terrarossa") with some flints and chalcedonies. They are used as fertilizers.

Also, it is necessary to mention the micaceous regolites that arises by weathering of mica schists and other schists. These regolites have abundant content of *micas* that represent important accumulations. Also, the *kainite* is contained in important weathering regolites on schists that also contain fragments of monokyanitic schists and abundant kyanite content as loose silt and they form potent deposits. *Magnesite*, *barite*, and *bentonite* deposits of residual type and practical importance exist, some have been used by the mining, but most remains little studied and the evaluation of their potentiality is pending.

9.5 Conclusions

The complexity of the geology of Cuba, in spite of being a relatively small country, contains an outstanding diversity of minerageodinamic scenarios, comparable with those of big territories. This has favored the formation and existence of a great readiness and potentiality in industrial minerals and rocks. At the present time, many of them are exploited and others are pending to complete their evaluation and use. This constitutes a task of great span for the Servicio Geológico de Cuba and entities related.

Acknowledgements The authors appreciate the most the geological information and map basis specially prepared by Dr. Cobiella Reguera JL and as well are grateful to Dr. Pardo Echarte M for the careful revision of the text and recommendations therein.

References

1. Atlas Nacional de Cuba (1970) Datos Físico-geográficos. Academia de Ciencias de Cuba y URSS
2. Bates RL (1959) Classification of the nonmetallics. Econ Geol 1 (54):248–253

3. Bates RL (1969) Geology of industrial rocks and minerals. Dover Publications, New York
4. Batista González R y otros (2001, actualizado al 2016). Sistema Informativo para los Recursos Minerales de la República de Cuba (INFOYAC), bases de datos. IGP, La Habana, Cuba
5. Batista González R y otros (2011) Mapa de rocas y minerales industriales de Cuba, escala 1:100000, CD-ROM. IGP, La Habana
6. Bauman L, Tishendorph G (1979). Введение в металлогению-минерагению (пер. с нем), М
7. Bilibin YA (1961) Металлогеническиепровинции и металлогеническиеэпохи. Избр. Тр. Т.3. -М.: Изд-во АН СССР: 67–130
8. Brito A, Coutín DP (2018) Las Zeolititas de Cuba: recuento de 45 años dedicados a su estudio. IGP, La Habana, Cuba
9. Cabral JrM y otros (2005) Minerais Industriais- orientaçâo para regularizaçâo e implementaçâo de empreendimentos. Sâo Paulo, IPT, 86p
10. Cobiella Reguera JL (2018) Texto explicativo del mapa tectónico de Cuba (inédito). IGP, La Habana, Cuba
11. Fisher WL (1969) The nonmetallic industrial minerals: examples of diversity and quality. Min Congr J 55(2):120–126
12. Iturralde Vinent MA y otros (1998) Sinopsis de la Constitución Geológica de Cuba. Acta Geológica Hispánica 33(1–4):9–56
13. Iturralde Vinent MA y otros (2006) Naturaleza Geológica de Cuba. Editorial Academia, ISBN 978-959-270-096-3
14. Iturralde Vinent MA and others (2016) The geology of Cuba: a brief overview and synthesis. Geol Soc Am 26(10)
15. Kuzvart M (1984) Industrial minerals and rocks. Developments: in economic geology 18. Elsevier, Amsterdam
16. Lavaut Copa W (1998) Tendencias geológicas del intemperismo de las rocas ultramáficas en Cuba Oriental (vol 1). Publicación Minería y Geología
17. Lavaut Copa W (2018) Geological classification for the rocks of weathering. Petrol Sci Eng 2(1):1–6. Disponible en: http://www.sciencepublishinggroup.com/journal/paperinfo?journalid=608&doi=10.11648/j.pse.20180201.11
18. Lingren W (1963) Mineral deposits (4th ed.). New York, McGraw-Hill Book Co., Inc
19. Luz AB y Lins FF (2004) Tratamento de Minérios, 4ªEdiçâo. AB Luz, JA Sampaio, SLM Almeida (eds) Rio de Janeiro, CETEM, cap 1, pp 1–16
20. Noetstaller R (1988) Industrial minerals: a technical review. Technical Paper No. 76. World Bank, Washington, DC
21. Pirajno F (2010) Hydrothermal processes and mineral systems. Geological Survey of Western Australia, Perth, WA, Australia
22. Ray DK (1963) Metallogenic-mineragenetic map of India [cartographic material]. Survey of India
23. Tabío E y Guarch JM (1966) Excavaciones en Arroyo del Palo, Mayarí, Oriente, Cuba. Dpto. de Antropología, Academia de Ciencias de Cuba, La Habana
24. Tijomirov IN (2016) Минерагенические карты: содержание, принципы составления. Uralskiy Geologicheskiy Zhurnal 3 (111):42–55, UDK 553.07
25. Vidal P y otros (2018) Informe de Prospección y Exploración Mármoles La Gran Piedra. Roca Real, Micons, La Habana (Inédito)
26. Wright LA, Burnett JL (1962) The search for industrial minerals. Min Inform Serv Calif Div Mines Geol 15(1):1–8
27. Zappettini EO (2018) Metallogenic potential of Argentina. In: 15th quadrennial IAGOD international association on the genesis of ore deposits symposium, Salta, Argentina

Printed in Great Britain
by Amazon